OPTIMAL, PREDICTIVE, AND ADAPTIVE CONTROL

PRENTICE HALL INFORMATION AND SYSTEM SCIENCES SERIES

Thomas Kailath, Editor

OPTIMAL, PREDICTIVE, AND ADAPTIVE CONTROL

EDOARDO MOSCA

Department of Systems and Computer Engineering
University of Florence
Florence, Italy

PRENTICE HALL, Englewood Cliffs, New Jersey 07632

Library of Congress Cataloging-in-Publication Data

Mosca, Edoardo.
   Optimal, predictive, and adaptive control / Edoardo Mosca.
      p.   cm.
   Includes bibliographical references and index.
   ISBN 0-13-847609-8
    1. Automatic control.   2. Predictive control.   3. Adaptive control
systems. I. Title.
TJ213.M558   1995
629.8—dc20   94-6082
                     CIP

Acquisitions Editor: Don Fowley
Project Manager: Bayani Mendoza de Leon
Copy Editor: Peter Zurita
Cover Designer: Karen Salzbach
Production Coordinator: Dave Dickey
Editorial Assistant: Jennifer Klein

© 1995 by Prentice-Hall, Inc.
A Paramount Communications Company
Englewood Cliffs, New Jersey 07632

The author and publisher of this book have used their best efforts in preparing this book. These efforts include the development, research, and testing of the theories and formulas to determine their effectiveness. The author and publisher shall not be liable in any event for incidental or consequential damages in connection with, or arising out of, the furnishing, performance, or use of these formulas.

Printed in the United States of America

10 9 8 7 6 5 4 3 2 1

ISBN 0-13-847609-8

Prentice-Hall International (UK) Limited, *London*
Prentice-Hall of Australia Pty. Limited, *Sydney*
Prentice-Hall Canada, Inc., *Toronto*
Prentice-Hall Hispanoamericana, S.A., *Mexico*
Prentice-Hall of India Private Limited, *New Delhi*
Prentice-Hall of Japan, Inc., *Tokyo*
Simon & Schuster Asia Pte. Ltd., *Singapore*
Editora Prentice-Hall do Brasil, Ltda., *Rio de Janeiro*

*To the colleagues whose work
kept me intellectually busy
over the years*

CONTENTS

PREFACE

This book is the result of my teaching a first-year graduate course on linear optimal systems and a third-year graduate course on stochastic estimation and control to applied mathematics and engineering students. The contents of the book have been strongly influenced by the great advances in the field and the related literature. However, no attempt was made to have a complete coverage, an unrealizable goal for a single book. On the contrary, a selection of topics is given to present a balanced blend of theoretical foundations, analysis and design methodologies, and application-oriented tools. Except for a few examples, only discrete-time linear plants are considered, even if some of the methods can be extended to some classes of nonlinear plants. The purpose of this book is to provide a systematic access to the main topics of linear quadratic control, predictive control, and adaptive predictive control.

My main motivation was to write a book aiming at presenting these related but yet different subjects within a common unifying framework. The tendency throughout the book is to adopt methods of analysis and design that appear significant from both the conceptual and application viewpoints. It is my hope that the book will be a useful reference for the second- or later-year graduate student.

It is assumed that the typical reader with an engineering background will have gone through a conventional undergraduate/first-year graduate single-input single-output linear systems course covering state–space descriptions, the dual notion of complete reachability and observability, and related canonical decompositions. An elementary course in control is not indispensable but may be useful for motivational purposes. For Chapters 6 and 7, and Secs. 8.2, 8.7, 8.8, 9.2–9.6, an elementary course on probability and stochastic processes—complemented with the material of Appendix D—is a prerequisite.

The problems are not always concentrated at the end of each section, but are distributed throughout the text in the places where the related questions apply or their solutions
are required. The problems mainly consist of extensions and completions of theoretical
material for which hints are often provided.

NOTE TO THE READER

Reference numbers are used to separately identify different items such as definitions, formulas, results, lemmas, propositions, and theorems. These reference numbers are started
at the beginning of each section, the latter being identified by two symbols, for example,
Sec. 3.4 and Appendix B.2. When referring to an item inside a section, we use a bare number for a formula, for example, (24), or a number preceded by the item specification for
the other items, for example, Theorem 2. When referring to an item outside a section but
inside the same chapter, we use two symbols, for example, (3-24) indicates equation (24)
of the third section inside the given chapter. When referring to an item outside the chapter,
we use three symbols, for example, (5.3-24) indicates equation (24) of the third section of
Chapter 5.

A list of abbreviations follows and an analytic index appears at the end of the book.

Edoardo Mosca
Florence, Italy

ABBREVIATIONS

a.s.	almost surely
ARE	algebraic Riccati equation
ARX	autoregressive with exogeneous input
CAR	controlled autoregressive
CARIMA	controlled autoregressive integrated moving average
CARMA	controlled autoregressive moving average
CC	cheap control
CMUSMAR	constrained multistep multivariable adaptive regulator
crd	common right divisor
CT-NRLS	constant-trace normalized recursive least squares
DOF	degree of freedom
DZ-CT-NRLS	constant-trace normalized recursive least squares with dead zone
ECE	enforced certainty equivalence
FARE	fake algebraic Riccati equation
FIR	finite impulse response
gcd	greatest common divisor
gcld	greatest common left divisor
gcrd	greatest common right divisor

GK	Gilbert–Kalman
GMV	generalized minimum variance
GPC	generalized predictive control
GPR	generalized predictive regulation
I/O	input/output
ISLM	indirect-sensing linear measurement
KF	Kalman filter
L.H.S.	left-hand side
LQ	linear quadratic
LQG	linear quadratic Gaussian
LQOR	linear quadratic output regulation
LQR	linear quadratic regulation
LQS	linear quadratic stochastic
LQSL	linear quadratic stochastic linear
LS	least squares
MFD	matrix-fraction description
MIMO	multiinput multioutput
MKI	modified Kleinman iterations
MMSE	minimum mean-square error
MPE	minimum prediction error
MRAC	model-reference adaptive control
MS	mean square
MUSMAR	multistep multivariable adaptive regulator
MV	minimum variance
ODE	ordinary differential equation
PR	positive real
PRBS	pseudo-random binary sequence
RDE	Riccati difference equation
RDZ	relative dead zone
RELS	recursive extended least squares
RELS(PO)	recursive extended least-squares with a posteriori prediction errors
RELS(PR)	recursive extended least-squares with a priori prediction errors
RH	receding horizon

RHC	receding-horizon control
RHR	receding-horizon regulation
R.H.S.	right-hand side
RLS	recursive least squares
RML	recursive maximum likelihood
RMPE	recursive minimum prediction error
SG	stochastic gradient
SIORHC	stabilizing input/output receding-horizon control
SIORHR	stabilizing input/output receding-horizon regulation
SISO	single-input single-output
SNR	signal-to-noise ratio
SPR	strictly positive real
ST	self-tuning
STC	self-tuning control(ler)
STCC	self-tuning cheap control(ler)
TCC	truncated cost control
TCI	truncated cost iterations
w.l.o.g.	without loss of generality
w.r.t.	with respect to
w.s.	weak sense

CHAPTER 1

INTRODUCTION

1.1 OPTIMAL, PREDICTIVE, AND ADAPTIVE CONTROL

This book covers various topics related to the design of discrete-time control systems via the *linear quadratic* (LQ) control methodology.

LQ control is an optimal control approach whereby the control law of a given dynamic *linear* system—the so-called plant—is analytically obtained by minimizing a performance index *quadratic* in the regulation/tracking error and control variables. LQ control is either *deterministic* or *stochastic* according to the deterministic or stochastic nature of the plant. To master LQ control theory is important for several reasons:

- LQ control theory provides a set of analytical design procedures that facilitate the synthesis of control systems with nice properties. These procedures, often implemented by commercial software packages, yield a solution that can be also used as a first cut in a trial-and-error iterative process, in case some specifications are not met by the initial LQ solution.

- LQ control allows us to design control systems under various assumptions on the information available to the controller. If this includes also the knowledge of the reference to be tracked, feedback as well as feedforward control laws— the so-called 2-DOF controllers—are jointly obtained analytically.

- More advanced control design methodologies, such as $\mathcal{H}_\infty$ control theory, can be regarded as extensions of LQ control theory.

- LQ control theory can be applied to nonlinear systems operating on a small-signal basis.

- There exists a relationship of duality between LQ control and minimum-mean-square linear prediction, filtering, and smoothing. Hence, any LQ control result has a direct counterpart in the latter areas.

LQ control theory is complemented in the book with a treatment of *multi-step predictive control* algorithms. With respect to LQ control, predictive control basically adds constraints in the tracking error, state and control variables, and uses the receding horizon control philosophy. In this way, relatively simple 2-DOF control laws can be synthesized. Their feature is that the profile of the reference over the prediction horizon can be made part of the overall control system design and dependent on both the current plant state and the desired set point. This extra freedom can be effectively used so as to guarantee a bumpless behavior and to avoid surpassing saturation bounds. These are aspects of such an importance that multistep predictive control has gained wide acceptance in industrial control applications.

Both LQ and the predictive control methodologies assume that a model of the physical system is available to the designer. When, as often happens in practice, this is not the case, we can combine on-line system identification and control design methods to build up *adaptive control* systems. Our study of such systems includes basically two classes of adaptive control systems: single-step-ahead self-tuning controllers and multistep predictive adaptive controllers. The controllers in the first class are based on simple control laws and the price paid for them originates the stability problems that they exhibit with nonminimum-phase and open-loop unstable plants. They also require that the plant I/O delay be known. The multistep predictive adaptive controllers have a substantially greater computational load and require a greater effort for convergence analysis, but overcome the just mentioned limitations.

1.2   ABOUT THIS BOOK

LQ control is by far the most thoroughly studied analytic approach to linear feedback control system design. In particular, several excellent textbooks exist on the topic. Considering also the already available books on adaptive control at various levels of rigor and generality, the question arises as to whether this new addition to the preexisting literature can be justified. We answer this question by listing some of the distinguishing features of this book.

1.2.1   The Dynamic Programming vs. the Polynomial Equation Approach

LQ control, either in a deterministic or in a stochastic setting, is customarily approached via dynamic programming by using a state–space or "internal" model of the physical system. This is a time-domain approach and yields the desired solution in terms of a Riccati difference equation. For time-invariant plants, the so-called steady-state LQ control law is obtained by letting the control horizon to become of infinite length, and, henceforth, can be computed by solving an algebraic Riccati equation. This steady-state solution can be also obtained via an alternative way, the so-called polynomial equation approach. This derives from a quite different way of looking at the LQ control problem. It uses transfer matrices or "external" models of the physical system, and turns out to be more akin to a frequency-domain methodology. It leads us to solve the steady-state LQ control problem by spectral factorization and a couple of linear Diophantine equations.

In this book, the dynamic programming and the polynomial equation approaches are thoroughly studied and compared, our experience being that mastering both can be highly beneficial for the student or practitioner. Both approaches play a synergetic role, providing two alternative ways of looking at the same problem and different sets of solving equations. As a consequence, our insight is enhanced and our ability in applying the theory strengthened.

1.2.2 Predictive vs. LQ Control

Multistep or long-range predictive control is an important topic for process control applications, some of the reasons having been outlined earlier. In this book, the emphasis is, however, on design techniques that are applicable when the plant is only partially known. Further, we study predictive control within the framework of LQ control theory. In fact, a predictive control law referred to as SIORHC (stabilizing input–output receding-horizon control) is singled out by addressing a dynamic LQ control problem in the presence of suitable constraints on its terminal regulation/tracking error and on control signals. SIORHC has the peculiarity of possessing a guaranteed stabilizing property, provided that its prediction horizon length exceeds or equals the order of the plant. This finite-horizon stabilizing property makes SIORHC particularly well suited for adaptive control wherein stabilization must be ensured irrespective of the actual value of the estimated plant parameters. SIORHC, as its prediction horizon becomes larger, tightly approximates the steady-state LQ control, when the latter optimally exploits the knowledge of the future reference profile. However, thanks to the finite length of its prediction horizon, SIORHC can be more easily computable than steady-state LQ control, because it does not require the solution of an algebraic Riccati equation or a spectral factorization problem.

1.2.3 Single-Step-Ahead vs. Multistep Predictive Adaptive Control

An entire chapter is devoted to single-step-ahead self–tuning control. This is done mainly to introduce the subject of adaptive control and the tools for analyzing more general schemes, our prevalent interest being in adaptive (multistep) predictive control systems because of their wider application potential. However, in going from single-step-ahead to more complicated control design procedures such as pole assignment, LQ, and some predictive control laws, a difficulty arises in that it may not be always feasible to evaluate the control law. A typical situation occurs when the estimated model has unstable pole-zero cancellations, that is, the estimated model becomes unstabilizable. We refer to this difficulty as the singularity problem. This has been one of the stumbling blocks for the construction of stable adaptive predictive control systems. The standard way to circumvent the singularity problem is to assume that the true plant parameter vector belongs to an a priori known convex set whose elements correspond to reachable plant models. Next, the recursive identification algorithm is modified so as to guarantee that the estimates belong to the set. For example, this can be achieved by embodying a projection facility in the identification algorithm. The alleged prior knowledge of such a convex set is instrumental to the development of (locally) convergent algorithms, but it does not appear to be justifiable in many instances. In contrast with the previous approach, in order to address convergence of adaptive multistep predictive control systems, an alternative technique is here adopted and analyzed. It consists of a self-excitation mechanism by which a dither signal is superimposed to the plant input whenever the estimated plant model is close to becoming unreachable. Under quite general conditions, the self-excitation mechanism turns off after a finite time and global convergence of the adaptive system is ensured.

1.2.4   Implicit Adaptive Predictive Control

One classic result in stochastic adaptive control is that an autoregressive moving average plant under minimum-variance control can be described in terms of a linear-regression model. This allows one to construct simple implicit self-tuning controllers based on the minimum-variance control law. In this book, it is shown that a similar property holds also when multistep predictive control laws are used. Hence, implicit adaptive predictive controllers can be constructed, wherein simple linear-regression identifiers are used. The fact that no global convergence proofs are generally available—or even feasible—does not deter us from considering implicit adaptive predictive control in view of its excellent local self-optimizing properties in the presence of neglected dynamics, and hence its possible use for autotuning reduced-complexity controllers of complex plants.

1.3   PART AND CHAPTER OUTLINES

In this section, we briefly describe the breakdown of the book into parts and chapters. The parts are three and they are described in what follows.

Part I: — Basic Deterministic Theory of LQ and Predictive Control.   This part consists of Chapters 2 to 5. Chapter 2 establishes the main facts on the deterministic LQ regulation problem. Dynamic programming is discussed and used to get the Riccati-based solution of the LQ regulator. Next, time-invariant LQ regulation is considered, and existence conditions and properties of the steady-state LQ regulator are established. Finally, two simple versions of LQ regulation, based on a control horizon comprising a single step, are analyzed and their limitations are pointed out.

Chapter 3 introduces the d–representation of a sequence, matrix-fraction descriptions of system transfer matrices and system polynomial representations. With these tools, a study follows on the characterization of stability of feedback linear systems and on the so-called YJBK parameterization of all stabilizing compensators. Finally, the asymptotic tracking problem is considered and formulated as a stability problem of a feedback system.

In Chapter 4, the polynomial approach to LQ regulation is addressed, the related solution found in terms of a spectral factorization problem and a couple of linear Diophantine equations, and its relationship with the Riccati-based solution is established. Some remarks follow on robust stability of LQ-regulated systems.

Chapter 5 introduces receding-horizon control. Zero-terminal-state regulation is first considered so as to develop dynamic receding-horizon regulation with a guaranteed stabilizing property. Within the same framework, generalized predictive regulation is treated. Next, receding-horizon iterations are introduced, our interest in them being motivated by their possible use in adaptive multistep predictive control. Finally, the tracking problem is discussed. In particular, predictive control is introduced as a 2-DOF receding-horizon control methodology, whereby the feedforward action is made dependent on the future reference evolution that, in turn, can be selected on-line so as to avoid saturation phenomena.

Part II: State Estimation, System Identification, LQ and Predictive Stochastic Control. This part consists of Chapters 6 and 7. Chapter 6 lays down the main results on recursive state estimation and system identification. The Kalman filter and various linear and pseudo-linear recursive system parameter estimators are considered and related to algorithms derived systematically via the prediction-error method. Finally, convergence properties of recursive identification algorithms are studied. The emphasis here is to prove convergence to the true system parameter vector under some strong conditions that typically cannot be guaranteed in adaptive control.

Chapter 7 extends LQ and predictive receding-horizon control to a stochastic setting. To this end, stochastic dynamic programming is used to yield the optimal LQG (linear quadratic Gaussian) control solution via the so-called certainty equivalence principle. Next, minimum-variance control and steady-state LQ stochastic control for CARMA (controlled autoregressive moving average) plants are tackled via the stochastic variant of the polynomial equation approach introduced in Chapter 3. Finally, 2-DOF tracking and servo problems are considered, and the stabilizing predictive control law, introduced in Chapter 5, is extended to a stochastic setting.

Part III: Adaptive Control. Chapters 8 and 9 combine recursive system identification algorithms with LQ and predictive control methods to build adaptive control systems for unknown linear plants. Chapter 8 describes the two basic groups of adaptive controllers, viz., model-reference and self-tuning controllers. Next, we point out the difficulties encountered in formulating adaptive control as an optimal stochastic control problem, and, in contrast, the possibility of adopting a simple suboptimal procedure by enforcing the certainty equivalence principle. We discuss the deterministic properties of the RLS (recursive least squares) identification algorithm not subject to persistency of excitation and, hence, applicable in the analysis of adaptive systems. These properties are used so as to construct a self-tuning control system, based on a simple one-step-ahead control law, for which global convergence can be established. Global convergence is also shown to hold true when a constant-trace RLS identifier with data normalization is used, the finite memory length of this identifier being important for time-varying plants. Self-tuning minimum-variance control is discussed by pointing out that implicit modeling of CARMA plants under minimum-variance control can be exploited so as to construct algorithms whose global convergence can be proved via the stochastic Lyapunov equation method. Further, it is shown that generalized minimum-variance control is equivalent to minimum-variance control of a modified plant, and, hence, globally convergent self-tuning algorithms based on the former control law can be developed by exploiting such an equivalence. Chapter 8 ends by discussing how to robustify self-tuning single-step-ahead controllers so as to deal with plants with bounded disturbances and neglected dynamics.

Chapter 9 studies various adaptive multistep predictive control algorithms, the main interest being in extending the potential applications beyond the restrictions inherent to single-step-ahead controllers. We start with considering an indirect adaptive version of the stabilizing predictive control (SIORHC) algorithm introduced in Chapter 5. We show that in order to avoid a singularity problem in the controller parameter evaluation, the notion of a self-excitation mechanism can be used. The resulting control philosophy is of dual control type in that the self-excitation mechanism switches on an input dither whenever the estimated plant parameter vector becomes close to singularity. The dither intensity must be suitably chosen, by taking into account the interaction between the dither and the feedforward signal, so as to ensure global convergence to the adaptive system. We next discuss how the indirect adaptive predictive control algorithm can be robustified in order to deal with plant bounded disturbances and neglected dynamics.

The second part of Chapter 9 deals with implicit adaptive predictive control. It first shows how the implicit modeling property of CARMA plants, previously derived under minimum-variance control, can be extended to more complex control laws, such as steady-state LQ stochastic control and variants thereof. Next, the possible use of implicit prediction models in adaptive predictive control is discussed and some examples of such controllers are given. One such controller, MUSMAR, which possesses attractive local self-optimizing properties, is studied via the ordinary differential equation (ODE) approach to analyze recursive stochastic algorithms. Two extensions of MUSMAR are finally studied. Such extensions are finalized to recover exactly the steady-state LQ stochastic regulation law as an equilibrium point of the algorithm, and to impose a mean-square input constraint to the controlled system.

Appendices. Results are given from linear system theory, polynomial matrix theory, linear Diophantine equations, probability theory, and stochastic processes.

PART
1

BASIC DETERMINISTIC
THEORY OF LQ
AND PREDICTIVE CONTROL

CHAPTER 2

DETERMINISTIC LQ REGULATION–I: RICCATI-BASED SOLUTION

The purpose of this chapter is to establish the main facts on the deterministic *linear quadratic* (LQ) regulator. After formulating the problem in Sec. 1, dynamic programming is discussed in Sec. 2 and used in Sec. 3 to get the Riccati-based solution of the LQ regulator. Section 4 discusses the time-invariant LQ regulation, the existence and properties of the steady-state regulator resulting asymptotically by letting the regulation horizon become infinitely large. Section 5 considers iterative methods for computing the steady-state regulator. In Secs. 6 and 7, two simple versions of LQ regulation, cheap control and single-step Control, are presented and analyzed.

2.1   THE DETERMINISTIC LQ REGULATION PROBLEM

The *plant* to be regulated consists of a discrete-time linear dynamic system represented as follows

$$x(k+1) = \Phi(k)x(k) + G(k)u(k) \tag{2.1-1}$$

where $k \in \mathbb{Z} := \{\cdots, -1, 0, 1, \cdots\}$; $x(k) \in \mathbb{R}^n$ denotes the plant *state* at time k; $u(k) \in \mathbb{R}^m$ the plant *input* or control at time k; and $\Phi(k)$ and $G(k)$ are matrices of compatible dimensions.

Assuming that the plant state at a given time t_0 is $x(t_0)$, the interest is to find a control sequence over the *regulation horizon* $[t_0, T]$, $t_0 \leq T - 1$,

$$u_{[t_0, T)} := \left\{ u(k) \right\}_{k=t_0}^{T-1} \tag{2.1-2}$$

that minimizes the *quadratic performance index* or *cost functional*

$$J\left(t_0,\, x(t_0),\, u_{[t_0,T)}\right) := \|x(T)\|^2_{\psi_x(T)}$$

$$+ \sum_{k=t_0}^{T-1} \left[\|x(k)\|^2_{\psi_x(k)} + 2u'(k)M(k)x(k) + \|u(k)\|^2_{\psi_u(k)} \right] \tag{2.1-3}$$

where $\|x\|^2_\psi := x'\psi x$, and the prime denotes matrix transposition. Without loss of generality, it will be assumed that $\psi_x(k)$, $\psi_u(k)$ and $\psi_x(T)$ are symmetric matrices.

Problem 2.1-1   Consider the quadratic form $x'\psi x$, $x \in \mathbb{R}^n$, with ψ any $n \times n$ matrix with real entries. Let $\psi_s = \psi'_s := (\psi + \psi')/2$. Show that $x'\psi x = x'\psi_s x$. (*Hint:* Use the fact that $\psi = \psi_s + \psi_{\bar{s}}$ if $\psi_{\bar{s}} := (\psi - \psi')/2$.)

$J(t_0, x(t_0), u_{[t_0,T)})$ quantifies the regulation performance of the plant (1), from the initial *event* $(t_0, x(t_0))$ when its input is specified by $u_{[t_0,T)}$. It is assumed that any nonzero input $u(k) \neq O_m$ is costly. This condition amounts to assuming that the symmetric matrix $\psi_u(k)$ is positive definite

$$\psi_u(k) = \psi'_u(k) > 0 \tag{2.1-4}$$

It is also assumed that the *instantaneous loss* at time k, viz., the term within brackets in (3), is nonnegative

$$\ell(k, x(k), u(k)) := \|x(k)\|^2_{\psi_x(k)} + 2u'(k)M(k)x(k) + \|u(k)\|^2_{\psi_u(k)} \geq 0 \tag{2.1-5}$$

Because by (4), $\psi_u(k)$ is nonsingular, (5) is equivalent to the following nonnegative definiteness condition:

$$\psi_x(k) - M'(k)\psi_u^{-1}(k)M(k) \geq 0 \tag{2.1-6}$$

Problem 2.1-2   Consider the quadratic form

$$\ell(x, u) := \|x\|^2_{\psi_x} + 2u'Mx + \|u\|^2_{\psi_u}$$

with $x \in \mathbb{R}^n$, $u \in \mathbb{R}^m$, and $\psi_u > 0$. Show that $\ell(x, u) \geq 0$ for every $(x, u) \in \mathbb{R}^n \times \mathbb{R}^m$ if and only if $\psi_x - M'\psi_u^{-1}M \geq 0$. (*Hint:* Find the vector $u^0(x) \in \mathbb{R}^m$ that minimizes $\ell(x, u)$ for any given x, viz., $\ell(x, u^0(x)) \leq \ell(x, u)$, $u \in \mathbb{R}^m$.)

The *terminal cost* $\|x(T)\|^2_{\psi_x(T)}$ is finally assumed nonnegative

$$\psi_x(T) = \psi'_x(T) \geq 0 \tag{2.1-7}$$

Let us consider the following as a formal statement of the deterministic LQ regulation problem.

Deterministic LQ Regulator (LQR) Problem. Consider the linear plant (1). Define the quadratic performance index (3) with $\psi_x(k)$, $\psi_u(k)$, and $\psi_x(T)$ symmetric matrices satisfying (4), (6), and (7). Find an *optimal input* $u^0_{[t_0, T)}$ to the plant (1), initialized from the event $(t_0, x(t_0))$, minimizing the performance index (3).

The general LQR problem can be transformed into an equivalent problem with no cross-product terms in its instantaneous loss. In order to see this, set

$$u(k) = \bar{u}(k) - K(k)x(k) \tag{2.1-8}$$

This means that the plant input $u(k)$ at time k is the sum of $-K(k)x(k)$, a state-feedback component, and a vector $\bar{u}(k)$.

Problem 2.1-3 Consider the instantaneous loss (5). Rewrite it as

$$\bar{\ell}(k, x(k), \bar{u}(k)) := \ell(k, x(k), \bar{u}(k) - K(k)x(k))$$

Show that the cross-product terms in $\bar{\ell}$ vanish provided that

$$K(k) = \psi_u^{-1}(k)M(k) \tag{2.1-9}$$

Show also that under the choice (9),

$$\bar{\ell}(k, x(k), \bar{u}(k)) = \|x(k)\|^2_{\bar{\psi}_x(k)} + \|\bar{u}(k)\|^2_{\psi_u(k)} \tag{2.1-10}$$

where $\bar{\psi}_x(k)$ equals the L.H.S. of (6):

$$\bar{\psi}_x(k) := \psi_x(k) - M'(k)\psi_u^{-1}(k)M(k) \tag{2.1-11}$$

Taking into account the solution of Problem 3, we can see that the general LQR problem is equivalent to the following. Given the plant

$$x(k+1) = [\Phi(k) - G(k)\psi_u^{-1}(k)M(k)]x(k) + G(k)\bar{u}(k) \tag{2.1-12}$$

find an optimal input $\bar{u}^0_{[t_0, T)}$ minimizing the performance index

$$J(t_0, x(t_0), \bar{u}_{[t_0, T)}) = \sum_{k=t_0}^{T-1} \bar{\ell}(k, x(k), \bar{u}(k)) + \|x(T)\|^2_{\psi_x(T)} \tag{2.1-13}$$

where the instantaneous loss is given by (10).

Problem 2.1-4 (LQ Tracking) Consider the plant (1) along with the n_w-dimensional linear system

$$x_w(k+1) = \Phi_w(k)x_w(k)$$

with $x_w(t_0) \in \mathbb{R}^{n_w}$ given. Let

$$\tilde{x}(k) := x(k) - x_w(k)$$

and

$$J\left(t_0, \tilde{x}(t_0), u_{[t_0,T)}\right) := \sum_{k=t_0}^{T-1} \ell(k, \tilde{x}(k), u(k)) + \|\tilde{x}(T)\|^2_{\psi_x(T)}$$

$$\ell(k, \tilde{x}(k), u(k)) := \|\tilde{x}(k)\|^2_{\psi_x(k)} + 2u'(k)M(k)\tilde{x}(k) + \|u(k)\|^2_{\psi_u(k)}$$

Show that the problem of finding an optimal input $u^0_{[t_0,T)}$ for the plant (1) that minimizes the given performance index can be cast into an equivalent LQR problem. (*Hint*: Consider the plant with "extended" state $\chi(k) := [\; x'(k) \quad x'_w(k)\;]'$.)

2.2 DYNAMIC PROGRAMMING

A solution method that exploits in an essential way the dynamic nature of the LQR problem is Bellman's technique of *dynamic programming*. Dynamic programming is discussed here only to the extent necessary to solve the LQR problem. In doing this, we consider a larger class of optimal regulation problems so as to better focus our attention on the essential features of dynamic programming.

Let the plant be described by a possibly nonlinear state–space representation:

$$x(k + 1) = f(k, x(k), u(k)) \tag{2.2-1}$$

As in (1-1), $x(k) \in \mathbb{R}^n$ and $u(k) \in \mathbb{R}^m$. Function f, referred to as the *local state-transition function*, specifies the rule according to which the *event* $(k, x(k))$ is transformed, by a given input $u(k)$ at time k, into the next *plant state* $x(k + 1)$ at time $k + 1$. By iterating (1), it is possible to define the *global state-transition function*:

$$x(j) = \varphi\left(j, k, x(k), u_{[k,j)}\right) \quad , \quad j \geq k \tag{2.2-2}$$

Function φ, for a given input sequence $u_{[k,j)}$, $j \geq k$, specifies the rule according to which the initial event $(k, x(k))$ is transformed into the final event $(j, x(j))$. For example,

$$\begin{aligned}
x(k + 2) &= f(k + 1, x(k + 1), u(k + 1)) \\
&= f(k + 1, f(k, x(k), u(k)), u(k + 1)) \qquad [(1)] \\
&=: \varphi(k + 2, k, x(k), u_{[k,k+2)})
\end{aligned}$$

For $j = k$, $u_{[k,j)}$ is empty, and, consequently, the system is left in the event $(k, x(k))$. This amounts to assuming that φ satisfies the following *consistency condition*:

$$\varphi(k, k, x(k), u_{[k,k)}) = x(k) \tag{2.2-3}$$

Problem 2.2-1 Show that for the linear dynamic system (1-1), the global state-transition function equals

$$\varphi(j, k, x(k), u_{[k,j)}) = \Phi(j, k)x(k) + \sum_{i=k}^{j-1} \Phi(j, i + 1)G(i)u(i)$$

where

$$\Phi(j,k) := \left\{ \begin{array}{ll} I_n, & j = k \\ \Phi(j-1)\cdots\Phi(k), & j > k \end{array} \right. \tag{2.2-4}$$

is the *state-transition matrix* of the linear system.

Along with the plant (1) initialized from the event $(t_0, x(t_0))$, we consider the following possibly nonquadratic performance index:

$$J(t_0, x(t_0), u_{[t_0,T)}) = \sum_{k=t_0}^{T-1} \ell(k, x(k), u(k)) + \psi(x(T)) \tag{2.2-5}$$

where again $\ell(k, x(k), u(k))$ stands for a nonnegative instantaneous loss incurred at time k, $\psi(x(T))$ for a nonnegative loss due to the terminal state $x(T)$, and $[t_0, T]$ for the regulation horizon. The problem is to find an optimal control $u_{[t_0,T)}^0$ for the plant (1), initialized from $(t_0, x(t_0))$, minimizing (5).

Hereafter, conditions on (1) and (5) will be implicitly assumed in order that each step of the adopted optimization procedure makes sense. For $t \in [t_0, T]$, consider the so called *Bellman function:*

$$\begin{aligned} V(t, x(t)) \quad := \quad & \min_{u_{[t,T)}} J\Big(t, x(t), u_{[t,T)}\Big) \\[2mm] = \quad & \min_{u_{[t,t_1)}} \left\{ \min_{u_{[t_1,T)}} \left[\sum_{k=t}^{t_1-1} \ell(k, x(k), u(k)) \right. \right. \\[2mm] & \left. \left. + J\Big(t_1, \varphi\Big(t_1, t, x(t), u_{[t,t_1)}\Big), \; u_{[t_1,T)}\Big) \right] \right\} \end{aligned} \tag{2.2-6}$$

The second equality follows because $u_{[t,T)} = u_{[t,t_1)} \otimes u_{[t_1,T)}$, for $t_1 \in [t, T)$, $\otimes$ denoting concatenation. Equation (6) can be rewritten as follows:

$$\begin{aligned} V(t, x(t)) \quad = \quad & \min_{u_{[t,t_1)}} \left\{ \sum_{k=t}^{t_1-1} \ell(k, x(k), u(k)) \right. \\[2mm] & \left. + \min_{u_{[t_1,T)}} J\Big(t_1, \varphi\Big(t_1, t, x(t), u_{[t,t_1)}\Big), \; u_{[t_1,T)}\Big) \right\} \\[2mm] = \quad & \min_{u_{[t,t_1)}} \left\{ \sum_{k=t}^{t_1-1} \ell(k, x(k), u(k)) + V\left(t_1, \varphi\left(t_1, t, x(t), u_{[t,t_1)}\right)\right) \right\} \end{aligned} \tag{2.2-7}$$

Suppose now that $u_{[t,T)}^0$ is an optimal input over the horizon $[t, T)$ for the initial event $(t, x(t))$, viz.

$$\begin{aligned} V(t, x(t)) \quad &= \quad J\Big(t, x(t), u_{[t,T)}^0\Big) \\[2mm] &\leq \quad J\Big(t, x(t), u_{[t,T)}\Big) \end{aligned}$$

for all control sequences $u_{[t,T)}$. Then, from (7), it follows that $u^0_{[t_1,T)}$, the restriction of $u^0_{[t,T)}$ to $[t_1, T)$, is again an optimal input over the horizon $[t_1, T)$ for the initial event $(t_1, x(t_1))$, $x(t_1) := \varphi(t_1, t, x(t), u^0_{[t,t_1)})$, viz.,

$$
\begin{aligned}
V(t_1, x(t_1)) &= J(t_1, x(t_1), u^0_{[t_1,T)}) \\
&\leq J(t_1, x(t_1), u_{[t_1,T)})
\end{aligned}
$$

The following statement is a way of expressing Bellman's *principle of optimality*.

> **Principle of Optimality.** This principle states that an optimal input sequence $u^0_{[t,T)}$ is such that, given an event $(t_1, x(t_1))$ along the corresponding optimal trajectory, $x(t_1) = \varphi(t_1, t_0, x(t_0), u^0_{[t_0,t_1)})$, the subsequent input sequence $u^0_{[t_1,T)}$ is again optimal for the cost to go over the horizon $[t_1, T]$.

For $t_1 = t + 1$, (7) yields the *Bellman equation:*

$$
V(t, x(t)) = \min_{u(t)} \left\{ \ell(t, x(t), u(t)) + V(t+1, f(t, x(t), u(t))) \right\} \tag{2.2-8}
$$

with the terminal-event condition:

$$
V(T, x(T)) = \psi(x(T)) \tag{2.2-9}
$$

The functional equation (8) can be used as follows. Equation (8) for $t = T - 1$ gives

$$
\begin{aligned}
V(T-1, x(T-1)) &= \min_{u(T-1)} \left\{ \ell(T-1, x(T-1), u(T-1)) + \psi(x(T)) \right\} \\
x(T) &= f(T-1, x(T-1), u(T-1)) \tag{2.2-10}
\end{aligned}
$$

If this can be solved w.r.t. $u(T-1)$ for any state $x(T-1)$, one finds an optimal input at time $T - 1$ in a state-feedback form:

$$
u^0(T-1) = u^0(T-1, x(T-1)) \tag{2.2-11}
$$

and hence determines $V(T-1, x(T-1))$. By iterating backward the preceding procedure, provided that at each step a solution can be found, one can determine an optimal control law in a state-feedback form:

$$
u^0(k) = u^0(k, x(k)) \qquad k \in [t_0, T) \tag{2.2-12}
$$

and $V(k, x(k))$.

Before proceeding any further, let us consolidate the discussion so far. We have used the principle of optimality of dynamic programming to obtain the Bellman equation (8). This suggests the procedure just outlined for obtaining an optimal control. It is remarkable that, if a solution can be obtained, it is in a state-feedback form. The next theorem shows that, provided that the procedure yields a solution, it solves the optimal regulation problem at hand.

Theorem 2.2-1. *Suppose that $\{V(t,x)\}_{t=t_0}^{T}$ satisfies the Bellman equation (8) with terminal condition (9). Suppose that a minimum as in (8) exists and is attained at*

$$\hat{u}(t) = \hat{u}(t,x)$$

viz.,

$$\ell(t,x,\hat{u}(t)) + V(t+1, f(t,x,\hat{u}(t))) \leq \ell(t,x,u) + V(t+1, f(t,x,u)) \ , \ \ \forall u \in \mathbb{R}^m$$

Define $x^0_{[t_0,T]}$ and $u^0_{[t_0,T)}$ recursively as follows:

$$x^0(t_0) \ = \ x(t_0) \tag{2.2-13}$$

$$\left.\begin{array}{rcl} u^0(t) & = & \hat{u}(t, x^0(t)) \\ x^0(t+1) & = & f(t, x^0(t), u^0(t)) \end{array}\right\} \ \ t = t_0, t_0+1, \cdots, T-1 \tag{2.2-14}$$

Then $u^0_{[t_0,T)}$ is an optimal control sequence, and the minimum cost equals $V(t_0, x(t_0))$.

Proof: It is to be shown that, if $u^0_{[t_0,T)}$ is defined as in (14),

$$\begin{array}{rcl} V(t_0, x(t_0)) & = & J\Big(t_0, x(t_0), u^0_{[t_0,T)}\Big) \\ & \leq & J\Big(t_0, x(t_0), u_{[t_0,T)}\Big) \end{array} \tag{2.2-15}$$

for all control sequences $u_{[t_0,T)}$.

Because for $x(t) = x^0(t)$, the R.H.S. of (8) attains its minimum at $u^0(t)$, one has

$$V(t, x^0(t)) = \ell(t, x^0(t), u^0(t)) + V(t+1, x^0(t+1)) \tag{2.2-16}$$

Hence,

$$V\Big(t_0, x(t_0)\Big) - V\Big(T, x^0(T)\Big) \tag{2.2-17}$$

$$= \sum_{t=t_0}^{T-1} \Big[V\Big(t, x^0(t)\Big) - V\Big(t+1, x^0(t+1)\Big)\Big]$$

$$= \sum_{t=t_0}^{T-1} \ell\Big(t, x^0(t), u^0(t)\Big)$$

$V(T, x^0(T)) = \psi(x(T))$, so the equality in (15) follows.

Now for every control sequence $u_{[t_0,T)}$ applied to the plant initialized from the event $(t_0, x(t_0))$, one has

$$V(t, x(t)) \leq \ell(t, x(t), u(t)) + V(t+1, x(t+1)) \tag{2.2-18}$$

if

$$\begin{array}{rcl} x(t+1) & = & f(t, x(t), u(t)) \\ & = & \varphi\big(t, t_0, x(t_0), u_{[t_0,T)}\big) \end{array}$$

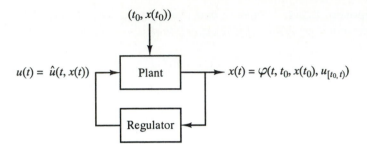

Figure 2.2-1: Optimal solution of the regulation problem in a state-feedback form as given by dynamic programming.

Using (18) instead of (16), one finds the next inequality in place of (17):

$$V(t_0, x(t_0)) \quad \leq \quad \sum_{t=t_0}^{T-1} \ell(t, x(t), u(t)) + \psi_x(x(T))$$

$$= \quad J\left(t_0, x(t_0), u_{[t_0, T)}\right) \qquad (2.2\text{-}19)$$

Main Points of the Section. The Bellman equation of dynamic programming, if solvable, yields, via backward iterations, the optimal regulator in a state-feedback form (Fig. 1).

2.3   RICCATI-BASED SOLUTION

The Bellman equation (2-8) is now applied to solve the deterministic LQR problem of Sec. 2.1. In this case, the plant is as in (1-1), the performance index as in (1-3) with the instantaneous loss as in (1-5). Taking into account (1-7), one sees that $V(T, x(T))$, the Bellman function at the terminal event, equals the quadratic function $x'(T)\psi_x(T)x(T)$ with the matrix $\psi_x(T)$ symmetric and nonnegative definite. By adopting the procedure outlined in Sec. 2.2 to compute backward $V(t, x(t)), t = T - 1, T - 2, \cdots, t_0$, the solution of the LQR problem is obtained.

Theorem 2.3-1.   *The solution to the deterministic LQR problem of Sec. 2.1 is given by the following linear state-feedback control law*

$$u(t) = F(t)x(t) , \quad t \in [t_0, T) \qquad (2.3\text{-}1)$$

where $F(t)$ is the LQR feedback-gain *matrix:*

$$F(t) = -\left[\psi_u(t) + G'(t)\mathcal{P}(t+1)G(t)\right]^{-1}\left[M(t) + G'(t)\mathcal{P}(t+1)\Phi(t)\right] \qquad (2.3\text{-}2)$$

*and $\mathcal{P}(t)$ is the symmetric nonnegative definite matrix given by the solution of the
following backward* Riccati difference equation (RDE):

$$
\begin{aligned}
\mathcal{P}(t) &= \Phi'(t)\mathcal{P}(t+1)\Phi(t) - \Big[M'(t) + \Phi'(t)\mathcal{P}(t+1)G(t)\Big] \\
&\quad \times \Big[\psi_u(t) + G'(t)\mathcal{P}(t+1)G(t)\Big]^{-1} \\
&\quad \times \Big[M(t) + G'(t)\mathcal{P}(t+1)\Phi(t)\Big] + \psi_x(t)
\end{aligned}
\tag{2.3-3}
$$

$$
\begin{aligned}
&= \Phi'(t)\mathcal{P}(t+1)\Phi(t) \\
&\quad - F'(t)\Big[\psi_u + G'(t)\mathcal{P}(t+1)G(t)\Big]F(t) + \psi_x(t)
\end{aligned}
\tag{2.3-4}
$$

$$
\begin{aligned}
&= \Big[\Phi(t) + G(t)F(t)\Big]'\mathcal{P}(t+1)\Big[\Phi(t) + G(t)F(t)\Big] \\
&\quad + F'(t)\psi_u(t)F(t) + M'(t)F(t) + F'(t)M(t) + \psi_x(t)
\end{aligned}
\tag{2.3-5}
$$

with terminal condition

$$
\mathcal{P}(T) = \psi_x(T)
\tag{2.3-6}
$$

Further,

$$
\begin{aligned}
V(t, x(t)) &= \min_{u_{[t,T)}} J\Big(t, x(t), u_{[t,T)}\Big) \\
&= x'(t)\mathcal{P}(t)x(t)
\end{aligned}
\tag{2.3-7}
$$

Proof (by induction): It is known that $V(T, x(T))$ is given by (7) if $\mathcal{P}(T)$ is as in
(6). Next, assume that $V(t+1, x(t+1)) = \|x(t+1)\|^2_{\mathcal{P}(t+1)}$ with $\mathcal{P}(t+1) = \mathcal{P}'(t+1) \geq 0$
and $x(t+1) = \Phi(t)x(t) + G(t)u(t)$. Show that $V(t, x(t)) = \|x(t)\|^2_{\mathcal{P}(t)}$ with $\mathcal{P}(t)$ satisfying
(3). One has

$$
\begin{aligned}
V(t, x(t)) &= \min_{u(t)} \hat{J}(t, x(t), u(t)) \\
&:= \min_{u(t)} \Big\{ \|x(t)\|^2_{\psi_x(t)} + 2u'(t)M(t)x(t) \\
&\qquad + \|u(t)\|^2_{\psi_u(t)} + \|\Phi(t)x(t) + G(t)u(t)\|^2_{\mathcal{P}(t+1)} \Big\}
\end{aligned}
\tag{2.3-8}
$$

Let $u(t) = [u_1(t) \cdots u_m(t)]'$. Set to zero the gradient vector of $\hat{J}$ w.r.t. $u(t)$

$$
\begin{aligned}
O_m = \frac{\partial \hat{J}(t, x(t), u(t))}{2\partial u(t)} &:= \frac{1}{2}\Big[\frac{\partial \hat{J}}{\partial u_1(t)} \cdots \frac{\partial \hat{J}}{\partial u_m(t)}\Big]' \\
&= [M(t) + G'(t)\mathcal{P}(t+1)\Phi(t)]x(t) \\
&\quad + [\psi_u(t) + G'(t)\mathcal{P}(t+1)G(t)]u(t)
\end{aligned}
\tag{2.3-9}
$$

This yields (1) and (2). That these two equations give uniquely the optimizing input $u(t)$,
it follows from invertibility of $[\psi_u(t) + G'(t)\mathcal{P}(t+1)G(t)]$ and positive definiteness of the

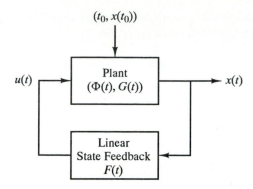

Figure 2.3-1: LQR solution.

Hessian matrix:

$$\frac{\partial^2 \hat{J}(t, x(t), u(t))}{\partial^2 u(t)} = 2[\psi_u(t) + G'(t)\mathcal{P}(t+1)G(t)] > 0$$

By substituting (1) and (2) into $J(t, x(t), u(t))$, (7) is obtained with $\mathcal{P}(t)$ satisfying (3). Equation (3) shows that $\mathcal{P}(t)$ is symmetric.

To complete the induction, it now remains to show that $\mathcal{P}(t)$ is nonnegative definite. Rewrite (4) as follows:

$$\mathcal{P}(t) = \Phi'(t)\mathcal{P}(t+1)\big[\Phi(t) + G(t)F(t)\big] + M'(t)F(t) + \psi_x(t) \qquad (2.3\text{-}10)$$

Further, premultiply both sides of (2) by $F'(t)[\psi_u(t) + G'(t)\mathcal{P}(t+1)G(t)]$ to get

$$F'(t)\psi_u(t)F(t) \quad = \quad -F'(t)M(t) - F'(t)G'(t)\mathcal{P}(t+1)\big[\Phi(t) + G(t)F(t)\big] \qquad (2.3\text{-}11)$$

Subtracting (11) from (10), we find (5). Next, by virtue of (1-6),

$$
\begin{aligned}
& F'(t)\psi_u(t)F(t) + M'(t)F(t) + F'(t)M(t) + \psi_x(t) \\
\geq \quad & F'(t)\psi_u(t)F(t) + M'(t)F(t) + F'(t)M(t) + M'(t)\psi_u^{-1}(t)M(t) \qquad (2.3\text{-}12) \\
= \quad & \big[F'(t) + M'(t)\psi_u^{-1}(t)\big]\psi_u(t)\big[F(t) + \psi_u^{-1}(t)M(t)\big]
\end{aligned}
$$

From (5), (12), and the nonnegative definiteness of $\mathcal{P}(t+1)$, $\mathcal{P}(t)$ is seen to be lower-bounded by the sum of two nonnegative definite matrices. Hence, $\mathcal{P}(t)$ is also nonnegative definite.

Main Points of the Section. For any horizon of finite length, the LQR problem is solved (Fig. 1) by a regulator consisting of a linear time-varying state-feedback gain matrix $F(t)$, computable by solving a Riccati difference equation.

2.4 TIME-INVARIANT LQR

It is of interest to make a detailed study of the LQR properties in the *time-invariant case*. In this case, the plant (1-1) and the weights in (1-3) are time-invariant, viz., $\Phi(k) = \Phi$; $G(k) = G$; $\psi_x(k) = \psi_x$; $M_k(k) = M$; and $\psi_u(k) = \psi_u$, for $k = t_0, \cdots, T-1$.

By time invariance, we have for the cost (1-3)

$$J\left(t_0, x, \hat{u}_{[t_0,T)}\right) = J\left(0, x, u_{[0,N)}\right) \tag{2.4-1}$$

where

$$u(\cdot) := \hat{u}(\cdot + t_0) \quad \text{and} \quad N := T - t_0$$

for any $x \in \mathbb{R}^n$ and input sequence $\hat{u}(\cdot)$. In (1), $\hat{u}(\cdot + t_0)$ indicates the sequence $\hat{u}(\cdot)$ anticipated in time by t_0 steps. The notation can be further simplified, by rewriting (1) as $J\left(x, u_{[0,N)}\right)$, where it is understood that x denotes the initial state of the plant to be regulated, and N the length of the regulation horizon. The following is a restatement of the deterministic LQR problem in the time-invariant case.

LQR Problem in the Time-Invariant Case. Consider the time-invariant linear plant

$$\begin{aligned} x(k+1) &= \Phi x(k) + Gu(k) \\ x(0) &= x \end{aligned} \left. \right\} \tag{2.4-2}$$

along with the quadratic performance index

$$J\left(x, u_{[0,N)}\right) := \sum_{k=0}^{N-1} \ell(x(k), u(k)) + \|x(N)\|^2_{\psi_x(N)} \tag{2.4-3}$$

$$\ell(x, u) := \|x\|^2_{\psi_x} + 2u'Mx + \|u\|^2_{\psi_u} \tag{2.4-4}$$

where ψ_x, ψ_u, and $\psi_x(N)$ are symmetric matrices satisfying

$$\psi_u = \psi_u' > 0 \tag{2.4-5}$$
$$\bar{\psi}_x := \psi_x - M'\psi_u^{-1}M \geq 0 \tag{2.4-6}$$
$$\psi_x(N) = \psi_x'(N) \geq 0 \tag{2.4-7}$$

Find an optimal input $u^0(\cdot)$ to the plant (2) with initial state x, minimizing the performance index (3) over an N-step regulation horizon.

For any *finite* N, Theorem 3-1 provides, of course, the solution to the problem (2) to (7). Here the solution depends on the matrix sequence $\{\mathcal{P}(t)\}_{t=0}^N$ that can be computed by iterating backward the matrix Riccati equation (3-3) to (3-5). Equivalently, by setting

$$P(j) := \mathcal{P}(N-j) \quad , \quad j = 0, 1, \cdots, N \tag{2.4-8}$$

we can express the solution via forward Riccati iterations as in the next theorem.

Theorem 2.4-1. *In the time-invariant case, the solution to the determin-istic LQR problem (2) to (7) is given by the following state-feedback control:*

$$u(N - j) = F(j)x(N - j) \quad , \quad j = 1, \cdots, N \tag{2.4-9}$$

where $F(j)$ is the LQR feedback matrix:

$$F(j) = -[\psi_u + G'P(j - 1)G]^{-1}[M + G'P(j - 1)\Phi] \tag{2.4-10}$$

and $P(j)$ is the symmetric nonnegative definite matrix solution of the following forward RDE:

$$
\begin{aligned}
P(j) &= \Phi'P(j - 1)\Phi - \Big[M' + \Phi'P(j - 1)G\Big] \\
&\quad \times \Big[\psi_u + G'P(j - 1)G\Big]^{-1}\Big[M + G'P(j - 1)\Phi\Big] + \psi_x \tag{2.4-11} \\
&= \Phi'P(j - 1)\Phi - F'(j)\Big[\psi_u + G'P(j - 1)G\Big]F(j) + \psi_x \tag{2.4-12} \\
&= \Big[\Phi + GF(j)\Big]'P(j - 1)\Big[\Phi + GF(j)\Big] \\
&\quad + F'(j)\psi_u F(j) + M'F(j) + F'(j)M + \psi_x \tag{2.4-13}
\end{aligned}
$$

with initial condition

$$P(0) = \psi_x(N) \tag{2.4-14}$$

Further, the Bellman function $V_j(x)$, relative to an initial state x and a j-step regulation horizon, with terminal state costed by $P(0)$, equals

$$
\begin{aligned}
V_j(x) : &= \min_{u_{[N-j,N)}} J\Big(x, u_{[N-j,N)}\Big) \\
&= \min_{u_{[0,j)}} J\Big(x, u_{[0,j)}\Big) \tag{2.4-15} \\
&= x'P(j)x
\end{aligned}
$$

Our interest will be now focused on the limit properties of the LQR solution (9) to (15) as $j \to \infty$, that is, as the length of the regulation horizon becomes infinite. The interest is motivated because if a limit solution exists, the corresponding state feedback may yield good transient as well as steady-state regulation properties to the controlled system.

We start by studying the convergence properties of $P(j)$ as $j \to \infty$. As Example 1 shows, the limit of $P(j)$ for $j \to \infty$ need not exist. In particular, we see that some stabilizability condition on the pair (Φ, G) must be satisfied if the limit has to exist.

Example 2.4-1 Consider the plant (2) with

$$\Phi = \begin{bmatrix} 1 & 1 \\ 0 & 2 \end{bmatrix} \qquad G = \begin{bmatrix} 1 \\ 0 \end{bmatrix} \tag{2.4-16}$$

For the pair (Φ, G), 2 is an unstable unreachable eigenvalue. Hence, (Φ, G) is not stabilizable. Let $x_2(k)$ be the second component of the plant state $x(k)$. It is seen that $x_2(k)$ is unaffected by $u(\cdot)$. In fact, it satisfies the following homogeneous difference equation

$$x_2(k+1) = 2x_2(k) \tag{2.4-17}$$

Consider the performance index (3) with $\psi_x(N) = O_{2\times 2}$ and instantaneous loss

$$\ell(x, u) = x_2^2 + u^2 \tag{2.4-18}$$

Assume that the corresponding matrix sequence $\{P(j)\}_{j=0}^{\infty}$ admits a limit as $j \to \infty$:

$$\lim_{j\to\infty} P(j) = P(\infty) \leq M \tag{2.4-19}$$

Then, according to (15), there is an input sequence for which

$$\begin{aligned}
\lim_{j\to\infty} J\big(x, u_{[0,j)}\big) &= \lim_{j\to\infty} \sum_{k=0}^{j-1} \left[x_2^2(k) + u^2(k) \right] \\
&= x'P(\infty)x \;<\; \infty
\end{aligned}$$

However, the last inequality contradicts the fact that the performance index (3), with $\ell(x, u)$ as in (18) and $x_2(k)$ satisfying (17), diverges as $j \to \infty$ for any initial state $x \in \mathbb{R}^2$ such that $x_2 \neq 0$, irrespective of the input sequence. Therefore, by contradiction, we conclude that the limit (19) does not exist.

Problem 1 applies the results of Theorem 1 to the plant (2) when $G = O_{n\times m}$ and Φ is a stability matrix.

Problem 2.4-1 Consider the sequence $\{x(k)\}_{k=0}^{N-1}$ satisfying the difference equation

$$x(k+1) = \Phi x(k) \tag{2.4-20}$$

Show that

$$\sum_{k=0}^{N-1} \|x(k)\|_{\psi_x}^2 = \|x(0)\|_{\mathcal{L}(N)}^2 \tag{2.4-21}$$

where $\mathcal{L}(N)$ is the symmetric nonnegative definite matrix obtained by the following *Lyapunov difference equation:*

$$\mathcal{L}(j+1) = \Phi' \mathcal{L}(j)\Phi + \psi_x \qquad j = 0, 1, \cdots \tag{2.4-22}$$

initialized from $\mathcal{L}(0) = O_{n\times m}$. Next, show that the following limits exist:

$$\lim_{N\to\infty} \sum_{k=0}^{N-1} \|x(k)\|_{\psi_x}^2 = \|x(0)\|_{\mathcal{L}(\infty)}^2 \tag{2.4-23}$$

$$\lim_{N\to\infty} \mathcal{L}(N) =: \mathcal{L}(\infty) \tag{2.4-24}$$

provided that Φ is a *stability matrix*, that is,

$$|\lambda(\Phi)| < 1 \tag{2.4-25}$$

if $\lambda(\Phi)$ denotes any eigenvalue of Φ. Finally, show that $\mathcal{L}(\infty)$ satisfies the following (algebraic) *Lyapunov equation:*

$$\mathcal{L}(\infty) = \Phi'\mathcal{L}(\infty)\Phi + \psi_x \tag{2.4-26}$$

That (26) has a unique solution under (25), it follows from a result of matrix theory [Fra64].

The next lemma will be used in the study of the limiting properties as $j \to \infty$ of the solution $P(j)$ of the Riccati equation (11) to (13)

Lemma 2.4-1.   *Let $\{P(j)\}_{j=0}^{\infty}$ be a sequence of matrices in $\mathbb{R}^{n \times n}$ such that:*

(i) *Every $P(j)$ is symmetric and nonnegative definite*

$$P(j) = P'(j) \geq 0 \tag{2.4-27}$$

(ii) *$\{P(j)\}_{j=0}^{\infty}$ is monotonically nondecreasing, viz.*

$$i \leq j \quad \Rightarrow \quad P(i) \leq P(j) \tag{2.4-28}$$

(iii) *$\{P(j)\}_{j=0}^{\infty}$ is bounded from above, viz., there exists a matrix $Q \in \mathbb{R}^{n \times n}$ such that, for every j,*

$$P(j) \leq Q \tag{2.4-29}$$

Then, $\{P(j)\}_{j=0}^{\infty}$ admits a symmetric nonnegative definite limit $\hat{P}$ as $j \to \infty$

$$\lim_{j \to \infty} P(j) = \hat{P} \tag{2.4-30}$$

Proof: For every $x \in \mathbb{R}^n$, the real-valued sequence $\{\alpha(j)\} := \{x'P(j)x\}$ is, by (ii), monotonically nondecreasing and, by (iii), upper bounded by $x'Qx$. Hence, there exists $\lim_{j \to \infty} \alpha(j) = \hat{\alpha}$. Now take $x = e_i$, where e_i is the ith vector of the natural basis of $\mathbb{R}^n$. Thus, with such a choice, $x'P(j)x = P_{ii}(j)$ if P_{ik} denotes the (i,k)th entry of P. Hence, we have established that there exist

$$\lim_{j \to \infty} P_{ii}(j) = \hat{P}_{ii} \quad i = 1, \cdots, n$$

Next, take $x = e_i + e_k$. Under such a choice, $x'P(j)x = P_{ii}(j) + 2P_{ik}(j) + P_{kk}(j)$. This admits a limit as $j \to \infty$. Because $\lim_{j \to \infty} P_{ii}(j) = \hat{P}_{ii}$ and $\lim_{j \to \infty} P_{kk}(j) = \hat{P}_{kk}$, there exists

$$\lim_{j \to \infty} P_{ik}(j) = \hat{P}_{ik}$$

Because we have established the existence of the limit as $j \to \infty$ of all entries of $P(j)$, and $P(j)$ satisfies (27), it follows that $\hat{P}$ exists symmetric and nonnegative definite.

We show next that the solution of the Riccati iterations (11) to (13) initialized from $P(0) = O_{n \times n}$ enjoys properties (i) to (iii) of Lemma 1, provided that the pair (Φ, G) is stabilizable.

Proposition 2.4-1. *Consider the matrix sequence $\{P(j)\}_{j=0}^{\infty}$ generated by the Riccati iterations (11) to (13) initialized from $P(0) = O_{n \times n}$. Then $\{P(j)\}_{j=0}^{\infty}$ enjoys properties (i) to (iii) of Lemma 1, provided that (Φ, G) is a stabilizable pair.*

Proof: Property (i) of Lemma 1 is clearly satisfied. To prove property (ii), we proceed as follows. Consider the LQ optimal input $u_{[0,j+1)}^0$ for the regulation horizon $[0, j+1]$ and an initial plant state x. Let $x_{[0,j+1]}^0$ be the corresponding state evolution. Then

$$
\begin{aligned}
x'P(j+1)x &= \sum_{k=0}^{j-1} \ell(x^0(k), u^0(k)) + \ell(x^0(j), u^0(j)) \\
&\geq \sum_{k=0}^{j-1} \ell(x^0(k), u^0(k)) \\
&\geq \min_{u_{[0,j)}} \sum_{k=0}^{j-1} \ell(x(k), u(k)) = x'P(j)x
\end{aligned}
\tag{2.4-31}
$$

Hence, $\{P(j)\}_{j=0}^{\infty}$ is monotonically nondecreasing.

To check property (iii), consider a feedback-gain matrix F that stabilizes Φ, viz. $\Phi + GF$ is a stability matrix. Let

$$
\hat{u}(k) = F\hat{x}(k) \qquad k = 0, 1, \cdots \tag{2.4-32}
$$

and correspondingly

$$
\begin{aligned}
\hat{x}(k+1) &= (\Phi + GF)\hat{x}(k) \\
x(0) &= x
\end{aligned}
\tag{2.4-33}
$$

Recall that by (3-12), $\psi_x + F'M + M'F + F'\psi_u F$ is a symmetric and nonnegative definite matrix. Then, by Problem 1, there exists a matrix $Q = Q' \geq 0$, a solution of the Lyapunov equation:

$$
Q = (\Phi + GF)'Q(\Phi + GF) + \psi_x + F'M + M'F + F'\psi_u F \tag{2.4-34}
$$

and such that

$$
\begin{aligned}
x'Qx &= \sum_{k=0}^{\infty} \ell(\hat{x}(k), \hat{u}(k)) \\
&\geq \sum_{k=0}^{j-1} \ell(\hat{x}(k), \hat{u}(k)) \\
&\geq \min_{u_{[0,j)}} \sum_{k=0}^{j-1} \ell(x(k), u(k)) = x'P(j)x
\end{aligned}
\tag{2.4-35}
$$

Hence, $\{P(j)\}_{j=0}^{\infty}$ is upper-bounded by Q.

Proposition 1, together with Lemma 1, enables us to establish a sufficient condition for the existence of the limit of $P(j)$ as $j \to \infty$.

Theorem 2.4-2.   *Consider the matrix sequence* $\{P(j)\}_{j=0}^{\infty}$ *generated by the Riccati iterations (11) to (13) initialized from* $P(0) = O_{n \times n}$. *Then, if* (Φ, G) *is a stabilizable pair, there exists the limit of* $P(j)$ *as* $j \to \infty$:

$$\hat{P} := \lim_{j \to \infty} P(j) \qquad (2.4\text{-}36)$$

$\hat{P}$ *is symmetric nonnegative definite and satisfies the* algebraic Riccati equation

$$
\begin{aligned}
\hat{P} &= \Phi' \hat{P} \Phi - & (2.4\text{-}37)\\
&\quad (M' + \Phi' \hat{P} G)(\psi_u + G' \hat{P} G)^{-1}(M + G' \hat{P} \Phi) + \psi_x \\
&= \Phi' \hat{P} \Phi - \hat{F}(\psi_u + G' \hat{P} G)^{-1} \hat{F} + \psi_x & (2.4\text{-}38)\\
&= (\Phi + G\hat{F})' \hat{P}(\Phi + G\hat{F}) + \hat{F}' \psi_u \hat{F} + M' \hat{F} + \hat{F}' M + \psi_x & (2.4\text{-}39)
\end{aligned}
$$

with

$$\hat{F} = -(\psi_u + G' \hat{P} G)^{-1}(M + G' \hat{P} \Phi) \qquad (2.4\text{-}40)$$

Under the foregoing circumstances, the infinite-horizon or steady-state LQR, *for which*

$$\min_{u_{[0,\infty)}} J\left(x, u_{[0,\infty)}\right) = x' \hat{P} x \qquad (2.4\text{-}41)$$

is given by the state-feedback control:

$$u(k) = \hat{F} x(k) \qquad k = 0, 1, \cdots \qquad (2.4\text{-}42)$$

It is to be pointed out that Theorem 1 does *not* give any insurance on the asymptotic stability of the resulting closed-loop system:

$$x(k+1) = (\Phi + G\hat{F})x(k) \qquad (2.4\text{-}43)$$

Stability has to be guaranteed in order to make the steady-state LQR applicable in practice. We now begin to study stability of the closed-loop system (43), should $P(j)$ admit a limit $\hat{P}$, for $j \to \infty$. For the sake of simplicity, this study will be carried out with reference to the *linear quadratic output regulation* (LQOR) problem defined as follows.

LQOR Problem in the Time–Invariant Case.   Here the plant is described by a linear time-invariant state–space representation:

$$
\left.
\begin{aligned}
x(k+1) &= \Phi x(k) + G u(k) \\
x(0) &= x \\
y(k) &= H x(t)
\end{aligned}
\right\} \qquad (2.4\text{-}44)
$$

where $y(k) \in \mathbb{R}^p$ is the *output* to be regulated at zero. A quadratic performance index as in (3) is considered with instantaneous loss:

$$\ell(x, u) := \|y\|_{\psi_y}^2 + \|u\|_{\psi_u}^2 \qquad (2.4\text{-}45)$$

where ψ_u satisfies (5) and

$$\psi_y = \psi_y' > 0 \qquad (2.4\text{-}46)$$

Because in view of (44), $\|y\|_{\psi_y}^2 = \|x\|_{\psi_x}^2$ whenever

$$\psi_x = H'\psi_y H \qquad (2.4\text{-}47)$$

it appears that the LQOR problem is an LQR problem with $M = 0$. However, we recall that, by (1-8) to (1-13), each LQR problem can be cast into an equivalent problem with no cross-product terms in the instantaneous loss. In turn, any state instantaneous loss such as $\|x\|_{\psi_x}^2$ can be equivalently rewritten as $\|y\|_{\psi_y}^2$, $y = Hx$ and $\psi_y = \psi_y' > 0$, if H and ψ_y are selected as follows. Let rank $\psi_x = p \le n$. Then there exist matrices $H \in \mathbb{R}^{p \times n}$ and $\psi_y = \psi_y' > 0$ such that the factorization (47) holds. Any such a pair (H, ψ_y) can be used for rewriting $\|x\|_{\psi_x}^2$ as $\|y\|_{\psi_y}^2$. Therefore, we conclude that, in principle, there is no loss of generality in considering the LQOR in place of the LQR problem.

For any finite regulation horizon, the solution of the LQOR problem in the time-invariant case is given by (9) to (15) of Theorem 1, provided that $M = 0$ and ψ_x is as in (47). An advantage of the LQOR formulation is that the limiting properties as $N \to \infty$ of the LQOR solution can be nicely related to the system-theoretic properties of the plant $\Sigma = (\Phi, G, H)$ in (44).

Problem 2.4-2   Consider the plant (44) in a Gilbert–Kalman (GK) canonical observability decomposition:

$$\Phi = \begin{bmatrix} \Phi_o & 0 \\ \Phi_{\bar{o}o} & \Phi_{\bar{o}} \end{bmatrix} \qquad G = \begin{bmatrix} G_o \\ G_{\bar{o}} \end{bmatrix}$$

$$H = \begin{bmatrix} H_o & 0 \end{bmatrix} \qquad x = \begin{bmatrix} x_o' & x_{\bar{o}}' \end{bmatrix}' \qquad (2.4\text{-}48)$$

It is to be remarked that this can be assumed w.l.o.g. because any plant (44) is algebraically equivalent to (48). With reference to (10) to (15) with $M = 0$ and $\psi_x = H'\psi_y H$, show that, if

$$P(0) = \begin{bmatrix} P_o(0) & 0 \\ 0 & 0 \end{bmatrix} \qquad (2.4\text{-}49)$$

then

$$P(j) = \begin{bmatrix} P_o(j) & 0 \\ 0 & 0 \end{bmatrix} \qquad (2.4\text{-}50)$$

with

$$P_o(j+1) = \Phi_o' P_o(j)\Phi_o - \qquad (2.4\text{-}51)$$
$$\Phi_o' P_o(j)G_o \Big[\psi_u + G_o' P_o(j)G_o\Big]^{-1} G_o' P_o(j)\Phi_o + H_o'\psi_y H_o$$

and

$$F(j) = \begin{bmatrix} F_o(j) & 0 \end{bmatrix} \qquad (2.4\text{-}52)$$

with

$$F_o(j) = -[\psi_u + G_o' P_o(j-1)G_o]^{-1} G_o' P_o(j-1)\Phi_o \qquad (2.4\text{-}53)$$

Expressing in words the conclusions of Problem 2, we can say that the solution of
the LQOR problem depends solely on the observable subsystem $\Sigma_o = (\Phi_o, G_o, H_o)$
of the plant, provided that only the observable component $x_o(N)$ of the final state

$$x(N) = \begin{bmatrix} x_o'(N) & x_{\bar{o}}'(N) \end{bmatrix}'$$

is costed.

For the time-invariant LQOR problem, next Theorem 2 gives a necessary and
sufficient condition for the existence of $\hat{P}$ in (36).

Theorem 2.4-3. *Consider the time-invariant LQOR problem and the corresponding matrix sequence $\{P(j)\}_{j=0}^{\infty}$ generated by the Riccati iterations (11) to (13), with $M = O_{n \times n}$, initialized from $P(0) = O_{m \times n}$. Let $\Sigma_o = (\Phi_o, G_o, H_o)$ be the completely observable subsystem obtained via a GK canonical observability decomposition of the plant (44) $\Sigma = (\Phi, G, H)$. Next, let $\Phi_{o\bar{r}}$ the state-transition matrix of the unreachable subsystem obtained via a GK canonical reachability decomposition of Σ_o. Then, there exists*

$$\hat{P} = \lim_{j \to \infty} P(j) \tag{2.4-54}$$

if and only if $\Phi_{o\bar{r}}$ is a stability matrix.

Proof: According to Problem 2, everything depends on Σ_o. Thus, w.l.o.g., we
can assume that the plant is Σ_o. To say that $\Phi_{o\bar{r}}$ is a stability matrix is equivalent to the
stabilizability of Σ_o. Then, by Theorem 1, the previous condition implies (54).

We prove that the condition is necessary by contradiction. Assume that $\Phi_{o\bar{r}}$ is
not a stability matrix. Therefore, there are observable initial states of the form $x_o = \begin{bmatrix} x_{or}' = 0 & x_{o\bar{r}}' \end{bmatrix}'$ such that $\sum_{k=0}^{j-1} \|y(k)\|_{\psi_x}^2$ diverges as $j \to \infty$, irrespective of the input
sequence. This contradicts (54).

The reader is warned of the right order for the GK canonical decompositions
that must be used to get $\Phi_{o\bar{r}}$ in Theorem 2.

Example 2.4-2 Consider the plant $\Sigma = (\Phi, G, H)$ with

$$\Phi = \begin{bmatrix} 1 & 1 \\ 0 & 2 \end{bmatrix} \qquad G = \begin{bmatrix} 1 \\ 0 \end{bmatrix} \qquad H = \begin{bmatrix} 1 & 0 \end{bmatrix}$$

Σ is seen to be completely observable. Hence, we can set $\Sigma = \Sigma_o$. Further, Σ is already
in a GK reachability canonical decomposition with $\Phi_{o\bar{r}} = 2$. Hence, we conclude that the
limit (54) does not exist.

If we reverse the order of the GK canonical decompositions, we first get the unreachable $\Sigma_{\bar{r}}$ of Σ. It equals $\Sigma_{\bar{r}} = (2, 0, 0)$ which is unobservable. Then, $\Phi_{\bar{r}o}$ is "empty"
(no unreachable and observable eigenvalue). Hence, we would erroneously conclude that
the limit (54) exists.

Problem 2.4-3 Consider the LQOR problem for the plant $\Sigma = (\Phi, G, H)$. Assume that
matrix $\Phi_{o\bar{r}}$, defined in Theorem 3, is a stability matrix. Then, by Theorem 3, there exists
$\hat{P}$ as in (54). Prove by contradiction that $\hat{P}$ is positive definite if and only if the pair
(Φ, H) is completely observable. (*Hint:* Use (50) and positive definiteness of ψ_y.)

Theorem 2.4-4.   *Consider the time-invariant LQOR problem and the corresponding matrix sequence $\{P(j)\}_{j=0}^{\infty}$ generated by the Riccati iterations (11) to (13), with $M = O_{m \times n}$, initialized from $P(0) = O_{n \times n}$. Then, there exists*

$$\hat{P} = \lim_{j \to \infty} P(j)$$

such that the corresponding feedback-gain matrix

$$\hat{F} = -(\psi_u + G'\hat{P}G)^{-1}G'\hat{P}\Phi \tag{2.4-55}$$

yields a state-feedback control law $u(k) = \hat{F}x(k)$ that stabilizes the plant, viz. $\Phi + G\hat{F}$ is a stability matrix, if and only if the plant $\Sigma = (\Phi, G, H)$ is stabilizable and detectable.

Proof: We first show that stabilizability and detectability of Σ is a necessary condition for the existence of $\hat{P}$ and the stability the corresponding closed-loop system.

First, $\Phi + G\hat{F}$ stable implies stabilizability of the pair (Φ, G). Second, necessity of detectability of (Φ, H) is proved by contradiction. Assume then that (Φ, H) is undetectable. By referring to Problem 2, w.l.o.g., (Φ, G, H) can be considered in a GK canonical observability decomposition and, according to (52), $\hat{F} = \begin{bmatrix} \hat{F}_o & 0 \end{bmatrix}$. Hence, the unobservable subsystem of Σ is left unchanged by the steady-state LQ regulator. This contradicts the stability of $\Phi + G\hat{F}$.

We next show that stabilizability and detectability of Σ is a sufficient condition. Because the pair (Φ, G) is stabilizable, by Theorem 1, there exists $\hat{P}$. Further, according to Problem 2, the unobservable eigenvalues of Σ are again eigenvalues of $\Phi + G\hat{F}$. Because by detectability of (Φ, H) they are stable, w.l.o.g., we can complete the proof by assuming that (Φ, H) is completely observable. Suppose now that $\Phi + G\hat{F}$ is not a stability matrix, and show that this contradicts (54). To see this, consider that complete observability of (Φ, H) implies complete observability of $\left(\Phi + GF, \begin{bmatrix} F' & H' \end{bmatrix}'\right)$ for any F of compatible dimensions. Then, if $\hat{\Phi} + G\hat{F}$ is not a stability matrix, there exists states $x(0)$ such that

$$\sum_{k=0}^{j-1}\left[\|y(k)\|_{\psi_y}^2 + \|u(k)\|_{\psi_u}^2\right] = \sum_{k=0}^{j-1}\left\{x'(k)\begin{bmatrix} \hat{F}' & H' \end{bmatrix}\begin{bmatrix} \psi_u & 0 \\ 0 & \psi_y \end{bmatrix}\begin{bmatrix} \hat{F} \\ H \end{bmatrix}x(k)\right\}$$

diverges as $j \to \infty$. This contradicts (54).

We next show that whenever the validity conditions of Theorem 4 are fulfilled, the Riccati iterations (11) to (13), with $M = O_{m \times n}$, initialized from *any* $P(0) = P(0) \geq 0$, yield the same limit as (54).

Lemma 2.4-2.   *Consider the time-invariant LQOR problem (44) to (46) with terminal-state cost weight $P(0) = P'(0) \geq 0$. Let the plant be stabilizable and detectable. Then, the corresponding matrix sequence $\{P(j)\}_{j=0}^{\infty}$ generated by the Riccati iterations (11) to (13) with $M = O_{m \times n}$, admits, as $j \to \infty$, a unique limit, no matter how $P(0)$ is chosen. Such a limit is the same as the one of (54). Further,*

$$x'\hat{P}x = \lim_{j \to \infty} \min_{u_{[0,j)}}\left\{\sum_{k=0}^{j-1}\left[\|y(k)\|_{\psi_y}^2 + \|u(k)\|_{\psi_u}^2\right] + \|x(j)\|_{P(0)}^2\right\} \tag{2.4-56}$$

and the optimal input sequence minimizing the performance index in (56) is given by the state-feedback control law $u(k) = \hat{F}x(k)$ with $\hat{F}$ as in (55).

Proof: Because the plant is stabilizable and detectable, Theorem 3 guarantees that, if we adopt the control law $u^0(k) = \hat{F}x^0(k)$, then

$$\lim_{j \to \infty} \|x^0(j)\|^2_{P(0)} = 0$$

and

$$\lim_{j \to \infty} \left\{ \sum_{k=0}^{j-1} \left[\|y^0(k)\|^2_{\psi_y} + |u^0(k)\|^2_{\psi_u} \right] + \|x^0(j)\|^2_{P(0)} \right\} = x'\hat{P}x$$

where the superscript denotes all system variables obtained by using the previous control law. Assume now that $\hat{F}$ is not the steady-state LQOR feedback-gain matrix for some $P(0)$ and initial state $x \in \mathbb{R}^n$. Then

$$\infty \; > \; x'\hat{P}x \; \underset{\neq}{\geq} \; \lim_{j \to \infty} \left\{ \sum_{k=0}^{j-1} \left[\|y(k)\|^2_{\psi_y} + \|u(k)\|^2 \right] + \|x(j)\|^2_{P(0)} \right\}$$

$$\geq \; \lim_{j \to \infty} \left\{ \sum_{k=0}^{j-1} \left[\|y(k)\|^2_{\psi_y} + \|u(k)\|^2_{\psi_u} \right] \right\}$$

This contradicts steady-state optimality of $\hat{F}$ for $P(0) = O_{n \times n}$.

Whenever the Riccati iterations (11) to (13) for the LQOR problem converge as $j \to \infty$ and $P = \lim_{j \to \infty} P(j)$, the limit matrix P satisfies the following *algebraic Riccati equation (ARE):*

$$P = \Phi'P\Phi - \Phi'PG(\psi_u + G'PG)^{-1}G'P\Phi + \psi_x \tag{2.4-57}$$

Conversely, all the solutions of (57) need not coincide with a limiting matrix of the Riccati iterations for the LQOR problem. The situation again simplifies under stabilizability and detectability of the plant.

Lemma 2.4-3. *Consider the time-invariant LQOR problem. Let the plant be stabilizable and detectable. Then, the ARE (57) has a unique symmetric non-negative definite solution that coincides with the matrix $\hat{P}$ in (54).*

Proof: Assume that, besides $\hat{P}$, (57) has a different solution $\tilde{P} = \tilde{P}' \geq 0$, $\tilde{P} \neq \hat{P}$. If the Riccati iterations (11) to (13) are initialized from $P(0) = \tilde{P}$, we get $P(j) = \tilde{P}$, $j = 1, 2, \cdots$. Then, $\hat{P}$ and $\tilde{P}$ are two different limits of the Riccati iterations. This contradicts Lemma 2.

Because the ARE is a nonlinear matrix equation, it has many solutions. Among these solutions P, the *strong solutions* are called the ones yielding a feedback-gain matrix $F = -(\psi_u + G'PG)^{-1}G'P\Phi$ for which the closed-loop transition matrix has eigenvalues in the closed unit disk. The following result completes Lemma 3 in this respect.

Result 2.4-1. *Consider the time-invariant LQOR problem and its associated ARE. Then:*

1. *The ARE has a unique strong solution if and only if the plant is stabilizable;*

2. *The strong solution is the only nonnegative definite solution of the ARE if and only if the plant is stabilizable and has no undetectable eigenvalue outside the closed unit disk.*

The most useful results of steady-state LQR theory are summed up in Theorem 5. Its conclusions are reassuming in that, under general conditions, they guarantee that the steady-state LQOR exists and stabilizes the plant. One important implication is that steady-state LQR theory provides a tool for systematically designing regulators that, while optimizing an engineering significant performance index, yield stable closed-loop systems.

Theorem 2.4-5. *Consider the time-invariant LQOR problem (44) to (46) and the related matrix sequence $\{P(j)\}_{j=0}^{\infty}$ generated via the Riccati iterations (11) to (13) with $M = O_{m \times n}$, initialized from any $P(0) = P'(0) \geq 0$. Then, there exists*

$$P = \lim_{j \to \infty} P(j) \tag{2.4-58}$$

such that

$$x'Px = V_\infty(x) \tag{2.4-59}$$

$$= \lim_{j \to \infty} \min_{u_{[0,j)}} \left\{ \sum_{k=0}^{j-1} \left[\|y(k)\|_{\psi_y}^2 + \|u(k)\|_{\psi_u}^2 \right] + \|x(j)\|_{P(0)}^2 \right\}$$

and the LQOR control law given by

$$u(k) = Fx(k) \tag{2.4-60}$$

$$F = -(\psi_u + G'PG)^{-1}G'P\Phi \tag{2.4-61}$$

stabilizes the plant, if and only if the plant (Φ, G, H) is stabilizable and detectable. Further, under such conditions, the matrix P in (58) coincides with the unique symmetric nonnegative definite solution of the ARE (57).

Main Points of the Section. The infinite-time or steady-state LQOR solution can be used so as to stabilize any time-invariant plant, while optimizing a quadratic performance index, provided that the plant is stabilizable and detectable. The steady-state LQOR consists of a time-invariant state-feedback whose gain (61) is expressed in terms of the limit matrix P (58) of the Riccati iterations (11) to (13). This also coincides with the unique symmetric nonnegative definite solution of the ARE (57). Although stabilizability appears as an obvious intrinsic property that cannot be enforced by the designer, on the contrary, detectability can be guaranteed by a suitable choice of the matrix H or the state-weighting matrix ψ_x (47).

Problem 2.4-4 Show that the zero eigenvalues of the plant, are also eigenvalues of the LQOR closed-loop system.

Problem 2.4-5 (Output Dynamic Compensator as an LQOR)   Consider the single-input single-output (SISO) plant described by the following difference equation:

$$y(t) + a_1 y(t-1) + \cdots + a_n y(t-n) = b_1 u(t-1) + \cdots + b_n u(t-n) \qquad (2.4\text{-}62)$$

Show that:

(i) $x(t) := \begin{bmatrix} y(t-n+1) & \cdots & y(t) & u(t-n+1) & \cdots & u(t-1) \end{bmatrix}'$ (2.4-63)
is a state vector for the plant (62).

(ii) The state-space representation (Φ, G, H) with state $x(t)$ is stabilizable and detectable if the polynomials

$$\left. \begin{array}{l} A(q) := q^n + a_1 q^{n-1} + \cdots + a_n \\ B(q) := b_1 q^{n-1} + \cdots + b_n \end{array} \right\} \qquad (2.4\text{-}64)$$

have a strictly Hurwitz greatest common divisor.

(iii) Under the assumption in (ii), the steady-state LQOR, obtained by using the triplet (Φ, G, H), consists of an *output dynamic compensator* of the form

$$\begin{array}{l} u(t) + r_1 u(t-1) + \cdots + r_{n-1} u(t-n+1) \\ = \sigma_0 y(t) + \sigma_1 y(t-1) + \cdots + \sigma_{n-1} y(t-n+1) \end{array} \qquad (2.4\text{-}65)$$

(*Hint:* Use the result [GS84] according to which (Φ, G) is completely reachable if and only if $A(q)$ and $B(q)$ are coprime.)

Problem 2.4-6 Consider the LQOR problem for the SISO plant

$$\Phi = \begin{bmatrix} 0 & 1 \\ -\frac{1}{2} & \frac{3}{2} \end{bmatrix} \qquad G = \begin{bmatrix} 0 \\ 1 \end{bmatrix} \qquad H = \begin{bmatrix} \alpha & 1 \end{bmatrix}, \qquad \alpha \in \mathbb{R} \qquad (2.4\text{-}66)$$

and the cost $\sum_{k=0}^{\infty} [y^2(k) + \rho u^2(k)]$, $\rho > 0$. Find the values of α for which the steady-state LQOR problem has no solution yielding an asymptotically stable closed-loop system. (*Hint:* The unobservable eigenvalues of (66) coincide with the common roots of $\bar{\chi}_\Phi(z) := \det(zI_2 - \Phi)$ and $H \operatorname{Adj}(zI_2 - \Phi)G$.)

Problem 2.4-7 Consider the plant

$$y(t) - \frac{1}{4} y(t-2) = u(t-1) + u(t-2)$$

with initial conditions

$$\begin{array}{rcl} x_1(0) & := & \frac{1}{4} y(t-1) + u(-1) \\ x_2(0) & := & y(0) \end{array}$$

and state-feedback control law $u(t) = -\frac{1}{4} y(t)$. Compute the corresponding cost $J = \sum_{k=0}^{\infty} [y^2(k) + \rho u^2(k)]$, $\rho \geq 0$. (*Hint:* Use the Lyapunov equation (26) with a suitable choice for $x(t)$.)

Problem 2.4-8   Consider the LQOR problem for the plant

$$\Phi = \begin{bmatrix} \frac{1}{2} & 0 \\ -2 & \alpha \end{bmatrix} \qquad G = \begin{bmatrix} g_1 \\ g_2 \end{bmatrix} \qquad H = \begin{bmatrix} 0 & 1 \end{bmatrix}$$

and performance index $J = \sum_{k=0}^{\infty}[y^2(k) + 10^{-4}u^2(k)]$. Give detailed answers to the following questions:

(a) Find the set of values for the parameters (α, g_1, g_2) for which there exists $\hat{P} = \lim_{j \to \infty} P(j)$ as in (54).

(b) Assuming that $\hat{P}$ as in (a) exists, find for which values of (α, g_1, g_2) the state-feedback control law $u(k) = -(10^{-4} + G'\hat{P}G)^{-1}G'\hat{P}\Phi x(k)$ makes the closed-loop system asymptotically stable.

Problem 2.4-9 (LQOR with a Prescribed Degree of Stability)   Consider the LQOR problem for a plant (Φ, G, H) and performance index

$$J = \sum_{k=0}^{\infty} r^{2k}\left[\|y(k)\|_{\psi_y}^2 + \|u(k)\|_{\psi_u}^2 \right] \tag{2.4-67}$$

with $r \geq 1$, $\psi_y = \psi_y' > 0$, and $\psi_u = \psi_u' > 0$. Show the following:

(a) This LQOR problem is equivalent to an LQOR problem with the following performance index:

$$\sum_{k=0}^{\infty}\left[\|\bar{y}(k)\|_{\psi_y}^2 + \|\bar{u}(k)\|_{\psi_u}^2 \right]$$

and a new plant $(\bar{\Phi}, \bar{G}, \bar{H})$ to be specified.

(b) Provided that $(\bar{\Phi}, \bar{H})$ is a detectable pair, the eigenvalues λ of the characteristic polynomial of the closed-loop system consisting of the initial plant optimally regulated according to (67) satisfy the inequality

$$|\lambda| < \frac{1}{r}$$

Problem 2.4-10 (Tracking as a Regulation Problem)   Consider a detectable plant (Φ, G, H) with input $u(t)$, state $x(t)$, and scalar output $y(t)$. Let r be any real number. Define $\varepsilon(t) := y(t) - r$. Prove that, if 1 is an eigenvalue of Φ, viz., $\chi_\Phi(1) := \det(I_n - \Phi) = 0$, there exist eigenvectors x_r of Φ associated with the eigenvalue 1 such that, for $\tilde{x}(t) := x(t) - x_r$, we have

$$\begin{aligned} \tilde{x}(t+1) &= \Phi\tilde{x}(t) + Gu(t) \\ \varepsilon(t) &= H\tilde{x}(t) \end{aligned} \right\} \tag{2.4-68}$$

This shows that, under the stated assumptions, the plant with input $u(t)$, state $\tilde{x}(t)$, and output $\varepsilon(t)$ has a description coinciding with the initial triplet (Φ, G, H). Then, if (Φ, G) is stabilizable, the LQ regulation law $u(t) = F\tilde{x}(t)$ minimizing

$$\sum_{k=0}^{\infty}[\varepsilon^2(k) + \|u(k)\|_{\psi_u}^2], \qquad \psi_u > 0 \tag{2.4-69}$$

for the plant (62) exists, and the corresponding closed-loop system is asymptotically stable.

Problem 2.4-11 (Tracking as a Regulation Problem) Consider again the situation described in Problem 2.4-12 where $u(t) \in \mathbb{R}$. Let

$$\left.\begin{array}{rcl} \delta x(t) & := & x(t) - x(t-1) \\ \delta u(t) & := & u(t) - u(t-1) \\ \xi(t) & := & \left[\begin{array}{cc} \delta x'(t) & \varepsilon(t) \end{array}\right]' \in \mathbb{R}^{n+1} \end{array}\right\} \tag{2.4-70}$$

(a) Show that the state–space representation of the plant with input $\delta u(t)$, state $\xi(t)$, and output $\varepsilon(t)$ is given by the triplet

$$\Sigma = \left(\left[\begin{array}{cc} \Phi & 0 \\ H\Phi & 1 \end{array}\right], \left[\begin{array}{c} G \\ HG \end{array}\right], \left[\begin{array}{cc} O'_n & 1 \end{array}\right] \right)$$

(b) Let Θ be the observability matrix of Σ. Show that by taking elementary row operations on Θ, we can get a matrix that can be factorized as follows:

$$\hat{\Theta}\left[\begin{array}{cc} \Phi & O_n \\ O'_n & 1 \end{array}\right], \qquad \hat{\Theta} = \left[\begin{array}{cc} O'_n & 1 \\ H & 0 \\ H\Phi & 0 \\ \vdots & \vdots \\ H\Phi^{n-1} & 0 \end{array}\right]$$

Show that (Φ, H) detectable implies detectability of Σ.

(c) Let R be the reachability matrix of Σ. Define

$$\tilde{R} := \left[\begin{array}{cc} I_n & O_n \\ -H & 1 \end{array}\right] R$$

Show that by taking elementary column operations on $\tilde{R}$, we can get a matrix that can be factorized as $L\hat{R}$, with

$$L = \left[\begin{array}{cc} G & \Phi - I_n \\ 0 & H \end{array}\right], \qquad \hat{R} = \left[\begin{array}{ccccc} 1 & 0 & 0 & \cdots & 0 \\ 0 & G & \Phi G & \cdots & \Phi^{n-1}G \end{array}\right]$$

(d) Prove that nonsingularity of L is equivalent to $H_{yu}(1) := H(I_n - \Phi)^{-1}G \neq 0$.

(e) Prove that (Φ, G) stabilizable and $H_{yu}(1) \neq 0$ implies that Σ is stabilizable.

(f) Conclude that if (Φ, G, H) is stabilizable and detectable, and $H_{yu}(1) \neq 0$, the LQ regulation law

$$\delta u(t) = F_x \delta x(t) + F_\varepsilon \varepsilon(t) \tag{2.4-71}$$

minimizing

$$\sum_{k=0}^{\infty} \left[\varepsilon^2(k) + \rho[\delta u(k)]^2\right], \qquad \rho > 0 \tag{2.4-72}$$

for the plant Σ exists, and the corresponding closed-loop system is asymptotically stable.

Note that (71) gives

$$u(t) - u(0) \quad = \quad \sum_{k=1}^{t} \delta u(t) \tag{2.4-73}$$

$$= \quad F_x[x(t) - x(0)] + F_\varepsilon \sum_{k=1}^{t} \varepsilon(t)$$

In other terms, (71) is a feedback-control law including an *integral action* from the tracking error.

Problem 2.4-12 (Fake ARE)   Consider the Riccati forward difference equation (11) with $M = O_{m \times n}$ and ψ_x as in (46) and (47):

$$P(j+1) = \Phi' P(j) \Phi - \Phi' P(j) G[\psi_u + G' P(j) G]^{-1} G' P(j) \Phi + H' \psi_y H \tag{2.4-74}$$

We note that this equation can be formally rewritten as follows:

$$P(j) \quad = \quad \Phi' P(j) \Phi - \Phi' P(j) G[\psi_u + G' P(j) G]^{-1} G' P(j) \Phi + Q(j) \tag{2.4-75}$$

$$Q(j) \quad := \quad H' \psi_y H + P(j) - P(j+1) \tag{2.4-76}$$

The latter has the same form as the ARE (57) and has been called [BGW90] *fake* ARE. Use Theorem 4 to show that the feedback-gain matrix

$$F(j+1) = -[\psi_u + G' P(j) G]^{-1} G' P(j) \Phi \tag{2.4-77}$$

stabilizes the plant, viz. $\Phi + G F(j+1)$ is a stability matrix, provided that (Φ, G, H) is stabilizable and detectable, and P has the property

$$P(j) - P(j+1) \geq 0 \tag{2.4-78}$$

(*Hint*: Show that (78) implies that $Q(j)$ can be written as $H' \psi_y H + \Gamma' \psi_\gamma \Gamma$, $\psi_\gamma = \psi_\gamma' > 0$, $\Gamma \in \mathbb{R}^{r \times n}$, and $r := \text{rank}[P(j) - P(j+1)]$. Next, prove that detectability of (Φ, H) implies detectability of $(\Phi, [\ H'\ \ \Gamma'\ ]')$. Finally, consider the fake ARE.)

2.5   STEADY-STATE LQR COMPUTATION

There are several numerical procedures available for computing matrix P in (4-58). We limit our discussion to the ones that will be used in this text. In particular, we shall not enter here into numerical factorization techniques for solving LQ problems, which will be touched upon for the dual estimation problem in Sec. 6.5.

2.5.1   Riccati Iterations

Equations (4-11) to (4-13), with $M = O_{m \times n}$, can be iterated once they are initialized from any $P(0) = P'(0) \geq 0$, for computing P as in (4-58). Of the three different forms, the third, viz.,

$$
\begin{aligned}
P(j+1) &= [\Phi + GF(j+1)]'P(j)[\Phi + GF(j+1)] \\
&\quad + F'(j+1)\psi_u F(j+1) + \psi_x & \text{(2.5-1)} \\
F(j+1) &= -[\psi_u + G'P(j)G]^{-1}G'P(j)\Phi & \text{(2.5-2)}
\end{aligned}
$$

is sometimes referred to as the *robustified form* of the Riccati iterations. The attribute is motivated by the fact that, unlike the other two remaining forms, it updates matrix $P(j)$ by adding symmetric nonnegative definite matrices. When computations with roundoff errors are considered, this is a feature that can help to obtain at each iteration step a new symmetric nonnegative definite matrix $P(j)$, as required by LQR theory.

The rate of convergence of the Riccati iterations is generally not very rapid, even in the neighborhood of steady-state solution P. The numerical procedure described next exhibits fast convergence in the vicinity of P.

2.5.2 Kleinman Iterations

Given a stabilizing feedback-gain matrix $F_k \in \mathbb{R}^{m \times n}$, let $\mathcal{L}_k$ be the solution of the Lyapunov equation

$$
\begin{aligned}
\mathcal{L}_k &= \Phi_k' \mathcal{L}_k \Phi_k + F_k' \psi_u F_k + \psi_x & \text{(2.5-3)} \\
\Phi_k &:= \Phi + GF_k & \text{(2.5-4)}
\end{aligned}
$$

The next feedback-gain matrix F_{k+1} is then computed:

$$
F_{k+1} = -(\psi_u + G'\mathcal{L}_k G)^{-1}G'\mathcal{L}_k \Phi \qquad \text{(2.5-5)}
$$

The iterative equations (3) to (5), $k = 0, 1, 2, \cdots$, enjoy the following properties.

Suppose that the ARE (4-57) has a unique nonnegative definite solution, for example, (Φ, G, H), with $\psi_x = H'\psi_y H$, $\psi_y = \psi_y' > 0$, stabilizable, and detectable. Then, provided that F_0 is such as to make Φ_0 a stability matrix:

1. the sequence $\{\mathcal{L}_k\}_{k=0}^{\infty}$ is monotonic nonincreasing and lower-bounded by the solution P of the ARE (4-57):

$$
\mathcal{L}_0 \geq \cdots \geq \mathcal{L}_k \geq \mathcal{L}_{k+1} \geq P \qquad \text{(2.5-6)}
$$

2.
$$
\lim_{k \to \infty} \mathcal{L}_k = P \qquad \text{(2.5-7)}
$$

3. the rate of convergence to P is quadratic, viz.,

$$
\|P - \mathcal{L}_{k+1}\| \leq c\|P - \mathcal{L}_k\|^2 \qquad \text{(2.5-8)}
$$

for any matrix norm and for a constant c independent of the iteration index k.

Equation (8) shows that the rate of convergence of the Kleinman iterations is fast in the vicinity of P. It is required, however, that the iterations be initialized from a *stabilizing* feedback-gain matrix F_0. In order to speed up convergence, [AL84] suggests to select F_0 via a direct Schur-type method.

The main problem with Kleinman iterations is that (3) must be solved at each iteration step. Although (3) is linear in $\mathcal{L}_k$, its solution cannot be obtained by simple matrix inversion. Actually, the numerical effort for solving it may be rather formidable because the number of linear equations that must be solved at each iteration step equals $n(n+1)/2$ if n denotes the plant order.

Kleinman iterations result from using Newton–Raphson's method [Lue69] for solving the ARE (4-57).

Problem 2.5-1   Consider the matrix function

$$N(P) := -P + \Phi'P\Phi - \Phi'PG[H(P)]^{-1}G'P\Phi + \psi_x$$

where

$$H(P) := (\psi_u + G'PG)$$

The aim is to find the symmetric nonnegative definite matrix P such that

$$N(P) = O_{n \times n}$$

Let $\mathcal{L}_{k-1} = \mathcal{L}'_{k-1} \geq 0$ be a given approximation to P. It is asked to find a next approximation $\mathcal{L}_k$ by increasing $\mathcal{L}_{k-1}$ by a "small" correction $\tilde{\mathcal{L}}$:

$$\mathcal{L}_k = \mathcal{L}_{k-1} + \tilde{\mathcal{L}}$$

$\tilde{\mathcal{L}}$, and hence $\mathcal{L}_k$, has to be determined in such a way that $N(\mathcal{L}_k) \approx O_{n \times n}$.

By omitting the terms in $\tilde{\mathcal{L}}$ of order higher than the first, show that

$$H^{-1}(\mathcal{L}_k) \approx H^{-1}(\mathcal{L}_{k-1}) - H^{-1}(\mathcal{L}_{k-1})G'\tilde{\mathcal{L}}GH^{-1}(\mathcal{L}_{k-1})$$

and, further, that $N(\mathcal{L}_k) \approx O_{n \times n}$ if $\mathcal{L}_k$ satisfies (3) and (4).

Control-Theoretic Interpretation.   It is of interest for its possible use in adaptive control to give a specific control-theoretic interpretation to the Kleinman iterations. To this end, consider the quadratic cost $J(x, u_{[0,\infty)})$ under the assumption that all inputs, except $u(0)$, are given by feeding back the current plant state by a stabilizing constant-gain matrix F_k, viz.,

$$u(j) = F_k x(j) , \qquad j = 1, 2, \cdots \tag{2.5-9}$$

The situation is depicted in Fig. 1, where $t = 0^+$ indicates that the switch commutes from position a to position b after $u(0)$ has been applied and before $u(1)$ is fed into the plant.

Let the corresponding cost be denoted as follows:

$$J\Big(x, u(0), F_k\Big) := J\Big(x, u_{[0,\infty)} \mid u(j) = F_k x(j), \quad j = 1, 2, \cdots\Big) \tag{2.5-10}$$

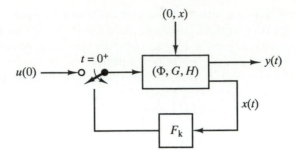

Figure 2.5-1: A control-theoretic interpretation of Kleinman iterations.

We show that, for given x and F_k,

$$u(0) = F_{k+1}x \tag{2.5-11}$$

minimizes (10) w.r.t. $u(0)$, if F_{k+1} is related to F_k via (3) to (6). To see this, rewrite (10) as follows:

$$
\begin{aligned}
J(x, u(0), F_k) &= \|x\|^2_{\psi_x} + \|u(0)\|^2_{\psi_u} + \sum_{j=1}^{\infty} \|x(j)\|^2_{(\psi_x + F'_k \psi_u F_k)} \\
&= \|x\|^2_{\psi_x} + \|u(0)\|^2_{\psi_u} + \|x(1)\|^2_{\mathcal{L}_k} \\
&= \|x\|^2_{\psi_x} + \|u(0)\|^2_{\psi_u} + \|\Phi x + Gu(0)\|^2_{\mathcal{L}_k} \tag{2.5-12}
\end{aligned}
$$

where the first equality follows from (9), the second from Problem 4-1 if $\mathcal{L}_k$ is the solution of the Lyapunov equation (3), and the third because $x(1) = \Phi x + Gu(0)$. Minimization of (10) w.r.t. $u(0)$ yields (11) with F_{k+1} as in (5).

Problem 2.5-2   Consider (12) and define the symmetric nonnegative definite matrix R_{k+1} implicitly via

$$x' R_{k+1} x := \min_{u(0)} J(x, u(0), F_k) \tag{2.5-13}$$

Show that R_{k+1} satisfies the recursions

$$R_{k+1} = \Phi' \mathcal{L}_k \Phi - \Phi' \mathcal{L}_k G(\psi_u + G' \mathcal{L}_k G)^{-1} G' \mathcal{L}_k \Phi + \psi_x \tag{2.5-14}$$

with $\mathcal{L}_k$ as in (3).

Problem 2.5-3   If R_{k+1} and $\mathcal{L}_k$ are as in Problem 2, show that

$$\mathcal{L}_k - R_{k+1} \geq 0 \tag{2.5-15}$$

Problem 2.5-4   Assume that $\psi_x = H' \psi_y H$, $\psi'_y = \psi_y > 0$, and (Φ, G, H) is stabilizable and detectable. Use (14) and (15) to prove that (5) is a stabilizing feedback-gain matrix, viz. $\Phi_{k+1} = \Phi + GF_{k+1}$ is a stability matrix. (*Hint*: Refer to Problem 4-12. From (14), form a fake ARE $\mathcal{L}_k = \Phi' \mathcal{L}_k \Phi - \Phi' \mathcal{L}_k G(\psi_u + G' \mathcal{L}_k G)^{-1} G' \mathcal{L}_k \Phi + Q_k$. And so on.)

2.6   CHEAP CONTROL

The performance index used in the LQR problem has to be regarded as a compromise between two conflicting objectives: to obtain good *regulation performance*, viz., small $\|y(k)\|$, as well as to prevent $\|u(k)\|$ from becoming too large. This compromise is achieved by selecting suitable values for the weights ψ_u, ψ_x, and M in the performance index. However, it is interesting to consider in the time-invariant case a performance index in which

$$\psi_u = O_{m \times m} \qquad M = O_{m \times n} \qquad \psi_x(N) = O_{n \times n}$$

This means that the plant input is allowed to take on even very large values, the control effort not being penalized in the resulting performance index:

$$J(x, u_{[0,N)}) = \sum_{k=0}^{N-1} \|x(k)\|_{\psi_x}^2 = \sum_{k=0}^{N-1} \|y(k)\|_{\psi_y}^2 \qquad (2.6\text{-}1)$$

This choice should hopefully yield a high regulation performance though at the expense of possibly large inputs. The LQR problem with performance index (1) will be referred to as the cheap control problem and, whenever it exists, the corresponding optimal input for $N \to \infty$ as *cheap control*.

It is to be noticed that, because in (1) $\psi_u = O_{m \times m}$, we have to check that for solving the cheap control problem, one can still use (4-9) to (4-15) where, on the opposite, it was assumed that $\psi_u > 0$. As can be seen from the proof of Theorem 3-1, for any finite N (4-9) to (4-15) hold true even for $\psi_u = O_{m \times m}$, provided that $G'P(j)G$ is nonsingular. However, cheap control is not comprised in the asymptotic theory of Sec. 2.4, which is crucially based on the assumption that $\psi_u > 0$. In particular, no stability property can be ensured by Theorem 4-5 to a cheap control regulated plant. Indeed, as we shall now show, the regulation law that for $N \to \infty$ minimizes (1) does not yield in general an asymptotically stable regulated system.

In order to hold this issue in focus, we avoid needless complications by restricting ourselves to SISO plants, viz., $m = p = 1$. Thus, w.l.o.g., we can set $\psi_y = 1$:

$$
\begin{aligned}
J(x, u_{[0,N)}) &= \sum_{k=0}^{N-1} y^2(k) \\[2mm]
&= \sum_{k=0}^{N-2} y^2(k) + \|x(N-1)\|_{H'H}^2
\end{aligned}
\qquad (2.6\text{-}2)
$$

We also assume that the first sample of the impulse response of the plant (4-44) is nonzero:

$$w_1 := HG \neq 0 \qquad (2.6\text{-}3)$$

We shall refer to this condition by saying that the plant has unit I/O delay. Then, we can solve the Riccati difference equation (4-11), initialized from $P(0) = \psi_x(N) =$

$O_{n \times n}$, or, according to the second of (2.6-2) from $P(1) = \psi_x(N-1) = H'H$, to find

$$P(2) = \Phi'H'H\Phi - \frac{\Phi'H'HGG'H'H\Phi}{G'H'HG} + H'H = H'H$$

Then, it follows that for every $j = 1, 2, \cdots$,

$$
\begin{aligned}
P(j) &= H'H & \text{(2.6-4)} \\
F(j) &= -(HG)^{-1}H\Phi \\
     &= -\frac{H\Phi}{w_1} =: F & \text{(2.6-5)}
\end{aligned}
$$

Correspondingly,

$$
\begin{aligned}
V_j(x) &= \min_{u_{[0,j)}} \sum_{k=0}^{j-1} y^2(k) \\
       &= x'H'Hx = y^2(0) & \text{(2.6-6)}
\end{aligned}
$$

This result shows that, whenever $w_1 \neq 0$, the constant-feedback row-vector (5) is such that the corresponding time-invariant cheap control regulator

$$u(k) = -\frac{H\Phi}{w_1}x(k), \qquad k = 0, 1, \cdots \qquad \text{(2.6-7)}$$

takes the plant output to zero at time $k = 1$ and holds it at zero thereafter. In fact, by (7),

$$
\begin{aligned}
x(k+1) &= \Phi x(k) + Gu(k) \\
       &= \Phi x(k) - G\frac{H\Phi}{w_1}x(k)
\end{aligned}
$$

Hence,

$$
\begin{aligned}
y(k+1) &= Hx(k+1) \\
       &= H\Phi x(k) - \frac{w_1}{w_1}H\Phi x(k) = 0
\end{aligned}
$$

In order to find conditions under which the cheap control regulated system is asymptotically stable, the plant is assumed to be stabilizable. Hence, because its unreachable modes are stable and are left unmodified by the input, w.l.o.g., we can restrict ourselves to a completely reachable plant in a canonical reachability representation:

$$\Phi = \begin{bmatrix} 0 & \vdots & I_{n-1} \\ \cdots & \cdots & \cdots \cdots \cdots \cdots \\ -a_n & \vdots & -a_{n-1} \cdots - a_1 \end{bmatrix} \qquad G = \begin{bmatrix} 0 \\ \cdots \\ 1 \end{bmatrix} \qquad H = \begin{bmatrix} b_n \cdots b_1 \end{bmatrix} \qquad \text{(2.6-8)}$$

It is known that the transfer function $\bar{H}_{yu}(z)$ from input u to output y of the system (8) equals

$$\bar{H}_{yu}(z) := H(zI_n - \Phi)^{-1}G = \frac{\bar{B}(z)}{\bar{A}(z)} \qquad (2.6\text{-}9)$$

where

$$\begin{aligned} \bar{A}(z) &:= \det(zI_n - \Phi) \\ &= z^n + a_1 z^{n-1} + \cdots + z^{n-1} + \cdots + a_n \qquad (2.6\text{-}10) \\ \bar{B}(z) &:= b_1 z^{n-1} + \cdots + b_n \qquad (2.6\text{-}11) \end{aligned}$$

Further,

$$w_1 := HG = b_1 \neq 0$$

Problem 2.6-1 Show that the closed-loop state-transition matrix Φ_{cl} of the unit I/O delay plant (8) regulated by the cheap control (7) equals

$$\Phi_{cl} := \Phi - \frac{GH\Phi}{w_1} = \left[\begin{array}{c|c} 0 & I_{n-1} \\ \hline 0 & -b_n/b_1 \cdots - b_2/b_1 \end{array} \right] \qquad (2.6\text{-}12)$$

Because Φ_{cl} is in companion form, its characteristic polynomial can be obtained by inspection:

$$\begin{aligned} \bar{\chi}_{cl}(z) &:= \det(zI_n - \Phi_{cl}) \qquad (2.6\text{-}13) \\ &= z^n + \frac{b_2}{b_1}z^{n-1} + \cdots + \frac{b_n}{b_1}z \\ &= \frac{1}{b_1}(b_1 z^n + b_2 z^{n-1} + \cdots + b_n z) \\ &= \frac{1}{b_1}z\bar{B}(z) \end{aligned}$$

We say that $\bar{H}_{yu}(z)$ or (Φ, G, H) is a *minimum-phase* transfer function, or plant, if the numerator polynomial $\bar{B}(z)$ in (9) is *strictly Hurwitz*, viz., it has no root in the complement of the open unit disc of the complex plane. Further, the control law $u(k) = Fx(k)$ is said to stabilize the plant (Φ, G, H) if the closed-loop state-transition matrix $\Phi_{cl} := \Phi + GF$ is a stability matrix, that is, $\bar{\chi}_{cl}(z)$ is strictly Hurwitz. Because the polynomial in (13) is strictly Hurwitz if and only if $\bar{B}(z)$ is such, we arrive at the following conclusion, which is a generalization (cf. Problem 2) of the prior analysis.

Theorem 2.6-1. *Let the plant (Φ, G, H) be time-invariant, SISO, and stabilizable. Then, the state-feedback regulator solving the cheap control problem yields an asymptotically stable closed-loop system if and only if the plant is minimum-phase. If the plant has I/O delay τ, $1 \leq \tau \leq n$,*

$$b_1 = b_2 = \cdots = b_{\tau-1} = 0$$

$$w_\tau := H\Phi^{\tau-1}G = b_\tau \neq 0$$

the cheap control law is given by

$$u(k) = -\frac{H\Phi^\tau}{b_\tau}x(k) , \qquad k = 0, 1, \cdots \qquad (2.6\text{-}14)$$

Further, provided that the plant is completely reachable, (14) yields a closed-loop characteristic polynomial given by

$$\bar{\chi}_{cl}(z) = \frac{1}{b_\tau}z^\tau \bar{B}(z) \qquad (2.6\text{-}15)$$

Finally the cheap control law (14) is output-deadbeat *in that, for any initial plant state $x(0) = x \in \mathbb{R}^n$,*

$$y(k) = 0 , \qquad k = \tau, \tau+1, \cdots$$

Correspondingly, for every $j \geq \tau$,

$$V_j(x) = \min_{u_{[0,j)}} \sum_{k=0}^{j-1} y^2(k) = \sum_{k=0}^{\tau-1} y^2(k)$$

$$= x' \sum_{k=0}^{\tau-1} (\Phi')^k H' H \Phi^k x$$

Problem 2.6-2 Consider the plant (8) with I/O delay τ, $1 \leq \tau \leq n$. Show that, similarly to (4),

$$P(j) = \sum_{k=0}^{j-1} (\Phi')^k H' H \Phi^k , \quad j \leq \tau$$

$$= P(\tau) , \quad j \geq \tau$$

Next, using (4-10), find (14). Finally, verify (15).

Naively, one might think that the cheap control regulator is obtained by letting $\psi_u \downarrow O_{m\times m}$ in the regulator solving the steady-state LQOR problem. Indeed Problem 5 shows that this is the case for minimum-phase SISO plants. However, for nonminimum-phase SISO plants, this is generally not the case (cf. Problem 6). In fact, in contrast with cheap control, the solution of the steady-state LQOR problem, provided that the plant is stabilizable and detectable, yields an asymptotically stable closed-loop system even for vanishingly small $\psi_u > 0$ [KS72] (cf. Problem 6).

Main Points of the Section. Cheap control regulation is obtained by setting $\psi_u = O_{m\times m}$ in the performance index of the LQOR problem. For SISO time-invariant stabilizable plants, cheap control regulation is achieved by a time-invariant state-feedback control law that is well-defined for each regulation horizon greater than or equal to the I/O delay of the plant. In this case, the cheap control law can be computed in a simple way (cf. (14)). In particular, in contrast with the $\psi_u > 0$ case, no Riccati-like equation has to be solved. However, applicability of cheap control is severely limited by the fact that it yields an unstable closed-loop system whenever the plant is nonminimum-phase.

Problem 2.6-3 Consider the polynomial $\bar{B}(z)$ in (11) with $b_1 \neq 0$. Show that $|b_n/b_1| \geq 1$ implies that $\bar{B}(z)$ is not a strictly Hurwitz polynomial. (*Hint:* If r_i, $i = 1, \cdots, n - 1$, denote the roots of $\bar{B}(z)$, $\bar{B}(z)/b_1 = \prod_{i=1}^{n-1}(z - r_i)$.)

Problem 2.6-4 Show that $L := P(\tau)$ of Problem 2 satisfies the following matrix equation:

$$L = \Phi' L \Phi + H'H - (\Phi')^\tau H'H(\Phi)^\tau$$

Show that, provided that Φ is a stability matrix, the foregoing equation becomes as $\tau \to \infty$ the Lyapunov equation (4-26) with $\psi_x = H'H$.

Problem 2.6-5 Consider the following SISO completely reachable and observable minimum-phase plant:

$$\Phi = \begin{bmatrix} 0 & 1 \\ 0 & 0 \end{bmatrix} \qquad G = \begin{bmatrix} 0 \\ 1 \end{bmatrix} \qquad H = \begin{bmatrix} 1 & 2 \end{bmatrix}$$

and the related steady-state LQOR problem with performance index

$$J(x, u_{[0,\infty)}) = \sum_{k=0}^{\infty} \left[y^2(k) + \rho u^2(k) \right]$$

with $\rho > 0$. Show the following:

(a) The corresponding ARE (4-37) has nonnegative definite solution

$$\hat{P}(\rho) \quad - \quad \begin{bmatrix} 1 & 2 \\ 2 & p(\rho) \end{bmatrix}$$

$$p(\rho) \quad := \quad \frac{5 - \rho}{2} + \frac{1}{2}(\rho^2 + 10\rho + 9)^{1/2}$$

(b) The steady-state LQOR row-vector feedback (4-40) equals

$$\hat{F}(\rho) = -[\rho + p(\rho)]^{-1} \begin{bmatrix} 0 & 2 \end{bmatrix}$$

(c) The cheap control row-vector feedback (5) equals

$$F = \begin{bmatrix} 0 & -\frac{1}{2} \end{bmatrix}$$

and the corresponding strictly Hurwitz closed-loop characteristic polynomial is

$$\bar{\chi}_{cl}(z) = z^2 + \frac{1}{2}z$$

(d) $\hat{F}(\rho) \to F$, as $\rho \downarrow 0$.

Problem 2.6-6   Consider the SISO plant (Φ, G, H) with Φ and G as in Problem 5 and

$$H = \begin{bmatrix} 2 & 2 \end{bmatrix}$$

Show the following:

(a) The plant is nonminimum-phase.

(b) $\hat{P}(\rho)$ and $\hat{F}(\rho)$ associated to the same performance index as in Problem 5 are given again as in (a) and (b) of Problem 5.

(c) The cheap control row-vector feedback (5) equals

$$F = \begin{bmatrix} 0 & -2 \end{bmatrix}$$

and yields the non-Hurwitz closed-loop characteristic polynomial

$$\bar{\chi}_{cl}(z) = z^2 + 2z$$

(d)
$$\hat{F}_0 := \lim_{\rho \downarrow 0} \hat{F}(\rho) = \begin{bmatrix} 0 & -\frac{1}{2} \end{bmatrix} \neq F$$

yields a strictly Hurwitz closed-loop characteristic polynomial whose roots are given by the stable roots and the reciprocal of the unstable roots of $\bar{\chi}_{cl}(z)$ in (c).

2.7   SINGLE-STEP REGULATION

Assume that the time-invariant plant (4-44) has I/O delay τ, $1 \leq \tau \leq n$, viz., its impulse-response sample matrices $W_k := H\Phi^{k-1}G$ are such that

$$W_k = O_{p \times m}, \quad k = 1, 2, \cdots, \tau - 1 \tag{2.7-1}$$

and

$$W_\tau \neq O_{p \times m} \tag{2.7-2}$$

Then, consider the performance index with $\psi_u > 0$:

$$J(x(0), u_{[0,\tau)}) = \sum_{k=0}^{\tau-1} \left[\|y(k)\|_{\psi_y}^2 + \|u(k)\|_{\psi_u}^2 \right] + \|y(\tau)\|_{\psi_y}^2 \tag{2.7-3}$$

Because of the I/O delay τ, $y_{[0,\tau)}$ is not affected by $u_{[0,\tau)}$. In fact,

$$y(k) = \begin{cases} H\Phi^k x(0), & k = 0, \cdots, \tau - 1 \\ W_\tau u(0) + H\Phi^\tau x(0), & k = \tau \end{cases} \tag{2.7-4}$$

Therefore, the optimal $u_{[0,\tau)}^0$ minimizing (3) is given by

$$u^0(k) = \begin{cases} Fx(0), & k = 0 \\ O_m, & k = 1, \cdots, \tau - 1 \end{cases} \tag{2.7-5}$$

$$F = -\left[\psi_u + W_\tau' \psi_y W_\tau\right]^{-1} W_\tau' \psi_y H\Phi^\tau \tag{2.7-6}$$

Problem 2.7-1   Verify (5) to (6) by using (4-9) to (4-11).

It is to be pointed out that, relative to the determination of the only nonzero input $u^0(0)$ in the optimal sequence $u^0_{[0,\tau)}$, (3) can be replaced by the following performance index comprising a *single regulation step*:

$$\tilde{J}(x(0), u(0)) := \|y(\tau)\|^2_{\psi_y} + \|u(0)\|^2_{\psi_u} \tag{2.7-7}$$

Problem 2.7-2   Verify (6) by direct minimization of (7) w.r.t. $u(0)$.

It follows that the time-invariant state-feedback control law

$$u(k) = Fx(k) \tag{2.7-8}$$

with F as in (6), minimizes for each $k \in \mathbb{Z}$ the *single-step performance index*:

$$\tilde{J}(x(k), u(k)) := \|y(k+\tau)\|^2_{\psi_y} + \|u(k)\|^2_{\psi_u} \tag{2.7-9}$$

for the plant (4-44) with I/O delay τ. This will be referred to as *single-step regulation*.

As in cheap control, the feedback-gain of the single-step regulator can be computed without solving any Riccati-like equation. Similarly to cheap control, this has negative consequences for the stability of the closed-loop system. In order to find out the intrinsic limitations of single-step regulated systems, it suffices to consider a SISO plant. We assume also that the plant is stabilizable. Using the same argument as for cheap control, we can restrict ourselves to a completely reachable plant (Φ, G, H) with I/O delay τ, Φ, and G as in (6-8), and

$$H = \begin{bmatrix} b_n & \cdots & b_\tau & 0 & \cdots & 0 \end{bmatrix} \tag{2.7-10}$$

being

$$w_\tau = b_\tau \neq 0$$

In this case, for $\psi_y = 1$ and $\psi_u = \rho > 0$, (6) becomes

$$\begin{aligned} F &= -\frac{b_\tau}{\rho + b_\tau^2} H\Phi^\tau \tag{2.7-11} \\[2mm] &= -\frac{b_\tau}{\rho + b_\tau^2} \begin{bmatrix} 0 & \cdots & 0 & b_n & \cdots & b_\tau \end{bmatrix} \Phi \end{aligned}$$

Problem 2.7-3   Verify that, for H as in (10) and Φ as in (6-8),

$$H\Phi^{\tau-1} = \begin{bmatrix} 0 \cdots 0 & b_n \cdots b_\tau \end{bmatrix}$$

Problem 2.7-4   Show that the closed-loop state-transition matrix Φ_{cl} of the plant (Φ, G, H) with I/O delay τ, Φ, and G as in (6-8), and H as in (10), is again in companion form with its last row as follows:

$$(1 - \gamma b_\tau) \begin{bmatrix} -a_n & \cdots & -a_{n-\tau+1} & -a_{n-\tau} & \cdots & -a_1 \end{bmatrix} + \gamma \begin{bmatrix} 0 & \cdots & 0 & -b_n & \cdots & -b_{\tau+1} \end{bmatrix} \tag{2.7-12}$$

with $\gamma := b_\tau/(\rho + b_\tau^2)$.

Because Φ_{cl} is in companion form, its characteristic polynomial can be obtained by inspection:

$$\bar{\chi}_{cl}(z) = \gamma \left[\frac{\rho}{b_\tau} \bar{A}(z) + z^\tau \bar{B}(z) \right] \qquad (2.7\text{-}13)$$

with $\bar{A}(z)$ as in (6-10) and

$$\bar{B}(z) := b_\tau z^{n-\tau} + \cdots + b_n \qquad (2.7\text{-}14)$$

Theorem 2.7-1. *Let the plant* (Φ, G, H) *be time-invariant, SISO, stabilizable, and with I/O delay* τ. *Then, the single-step regulator* $u(k) = Fx(k)$ *with feedback gain (11) yields a closed-loop system with characteristic polynomial (13).*

Stability of a SISO single-step regulated plant can be investigated by the root-locus method. The eigenvalues of the closed-loop system are close to the roots of $z^\tau \bar{B}(z)$ if ρ is small, and close to the roots of $\bar{A}(z)$ if ρ is large. If the plant is minimum-phase, there is an upper bound on ρ, call it ρ_M, such that for $\rho < \rho_M$ the closed-loop system is stable. If the plant is open-loop stable, there is a lower bound on ρ, call it ρ_m, such that for $\rho > \rho_m$, the closed-loop system is stable. If the plant is either nonminimum-phase or open-loop unstable, there are, however, critical control weights that yield an unstable closed-loop system. If the plant is both nonminimum-phase and open-loop unstable, there may be *no* value of ρ that makes the single-step regulated plant asymptotically stable.

Main Points of the Section. The single-step regulation law for a SISO plant can be easily computed (cf. (11)). The price paid for it is that applicability of single-step regulation is limited in that it yields an asymptotically stable closed-loop system only under restrictive assumptions on the plant and the input weight ρ. In particular, there may be *no* single-step regulator capable of stabilizing a nonminimum-phase open-loop unstable plant.

Problem 2.7-5   Consider the open-loop unstable nonminimum-phase plant

$$\Phi = \begin{bmatrix} 0 & 1 \\ 1 & 2 \end{bmatrix} \qquad G = \begin{bmatrix} 0 \\ 1 \end{bmatrix} \qquad H = \begin{bmatrix} 2 & 1 \end{bmatrix}$$

Compute the single-step feedback-gain row-vector (11). Show that the closed-loop eigenvalues of the single-step regulated systems equal $1 \pm \sqrt{1 + \rho}$. Conclude that there is no ρ, $\rho \in [0, \infty)$, yielding a stable closed-loop system.

NOTES AND REFERENCES

LQR is a topic widely and thoroughly discussed in standard textbooks: [AF66], [AM71], [AM90], [BH75], [DV85], [KS72], and [Lew86].

Dynamic programming was introduced by Bellman [Bel57]. More recent texts include [Ber76] and [Whi81].

The role of the Riccati equation in the LQR problem was emphasized by Kalman [Kal60a]. See also [Wil71]. Strong solutions of the ARE are discussed in [CGS84] and [dSGG86]. The literature on the Riccati equation is now immense, for example, [Ath71] and [Bit89].

Numerical factorization techniques for solving LQ problems were addressed by [Bie77] and [LH74]. See also [KHB$^+$85] and [Pet86]. Kleinman iterations for solving the ARE were analyzed in [Kle68], [McC69], and for the discrete-time case in [Hew71]. The control-theoretic interpretation of Kleinman iterations depicted in Fig. 5-1 appeared in [MM80].

Cheap control is the deterministic state-space version of the minimum-variance regulation of stochastic control [Åst70]. Similarly, in a deterministic state-space framework, single-step regulation corresponds to the generalized minimum-variance regulation discussed in [CG75], [CG79], and [WR79].

CHAPTER 3

I/O DESCRIPTIONS
AND FEEDBACK SYSTEMS

This chapter introduces representations for signals and linear time-invariant dynamic systems that are alternative to the state-space ones. These representations basically consist of system transfer matrices and matrix-fraction descriptions. They will be generically referred to as I/O or *external* descriptions in contrast with the state-space or *internal* descriptions. The experience has led us to appreciate one's advantage of being able to use both kinds of descriptions in a cooperative fashion, and exploit their relative merits.

The chapter is organized as follows. In Sec. 1, we introduce the *d*-representation of a sequence and matrix-fraction descriptions of system transfer matrices. Section 2 shows how stability of a feedback linear system can be studied by using matrix-fraction descriptions of the plant and the compensator. These tools allow us to characterize all feedback compensators, in a suitably parameterized form, that stabilize a given plant. The issue of robust stability is addressed in Sec. 3. After introducing in Sec. 4 system polynomial representations in the unit backward operator, Sec. 5 discusses how the asymptotic tracking problem can be formulated as a stability problem of a feedback system.

3.1 SEQUENCES AND MATRIX-FRACTION DESCRIPTIONS

Consider a time-indexed matrix-valued sequence

$$
\begin{aligned}
u(\cdot) &= \{u(k)\}_{k=-\infty}^{\infty} && \text{(3.1-1)} \\
&= \{\cdots, u(-1) \; ; \; u(0), \; u(1), \cdots\}
\end{aligned}
$$

where: $u(k) \in \mathbb{R}^{p \times m}$; $k \in \mathbb{Z}$; and the semicolon separates the samples at negative times on the left from the ones at nonnegative times on the right. Another possibility is to write

$$\hat{u}(d) := \sum_{k=-\infty}^{\infty} u(k) d^k \qquad (3.1\text{-}2)$$

This has to be interpreted as a representation of the given sequence where the symbol d^k, the kth power of d, indicates that the associated matrix $u(k)$ is the value taken on by $u(\cdot)$ at the integer k along the time–axis. *E.g.*, for the real–valued sequence

$$u(\cdot) = \{-1; 1, 2, -3\} \qquad (3.1\text{-}3)$$

we have

$$\hat{u}(d) = -d^{-1} + 1 + 2d - 3d^2 \qquad (3.1\text{-}4)$$

In $\hat{u}(d)$ the powers of d are instrumental for identifying the positions of the numbers -1,1,2,-3 along the time–axis. From this viewpoint, d is an *indeterminate* and no numerical value, either real or complex, pertains to it. It is only a time–marker. In particular, the power series (2) is a *formal series* in that it is not to be interpreted as a function of d, and there is no question of convergence whatsoever. We shall refer to $\hat{u}(d)$ as the *d-representation* of $u(\cdot)$.

Consider now

$$u_{-1}(\cdot) := u(\cdot - 1) \qquad (3.1\text{-}5)$$

a copy of the sequence $u(\cdot)$ delayed by one step. We have

$$\hat{u}_{-1}(d) = \sum_{k=-\infty}^{\infty} u(k-1) d^k = d\hat{u}(d) \qquad (3.1\text{-}6)$$

We see then that d applied to $\hat{u}(d)$ yields the d–representation of the sequence $u(\cdot)$ delayed by one–step.

Consider next the sequence $v(\cdot)$ obtained by convolving $w(\cdot)$ with $u(\cdot)$

$$
\begin{aligned}
v(k) &= \sum_{i=-\infty}^{\infty} w(i) u(k-i) \qquad (3.1\text{-}7) \\
&= \sum_{r=-\infty}^{\infty} w(k-r) u(r)
\end{aligned}
$$

we have

$$
\begin{aligned}
\hat{v}(d) &= \sum_{k=-\infty}^{\infty} \sum_{i=-\infty}^{\infty} w(i) u(k-i) d^k \qquad (3.1\text{-}8) \\
&= \sum_{i=-\infty}^{\infty} w(i) \sum_{k=-\infty}^{\infty} u(k-i) d^k
\end{aligned}
$$

$$= \sum_{i=-\infty}^{\infty} w(i)d^i \hat{u}(d) \qquad [(6)]$$

$$= \hat{w}(d)\hat{u}(d)$$

We then see that the d-representation of the convolution of two sequences $w(\cdot)$ and $u(\cdot)$ is the product of the d-representations of $w(\cdot)$ and $u(\cdot)$.

We insist again on pointing out that the operations under (8) are formal. In particular, the two infinite summations in (8) always commute as long as (7) makes sense.

Given $\hat{u}(d)$, we define as its *adjoint*

$$\hat{u}^*(d) := \hat{u}'(d^{-1}) \qquad (3.1\text{-}9)$$

Then, $\hat{u}^*(d)$ is the d-representation of a sequence obtained by taking the transpose of each sample of $u(\cdot)$ and reversing the time axis. We say that $\hat{u}(d)$ has *order* ℓ whenever ℓ is the minimum d-power present in (2). In such a case, we shall write

$$\text{ord}\,\hat{u}(d) = \ell \qquad (3.1\text{-}10)$$

For example, the order of $\hat{u}(d)$ in (4) equals -1. Because $u(\cdot)$ and $\hat{u}(d)$ identify the same entity, in the sequel, for the sake of conciseness, we shall simply refer to $\hat{u}(d)$ as a sequence, whenever no ambiguity can arise.

We say that $\hat{u}(d)$ is a *causal* sequence if $\text{ord}\,\hat{u}(d) \geq 0$, and *strictly causal* if $\text{ord}\,\hat{u}(d) > 0$. $\hat{u}(d)$ is called *anticausal* (*strictly anticausal*) if $\hat{u}^*(d)$ is causal (strictly causal). A sequence is called *one-sided* if it is either causal or anticausal. Otherwise, it is called *two-sided*.

We write $u(\cdot) \in \ell_2$, or, equivalently, $\hat{u}(d) \in \ell_2$, whenever the corresponding sequence has finite *energy*, viz.,

$$\|u(\cdot)\|^2 \quad := \quad \text{Tr} \sum_{k=-\infty}^{\infty} u'(k)u(k) \qquad (3.1\text{-}11)$$

$$= \quad \text{Tr} \sum_{k=-\infty}^{\infty} u(k)u'(k) < \infty$$

The previous quantity can be also computed as follows:

$$\|\hat{u}(d)\|^2 := \text{Tr}\langle \hat{u}^*(d)\hat{u}(d)\rangle \qquad (3.1\text{-}12)$$

where the symbol $\langle\,\rangle$ denotes extraction of the 0-power term. For example, $\langle -d^{-1} + 2 + d - 3d^2 \rangle = 2$. It is easy to verify that

$$\|u(\cdot)\|^2 = \|\hat{u}(d)\|^2 \qquad (3.1\text{-}13)$$

whenever $u(\cdot) \in \ell_2$.

Consider now temporarily the series (1) as a numerical series, viz., as a function of $d \in \mathbb{C}$, $\mathbb{C}$ denoting the field of complex numbers. Assume that the series converges for d in some subset $\mathcal{D}$ of $\mathbb{C}$ and its sum can be written in a closed form $S(d)$, viz.,

$$S(d) = \sum_{k=-\infty}^{\infty} u(k)d^k, \qquad d \in \mathcal{D} \subset \mathbb{C} \tag{3.1-14}$$

In such a case, we shall equal the formal series (2) to $S(d)$:

$$\hat{u}(d) = S(d) \tag{3.1-15}$$

and $S(d)$ in (15) will be called the *formal sum* of (1).

Example 3.1-1   For any square matrix $\Phi \in \mathbb{R}^{n \times n}$, consider the causal sequence

$$u(\cdot) = \{u(k) = \Phi^k\}_{k=0}^{\infty} = \{; I_n, \Phi, \Phi^2, \cdots\} \tag{3.1-16}$$

Then

$$\hat{u}(d) = \sum_{k=0}^{\infty} (\Phi d)^k = (I_n - d\Phi)^{-1} = S(d) \tag{3.1-17}$$

In fact, $(I_n - d\Phi)^{-1}$ is the sum of the numerical series $\sum_{k=0}^{\infty} (\Phi d)^k$ for every complex number d such that $|d| < 1/|\lambda_{\max}(\Phi)|$, where $\lambda_{\max}(\Phi)$ is the eigenvalue of Φ with maximum modulus.

In accordance with interpreting d as an indeterminate, we point out that (15) is formal and there is no question of convergence of $\hat{u}(d)$ to $S(d)$ whatsoever.

Another point that has to be brought out is the following. Given a two-sided sequence, its formal sum, whenever it exists, is well defined. Conversely, given a formal sum $S(d)$, in general, the corresponding series $\hat{u}(d)$ can be unambiguously identified only by specifying its order, for example, whether $\hat{u}(d)$ is either causal or anticausal.

Example 3.1-2   Consider the formal sum $S(d) = (I_n - d\Phi)^{-1}$ found in Example 1 for the sequence (16) with ord $\hat{u}(d) = 0$. Assuming Φ nonsingular, we can write formally

$$\begin{aligned} S(d) &= (I_n - d\Phi)^{-1} \\ &= -(d\Phi)^{-1}(I_n - d^{-1}\Phi^{-1})^{-1} \\ &= -\{\Phi^{-1}d^{-1} + \Phi^{-2}d^{-2} + \cdots\} \end{aligned} \tag{3.1-18}$$

Then, $S(d)$ is also the formal sum of the previous series, whose order is $-\infty$, and corresponds to the strictly anticausal sequence

$$v(\cdot) = \{ \quad \cdots \quad , -\Phi^{-2}, -\Phi^{-1}; \} \tag{3.1-19}$$

In fact, $(I_n - d\Phi)^{-1}$ is the sum of the numerical series $-\sum_{k=-\infty}^{-1} (\Phi d)^k$ for every complex number d such that $|d| > 1/|\lambda_{\min}(\Phi)|$, where $\lambda_{\min}(\Phi)$ is the eigenvalue of Φ with minimum modulus.

Problem 3.1-1   Convince yourself that there is no formal sum for the d-representations of the two–sided sequences:

$$z(\cdot) \quad := \quad \{\cdots, -\Phi^{-2}, -\Phi^{-1}; I_n, \Phi, \Phi^2, \cdots\} \tag{3.1-20}$$
$$h(\cdot) \quad := \quad \{\cdots, \Phi^{-2}, \Phi^{-1}; I_n, \Phi, \Phi^2, \cdots\} \tag{3.1-21}$$

In this book, it will be made always clear from the context whether a formal sum corresponds to either a causal or anticausal matrix sequence. The following example consolidates the point.

Example 3.1-3   Consider the linear time-invariant state-space representation

$$\begin{cases} x(k+1) &= \Phi x(k) + Gu(k) \\ y(k) &= Hx(k) \end{cases} \quad \text{for } k = 0, 1, 2, \cdots \tag{3.1-22}$$

We have

$$x(k) = \Phi^k x(0) + \sum_{i=-\infty}^{\infty} g(i)u(k-1) \tag{3.1-23}$$

where $u(\cdot) = \{; u(0), u(1), \cdots\}$ is a causal input sequence and

$$g(i) := \begin{cases} \Phi^{i-1}G & , i \geq 1 \\ O_{n\times m} & , \text{elsewhere} \end{cases} \tag{3.1-24}$$

is the ith sample of the impulse-response matrix of (Φ, G, I_n). Because by Example 1 and (6),

$$\hat{g}(d) = (I_n - d\Phi)^{-1}dG \tag{3.1-25}$$

using (8), we find

$$\hat{x}(d) = (I_n - d\Phi)^{-1}[x(0) + dG\hat{u}(d)] \tag{3.1-26}$$

Here, because $\hat{x}(d)$ and $\hat{u}(d)$ are causal sequences and the system (22) dynamic, $(I_n - d\Phi)^{-1}$ must be interpreted as the formal sum of the causal matrix sequence of Example 1. Further,

$$\hat{y}(d) = H(I_n - d\Phi)^{-1}x(0) + H_{yu}(d)\hat{u}(d) \tag{3.1-27}$$

where the rational matrix

$$H_{yu}(d) := H(I_n - d\Phi)^{-1}dG \tag{3.1-28}$$

is the *transfer matrix* of the system (22). This must be regarded as the formal sum of the d-representation of the sequence of the samples of the *impulse-response matrix* of the system (22):

$$h_{yu}(\cdot) := \{; O_{p\times m}, HG, H\Phi G, \cdots\} \tag{3.1-29}$$

We say that $\Sigma = (\Phi, G, H)$ is *free of hidden modes* if it is completely reachable and completely observable. Σ is *free of nonzero hidden eigenvalues* if it is completely controllable and completely reconstructible. In Appendix B, it is shown that Σ is completely controllable if and only if the polynomial matrices $(I_n - d\Phi)$ and dG are *left coprime*, and completely reconstructible if and only if $(I_n - d\Phi)$ and H are *right coprime*. Σ will be said to be *free of unstable hidden modes* if it is stabilizable and detectable. In Appendix B, it is shown that a necessary and sufficient condition for Σ to be stabilizable is that the *greatest common left divisors* (gcld) $\Delta(d)$ of $(I_n - d\Phi)$ and dG are *strictly Hurwitz*, viz.,

$$\det \Delta(d) \neq 0 \quad , \quad \forall \, |d| \leq 1 \qquad (3.1\text{-}30)$$

It is also shown that a necessary and sufficient condition for Σ to be detectable is that *the greatest common right divisors* (gcrd) of $(I_n - d\Phi)$ and H are strictly Hurwitz.

$H_{yu}(d)$ in (28) can be represented in terms of *matrix-fraction descriptions* (MFDs):

$$\hat{H}_{yu}(d) \;=\; A_1^{-1}(d)B_1(d) \qquad (3.1\text{-}31)$$
$$=\; B_2(d)A_2^{-1}(d) \qquad (3.1\text{-}32)$$

where $A_1(d)$ and $B_1(d)$ are polynomial matrices of dimensions $p \times p$ and $p \times m$, respectively; $A_2(d)$ and $B_2(d)$ are polynomial matrices of dimensions $m \times m$ and $p \times m$, respectively. $A_1^{-1}(d)B_1(d)$ is called a *left* MFD. Further, this is said to be an *irreducible* left MFD, whenever $A_1(d)$ and $B_1(d)$ are left coprime. *Mutatis mutandis* a similar terminology is used for the *right* MFD, $B_2(d)A_2^{-1}(d)$. For an irreducible left MFD, $A_1^{-1}(d)B_1(d)$, to represent the transfer matrix of a strictly causal dynamic system (22), it is necessary and sufficient that

1. $A_1(0)$ is nonsingular $\qquad\qquad\qquad\qquad\qquad\qquad\qquad$ (3.1-33)

2. $d \mid B_1(d)$ $\qquad\qquad\qquad\qquad\qquad\qquad\qquad\qquad\qquad\qquad$ (3.1-34)

The latter condition is expressed in words by saying that d divides $B_1(d)$, viz. $B_1(0) = O_{p \times m}$, that is, ord $B_1(d) > 0$. Then, $B_1(d)$ is a strictly causal matrix sequence of finite length. Condition 1. is necessary and sufficient for $A_1(d)$ to be causally invertible, viz., for the existence of a causal-matrix sequence $A_1^{-1}(d)$ such that

$$I_p = A_1(d)A_1^{-1}(d) = A_1^{-1}(d)A_1(d) \qquad (3.1\text{-}35)$$

Problem 3.1-2 Let $A(d)$ be a polynomial matrix:

$$A(d) = A_0 + A_1 d + \cdots + A_n d^n$$

with $A_i \in \mathbb{R}^{p \times p}$. Show that there exists a causal-matrix sequence

$$A^{-1}(d) = \sum_{k=0}^{\infty} V_k d^k \quad , \quad V_k \in \mathbb{R}^{p \times p}$$

such that

$$I_p = A(d)A^{-1}(d) = A^{-1}(d)A(d)$$

if and only if

$$A(0) = A_0$$

is nonsingular.

It is crucial to appreciate that, in general, the structural properties of Σ cannot be inferred from the MFDs of $H_{yu}(d)$. In particular, an irreducible MFD of $H_{yu}(d)$ does not provide information on the hidden modes of Σ. In order to make this precise, let us define the *d-characteristic polynomial* $\chi_\Phi(d)$ of Φ:

$$\chi_\Phi(d) := \det(I_n - d\Phi) \tag{3.1-36}$$

Then, from Appendix B, we have the following result.

Fact 3.1-1.   *Consider the system $\Sigma = (\Phi, G, H)$. Let the MFDs (31) and (32) of its transfer matrix be irreducible. Then,*

$$\chi_\Phi(d) = \frac{\det A_1(d)}{\det A_1(0)} = \frac{\det A_2(d)}{\det A_2(0)} \tag{3.1-37}$$

if and only if Σ is free of nonzero hidden eigenvalues, that is, controllable and reconstructible.

Problem 3.1-3   Consider the characteristic polynomial of Φ:

$$\bar\chi_\Phi(z) := \det(zI_n - \Phi) \tag{3.1-38}$$

Clearly, if $\partial\bar\chi_\Phi(z)$ denotes the degree of $\bar\chi_\Phi(z)$, we have $\partial\bar\chi_\Phi(z) = \dim\Phi = n$. Show that the d-characteristic polynomial of Φ is the *reciprocal polynomial* of $\bar\chi_\Phi$, viz.,

$$\begin{aligned} \chi_\Phi(d) &= d^{\partial\bar\chi_\Phi}\,\bar\chi_\Phi^*(d) \\ &= d^n\bar\chi_\Phi(d^{-1}) \end{aligned} \tag{3.1-39}$$

and that

$$\partial\chi_\Phi(d) = \partial\bar\chi_\Phi(z) \;-\; \text{number of zero roots of } \bar\chi_\Phi(z) \tag{3.1-40}$$

Further,

$$\bar\chi_\Phi(d) = d^n\chi_\Phi(d^{-1}) = d^n\chi_\Phi^*(d) \tag{3.1-41}$$

Note that if $p(d)$ is a polynomial, then the roots of its reciprocal polynomial, defined as $d^{\partial p}p^*(d)$, equal the reciprocal of the nonzero roots of $p(d)$.

Problem 3 shows that a necessary and sufficient condition for Σ to be asymptotically stable is that $\chi_\Phi(d)$ be strictly Hurwitz. Because, according to Fact 1, the determinants of the denominators of the irreducible MFDs of $H_{yu}(d)$ capture only the nonzero unhidden eigenvalues of Φ, a condition for the asymptotic stability of Σ can be stated as follows.

Proposition 3.1-1.   *Let the MFDs (31) and (32) of the transfer matrix of the system $\Sigma = (\Phi, G, H)$ be irreducible. Let Σ be free of unstable hidden modes. Then, Σ is asymptotically stable if and only if $A_1(d)$, or, equivalently, $A_2(d)$, is a strictly Hurwitz polynomial matrix.*

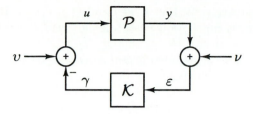

Figure 3.2-1: The feedback system.

It is customary to speak about the stability of transfer matrices in contrast with the asymptotic stability of state-space representations. A rational function $H(d)$ is said to be *stable* if the denominator polynomial $a(d)$ of its irreducible form $H(d) = b(d)/a(d)$ is strictly Hurwitz. We note that, according to this definition, $H(d) = \frac{b(d)\varphi(d)}{a(d)\varphi(d)}$, where $a(d)$, $b(d)$, are $\varphi(d)$ are polynomials in d with $a(d)$ and $b(d)$ coprime, is stable if and only if $a(d)$ is strictly Hurwitz, irrespective of $\varphi(d)$. Likewise, a rational matrix $H(d) = \{H_{ij}(d)\}$ is said to be stable if all the denominator polynomials $a_{ij}(d)$ of its irreducible elements $H_{ij}(d) = b_{ij}(d)/a_{ij}(d)$ are strictly Hurwitz. This is the same as requiring that the irreducible MFDs of $H(d)$ have strictly Hurwitz denominator matrices. For this reason, there is no difference between stability of a rational matrix $H(d)$ and stability of its irreducible MFDs. This will be consequently reflected in our language in that we shall talk indifferently about stability of either rational matrices or irreducible MFDs.

Main Points of the Section. Sequences can be described in terms of d-representations. Likewise, time-invariant linear dynamic systems can be described in terms of transfer matrices and matrix fraction descriptions. Equation (12) is central in the polynomial equation approach to least-squares optimization and replaces the usual complex integral

$$\|u(\cdot)\|^2 = \frac{1}{2\pi j} \operatorname{Tr} \oint_{|z|=1} \hat{u}'\left(\frac{1}{z}\right) \hat{u}(z) \frac{dz}{z} \tag{3.1-42}$$

or more generally, for $u(\cdot), v(\cdot) \in \ell_2$,

$$\operatorname{Tr}\langle \hat{u}^*(d)\hat{v}(d)\rangle = \frac{1}{2\pi j} \operatorname{Tr} \oint_{|z|=1} \hat{u}'\left(\frac{1}{z}\right) \hat{v}(z) \frac{dz}{z}$$
$$= \frac{1}{2\pi} \operatorname{Tr} \int_{-\pi}^{\pi} \hat{u}'\left(e^{-j\omega}\right) \hat{v}\left(e^{j\omega}\right) d\omega \tag{3.1-43}$$

3.2 FEEDBACK SYSTEMS

Consider the feedback system of Fig. 1, where $\mathcal{P}$ and $\mathcal{K}$ denote two discrete-time finite-dimensional linear time-invariant dynamic systems with transfer matrices

$P(d)$ and $K(d)$, respectively. In Fig. 1, $v(\cdot)$ and $\nu(\cdot)$, $v(k) \in \mathbb{R}^m$ and $\nu(k) \in \mathbb{R}^p$, represent two exogenous input sequences, and $u(\cdot)$ and $\varepsilon(\cdot)$ the sequences at the input of $\mathcal{P}$ and $\mathcal{K}$, respectively.

We say that the feedback system is *well posed* if given any bounded input pair $\hat{w}(d) := [\ \hat{\nu}'(d) \quad \hat{v}'(d)\ ]'$ such that

$$\text{ord}\,\hat{w}(d) > -\infty \tag{3.2-1}$$

the response of the feedback system as given by $\hat{z}(d) := [\ \hat{\varepsilon}'(d) \quad \hat{u}'(d)\ ]'$ can be uniquely determined. To this end, it is immaterial to specify from which initial states for $\mathcal{P}$ and $\mathcal{K}$ input $w(\cdot)$ is first applied. Whenever these initial states are unspecified, by default, they will be taken to be zero. Accordingly,

$$\hat{z}(d) = \hat{w}(d) + \begin{bmatrix} O_p & P(d) \\ -K(d) & O_m \end{bmatrix} \hat{z}(d) \tag{3.2-2}$$

It follows that the *well–posed condition* for the feedback system is that the following determinant be not identically zero:

$$\begin{aligned} \det \begin{bmatrix} I_p & -P(d) \\ K(d) & I_m \end{bmatrix} &= \det \begin{bmatrix} I_p & -P(d) \\ O_{m\times p} & I_m + K(d)P(d) \end{bmatrix} \\ &= \det \begin{bmatrix} I_p + P(d)K(d) & O_{p\times m} \\ K(d) & I_m \end{bmatrix} \\ &= \det[I_m + K(d)P(d)] \\ &= \det[I_p + P(d)K(d)] \neq 0 \end{aligned} \tag{3.2-3}$$

From now on, it is assumed that the $p \times m$ rational transfer matrix $P(d)$ is such that

$$\text{ord}\,P(d) > 0 \tag{3.2-4}$$

In words, $P(d)$ is a strictly causal matrix sequence. Further, the $m \times p$ rational transfer matrix $K(d)$ is assumed to be a causal-matrix sequence, viz.

$$\text{ord}\,K(d) \geq 0 \tag{3.2-5}$$

It follows that ord $K(d)P(d) > 0$ and ord $P(d)K(d) > 0$. Consequently, (4) and (5) imply a well–posed feedback system.

DEFINITIONS

1. The system of Fig. 1 with $\mathcal{P}$ and $\mathcal{K}$ satisfying (4) and (5), respectively, will be called the *feedback system* with *plant* $\mathcal{P}$ and *compensator* $\mathcal{K}$.

2. The feedback system is *internally stable* if the transfer matrix

$$H_{zw}(d) = \begin{bmatrix} H_{\varepsilon\nu}(d) & H_{\varepsilon v}(d) \\ H_{u\nu}(d) & H_{uv}(d) \end{bmatrix} \qquad (3.2\text{-}6)$$

is stable.

3. The feedback system is *asymptotically stable* if the dynamical system resulting from the feedback interconnection of the dynamical systems $\mathcal{P}$ and $\mathcal{K}$ is such.

We see that, in contrast with asymptotic stability, internal stability is the same as stability of the four transfer matrices:

$$\begin{aligned} H_{\varepsilon\nu}(d) &= [I_p + P(d)K(d)]^{-1} \\ &= I_p - P(d)[I_m + K(d)P(d)]^{-1}K(d) \end{aligned} \qquad (3.2\text{-}7)$$

$$\begin{aligned} H_{\varepsilon v}(d) &= [I_p + P(d)K(d)]^{-1}P(d) \\ &= P(d)[I_m + K(d)P(d)]^{-1} \end{aligned} \qquad (3.2\text{-}8)$$

$$\begin{aligned} H_{u\nu}(d) &= -[I_m + K(d)P(d)]^{-1}K(d) \\ &= -K(d)[I_p + P(d)K(d)]^{-1} \end{aligned} \qquad (3.2\text{-}9)$$

$$\begin{aligned} H_{uv}(d) &= [I_m + K(d)P(d)]^{-1} \\ &= I_m - K(d)[I_p + P(d)K(d)]^{-1}P(d) \end{aligned} \qquad (3.2\text{-}10)$$

In (7) to (10), all the equalities can be verified by inspection. In fact, (9) can be easily established. Then, the last expression in (7) equals $I_p - P(d)K(d)[I_p + P(d)K(d)]^{-1}$, which, in turn, coincides with $[I_p + P(d)K(d)]^{-1}$. Along the same line, we can check (9) and (10).

A simplification takes place whenever $K(d)$ is a stable transfer matrix. In fact, in such a case, instead of checking that four different transfer matrices are stable, internal stability of the feedback system can be ascertained by only checking the stability of $H_{\varepsilon v}(d)$. The latter is sometimes referred to as *external stability* of the feedback system.

Proposition 3.2-1. *Consider the feedback system of Fig. 1. Let the compensator transfer matrix $K(d)$ be stable. Then, a necessary and sufficient condition for internal stability of the feedback system is that $H_{\varepsilon v}(d)$ be stable.*

Proof: Stability of $H_{\varepsilon v}(d)$ is obviously necessary. Sufficiency follows because

$$H_{\varepsilon\nu}(d) = I_p - H_{\varepsilon v}(d)K(d) \qquad (3.2\text{-}11)$$
$$H_{uv}(d) = I_m - K(d)H_{\varepsilon v}(d) \qquad (3.2\text{-}12)$$
$$H_{u\nu}(d) = -H_{uv}(d)K(d) \qquad (3.2\text{-}13)$$

are stable transfer matrices, provided that $K(d)$ is stable.

If $K(d)$ is unstable, internal stability of the feedback system does not follow from external stability, viz. the *stability* of $H_{\varepsilon v}(d)$.

Problem 3.2-1   Consider a SISO feedback system with $P(d) = B(d)/A(d)$, $A(d)$ and $B(d)$ coprime, $B(d) = d(1 - 2d)$, $K(d) = S(d)/R(d)$, $R(d)$ and $S(d)$ coprime, $R(d) = (1 - 2d)$. Assume that $A(d) + dS(d)$ is strictly Hurwitz. Show that, though $H_{\varepsilon v}(d)$ is stable, $H_{uv}(d)$ is unstable.

The following Fact 1 shows that internal stability and asymptotic stability of the feedback system are equivalent, whenever $\mathcal{P}$ and $\mathcal{K}$ are free of unstable hidden modes [Vid85].

Fact 3.2-1.   *Let the plant $\mathcal{P}$ and the compensator $\mathcal{K}$ be free of unstable hidden modes. Then, the feedback system is asymptotically stable if and only if it is internally stable.*

For the sake of brevity, keeping in mind Fact 1, throughout this book, we shall simply say that the feedback system is *stable* whenever it is internally stable and $\mathcal{P}$ and $\mathcal{K}$ are understood to be free of unstable hidden modes.

Fact 1-1 holds true for the feedback system. Namely, the d-characteristic polynomial of any realization of the feedback system with $\mathcal{P}$ and $\mathcal{K}$ free of nonzero hidden eigenvalues, is proportional, according to (1-37), to the determinant of the denominator polynomial of any irreducible MFD of $H_{zw}(d)$. From (2), we find

$$H_{zw}(d) = \begin{bmatrix} I_p & -P(d) \\ K(d) & I_m \end{bmatrix}^{-1} \tag{3.2-14}$$

Consider the following irreducible MFDs of $P(d)$ and $K(d)$, respectively,

$$P(d) = A_1^{-1}(d)B_1(d) = B_2(d)A_2^{-1}(d) \tag{3.2-15}$$

$$K(d) = R_1^{-1}(d)S_1(d) = S_2(d)R_2^{-1}(d) \tag{3.2-16}$$

We find

$$\begin{bmatrix} I_p & -P(d) \\ K(d) & I_m \end{bmatrix} = \begin{bmatrix} I_p & -A_1^{-1}(d)B_1(d) \\ R_1^{-1}(d)S_1(d) & I_m \end{bmatrix} \tag{3.2-17}$$

$$= \begin{bmatrix} A_1^{-1}(d) & 0 \\ 0 & R_1^{-1}(d) \end{bmatrix} \begin{bmatrix} A_1(d) & -B_1(d) \\ S_1(d) & R_1(d) \end{bmatrix}$$

Then

$$H_{zw}(d) = \begin{bmatrix} A_1(d) & -B_1(d) \\ S_1(d) & R_1(d) \end{bmatrix}^{-1} \begin{bmatrix} A_1(d) & 0 \\ 0 & R_1(d) \end{bmatrix} \tag{3.2-18}$$

$$= \begin{bmatrix} R_2(d) & 0 \\ 0 & A_2(d) \end{bmatrix} \begin{bmatrix} R_2(d) & -B_2(d) \\ S_2(d) & A_2(d) \end{bmatrix}^{-1}$$

where the first equality follows from (14), and the second equality is obtained in a similar way by using the right coprime MFDs of $P(d)$ and $K(d)$.

Problem 3.2-2 Show that the two MFDs of $H_{zw}(d)$ in (18) are irreducible. (*Hint:* The left MFD $A^{-1}(d)B(d)$ is irreducible if and only if $A(d)$ and $B(d)$ satisfy the *Bezout identity* (B.1-10).)

According to Problem 2, internal stability of the feedback system is equivalent to strict Hurwitzianity of the polynomial denominators of (18). We have

$$\det \begin{bmatrix} A_1(d) & -B_1(d) \\ S_1(d) & R_1(d) \end{bmatrix}$$

$$= \det R_1(d) \det \left[A_1(d) + B_1(d)R_1^{-1}(d)S_1(d) \right]$$

$$= \det R_1(d) \det \left[A_1(d) + B_1(d)S_2(d)R_2^{-1}(d) \right]$$

$$= \frac{\det R_1(d)}{\det R_2(d)} \det \left[A_1(d)R_2(d) + B_1(d)S_2(d) \right]$$

$$= \frac{\det R_1(0)}{\det R_2(0)} \det \left[A_1(d)R_2(d) + B_1(d)S_2(d) \right] \qquad (3.2\text{-}19)$$

where the first equality holds because $R_1(d)$ is nonsingular [Kai80, p. 650], and the last follows from (1-37). The conclusions of the next theorem then follow at once.

Theorem 3.2-1. *Consider the feedback system of Fig. 1 with plant and compensator having the irreducible MFDs (15) and (16). Then, the feedback system is internally stable if and only if*

$$P_1(d) := A_1(d)R_2(d) + B_1(d)S_2(d) \qquad (3.2\text{-}20)$$

or, equivalently,

$$P_2(d) := R_1(d)A_2(d) + S_1(d)B_2(d) \qquad (3.2\text{-}21)$$

are strictly Hurwitz. Further, the d-characteristic polynomial $\chi_\Phi(d)$ of any realization of the feedback system with $\mathcal{P}$ and $\mathcal{K}$ both free of nonzero hidden eigenvalues, is given by

$$\chi_\Phi(d) = \frac{\det P_1(d)}{\det P_1(0)} = \frac{\det P_2(d)}{\det P_2(0)} \qquad (3.2\text{-}22)$$

Problem 3.2-3 Consider the plant

$$A(d)\hat{y}(d) = B(d)\hat{z}(d) + C(d)\hat{e}(d)$$

where $\begin{bmatrix} z' & e' \end{bmatrix} =: u$ denotes a partition of input u into two separate vectors z and e. Let $A^{-1}(d) \begin{bmatrix} B(d) & C(d) \end{bmatrix}$ be an irreducible left MFD. Show that a necessary condition for the existence of a compensator

$$\begin{bmatrix} \hat{z}(d) \\ \hat{e}(d) \end{bmatrix} = - \begin{bmatrix} N_{2z}(d) \\ O \end{bmatrix} M_2^{-1}(d)\hat{y}(d)$$

that makes the feedback system internally stable is that the greatest common left divisors of $A(d)$ and $B(d)$ are strictly Hurwitz. Further, show that, under this condition, for the feedback system, internal stability coincides with asymptotic stability if and only if the compensator is realized without unstable hidden modes and all the unstable eigenvalues of the actual plant realization are the same, counting their multiplicity, as those of the minimal realization of $H_{yu}(d)$ or, equivalently, the reciprocals of the roots of $\det A(d)$.

We see from Theorem 1 that if the feedback system is internally stable, (20) and (21) can be rewritten as follows:

$$I_p = A_1(d)M_2(d) + B_1(d)N_2(d) \qquad\qquad (3.2\text{-}23)$$

$$I_m = M_1(d)A_2(d) + N_1(d)B_2(d) \qquad\qquad (3.2\text{-}24)$$

where

$$M_2(d) := R_2(d)P_1^{-1}(d) \qquad N_2(d) := S_2(d)P_1^{-1}(d) \qquad\qquad (3.2\text{-}25)$$

$$M_1(d) := P_2^{-1}(d)R_1(d) \qquad N_1(d) := P_2^{-1}(d)S_1(d) \qquad\qquad (3.2\text{-}26)$$

are stable transfer matrices. We note that the transfer matrix of the controller can be written as the ratio of the previous transfer matrices:

$$\begin{aligned} K(d) &= N_2(d)M_2^{-1}(d) &\qquad (3.2\text{-}27)\\ &= M_1^{-1}(d)N_1(d) &\qquad (3.2\text{-}28) \end{aligned}$$

This representation for the controller transfer matrix is not only necessary for the feedback system to be internally stable. It also turns out to be sufficient as well.

Theorem 3.2-2. *Consider the feedback system of Fig. 1 and the irreducible MFDs (15) of the plant. Then, a necessary and sufficient condition for the feedback system to be internally stable is that the compensator transfer matrix be factorizable as in (27) (equivalently, (28)) in terms of the ratio of two stable transfer matrices $M_2(d)$ and $N_2(d)$ ($M_1(d)$ and $N_1(d)$) satisfying the identity (23) ((24)). Conversely, a compensator with a transfer matrix factorizable as in (27) ((28)) in terms of $M_2(d)$ and $N_2(d)$ ($M_1(d)$ and $N_2(d)$) satisfying (23) ((24)) makes the feedback system internally stable if and only if $M_2(d)$ and $N_2(d)$ ($M_1(d)$ and $N_1(d)$) are both stable transfer matrices.*

Proof: That the condition is necessary is proved by (23) to (28). Sufficiency is proved next by showing that the condition implies stability of the transfer matrices $H_{uv}(d)$, $H_{\varepsilon v}(d)$, $H_{u\nu}(d)$, and $H_{\varepsilon\nu}(d)$.

Using (7) to (10), we find

$$\begin{aligned} H_{uv}(d) &= [I_m + K(d)P(d)]^{-1}\\ &= [I_m + M_1^{-1}(d)N_1(d)B_2(d)A_2^{-1}(d)]^{-1}\\ &= A_2(d)[M_1(d)A_2(d) + N_1(d)B_2(d)]^{-1}M_1(d) &\qquad (3.2\text{-}29)\\ &= A_2(d)M_1(d) &\qquad [(16)] \end{aligned}$$

$$
\begin{aligned}
H_{u\nu}(d) &= -[I_m + K(d)P(d)]^{-1}K(d) \\
&= -A_2(d)M_1(d)M_1^{-1}(d)N_1(d) \qquad [(29)] \\
&= -A_2(d)N_1(d)
\end{aligned}
\qquad (3.2\text{-}30)
$$

$$
\begin{aligned}
H_{\varepsilon\nu}(d) &= [I_p + P(d)K(d)]^{-1} \\
&= [I_p + A_1^{-1}(d)B_1(d)N_2(d)M_2^{-1}(d)]^{-1} \\
&= M_2(d)[A_1(d)M_2(d) + B_1(d)N_2(d)]^{-1}A_1(d) \\
&= M_2(d)A_1(d)
\end{aligned}
\qquad
\begin{aligned}
&\\
&\\
&(3.2\text{-}31)\\
&[(23)]
\end{aligned}
$$

$$
\begin{aligned}
H_{\varepsilon v}(d) &= [I_p + P(d)K(d)]^{-1}P(d) \\
&= M_2(d)A_1(d)A_1^{-1}(d)B_1(d) \qquad [(31)] \qquad (3.2\text{-}32)\\
&= M_2(d)B_1(d)
\end{aligned}
$$

We see that because $M_1(d)$, $N_1(d)$, and $M_2(d)$ are stable transfer matrices, (29) to (32) are such.

Problem 3.2-4   Consider the feedback system of Fig. 1 and (29) to (32). Show that $H_{\gamma\nu}(d) = -H_{u\nu}(d)$ and $H_{yv}(d) = H_{\varepsilon v}(d)$ imply

$$N_2(d)A_1(d) = A_2(d)N_1(d) \qquad (3.2\text{-}30a)$$

$$B_2(d)M_1(d) = M_2(d)B_1(d) \qquad (3.2\text{-}32a)$$

Further, verify that $H_{\varepsilon\nu}(d) = H_{y\nu}(d) + I_p$ and $H_{u\nu}(d) = -H_{\gamma v}(d) + I_m$ imply

$$I_p = M_2(d)A_1(d) + B_2(d)N_1(d) \qquad (3.2\text{-}23a)$$

$$I_m = A_2(d)M_1(d) + N_2(d)B_1(d) \qquad (3.2\text{-}24a)$$

It is to be pointed out that given irreducible MFDs of $P(d)$ as in (15), (23) and (24) can be always solved w.r.t. $(M_2(d), N_2(d))$ and $(M_1(d), N_1(d))$, respectively. In fact, because $A_1(d)$ and $B_1(d)$ are left coprime, there are *polynomial* matrices $(M_2(d), N_2(d))$ solving the Bezout identity (23). Now polynomial matrices in d are stable transfer matrices representing impulse-response matrix sequences of finite length. Let, then, $(M_{20}(d)), N_{20}(d))$ and $(M_{10}(d), N_{10}(d))$ be two pairs of stable transfer matrices solving (23) and (24), respectively. It follows from the pertinent results of Appendix C that *all* other stable transfer matrices solving (23) and (24), respectively, are given by

$$
\left.
\begin{aligned}
M_2(d) &= M_{20}(d) - B_2(d)Q(d) \\
N_2(d) &= N_{20}(d) + A_2(d)Q(d)
\end{aligned}
\right\}
\qquad (3.2\text{-}33)
$$

and

$$
\left.
\begin{aligned}
M_1(d) &= M_{10}(d) - Q(d)B_1(d) \\
N_1(d) &= N_{10}(d) + Q(d)A_1(d)
\end{aligned}
\right\}
\qquad (3.2\text{-}34)
$$

where $Q(d)$ is any $m \times p$ stable transfer matrix. Summing up, we have the following result.

Theorem 3.2-3.   (YJBK Parameterization) *Consider the feedback system of Fig. 1 and the irreducible MFDs (15) of the plant. Then, there exist compensator transfer matrices $K(d)$ as in (27) and (28) which make the feedback system* *internally stable. Given one such a transfer matrix*

$$K_0(d) \quad = \quad N_{20}(d)M_{20}^{-1}(d) \tag{3.2-35}$$

$$= \quad M_{10}^{-1}(d)N_{10}(d) \tag{3.2-36}$$

with $(M_{20}(d)$, $N_{20}(d))$ and $(M_{10}(d)$, $N_{10}(d))$ two pairs of stable transfer matrices satisfying (23) and, respectively, (24), all other transfer matrices $K(d)$ which make the feedback system internally stable are given by

$$K(d) \quad = \quad [N_{20}(d) + A_2(d)Q(d)] \, [M_{20}(d) - B_2(d)Q(d)]^{-1} \tag{3.2-37}$$

$$= \quad [M_{10}(d) - Q(d)B_1(d)]^{-1} \, [N_{10}(d) + Q(d)A_1(d)] \tag{3.2-38}$$

where $Q(d)$ is any $m \times p$ stable transfer matrix.

Equations (37) and (38) give the *YJBK parameterization* of all $K(d)$ that make the feedback system internally stable. Here, the acronym YJBK stands for Youla, Jabr, and Bongiorno [YJB76] and Kučera [Kuč79], who first proposed the set of all stabilizing compensators in Q-parametric form.

Example 3.2-1 Let

$$P(d) = \frac{4d}{1 - 4d^2} \tag{3.2-39}$$

Here, $A(d) = 1 - 4d^2$ and $B(d) = 4d$ are coprime polynomials. Equation (23), or (24), becomes

$$(1 - 4d^2)M(d) + 4dN(d) = 1 \tag{3.2-40}$$

This is a Bezout identity having polynomial solutions $(M_0(d)$, $N_0(d))$. This solution can be made unique by requiring that either $\partial M_0(d) < 1$ or $\partial N_0(d) < 2$, $\partial p(d)$ denoting the degree of polynomial $p(d)$. The minimum-degree solution w.r.t. $M_0(d)$, viz. the one with $\partial M_0(d) < 1$, can be easily computed by equating the coefficients of equal powers of the polynomials on both sides of (40). We get

$$M_0(d) = 1 \quad \text{and} \quad N_0(d) = d \tag{3.2-41}$$

Then, (37), or (38), gives the YJBK parametric form of all compensator transfer functions, making, for the given $P(d)$, the feedback system internally stable:

$$K(d) \quad = \quad \frac{d + (1 - 4d^2)Q(d)}{1 - 4dQ(d)} \tag{3.2-42}$$

$$= \quad \frac{d\delta(d) + (1 - 4d^2)n(d)}{\delta(d) - 4dn(d)}$$

where $Q(d) = n(d)/\delta(d)$ with $n(d)$ any polynomial and $\delta(d)$ any strictly Hurwitz polynomial.

The next problem shows that the characteristic polynomial of the feedback system can be freely assigned by suitably selecting the denominator matrix of the MFDs of $Q(d)$.

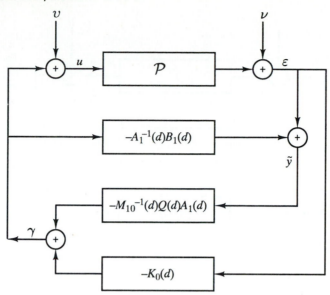

Figure 3.2-2: The feedback system with a Q-parameterized compensator.

Problem 3.2-5   Consider a YJBK parameterized compensator with transfer matrix $K(d)$ as in (37). Let $(M_{20}(d), N_{20}(d))$ be a pair of polynomial matrices satisfying (23). Write $Q(d)$ in the form of a right coprime MFD $Q(d) = L_2(d)D_2^{-1}(d)$. Show then that

$$K(d) = [N_{20}(d)D_2(d) + A_2(d)L_2(d)] \, [M_{20}(d)D_2(d) - B_2(d)L_2(d)]^{-1}$$

and $P_1(d)$ in (20) equals $D_2(d)$.

   Another feature of the YJBK parameterization that turns out to be useful in optimization problem [Vid85] is that transfer matrices $H_{uv}(d)$, $H_{u\nu}(d)$, $H_{\varepsilon\nu}(d)$, and $H_{\varepsilon v}(d)$, as shown in the proof of Theorem 2, are linear in $M_1(d)$, $N_1(d)$, and $M_2(d)$, respectively. It follows from (33) and (34) that all the previous transfer matrices are affine in the $Q(d)$ parameter.

   By (38), all control laws making the feedback system internally stable can be written as follows

$$\begin{aligned} \hat{\gamma}(d) \;=\; & -K_0(d)\hat{\varepsilon}(d) \\ & -M_{10}^{-1}(d)Q(d)A_1(d)[\hat{\varepsilon}(d) - A_1^{-1}(d)B_1(d)\hat{\gamma}(d)] \end{aligned} \qquad (3.2\text{-}43)$$

Figure 2 depicts the feedback system with a YJBK parameterized compensator as in (43). As can be seen, the "outer" loop with output feedback $-K_0(d)$ is increased by an "inner" loop where $\tilde{y}$ is the difference between the (disturbed) plant output ε and the output from the plant model $A_1^{-1}(d)B_1(d)$. If $A_1(d)$ is strictly Hurwitz, (24) is solved by $N_{10}(d) = O_{m \times p}$ and $M_{10}(d) = A_2^{-1}(d)$. Then, the scheme of Fig. 2 can be simplified as in Fig. 3, where we set

$$\hat{Q}(d) := A_2^{-1}(d)Q(d)A_1(d) \qquad (3.2\text{-}44)$$

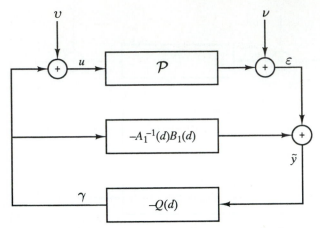

Figure 3.2-3: The feedback system with a Q-parameterized compensator for $P(d)$ stable.

Note that because $Q(d)$ is any stable $m \times p$ transfer matrix, $\hat{Q}(d)$ is any $m \times p$ stable transfer matrix as well. The scheme of Fig. 2 has been advocated [MZ89a] to be particularly advantageous in process control applications, where $P(d)$ turns to be stable.

   We now turn from internal to asymptotic stability. From Fact 1 and Theorem 1, the following result follows at once.

   Theorem 3.2-4.   *There exist compensators making the feedback system of Fig. 1 asymptotically stable if and only if the plant is free of unstable hidden modes. All such compensators are realizations free of unstable hidden modes of the YJBK parameterized transfer matrices $K(d)$ (37) or (38).*

Main Points of the Section.   Feedback systems can be studied by using matrix-fraction descriptions and properties of polynomial matrices. These tools can be used so as to nicely identify all feedback compensators, in YJBK parameterized form, that stabilize the plant.

3.3   ROBUST STABILITY

We consider again the feedback configuration of Fig. 2-1, where the plant is described by

$$\hat{y}(d) = P(d)\hat{u}(d) \tag{3.3-1}$$

with $P(d)$ the *true* or *actual* plant transfer matrix, and

$$\hat{u}(d) = -K(d)\hat{y}(d) \tag{3.3-2}$$

with $K(d)$ the compensator transfer matrix. We assume that the compensator is designed based upon the *nominal* plant transfer matrix $P^0(d)$ but applied to the actual plant $P(d)$.

The *robust stability* issue is to establish conditions under which stability of the designed closed loop implies stability of the actual closed loop. In order to address the point, it is convenient to let

$$
\begin{aligned}
G(d) &:= K(d)P(d) & (3.3\text{-}3) \\
G^0(d) &:= K(d)P^0(d) & (3.3\text{-}4)
\end{aligned}
$$

$G(d)$ $(G^0(d))$ is referred to as the actual (nominal) loop transfer matrix, viz., the transfer matrix of the plant/compensator cascade. We also assume that the actual plant transfer matrix equals the nominal plant transfer matrix postmultiplied by a *multiplicative perturbation* matrix, that is, we have

$$
P(d) = P^0(d)M(d) \tag{3.3-5}
$$

Hence,

$$
G(d) = G^0(d)M(d) \tag{3.3-6}
$$

We note that the relative, or percentage, error of the loop transfer matrix can be expressed in terms of the multiplicative perturbation matrix $M(d)$. In fact,

$$
\begin{aligned}
G^{-1}(d)[G_0(d) - G(d)] &= M^{-1}(d)\left[G^0(d)\right]^{-1}\left[G^0(d) - G^0(d)M(d)\right] \\
&= M^{-1}(d) - I_m \tag{3.3-7}
\end{aligned}
$$

For a complex square matrix A, denote by $\bar{\sigma}(A)$ and $\underline{\sigma}(A)$ the maximum and the minimum singular value of A, respectively, that is,

$$
\sigma(A) := +\lambda_{\max}^{1/2}(A^*A) \qquad \text{and} \qquad \underline{\sigma}(A) := +\lambda_{\min}^{1/2}(A^*A) \tag{3.3-8}
$$

where the star denotes Hermitian, that is, A^* is the complex-conjugate transpose of A. Further, we denote by Ω the contour in the complex plane consisting of the unit circle, suitably indented outwards around the roots of the open-loop poles of $G^0(d)$ on the circle.

We can now state the result on robust stability [BGW90], which will be used in Sec. 4.6.

Fact 3.3-1. *Denote by $\chi_{o\ell}(d)$ and $\chi_{o\ell}^0(d)$ the d-characteristic polynomial of the actual and nominal open-loop system, respectively; and $\chi_{c\ell}(d)$ and $\chi_{c\ell}^0(d)$ the characteristic polynomial of the actual and nominal closed-loop system, respectively. Let*

- *$\chi_{o\ell}(d)$ and $\chi_{o\ell}^0(d)$ have the same number of roots inside the unit circle*

- *$\chi_{o\ell}(d)$ and $\chi_{o\ell}^0(d)$ have the same unit circle roots*

- $\chi_{cl}^0(d)$ *is strictly Hurwitz*

Then $\chi_{cl}(d)$ is strictly Hurwitz provided that at each $d \in \Omega$,

$$\bar{\sigma}(M^{-1}(d) - I_m) < \min(\alpha(d), 1) \qquad (3.3\text{-}9)$$

$$\alpha(d) := \underline{\sigma}(I_m + G^0(d)) \qquad (3.3\text{-}10)$$

Transfer matrix $I_m + G^0(d)$ is called the *return difference* of the nominal loop, and (9) and (10) show that it plays an important role in robust stability. The importance of Fact 1 is that it shows that the feedback system of Fig. 2-1 remains stable when the relative error of the loop transfer matrix caused by multiplicative perturbations is small as compared to the nominal return difference.

We shall now use a different argument to point out a more generic and somewhat less direct result on robust stability. Namely, we shall show that *any nominally stabilizing compensator yields robust stability,* because it is capable of stabilizing all plants in a nontrivial neighborhood of the nominal one. This is a result of paramount interest that is worth studying in some detail for its far-reaching implications. To see this, we first point out (Problem 2.4-5) that the output dynamic compensation (2) can be looked at as a state-feedback compensation:

$$u(k) = Fx(k) \qquad (3.3\text{-}11)$$

provided that $x(k)$ be a plant state made up by a sufficient number of past input/output pairs:

$$x(k) := [y'(t - n + 1) \cdots y'(t) \; u'(t - n + 1) \cdots u'(t - 1)]' \qquad (3.3\text{-}12)$$

Further, if the nominal plant is stabilizable and detectable, neglecting its possible stable hidden modes of nonconcern to us, it can be described as follows:

$$\begin{cases} x(k+1) & = & \Phi^0 x(k) + G^0 u(k) \\ y(k) & = & Hx(k) \\ H & = & \begin{bmatrix} O_{p \times (n-1)p} & I_p & O_{p \times (n-1)m} \end{bmatrix} \end{cases} \qquad (3.3\text{-}13)$$

We assume that the nominal closed-loop system (11) to (13) is asymptotically stable, viz.,

$$\Phi_{cl}^0 := \Phi^0 + G^0 F \qquad (3.3\text{-}14)$$

is a stability matrix. Then, there are positive reals $\gamma \geq 1$ and $0 \leq \lambda < 1$ such that, for $k = 0, 1, \cdots$,

$$\|(\Phi_{cl}^0)^k\| \leq \gamma \lambda^k =: w_{cl}^0(k) \qquad (3.3\text{-}15)$$

where $\| \cdot \|$ denotes any matrix norm. Consider time-varying perturbed plants

$$x(k+1) = [\Phi^0 + \tilde{\Phi}(k)]x(k) + [G^0 + \tilde{G}(k)]u(k) \qquad (3.3\text{-}16)$$

such that the perturbations $\tilde{\Phi}(k)$ and $\tilde{G}(k)$ belong to the sets

$$\tilde{\Phi}(k) \quad \in \quad \left\{ \tilde{\Phi}(k) \in \mathbb{R}^{N \times N} \,\middle|\, \|\tilde{\Phi}(k)\| \leq \tilde{\varphi} \right\} \qquad (3.3\text{-}17)$$

$$\tilde{G}(k) \quad \in \quad \left\{ \tilde{G}(k) \in \mathbb{R}^{N \times m} \,\middle|\, \|\tilde{G}(k)\| \leq \tilde{g} \right\} \qquad (3.3\text{-}18)$$

where $N := \dim x$, and $\tilde{\varphi}$ and $\tilde{g}$ are positive reals. Then, we have the following result.

Theorem 3.3-1. (**Robust Stability of State-Feedback Systems**) *Consider the time-varying perturbed plants (16) with a fixed state-feedback compensation (11) such that the nominal closed-loop system with transition matrix (14) is asymptotically stable. Then, for all perturbed plants in the neighborhood of the nominal plant specified by the sets (17) and (18), the closed-loop system remains exponentially stable, whenever*

$$\frac{\gamma}{1-\lambda}(\tilde{\varphi} + \tilde{g}\|F\|) < 1 \qquad (3.3\text{-}19)$$

or

$$\|w_{cl}^0(\cdot)\|_1\, (\tilde{\varphi} + \tilde{\gamma}\|F\|) < 1 \qquad (3.3\text{-}20)$$

where $\|w_{cl}^0(\cdot)\|_1 := \sum_{k=0}^{\infty} |w_{cl}^0(k)|$.

In order to prove Theorem 1, we avail of the following lemma.

Lemma 3.3-1. (**Bellman–Gronwall** [Des70a]) *Let* $\{z(k)\}_{k=0}^{\infty}$ *be a nonnegative sequence and* m *and* c *two nonnegative reals such that*

$$z(k) \leq c + \sum_{i=0}^{k-1} mz(i) \qquad (3.3\text{-}21)$$

Then,

$$z(k) \leq c(1+m)^k \qquad (3.3\text{-}22)$$

Proof: Let

$$h(k) := c + \sum_{i=0}^{k-1} mz(i) \qquad (3.3\text{-}23)$$

It follows that

$$\begin{aligned} h(k) - h(k-1) \quad &= \quad mz(k-1) \\ &\leq \quad mh(k-1) \qquad \text{[(21), (23)]} \end{aligned}$$

Then

$$\begin{aligned} h(k) \quad &\leq \quad (1+m)h(k-1) \\ &\leq \quad (1+m)^k h(0) \\ &= \quad c(1+m)^k \end{aligned}$$

Proof of Theorem 1: By defining $\tilde{\Phi}_{cl}(k) := \tilde{\Phi}(k) + \tilde{G}(k)F$, the perturbed closed-loop system is

$$x(k+1) = \Phi_{cl}^0 x(k) + \tilde{\Phi}_{cl}(k)x(k) \quad , \quad k = 0, 1, \cdots$$

Then,

$$x(k) = \left(\Phi_{cl}^0\right)^k x(0) + \sum_{i=0}^{k-1} (\Phi_{cl}^0)^{k-1-i} \tilde{\Phi}_{cl}(i)x(i)$$

From (15), (17), and (18), it follows that

$$\|x(k)\| \leq \gamma\lambda^k \|x(0)\| + \gamma\lambda^k \lambda^{-1}(\tilde{\varphi} + \tilde{g}\|F\|) \sum_{i=0}^{k-1} \lambda^{-i}\|x(i)\|$$

or

$$z(k) \leq c + \sum_{i=0}^{k-1} m z(i)$$

with $z(i) := \lambda^{-i}\|x(i)\|$, $c := \gamma\|x(0)\|$, and $m := \gamma\lambda^{-1}(\tilde{\varphi} + \tilde{\gamma}\|F\|)$. Then, by virtue of the Bellman–Gronwall Lemma,

$$\begin{aligned} z(k) &= \lambda^{-k}\|x(k)\| \leq c(1+m)^k \\ &= \gamma\|x(0)\|[1 + \gamma\lambda^{-1}(\tilde{\varphi} + \tilde{g}\|F\|)]^k \end{aligned}$$

or

$$\|x(k)\| \leq \gamma\|x(0)\| \ [\lambda + \gamma(\tilde{\varphi} + \tilde{\gamma}\|F\|)]^k$$

The conclusion is that, as $k \to \infty$, $\|x(k)\|$ tends exponentially to zero, whenever (19) is fulfilled.

Note that the nonnegative sequence $\{w_{cl}^0(k)\}_{k=0}^{\infty}$ in (15) can be interpreted as the impulse response of the SISO first-order system:

$$\begin{aligned} \xi(k+1) &= \lambda\xi(k) + \gamma\lambda\nu(k) \\ \ell(k) &= \xi(k) + \gamma\nu(k) \end{aligned}$$

According to (15), $w_{cl}^0(k)$ upper-bounds the norm of the kth power of the state-transition matrix of the nominal closed-loop system. The more damped the latter, the smaller $\|w_{cl}^0(\cdot)\|_1$ can be made, and, by (20), the larger the size, as measured by $\tilde{\varphi} + \tilde{g}\|F\|$, of plant perturbations that do not destabilize the feedback system.

Main Points of the Section. For multiplicative perturbations affecting the plant nominal transfer matrix, robust stability of the feedback system can be analyzed via (9) and (10) by comparing the size of the relative error (7) of the loop transfer matrix with the size of the return difference of the nominal loop. A more generic and qualitative result, based on the bare fact that any output dynamic compensation can be looked at as a state-feedback compensation, is that any nominally stabilizing compensator yields robust stability, because it is capable of stabilizing all plants in a nontrivial neighborhood of the nominal one.

3.4   STREAMLINED NOTATIONS

It is convenient in several instances to depart from the notational conventions adopted so far for representing time-invariant linear systems in I/O form. An I/O description of such systems was given in terms of d-representations of sequences and MFDs of transfer functions as $\hat{y}(d) = A^{-1}(d)\left[B(d)\hat{u}(d) + \Gamma(d)\right]$ or

$$A(d)\hat{y}(d) = B(d)\hat{u}(d) + \Gamma(d) \tag{3.4-1}$$

where $\Gamma(d)$ is a polynomial vector depending upon the initial conditions (cf. (1-27)). This is an I/O *global* description, in that it allows us to compute the whole system output response $\hat{y}(d)$ once the whole input sequence $\hat{u}(d)$ and the initial conditions are assigned.

However, (1) bears upon an I/O *local* representation as well. This can be seen as follows. Let

$$A(d) = I_p + A_1 d + \cdots + A_{\partial A}d^{\partial A} \tag{3.4-2}$$

$$B(d) = \qquad B_1 d + \cdots + B_{\partial B}d^{\partial B} \tag{3.4-3}$$

Then, if $\hat{y}(d) = \sum_{k=0}^{\infty} y(k)d^k$ and $\hat{u}(d) = \sum_{k=0}^{\infty} u(k)d^k$, (1) can be written as the following difference equation:

$$y(t) + A_1 y(t-1) + \cdots + A_{\partial A}y(t-\partial A) = B_1 u(t-1) + \cdots + B_{\partial B}u(t-\partial B) \tag{3.4-4}$$

for every t, such that $t > \max\left(\partial A(d), \partial B(d), \partial\Gamma(d)\right)$.

This, in turn, can be rewritten in shorthand form as follows:

$$A(d)y(t) = B(d)u(t) \tag{3.4-5}$$

The last equation is the same as (4), rewritten in streamlined notation form.

Instead of interpreting d as a position marker as in (1), in (5), d is used as the *unit backward shift operator*, viz., $dy(t) = y(t-1)$. The reader is warned not to believe that (5) implies that the output vector $y(t)$ can be computed by premultiplying input vector $u(t)$ by matrix $A^{-1}(d)B(d)$. The I/O representation (5) is handy in that, though it gives no account of the initial conditions, it is appropriate for both stability analysis and synthesis purposes.

Main Points of the Section. For both study and design purposes, it is convenient to adopt system local I/O representations in a streamlined notation form as (5), where d plays the role of the unit backward operator.

Problem 3.4-1   Consider the difference equation

$$G(d)r(t) = 0$$

with

$$r(t) \in \mathbb{R} \quad \text{and} \quad G(d) = 1 + g_1 d + \cdots + g_n d^n$$

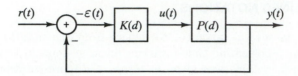

Figure 3.5-1: Unity-feedback configuration of a closed-loop system with a 1-DOF controller.

Set

$$\hat{r}(d) = \sum_{k=0}^{\infty} r(k)d^k \quad \text{and} \quad x(0) := [r(-n+1) \cdots r(-1)\ r(0)]'$$

Show that

$$\hat{r}(d) = \Gamma(d)/G(d)$$

for a suitable polynomial $\Gamma(d)$. (*Hint*: Put $r(t)$ in state-space representation.)

3.5   1-DOF TRACKERS

The *tracking* problem consists of finding inputs to a given plant so as to make its output $y(t) \in \mathbb{R}^p$ as closest as possible to a *reference* variable $r(t) \in \mathbb{R}^p$. Specifically, it is required to design a feedback control law that, while stabilizing the closed-loop system, as $t \to \infty$ reduces the *tracking error* to zero:

$$\varepsilon(t) := y(t) - r(t) \tag{3.5-1}$$

irrespective of the initial conditions. Whenever this happens, we say that *asymptotic tracking* is achieved or, in case of a constant reference, that the controller is *offset-free*. Typically, the design is carried out by assuming that $r(t)$ either belongs to a family of possible references or is preassigned.

Basically, there are two alternative approaches to solve the tracking problem. In the first, $\varepsilon(t)$ is the only input to the controller. For this reason, the latter is sometimes referred to as a *1-degree-of-freedom* (1-DOF) controller or tracker. As shown in Fig. 1, in such a case, the closed-loop system results in a unity-feedback configuration.

In the second approach, depicted in Fig. 2, the controller processes two separate inputs, viz., $y(t)$ and $r(t)$, in an independent fashion. For this reason, it is sometimes referred to as a *2-degrees-of-freedom* (2-DOF) controller or tracker.

Although 2-DOF controllers will be discussed in future chapters, we focus hereafter on how to solve the asymptotic tracking problem by 1-DOF controllers. Specifically, we show how to embed such a problem in the one of stabilizing a feedback system.

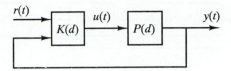

Figure 3.5-2: Closed-loop system with a 2-DOF controller.

We consider a plant with the same number of inputs and outputs:

$$\left.\begin{array}{rcl} A(d)y(t) & = & B(d)u(t) \\ \dim y(t) & = & \dim u(t) \;=\; p \end{array}\right\} \tag{3.5-2}$$

with $A^{-1}(d)B(d)$ a left-coprime MFD of the plant transfer matrix $P(d)$. Note that in (2), we use the streamlined notation of Sec. 4. Further, we assume that reference $r(t)$ is such that

$$D(d)r(t) = O_p \tag{3.5-3}$$

for some polynomial diagonal matrix $D(d)$ such that $D(0) = I_p$ and whose elements have only simple roots on the unit circle. This amounts to assuming that $r(t)$ is a bounded periodic sequence. For example, if $D(d) = 1 - d$, (3) yields $r(t) - r(t-1) = 0$, that is, $r(\cdot)$ is a constant sequence.

Problem 3.5-1 (Sinusoidal Reference)   Consider a polynomial $D(d)$ with roots $e^{j\theta}$ and $e^{-j\theta}$, $\theta \in [0, \pi]$. Find the corresponding difference equation (3) for $r(t)$. Plot $r(t)$ as a function of t, $t = 0, 1, \cdots$, for various θ, assuming that $r(-2) = r(-1) = 1$.

We now combine (2) and (3) so as to get a representation for $\varepsilon(t)$ in terms of $u(t)$. This can be achieved by, first, premultiplying (2) by $D(d)$ and (3) by $A(d)$ and, next, subtracting the second from the first. Accordingly,

$$A(d)D(d)\varepsilon(t) = B(d)\nu(t) \tag{3.5-4}$$

$$\nu(t) := D(d)u(t) \tag{3.5-5}$$

Equation (4) defines a new plant with input $\nu(t)$ and output $\varepsilon(t)$. If we can find a compensator that stabilizes (4), then

$$\varepsilon(t) \underset{(t\to\infty)}{\longrightarrow} O_p \quad \text{and} \quad \nu(t) \underset{(t\to\infty)}{\longrightarrow} O_p$$

Assuming that $D(d)$ and $B(d)$ are left-coprime, there exist right-coprime polynomial matrices $R_2(d)$ and $S_2(d)$ such as to make

$$\det\left[A(d)D(d)R_2(d) + B(d)S_2(d)\right] \tag{3.5-6}$$

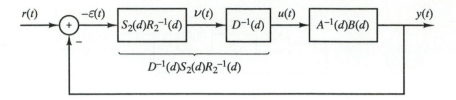

Figure 3.5-3: Unity-feedback closed-loop system with a 1-DOF controller for asymptotic tracking.

strictly Hurwitz. According to Theorem 2-1, (6) gives, apart from a multiplicative constant, the characteristic polynomial of the feedback system of Fig. 3 consisting of the plant (4) and the dynamic compensator:

$$D(d)u(t) = \nu(t) = -S_2(d)R_2^{-1}(t)\varepsilon(t) \qquad (3.5\text{-}7)$$

Note that the transfer matrix of the compensator is $D^{-1}(d)S_2(d)R_2^{-1}(d)$. Hence, it embodies the reference model (7). This result is in agreement with the so-called *internal model principle* [FW76] according to which, in order to possibly achieve asymptotic tracking, the compensator has to incorporate the model of the reference to be tracked. In the case of a constant reference, one should have $(1 - d)$ at the denominator of the compensator transfer function, so as to ensure offset-free behavior. This yields integral action as commonly employed in control system design. In this connection, see also Problem 2.4-11.

Problem 3.5-2 Show that if the feedback system of Fig. 1 is internally stable, the plant steady-state output response to a constant input u equals $P(1)u$. Conclude then that if the plant has no unstable hidden modes, asymptotic tracking of an arbitrary constant-reference vector is possible if and only if rank $P(1) = p$. In turn, this implies that $\dim u(t) \geq p$. Then, conclude that the condition $\dim u(t) = \dim y(t)$ in (2) entails no limitation.

Problem 3.5-3 Show that if polynomial matrix $D(d)$ in (3) is a divisor of $A(d)$ in (2), one can write $A(d)\varepsilon(t) = B(d)u(t)$. Conclude that in such a case, because all modes of the reference also belong to the plant, any stabilizing 1-DOF compensator yields asymptotic tracking.

Theorem 3.5-1.   (1-DOF Tracking) *Consider a square plant, viz.,*

$$\dim y(t) = \dim u(t)$$

free of unstable hidden modes and with a left-coprime MFD $A^{-1}(d)B(d)$. Let $r(t)$ be a bounded periodic reference modeled as in (3). Then, asymptotic tracking is achieved by any stabilizing compensator of the form (7), making the closed-loop system internally stable. Such compensators exist if and only if $D(d)$ and $B(d)$ are coprime. Further, if $D(d)$ is a divisor of $A(d)$, any stabilizing 1-DOF compensator yields asymptotic tracking.

The next problem shows that the output disturbance-rejection problem is isomorphic to the output reference-tracking problem.

Problem 3.5-4 Consider the square plant $y(t) = A^{-1}(d)B(d)u(t) + n(t)$ with $A(d)$ and $B(d)$ left-coprime, and $n(t)$ an output disturbance such that $D(d)n(t) = O_p$, with $D(d)$ as in (3). Show that if the feedback compensator

$$\nu(t) := D(d)u(t) = -S_2(d)R_2^{-1}(d)y(t)$$

is such that (6) is strictly Hurwitz, then it yields *asymptotic disturbance rejection*, viz.,

$$y(t) \underset{(t\to\infty)}{\longrightarrow} O_p \qquad \text{and} \qquad \nu(t) \underset{(t\to\infty)}{\longrightarrow} O_p$$

whenever $D(d)$ and $B(d)$ are coprime.

Main Points of the Section. The asymptotic tracking problem can be formulated as an internal stability problem of a feedback system. Under general conditions, asymptotic tracking can be achieved by 1-DOF controllers embodying the modes of the reference model.

Problem 3.5-5 (Joint Tracking and Disturbance Rejection) Consider the plant of Problem 4. Assume that the disturbed output $y(t)$ has to follow a reference $r(t)$ such that $G(d)r(t) = O_p$, with $G(d)$ a polynomial matrix with the same properties as $D(d)$ in (3). Let $L(d)$ be the least-common multiple of $G(d)$ and $D(d)$. Determine the conditions under which the feedback compensator

$$\nu(t) := L(d)u(t) = -S_2(d)R_2^{-1}(d)\varepsilon(t)$$

yields both asymptotic tracking and disturbance rejection.

NOTES AND REFERENCES

Appendix B provides a quick review of the results of polynomial matrix theory used throughout the chapter. Our notations and definitions mainly follow [Kuč79].

The formulation of the feedback stability problem first appeared in [DC75]. It has been widely used since, for example, [Vid85]. [DC75] also includes examples showing that any three of the blocks of $H_{zw}(d)$ can be stable and the fourth unstable.

The parametric form of all stabilizing controllers was introduced by [YJB76] for the continuous-time case, and by [Kuč75] and [Kuč79] for the discrete-time case. A subsequent version of the parameterization using factorizations in terms of stable transfer matrices was introduced by [DLMS80], and used since, for example, [Vid85].

The robust stability result of Sec. 3.3 follows the adaptation of the approach in [LSA81] to the discrete-time case as reported in [BGW90]. The generic robust stability result for discrete-time feedback systems is reported in [CMS91]. A similar result for continuous-time state-feedback systems is presented in [CD89].

CHAPTER 4

DETERMINISTIC LQ REGULATION–II: SOLUTION VIA POLYNOMIAL EQUATIONS

In this chapter, another approach for solving the steady-state LQOR problem is described. This, which will be referred to as the *polynomial equation approach* [Kuč79], can be regarded as an alternative to the one based on dynamic programming and leading to Riccati equations.

The reader may wonder why we should get bogged down with an alternative approach once we have found that the one using Riccati equations leads to efficient numerical solution routines. The answer is manifold. First, from a conceptual viewpoint, it is beneficial to appreciate that the Riccati-based solution is not the only way for solving the steady-state LQOR problem. As we shall see soon, this holds true even if the plant is given in a state-space representation as in (2.4-44). Second, the polynomial equation approach backs up the one based on dynamic programming, offering complementary insights. For example, in the polynomial equation approach, the eigenvalues of the closed-loop system show up in a direct way. This can be seen because the polynomial equation approach is basically a frequency-domain methodology in contrast with the time-domain nature of dynamic programming. Third, the polynomial equation approach can be extended to cover steady-state LQ stochastic control as well as filtering problems. This also yields additional insights to the ones offered by stochastic dynamic programming. For example, as will be seen in due time, the polynomial solution to the LQ stochastic servo problem provides a nice clue on how to realize high-performance 2-degree-of-freedom servocontrollers.

The polynomial equation approach to the steady-state deterministic LQOR problem will be discussed by disaggregating the required steps so as to emphasize their specific role. One reason is that, *mutatis mutandis*, the same steps can be followed to solve via polynomial equations other LQ optimization problems, such as linear minimum mean-square error filtering and steady-state LQ stochastic regulation. The advantage of introducing the basic tools of the polynomial equation approach to LQ optimization at this stage is twofold: first, we can nicely relate them to those pertaining to dynamic programming; and, second, presenting them within the simplest possible framework, maximize our understanding of their main features.

The chapter is organized as follows. Section 1 shows that the polynomial approach to the steady-state deterministic LQ regulation amounts to solving a spectral factorization problem, and decomposing a two-sided sequence into two additive sequences of which one is causal and the other is strictly anticausal and possibly of finite energy. The latter problem is addressed in Sec. 2 and consists of finding the solution to a bilateral Diphantine equation with a degree constraint. In Sec. 3, we show that stability of the optimally regulated system requires to solve a second bilateral Diophantine equation along with the one referred before. Section 4 proves the solvability of these two bilateral Diophantine equations under the stabilizability assumption of the plant.

4.1 POLYNOMIAL FORMULATION

Consider a time-invariant linear state representation of the plant to be output-regulated:

$$\begin{cases} x(t+1) = \Phi x(t) + Gu(t) \\ y(t) = Hx(t) \end{cases} \tag{4.1-1}$$

The problem is to find, whenever it exists, an input sequence

$$u(\cdot) = \{ \ ; \ u(0), u(1), \cdots \} \tag{4.1-2}$$

minimizing the quadratic performance index:

$$\begin{aligned} J \ &:= \ J(x(0), u_{[0,\infty)}) \ = \ \sum_{k=0}^{\infty} \Big[\|y(k)\|_{\psi_y}^2 + \|u(k)\|_{\psi_u}^2 \Big] \\ &= \ \sum_{k=0}^{\infty} \Big[\|x(k)\|_{\psi_x}^2 + \|u(k)\|_{\psi_u}^2 \Big] \end{aligned} \tag{4.1-3}$$

for any initial state $x(0)$. In (3), $\psi_x := H'\psi_y H$ and

$$\psi_y = \psi_y' > 0 \quad \text{and} \quad \psi_u = \psi_u' \geq 0 \tag{4.1-4}$$

As in Sec. 3.1, we use the d-representations $\hat{u}(d)$ and $\hat{x}(d)$ of sequences $u(\cdot)$ and $x(\cdot)$, respectively. In particular, (3.1-26) gives

$$\hat{x}(d) = A^{-1}(d) \left[x(0) + B(d)\hat{u}(d) \right] \tag{4.1-5}$$

where $A(d)$ and $B(d)$ are the following polynomial matrices:

$$A(d) \quad := \quad I - d\Phi \tag{4.1-6}$$

$$B(d) \quad := \quad dG \tag{4.1-7}$$

From (3.1-12), (3) can be rewritten as

$$
\begin{aligned}
J \quad &= \quad \langle \hat{y}^*(d)\psi_y \hat{y}(d) + \hat{u}^*(d)\psi_u \hat{u}(d) \rangle \\
&= \quad \langle \hat{x}^*(d)\psi_x \hat{x}(d) + \hat{u}^*(d)\psi_u \hat{u}(d) \rangle
\end{aligned} \tag{4.1-8}
$$

Let us introduce also a right-coprime MFD $B_2(d)A_2^{-1}(d)$ of transfer matrix $H_{xu}(d)$:

$$H_{xu}(d) = A^{-1}(d)B(d) = B_2(d)A_2^{-1}(d) \tag{4.1-9}$$

Then, (5) can be rewritten as

$$\hat{x}(d) = A^{-1}(d)x(0) + B_2(d)A_2^{-1}(d)\hat{u}(d) \tag{4.1-10}$$

Substituting (10) into (8), we get

$$
\begin{aligned}
J \quad = \quad &\langle \hat{u}^*(d)A_2^{-*}(d)\Big[A_2^*(d)\psi_u A_2(d) + B_2^*(d)\psi_x B_2(d)\Big]A_2^{-1}(d)\hat{u}(d) \\
&+ \hat{u}^*(d)A_2^{-*}(d)B_2^*(d)\psi_x A^{-1}(d)x(0) \\
&+ x'(0)A^{-*}(d)\psi_x B_2(d)A_2^{-1}(d)\hat{u}(d) \\
&+ x'(0)A^{-*}(d)\psi_x A^{-1}(d)x(0) \rangle
\end{aligned} \tag{4.1-11}
$$

where we used the shorthand notation $A_2^{-*}(d) := [A_2^{-1}(d)]^*$. Equation (11) can be simplified by considering an $m \times m$ Hurwitz polynomial matrix $E(d)$ solving the following *right-spectral factorization problem*

$$E^*(d)E(d) = A_2^*(d)\psi_u A_2(d) + B_2^*(d)\psi_x B_2(d) \tag{4.1-12}$$

$E(d)$ is then called a *right-spectral factor* of the R.H.S. of (12). $E(d)$ exists if and only if

$$\mathrm{rank} \left[\begin{array}{c} \psi_u A_2(d) \\ \psi_x B_2(d) \end{array} \right] = m := \dim u \tag{4.1-13}$$

The spectral factors are determined uniquely up to an orthogonal matrix multiple. If $E(d)$ and $\Gamma(d)$ are two right-spectral factors of the R.H.S. of (12), then

$$\Gamma(d) = UE(d) \tag{4.1-14}$$

where U is an orthogonal matrix, viz. $U'U = I_m$. In particular, if $m = 1$, the right-spectral factor $E(d)$ is a Hurwitz polynomial, and U represents just a change of sign.

We see that the R.H.S. of (12), call it $M(d, d^{-1})$, is a polynomial matrix in d and d^{-1} that, according to Sec. 3.1, can be considered as a two-sided matrix sequence of finite length. Further, it is symmetric about time 0, viz., $M^*(d, d^{-1}) = M(d, d^{-1})$. In the single-input case, $m = 1$, it is a polynomial symmetric in d and d^{-1}. Hence, if $d = a$ is a root, $d = 1/a$ is a root as well. Further, because its coefficients are real, if $d = a$ is a root, $d = a^*$ is also a root, a^* denoting the complex conjugate of a. It follows that $M(d, d^{-1})$ has an even number of inverse/Hermitian-symmetric complex roots. Each root on the unit disc must have an even multiplicity. Therefore, $E(d)$ can be constructed by collecting all the d-roots of $M(d, d^{-1})$ such that $|d| > 1$, along with every root such that $|d| = 1$ with multiplicity equal to one-half the corresponding multiplicity pertaining to $M(d, d^{-1})$.

Example 4.1-1 Let

$$\Phi = \begin{bmatrix} 1 & 0 \\ 0 & \frac{1}{2} \end{bmatrix} ; \quad G = \begin{bmatrix} 1 \\ 0 \end{bmatrix} \tag{4.1-15}$$

$$H = \begin{bmatrix} 1 & 1 \end{bmatrix} ; \quad \psi_y = 1 ; \quad \psi_u = 2$$

We find

$$A(d) = \begin{bmatrix} 1 - d & 0 \\ 0 & 1 - \frac{d}{2} \end{bmatrix} \quad B(d) = \begin{bmatrix} d \\ 0 \end{bmatrix} \tag{4.1-16}$$

$$H_{xy}(d) = A^{-1}(d)B(d) = \begin{bmatrix} \frac{d}{(1-d)} \\ 0 \end{bmatrix} = \begin{bmatrix} d \\ 0 \end{bmatrix} \frac{1}{1 - d} \tag{4.1-17}$$

Hence,

$$A_2(d) = 1 - d \qquad B_2(d) = \begin{bmatrix} d \\ 0 \end{bmatrix} \tag{4.1-18}$$

For the R.H.S. of (12), we find

$$-2d^{-1} \left(1 - \frac{5}{2}d + d^2 \right) = -2d^{-1} \left(d - \frac{1}{2} \right)(d - 2) = 4 \left(1 - \frac{d^{-1}}{2} \right) \left(1 - \frac{d}{2} \right)$$

Hence,

$$E(d) = \pm 2 \left(1 - \frac{d}{2} \right) \tag{4.1-19}$$

Using (12) into (11), we obtain

$$J = J_1 + J_2 \tag{4.1-20}$$

with

$$J_1 := \langle \ell^*(d)\ell(d) \rangle \tag{4.1-21}$$

$$\ell(d) := E^{-*}(d)B_2^*(d)\psi_x A^{-1}(d)x(0) + E(d)A_2^{-1}(d)\hat{u}(d) \tag{4.1-22}$$

$$J_2 := \left\langle x'(0)A^{-*}(d) \left[\psi_x - \psi_x B_2(d)E^{-1}(d)E^{-*}(d)B_2^*(d)\psi_x \right] \right.$$
$$\left. A^{-1}(d)x(0) \right\rangle \tag{4.1-23}$$

Note that J_2 is not affected by $\hat{u}(d)$. Then, the problem amounts to finding causal input sequences $\hat{u}(d)$ minimizing (21). According to (3.1-12), (21) equals the square of the ℓ_2 norm of the m-vector sequence $\ell(d)$ in (22). In turn, (22) has two additive components. One, $E(d)A_2^{-1}(d)\hat{u}(d)$ is the d-representation of a causal sequence. The other results from premultiplying the d-representation of the causal sequence $\psi_x A^{-1}(d)x(0)$ by $E^{-*}(d)B_2^*(d) = [B_2(d)E^{-1}(d)]^*$. This, considering (3.1-33) and (3.1-34), can be interpreted as the d-representation of a strictly anticausal sequence being $E(0)$ nonsingular by Hurwitzianity of $E(d)$. By (3.1-8), the first term on the R.H.S. of (22) can be thus interpreted as the d-representation of a sequence obtained by convolving a causal sequence with a strictly anticausal sequence. Hence, the first additive term on the R.H.S. of (22) is a two-sided m-vector sequence.

Taking into account the previous interpretation in order to find the optimal causal input sequences $\hat{u}(d)$, we try to additively decompose $\ell(d)$ in terms of a causal sequence $\ell_+(d)$ plus a strictly anticausal sequence $\ell_-(d)$:

$$\ell(d) = \ell_+(d) + \ell_-(d) \tag{4.1-24}$$

$$\operatorname{ord}\ell_+(d) \geq 0 \ , \ \operatorname{ord}\ell_-^*(d) > 0 \tag{4.1-25}$$

Because from (25),

$$\operatorname{ord}[\ell_-^*(d)\ell_+(d)] > 0 \tag{4.1-26}$$

it follows that

$$\langle \ell_-^*(d)\ell_+(d)\rangle = \langle \ell_+^*(d)\ell_-(d)\rangle = 0 \tag{4.1-27}$$

Consequently, with decomposition (24), we would have

$$J_1 = \langle \ell_+^*(d)\ell_+(d)\rangle + \langle \ell_-^*(d)\ell_-(d)\rangle \tag{4.1-28}$$

Because the two additive terms on the R.H.S. of (28) are nonnegative, boundedness of J_1 requires that each of them be such. Therefore, we must possibly ensure that in (24) $\ell_-(d)$ be an ℓ_2 sequence. Further, boundedness of $\langle \ell_+^*(d)\ell_+(d)\rangle$ will follow by restricting $\hat{u}(d)$ so as to make $\ell_+(d)$ an ℓ_2 sequence.

Main Points of the Section. The polynomial equation approach to the steady-state deterministic LQOR problem amounts to finding a right-spectral factor $E(d)$ as in (12), and decomposing the two-sided sequence $\ell(d)$ in (22) into two additive sequences of which one is causal and the other is strictly anticausal and possibly of finite energy.

Problem 4.1-1   Consider (12) for a SISO plant (1) with transfer function $H_{yu}(d) = HB_2(d)/A_2(d)$. Assume that $\psi_u > 0$, $\psi_y > 0$, and polynomials $HB_2(d)$ and $A_2(d)$ have no common root on the unit circle. Show that the (two) spectral factors $E(d)$ solving (12) are strictly Hurwitz polynomials. (*Hint:* It is enough to check that $E(d)$ has no root for $|d| = 1$. Note also that $A^*(e^{i\theta})A(e^{j\theta}) = |A(e^{j\theta})|^2$.)

4.2   CAUSAL–ANTICAUSAL DECOMPOSITION

It is convenient to transform the right spectral factorization (12) into an equation involving solely polynomial matrices in the indeterminate d.
Let

$$q := \max\{\partial A_2(d), \partial B_2(d)\} \tag{4.2-1}$$

$$\bar{A}_2(d) := d^q A_2^*(d) \ ; \ \ \bar{B}_2(d) := d^q B_2^*(d) \ ; \ \ \bar{E}(d) := d^q E^*(d) \tag{4.2-2}$$

Then, (1-12) can be rewritten as

$$\bar{E}(d)E(d) = \bar{A}_2(d)\psi_u A_2(d) + \bar{B}_2(d)\psi_x B_2(d) \tag{4.2-3}$$

Likewise, the first additive term on the R.H.S. of (1-22) can be rewritten as

$$\tilde{\ell}(d) := \bar{E}^{-1}(d)\bar{B}_2(d)\psi_x A^{-1}(d)x(0) \tag{4.2-4}$$

Suppose now that we can find a pair of polynomial matrices Y and Z fulfilling the following *bilateral Diophantine equation:*

$$\bar{E}(d)Y(d) + Z(d)A(d) = \bar{B}_2(d)\psi_x \tag{4.2-5}$$

with the degree constraint

$$\partial Z(d) < \partial \bar{E}(d) = q \tag{4.2-6}$$

The last equality follows because $E(d)$ is Hurwitz, so $E(0)$ is nonsingular. Using (5) in (4), we find

$$\tilde{\ell}(d) := \tilde{\ell}_+(d) + \tilde{\ell}_-(d) \tag{4.2-7}$$

where

$$\tilde{\ell}_+(d) := Y(d)A^{-1}(d)x(0) \tag{4.2-8}$$

$$\tilde{\ell}_-(d) := \bar{E}^{-1}(d)Z(d)x(0) \tag{4.2-9}$$

are respectively a causal and a strictly anticausal sequence, the latter possibly with finite energy. Causality of the first follows from causality of $A^{-1}(d)$. Strict anticausality and possibly finite energy of $\tilde{\ell}_-(d)$ is proved by showing that

$$
\begin{aligned}
\tilde{\ell}_-^*(d) &= x'(0)Z^*(d)\bar{E}^{-*}(d) \\
&= x'(0)\left[Z_0' + Z_1'd^{-1} + \cdots + Z_{\partial Z}'d^{-\partial Z}\right]\left\{[d^q E^*(d)]^*\right\}^{-1} \\
&= x'(0)\left[Z_0' + Z_1'd^{-1} + \cdots + Z_{\partial Z}'d^{-\partial Z}\right]d^q E^{-1}(d) \\
&= x'(0)\left[Z_{\partial Z}'d^{(q-\partial Z)} + \cdots + Z_0'd^q\right]E^{-1}(d)
\end{aligned}
\tag{4.2-10}
$$

where we set

$$Z(d) = Z_0 + Z_1 d + \cdots + Z_{\partial Z}d^{\partial Z} \tag{4.2-11}$$

From (10), we see that strict causality and possibly finite energy of $\tilde{\ell}_*^*(d)$ follows from (6) and the Hurwitz property of $E(d)$. Should $E(d)$ be strictly Hurwitz, the finite energy of $\tilde{\ell}_-(d)$ would follow at once.

In conclusion, provided that we can find a pair $(Y(d), Z(d))$ solving (5) with the degree constraint (6), a decomposition (1-24) is given by

$$\ell_+(d) = Y(d)A^{-1}(d)x(0) + E(d)A_2^{-1}(d)\hat{u}(d) \qquad (4.2\text{-}12)$$
$$\ell_-(d) = \bar{E}^{-1}(d)Z(d)x(0) \qquad (4.2\text{-}13)$$

Let

$$J_3 = \langle \ell_-^*(d)\ell_-(d) \rangle \qquad (4.2\text{-}14)$$

With J_2 as in (1-23), assume that $J_2 + J_3$ is bounded. Then, an optimal input sequence is obtained by setting $\ell_+(d) = O_m$, that is,

$$\hat{u}(d) = -A_2(d)E^{-1}(d)Y(d)A^{-1}(d)x(0) \qquad (4.2\text{-}15)$$
$$= -A_2(d)E^{-1}(d)Y(d)\left[\hat{x}(d) - A^{-1}(d)B(d)\hat{u}(d)\right] \qquad [(1\text{-}5)]$$

Equivalently,

$$\hat{u}(d) = -M_1^{-1}(d)N_1(d)\hat{x}(d) \qquad (4.2\text{-}16)$$

where, for reasons that will become clearer in the next section, we have introduced the following transfer matrices:

$$M_1(d) := E^{-1}(d)\left[E(d) - Y(d)B_2(d)\right]A_2^{-1}(d) \qquad (4.2\text{-}17)$$
$$N_1(d) := E^{-1}(d)Y(d) \qquad (4.2\text{-}18)$$

The following lemma sums up the results obtained so far.

Lemma 4.2-1. *Provided that:*

(i) condition (1-13) is satisfied;

(ii) Equation (5) admits solutions $(Y(d), Z(d))$ with the degree constraint (6);

(iii) $J_2 + J_3$ is bounded;

the LQOR problem has either open-loop solutions (15) or linear state-feedback solutions (16) to (18).

Main Points of the Section. The bilateral Diophantine equation (5) along with the degree constraint (6), if solvable, allows one to obtain a causal/strictly anticausal decomposition of (4) and, hence, an optimal control sequence.

4.3 STABILITY

From Chapter 2, we already know that LQR optimality does not imply in general stability of the optimally regulated closed-loop system. In Sec. 2, we found that optimal LQOR laws can be obtained by solving a bilateral Diophantine equation. Even if solvable, the latter need not have a unique solution. By imposing stability of the closed-loop system, we obtain, under fairly general conditions, uniqueness of the solution. To do this, we resort to Theorem 3.2-1. First, for $M_1(d)$ and $N_1(d)$ as in (2-17) and (2-18), respectively,

$$
\begin{aligned}
M_1(d)A_2(d) \; & + \; N_1(d)B_2(d) \\
& = \; E^{-1}(d)[E(d) - Y(d)B_2(d)] - E^{-1}(d)Y(d)B_2(d) \\
& = \; I_m
\end{aligned}
\tag{4.3-1}
$$

Hence, (3.2-24) is satisfied. Then, internal stability is obtained if and only if both $M_1(d)$ and $N_1(d)$ are stable transfer matrices. We begin by finding a necessary condition for stability of $M_1(d)$. To this end, we write

$$
\begin{aligned}
M_1(d) \; = \; & E^{-1}(d)\bar{E}^{-1}(d)\Big[\bar{E}(d)E(d) - \bar{E}(d)Y(d)B_2(d)\Big]A_2^{-1}(d) \\
= \; & E^{-1}(d)\bar{E}^{-1}(d)\Big\{\bar{A}_2(d)\psi_u A_2(d) \\
& + \bar{B}_2(d)\psi_x B_2(d) - \Big[\bar{B}_2(d)\psi_x - Z(d)A(d)\Big]B_2(d)\Big\}A_2^{-1}(d) \\
= \; & E^{-1}(d)\bar{E}^{-1}(d)\Big[\bar{A}_2(d)\psi_u + Z(d)A(d)B_2(d)A_2^{-1}(d)\Big] \\
= \; & E^{-1}(d)\bar{E}^{-1}(d)\Big[\bar{A}_2(d)\psi_u + Z(d)B(d)\Big] \qquad [(1\text{-}9)]
\end{aligned}
\tag{4.3-2}
$$

where the second equality follows from (1-12) and (2-5). We note that, being $E(d)$ Hurwitz, $\bar{E}(d)$ turns out to be *anti-Hurwitz*, viz.,

$$
\det \bar{E}(d) = 0 \;\Rightarrow\; |d| \leq 1
\tag{4.3-3}
$$

Then, a necessary condition for stability of $M_1(d)$ is that the polynomial matrix within brackets in (2) be divided on the left by $\bar{E}(d)$. That is, there must be a polynomial matrix $X(d)$ such as to satisfy the following equation:

$$
\bar{E}(d)X(d) - Z(d)B(d) = \bar{A}_2(d)\psi_u
\tag{4.3-4}
$$

Recalling (2-5), we conclude that in order to solve the steady-state LQOR problem, in addition to the spectral factorization problem (1-12), we have to find a solution $(X(d), Y(d), Z(d))$ with $\partial Z(d) < \partial \bar{E}(d)$ of the two bilateral Diophantine equations (2-5) and (4).

Using (4) in (2), we find

$$
M_1(d) = E^{-1}(d)X(d)
\tag{4.3-5}
$$

This, along with (2-18), yields, for (2-16),

$$\hat{u}(d) = -X^{-1}(d)Y(d)\hat{x}(d) \qquad\qquad (4.3\text{-}6)$$

We then see that $Z(d)$ in (2-5) and (4) plays the role of a "dummy" polynomial matrix. By eliminating $Z(d)$ in (2-5) and (4), we get

$$X(d)A_2(d) + Y(d)B_2(d) = E(d) \qquad\qquad (4.3\text{-}7)$$

Problem 4.3-1   Derive (7) from (2-5), (4), and (1.12) by eliminating the "dummy" polynomial matrix $Z(d)$.

Problem 4.3-2   Show that a triplet $(X(d), Y(d), Z(d))$ is a solution of (2-5) and (4) if and only if it solves (2-5) and (7). (*Hint:* Prove sufficiency by using (2-3).)

It follows from (5) that $X(d)$ is nonsingular. This can be seen also by setting $d = 0$ in (7) to find $X(0)A_2(0) = E(0)$. In fact, recall that, by (9) and (7), $B_2(0) = O_{n\times m}$. Both $A_2(0)$ and $E(0)$ are nonsingular, so nonsingularity of $X(0)$, and hence of $X(d)$, follows.

It also follows that $X(d)$ and $Y(d)$ are constant matrices. This is proved in the following lemma.

Lemma 4.3-2.   *Let (2-5) and (4) [or (2-5) and (7)] have a solution $(X(d)$, $Y(d)$, $Z(d))$ with $\partial Z < \partial \bar{E}$. Then $X(d) = X$ and $Y(d) = Y$ are constant matrices, viz., $\partial X = \partial Y = 0$.*

Proof: Consider (2-5). $\bar{E}(d)$ is a *regular* polynomial matrix, viz. the coefficient matrix of its highest power is nonsingular. Further, by (2-2), $\partial\bar{E}(d) = q$. Then, it follows that $\partial[\bar{E}(d)Y(d)] = q + \partial Y(d)$. Next, $Z(d)A(d) = Z(d) - dZ(d)\Phi$. Hence, $\partial[Z(d)A(d)] \le \partial\bar{E}(d) - 1 + 1 = q$. Further, from (2-2) $\partial\bar{B}_2(d) \le q - 1$. Hence, $\partial[\bar{B}_2(d)\psi_x] \le q - 1$. Therefore,

$$
\begin{aligned}
q + \partial Y(d) &= \partial[\bar{E}(d)Y(d)] \\
&= \partial[\bar{B}_2(d)\psi_x - Z(d)A(d)] \\
&\le \max\left\{\partial[\bar{B}_2(d)\psi_x],\ \partial[Z(d)A(d)]\right\} \le q
\end{aligned}
$$

Hence, $\partial Y(d) = 0$.
Similarly, with reference to (4),

$$
\begin{aligned}
q + \partial X(d) &= \partial[\bar{E}(d)X(d)] \\
&= \partial[\bar{A}_2(d)\psi_u + Z(d)B(d)] \\
&\le \max\left\{\partial[\bar{A}_2(d)\psi_u],\ \partial[Z(d)B(d)]\right\} \le q
\end{aligned}
$$

Hence, $\partial X(d) = 0$.

From $\partial X(d) = 0$, it follows that X and Y are left-coprime. Then, using Theorem 3.2-1, we find that the d-characteristic polynomial $\chi_{cl}(d)$ of the closed-loop system (1-9) and (6), with plant and regulator free of nonzero hidden eigenvalues, is given by

$$\chi_{cl}(d) = \det E(d) / \det E(0) \tag{4.3-8}$$

$$E(d) = X A_2(d) + Y B_2(d) \tag{4.3-9}$$

The following lemma sums up the foregoing results.

Lemma 4.3-2. *Provided that:*

(i) condition (1-13) is satisfied;

(ii) Equations (2-5) and (4) [or (2-5) and (7)] admit a solution $(X, Y, Z(d))$ with $\partial Z(d) < \partial \bar{E}(d)$;

(iii) $J_2 + J_3$ is bounded;

the LQOR problem is solved by the linear state–feedback law

$$u(t) = -X^{-1} Y x(t) \tag{4.3-10}$$

where X and Y are the constant matrices in (ii), and, correspondingly,

$$J_{\min} = J_2 + J_3$$

Further, the optimal feedback system is internally stable if and only if the left spectral factor $E(d)$ in (1-12) is strictly Hurwitz.

Main Points of the Section. Stability of the optimally regulated system lead us to consider a second bilateral Diophantine equation to be jointly solved with the one related to optimality. Internal stability is achieved if and only if the spectral factorization problem (1-12) yields a strictly Hurwitz spectral factor.

4.4 SOLVABILITY

It remains to establish conditions under which (2-5) and (3-4) are solvable.

Lemma 4.4-1. *Let the greatest common left divisors of $A(d) = I - d\Phi$ and $B(d) = dG$ be strictly Hurwitz. Then, there is a unique solution $(X, Y, Z(d))$ of (2-5) and (3-4) [or (2-5) and (3-7)] such that $\partial Z(d) < \partial \bar{E}(d)$. Such a solution is called the minimum-degree solution w.r.t. $Z(d)$.*

Proof: Let $D(d)$ be a greatest common left divisor *(gcld)* of $A(d)$ and $B(d)$. Then, according to (B.1-9), there exists a unimodular matrix $P(d)$ such that

$$\begin{bmatrix} A(d) & -B(d) \end{bmatrix} P(d) = \begin{bmatrix} \overbrace{D(d)}^{n} & O_{n \times m} \end{bmatrix} \tag{4.4-1}$$

$$P(d) = \begin{bmatrix} P_{11}(d) & P_{12}(d) \\ P_{21}(d) & P_{22}(d) \end{bmatrix} \begin{matrix} \}n \\ \}m \end{matrix} \tag{4.4-2}$$
$$\underbrace{\phantom{P_{11}(d)}}_{n} \quad \underbrace{\phantom{P_{12}(d)}}_{m}$$

Consider now (2-5) and (3-4). Rewrite them as follows:

$$\bar{E}(d) \begin{bmatrix} Y(d) & X(d) \end{bmatrix} + Z(d) \begin{bmatrix} A(d) & -B(d) \end{bmatrix} = \begin{bmatrix} \bar{B}_2(d)\psi_x & \bar{A}_2(d)\psi_u \end{bmatrix} \tag{4.4-3}$$

Postmultiplying (3) by $P(d)$, setting

$$\begin{bmatrix} \tilde{Y}(d) & \tilde{X}(d) \end{bmatrix} := \begin{bmatrix} Y(d) & X(d) \end{bmatrix} P(d) \tag{4.4-4}$$

and using (1), we find

$$\bar{E}(d)\tilde{Y}(d) + Z(d)D(d) = \bar{A}_2(d)\psi_u P_{21}(d) + \bar{B}_2(d)\psi_x P_{11}(d) \tag{4.4-5}$$
$$\bar{E}(d)\tilde{X}(d) = \bar{A}_2(d)\psi_u P_{22}(d) + \bar{B}_2(d)\psi_x P_{12}(d) \tag{4.4-6}$$

Now, observe that

$$O_{n \times m} = -B(d)P_{22}(d) + A(d)P_{12}(d) \tag{4.4-7}$$
$$= -B(d)A_2(d) + A(d)B_2(d)$$

where the first equality follows from (1) and (2), and the second from (1-9). Because $A_2(d)$ and $B_2(d)$ are right-coprime, there is a polynomial matrix $V(d)$ such that

$$P_{12}(d) = B_2(d)V(d) \qquad P_{22}(d) = A_2(d)V(d) \tag{4.4-8}$$

Then, using (2-3), the R.H.S. of (6) can be rewritten as $\bar{E}(d)E(d)V(d)$. Hence, (6) reduces to

$$\tilde{X}(d) = E(d)V(d) \tag{4.4-9}$$

Further, because $D(d)$ is strictly Hurwitz and $\bar{E}(d)$ is anti–Hurwitz, $\det G(d)$ and $\det \bar{E}(d)$ are coprime. Then, if follows from Result C.2-1 that (5) is solvable. If $(\tilde{Y}_0(d), Z_0(d))$ solves (5), all solutions of (5) are given by

$$\tilde{Y}(d) = \tilde{Y}_0(d) + L(d)D(d) \tag{4.4-10}$$
$$Z(d) = Z_0(d) - \bar{E}(d)L(d) \tag{4.4-11}$$

with $L(d)$ any polynomial matrix of compatible dimensions. Because $\bar{E}(d)$ is regular, the minimum-degree solution w.r.t. $Z(d)$ can be found by left dividing $Z_0(d)$ by $\bar{E}(d)$, viz.,

$$Z_0(d) = \bar{E}(d)Q(d) + R(d) \tag{4.4-12}$$

with $\partial R(d) < \partial \bar{E}(d)$. Then, (11) becomes $Z(d) = R(d) + \bar{E}(d)[Q(d) - L(d)]$. Choosing $L(d) = Q(d)$, we obtain the minimum-degree solution w.r.t. $Z(d)$:

$$\left. \begin{aligned} \tilde{Y}(d) &= \tilde{Y}_0 + Q(d)D(d) \\ \tilde{X}(d) &= E(d)V(d) \\ Z(d) &= R(d) \end{aligned} \right\} \qquad (4.4\text{-}13)$$

or

$$\begin{aligned} \left[\begin{array}{cc} Y(d) & X(d) \end{array} \right] & P(d) \\ &= \left[\begin{array}{cc} \tilde{Y}_0(d) + Q(d)D(d) & E(d)V(d) \end{array} \right] & [(4)] \\ &= \left[\begin{array}{cc} \tilde{Y}_0(d) & E(d)V(d) \end{array} \right] + Q(d) \left[\begin{array}{cc} D(d) & O_{n\times m} \end{array} \right] \\ &= \left[\begin{array}{cc} \tilde{Y}_0(d) & E(d)V(d) \end{array} \right] + Q(d) \left[\begin{array}{cc} A(d) & -B(d) \end{array} \right] P(d) & [(1)] \end{aligned}$$

Hence,

$$\left. \begin{aligned} Y(d) &= Y_0(d) + Q(d)A(d) \\ X(d) &= X_0(d) - Q(d)B(d) \\ Z(d) &= R(d) \end{aligned} \right\} \qquad (4.4\text{-}14)$$

with

$$\left[\begin{array}{cc} Y_0(d) & X_0(d) \end{array} \right] := \left[\begin{array}{cc} \tilde{Y}_0(d) & E(d)V(d) \end{array} \right] P^{-1}(d) \qquad (4.4\text{-}15)$$

is the desired minimum degree solution w.r.t. $Z(d)$ of (2-5) and (3-4).

As pointed out in Sec. 3.1, the fact that the gcld's of $A(d)$ and $B(d)$ are strictly Hurwitz is equivalent to stabilizability of the pair (Φ, G). Therefore, Lemma 1 is the counterpart of Theorem 2.4-1 in the polynomial equation approach.

Example 4.4-1 Consider again the LQOR problem of Example 1-1. We see that the pair (Φ, G) is stabilizable, though not completely reachable. Consequently, $A(d) = I - d\Phi$ and $B(d) = dG$ have strictly Hurwitz gcld's. From (1-18), it follows that $q = 1$. Therefore, if we take (cf. (1-19)) $E(d) = 2(1 - d/2)$, we find

$$\begin{aligned} \bar{A}_2(d) &= d(1 - d^{-1}) = -1 + d \\ \bar{B}_2(d) &= d \left[\begin{array}{cc} d^{-1} & 0 \end{array} \right] = \left[\begin{array}{cc} 1 & 0 \end{array} \right] \\ \bar{E}(d) &= 2d(1 - d^{-1}/2) = -1 + 2d \end{aligned} \qquad (4.4\text{-}16)$$

We have to solve (2-5) and (3-4) with the degree constraint

$$\partial Z(d) < \partial \bar{E}(d) = 1 \qquad (4.4\text{-}17)$$

We find that the minimum-degree solution w.r.t. $Z(d)$ of (2-5) and (3-4) is, in agreement with Lemma 1, unique and equals

$$X = 2, \quad Y = \left[\begin{array}{cc} 1 & \frac{1}{3} \end{array} \right], \quad Z = \left[\begin{array}{cc} 2 & \frac{4}{3} \end{array} \right] \qquad (4.4\text{-}18)$$

Therefore, (3-10) becomes

$$\begin{aligned} u(t) &= -X^{-1}Yx(t) \\ &= -\left[\begin{array}{cc} \frac{1}{2} & \frac{1}{6} \end{array} \right] x(t) \end{aligned} \qquad (4.4\text{-}19)$$

Further, the transition matrix of the corresponding closed-loop system equals

$$\Phi_{cl} = \Phi - GX^{-1}Y = \begin{bmatrix} \frac{1}{2} & -\frac{1}{6} \\ 0 & \frac{1}{2} \end{bmatrix} \tag{4.4-20}$$

Therefore,

$$\begin{aligned} \chi_{cl}(d) &= \det(I - d\Phi_{cl}) \\ &= \left(1 - \frac{d}{2}\right)^2 \\ &= \frac{E(d)}{E(0)}\left(1 - \frac{d}{2}\right) \end{aligned} \tag{4.4-21}$$

Hence, the closed-loop system is asymptotically stable. Further, the last equality shows that the closed-loop eigenvalues are the reciprocal of the roots of the spectral factor $E(d)$, together with the unreachable eigenvalue of (Φ, G).

Problem 4.4-1   Consider again the LQOR problem of Example 1-1 except for the matrix Φ whose (2,2) entry is now 2 instead of 1/2. Consequently, the pair (Φ, G) is not stabilizable. Show the following:

(a) $A_2(d)$, $B_2(d)$, and $E(d)$ are again as in (1-18) and (1-19).

(b) Equations (2-5) and (3-4) have no solution.

It is of interest to establish conditions under which the constant pair (X, Y) can be computed by using (3-9) alone. Whenever possible, this would provide the matrices that are needed to construct the state-feedback gain matrix $-X^{-1}Y$, without the extra effort required to compute the "dummy" polynomial matrix $Z(d)$.

Lemma 4.4-2.   *Let the pair (Φ, G) be completely reachable. Then, (3-9) admits a unique constant solution (X, Y) that is the same as the one yielded by the minimum degree solution w.r.t. $Z(d)$ of (2-5) and (3-4) [or (2-5) and (3-9)].*

Proof: First, we show that, under the stated condition, (3-9) has a unique constant solution (X, Y). In fact, Lemma 1 guarantees that a constant solution of (3-9) exists. To see that it is unique, consider that any constant pair (X, Y) solves (3-9) if and only if it solves

$$\bar{A}_2(d)X' + \bar{B}_2(d)Y' = \bar{E}(d) \tag{4.4-22}$$

Further,

$$\bar{A}_2^{-1}(d)\bar{B}_2(d) = \bar{B}(d)\bar{A}^{-1} \tag{4.4-23}$$

with $\bar{A}(d) := dA^*(d) = dI - \Phi'$ and $\bar{B}(d) =: dB^*(d) = G'$. Because (Φ, G) is completely reachable, it follows from the PBH reachability test (B.5-1) that $\bar{A}(d)$ and $\bar{B}(d)$ are right-coprime. In addition, $\bar{A}(d)$ is regular. It then follows that (22) has a unique solution with $\partial Y' < \partial \bar{A}(d) = 1$.

Problem 4.4-2   Show that if in (22) Y' is a constant matrix, X' is also a constant matrix. (*Hint:* Recall that $\partial \bar{B}_2(d) < \partial \bar{A}_2(d) = \partial \bar{E}(d) = q$ and $A_2(0)$ and $E(0)$ are nonsingular.)

Problem 4.4-3 Consider the LQOR problem of Example 1-1 where (Φ, G) is not completely reachable. Show that (3-9) does not have a unique constant solution.

Problem 4.4-4 Consider the LQOR of Example 1-1. Check whether it is possible to use (3-9) only to find the constant matrices X and Y.

> **Theorem 4.4-1.** *Let (Φ, G) be a stabilizable pair, or, equivalently, $A(d) := I - d\Phi$ and $B(d) := dG$ have strictly Hurwitz gcld's. Let (1-13) be fulfilled. Let $(X, Y, Z(d))$ be the minimum-degree solution w.r.t. $Z(d)$ of the bilateral Diophantine equations (2-5) and (3-4) [or (2-5) and (3-7)]. Then, the constant state-feedback control*
>
> $$u(t) = -X^{-1}Yx(t) \qquad (4.4\text{-}24)$$
>
> *makes the closed–loop system internally stable if and only if the spectral factor $E(d)$ in (1-12) is strictly Hurwitz. In such a case, (24) yields the steady-state LQOR law. If (Φ, G) is controllable, the d-characteristic polynomial $\chi_{cl}(d)$ of the optimally regulated system is given by*
>
> $$\chi_{cl}(d) = \frac{\det E(d)}{\det E(0)} \qquad (4.4\text{-}25)$$
>
> *Finally, if (Φ, G) is a reachable pair, the matrix pair (X, Y) in (24) is the constant solution of the unilateral Diophantine equation (3-7).*

It is to be pointed out that Theorem 1 holds true for every $\psi_u = \psi_u' \geq 0$. If ψ_u is positive definite, $E(d)$ turns out to be strictly Hurwitz and the involved polynomial equations solvable if (Φ, G, H) is stabilizable and detectable [CGMN91]. This result agrees with the conclusions of Theorem 2.4-4 obtained via the Riccati equation approach.

Main Points of the Section. Solvability of the two linear Diophantine equations relevant to the LQOR problem is guaranteed by stabilizability of the pair (Φ, G). The strict Hurwitz property of the right spectral factor $E(d)$ yields internal stability of the optimally regulated closed-loop system. Whenever (Φ, G) is a reachable pair, the steady-state LQOR feedback gain can be computed via a single Diophantine equation.

Problem 4.4-5 (Stabilizing Cheap Control) Consider the polynomial solution of the steady-state LQOR problem when the control variable is not costed, viz., $\psi_u = O_{m \times m}$. The resulting regulation law will be referred to as *stabilizing cheap control* because the polynomial solution ensures closed-loop asymptotic stability if no unstable hidden modes are present and $E(d)$ is strictly Hurwitz. Find the stabilizing cheap control for the plant of both Problems 2.6-5 and 2.6-6. Finally, draw general conclusions on the location of the eigenvalues of SISO plants, either minimum or nonminimum phase, regulated by stabilizing cheap control. Contrast stabilizing cheap control with cheap control.

4.5   RELATIONSHIP WITH THE RICCATI-BASED SOLUTION

In order to find a direct relationship between the polynomial and the Riccati-based solution, we proceed as follows. Assume that (Φ, G, H) is stabilizable and detectable. Let P be the symmetric nonnegative definite solution of the ARE (2.4-57). Then, for every $x(0) \in \mathbb{R}^n$, we have

$$x'(0)Px(0) = J_2 + J_3$$

or

$$
\begin{aligned}
P = {} & \langle A^{-*}(d) \Big[\psi_x - \psi_x B_2(d) E^{-1}(d) E^{-*}(d) B_2^*(d) \psi_x \\
& + A^*(d) Z^*(d) \bar{E}^{-*}(d) \bar{E}^{-1}(d) Z(d) A(d) \Big] A^{-1}(d) \rangle
\end{aligned}
$$

Letting

$$
\begin{aligned}
E^{-*}(d) B_2^*(d) \psi_x &= \bar{E}^{-1}(d) \bar{B}_2(d) \psi_x \\
&= Y + \bar{E}^{-1}(d) Z(d) A(d) \qquad [(2\text{-}5)]
\end{aligned}
$$

we get

$$
\begin{aligned}
P &= \langle A^{-*}(d)(\psi_x - Y'Y)A^{-1}(d) - d^{-q} A^{-*}(d) Y' E^{-*}(d) Z(d) \\
& \quad - d^q Z^*(d) E^{-1}(d) Y A^{-1}(d) \rangle \\
&= \langle A^{-*}(d)(\psi_x - Y'Y)A^{-1}(d) \rangle \\
&= \langle \sum_{r,k=0}^{\infty} [d^{-r} d^k \Phi'^r (\psi_x - Y'Y)\Phi^k] \rangle \\
&= \sum_{k=0}^{\infty} \Phi'^k (\psi_x - Y'Y)\Phi^k \\
&= \Phi'P\Phi - Y'Y + \psi_x \qquad\qquad\qquad\qquad (4.5\text{-}1)
\end{aligned}
$$

In (1), the second equality follows because $d^{-q} E^{-*}(d) Z(d) = \bar{E}^{-1}(d) Z(d)$ is strictly anticausal and $A^{-*}(d) Y'$ is anticausal; the third recalling (3.1-17); and the fifth by a formal identity (cf. also Problem 2.4-1). Comparing (1) with the ARE (2.4-57), we find

$$Y'Y = \Phi'PG(\psi_u + G'PG)^{-1}G'P\Phi \qquad\qquad (4.5\text{-}2)$$

Further, comparing (2.4-60) with (4-24), find

$$(\psi_u + G'PG)^{-1}G'P\Phi = X^{-1}Y \qquad\qquad (4.5\text{-}3)$$

Let

$$
\begin{aligned}
Y'Y &= Y'X'^{-1}X'Y \\
&= (X^{-1}Y)'X'XX^{-1}Y
\end{aligned}
$$

Taking into account (2) and (3), we get

$$X'X = \psi_u + G'PG \tag{4.5-4}$$

From (4) and (3), it follows that

$$X'Y = G'P\Phi \tag{4.5-5}$$

Finally,

$$Z(d) = \bar{B}_2(d)P \tag{4.5-6}$$

To establish (6), we can proceed as follows. Rewrite (3-7) as

$$\bar{E}(d) = \bar{A}_2(d)X' + \bar{B}_2'Y'$$

Using this into (2-5), we get

$$\begin{aligned}
Z(d)A(d) &= \bar{B}_2(d)(\psi_x - Y'Y) - \bar{A}_2(d)X'Y \\
&= \bar{B}_2(d)(P - \Phi'P\Phi) - \bar{A}_2(d)G'P\Phi \qquad [(1) \text{ and } (5)]
\end{aligned}$$

Recalling that $A(d) = I - d\Phi$, we have

$$\begin{aligned}
Z(d) - \bar{B}_2(d)P &= \{dZ(d) - [\bar{B}_2(d)\Phi' + \bar{A}_2(d)G']P\}\Phi \\
&= d[Z(d) - \bar{B}_2(d)P]\Phi \tag{4.5-7}
\end{aligned}$$

The last equality follows because

$$\begin{aligned}
\bar{B}_2(d)\Phi' + \bar{A}_2(d)G' &= [\Phi B_2(d) + GA_2(d)]^*d^q \\
&= [d^{-1}B_2(d)]^*d^q \qquad [A(d)B_2(d) = dGA_2(d)] \\
&= d\bar{B}_2(d)
\end{aligned}$$

Equation (7) can be rewritten as follows:

$$[Z(d) - \bar{B}_2(d)P]A(d) = O_{m\times n}$$

This yields (6).

Main Points of the Section. Equations (1) to (2) and (4) to (6) give the relationship between the polynomial and the Riccati-based solution of the steady-state LQOR problem.

4.6   ROBUST STABILITY OF LQ-REGULATED SYSTEMS

We shall use the results of Sec. 3.3 to analyze robust stability properties of optimally LQ output-regulated systems. In this respect, the first comment is that, in view

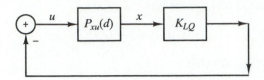

Figure 4.6-1: Plant/compensator cascade unity feedback for an LQ regulated system.

of Theorem 3.3-1, stability robustness is guaranteed from the outset by the state-feedback nature of the LQOR solution and asymptotic stability of the nominal closed-loop system. Nevertheless, we also intend to use Fact 3.3-1 so as to point out the connection in LQ-regulated systems between robust stability and the so-called return difference equality.

From the spectral factorization (1-12) and (3-9), we have

$$\psi_u + A_2^{-*}(d)B_2^*(d)\psi_x B_2(d)A_2^{-1}(d)$$

$$= [I_m + A_2^{-*}(d)B_2^*(d)Y'X'^{-1}]X'X[I_m + X^{-1}YB_2(d)A_2^{-1}(d)]$$

By recalling (1-9) and (4-32), this equation can be rewritten as follows:

$$\psi_u + P_{xu}^*(d)\psi_x P_{xu}(d) \qquad\qquad\qquad\qquad (4.6\text{-}1)$$

$$= [I_m + K_{\mathrm{LQ}}P_{xu}(d)]^*(\psi_u + G'PG)[I_m + K_{\mathrm{LQ}}P_{xu}(d)]$$

where

$$K_{\mathrm{LQ}} := X^{-1}Y \qquad\qquad\qquad\qquad (4.6\text{-}2)$$

denotes the constant transfer matrix (state-feedback-gain matrix) of the LQOR in the plant/compensator cascade unity feedback system, as in Fig. 1. Equation (1) is known as the *return difference equality* of the LQ-regulated system. Similarly to (3.3-4), we denote the loop transfer matrix of the plant/compensator cascade of Fig. 1 by

$$\mathcal{G}_{\mathrm{LQ}}(d) := K_{\mathrm{LQ}}P_{xu}(d) \qquad\qquad\qquad\qquad (4.6\text{-}3)$$

and the corresponding return difference by $I_m + \mathcal{G}_{\mathrm{LQ}}(d)$. In light of Sec. 3.3, we interpret $\mathcal{G}_{\mathrm{LQ}}(d)$ as the nominal loop transfer matrix. In fact, we suppose that K_{LQ} has been designed based upon the nominal plant transfer matrix $P_{xu}(d)$. Similarly to (3.3-6), we assume that the actual plant transfer matrix equals $P_{xu}(d)$ postmultiplied by a multiplicative perturbation matrix $M(d)$. Consequently, the actual loop transfer matrix $\mathcal{G}(d)$ equals

$$\mathcal{G}(d) = \mathcal{G}_{\mathrm{LQ}}(d)M(d) \qquad\qquad\qquad\qquad (4.6\text{-}4)$$

In order to use the robust stability result in Fact 3.3-1 here, we exploit (1) so as to find a lower bound for the R.H.S. of (3.3-10). We note that for d taking values on the unit circle of the complex plane, $H^*(d)$ is the Hermitian of $H(d)$, whenever the latter is a rational matrix with real coefficients. Consequently, for $d \in \mathbb{C}$ and $|d| = 1$, $H^*(d)H(d)$ is Hermitian symmetric nonnegative definite. The other point that we shall use to lower bounding the R.H.S. of (3.3-10) via (1) is that Theorem 2.4-4 ensures that the matrix P in (1) is a bounded symmetric nonnegative definite matrix, provided that the (Φ, G, H) is stabilizable and detectable. Hence, under such conditions, there exists a positive real β^2 such that

$$\psi_u \leq \psi_u + G'PG \leq \beta^2 I_m \tag{4.6-5}$$

Consequently, remembering the definition of the contour Ω in $\mathbb{C}$ given after (3.3-8), we find from (1), (2), and (3), for $d \in \Omega$,

$$\begin{aligned} \beta^2 [I_m + \mathcal{G}_{\mathrm{LQ}}^*(d)][I_m + \mathcal{G}_{\mathrm{LQ}}(d)] &\geq \\ \psi_u + P_{xu}^*(d)\psi_x P_{xu}(d) &\geq \quad \psi_u \end{aligned} \tag{4.6-6}$$

Hence,

$$\underline{\sigma}(I_m + \mathcal{G}_{\mathrm{LQ}}(d)) \geq \frac{\lambda_{\min}^{1/2}(\psi_u)}{\beta} =: \bar{\alpha} \leq 1 \tag{4.6-7}$$

Proposition 4.6-1.   *Consider the steady-state LQ output regulated system of Fig. 1, where (Φ, G, H) is stabilizable and detectable and $\psi_u > 0$. Then, there exists a positive real $\bar{\alpha} \leq 1$ lower-bounding the minimum singular value of the return difference matrix as in (7).*

We note that the bound $\bar{\alpha}$ in (7) is in general quite conservative. In [Sha86], a sharper bound is given. Our interest in (7) is that it shows that if an LQ regulator is designed on the basis of a nominal pair (Φ, G), then the corresponding nominal return difference $(I_m + \mathcal{G}_{\mathrm{LQ}}(d))$ has its smallest singular value lower-bounded by $\bar{\alpha} > 0$. Consequently, according to Fact 3.3-1, this LQ regulator will be capable of stabilizing all plants in a neighborhood of the nominal one.

Theorem 4.6-1.   *With reference to the notations in Fact 3.3-1, let:*

- *$\chi_{o\ell}(d)$ and $\chi_{o\ell}^o(d)$ have the same number of roots inside the unit circle;*

- *$\chi_{o\ell}(d)$ and $\chi_{o\ell}^o(d)$ have the same unit circle roots;*

- *(Φ, G, H) be stabilizable and detectable and $\psi_u > 0$;*

then the steady-state LQ output-regulated system designed for the nominal pair (Φ, G) remains asymptotically stable for all plants such that at each $d \in \Omega$,

$$\bar{\sigma}(M^{-1}(d) - I_m) < \bar{\alpha} \tag{4.6-8}$$

with $\bar{\alpha} > 0$ as in (7).

Main Points of the Section. The return difference equality of LQ regulation allows lower-bounding away from zero the minimum singular value of the return difference of the nominal loop and hence to show robust stability of LQ-regulated systems against plant multiplicative perturbations. Alternatively, stability robustness of LQ-regulated systems follows at once from the state-feedback nature of the LQOR solution, and asymptotic stability of the corresponding nominal closed-loop system.

NOTES AND REFERENCES

Appendix C gives a quick review of the results on linear Diophantine equations used throughout this chapter. The polynomial equation approach to LQ regulation was ushered by the fundamental monograph [Kuč79]. See also [Kuč91].

The deterministic LQOR problem in the discrete-time case seems to have been first directly tackled and constructively solved in its full generality by polynomial tools in [MN89]. Earlier related results also appeared in [Kuč83] and [Gri87].

The polynomial solution of the deterministic LQOR problem in the continuous-time case appeared in [CGMN91]. The dual problem of stochastic linear minimum mean–square-error state filtering, that is, Kalman filtering, was also solved in its full generality by polynomial equations in [CM91]. Earlier results on the subject were reported in [Kuč81] and [Gri85].

Spectral factorization is fundamental in LQ optimization, for example, in finding polynomial solutions to optimal filtering problems [CM91]. For a discussion on the scalar factorization problem, the reader is referred to [ÅW84]. In [Kuč79] and [JK85], algorithms are described that are applicable to spectral factorization of matrices. Spectral factorization was introduced by [Wie49]. In [You61], a spectral factorization for rational matrices was developed. See also [Kai68]. [And67] showed how spectral factorization for rational matrices can be computed using state-space methods, by solving an ARE.

For continuous-time plants, many robustness results are available [AM90], [Pet89], [POF89]. These results in general cannot be easily extended to the discrete-time case. For the latter, see [MZ88], [GdC87], and [NDD92].

CHAPTER 5

DETERMINISTIC RECEDING-HORIZON CONTROL

Receding-horizon control (RHC) is a conceptually simple method to synthesize feedback control laws for linear and nonlinear plants. Although the method, if desired, can be also used to synthesize approximations to the steady-state LQR feedback with a guaranteed stabilizing property, it has extra features that make it particularly attractive in some application areas. In fact, because it involves a horizon made up by only a finite number of time steps, the RHC input can be sequentially calculated on line by existing optimization routines so as to minimize a performance index and fulfill hard constraints, for example, bounds on the time evolutions of the input and state. This is of paramount importance whenever the previously mentioned constraints are part of the control design specifications. In contrast, in the LQ control problem over a semiinfinite horizon, hard constraints cannot be managed by standard optimization routines. Consequently, in steady-state LQ control, we are forced to replace, typically at performance-degradation expense, hard with soft constraints, for example, instantaneous input hard limits with input mean-square upper bounds.

Clearly, RHC is most suitable for slow linear and nonlinear systems, such as chemical batch processes, where it is possible to solve constrained optimization control problems on line. Another direction is to use simple RHC rules that yield easy feedback computations while guaranteeing a stable closed-loop system for generic linear plants or plants satisfying crude open-loop properties. In view of adaptive and self-tuning control applications, in this chapter, we focus our attention mainly on the latter use of RHC.

After a general formulation of the method in Sec. 1, specific receding-horizon regulation laws are considered and analyzed in Secs. 2 to 7. In Sec. 8, it is shown how the results of the previous sections can be suitably extended, by adding a feedforward action to the basic regulation laws, so as to cover the 2-DOF tracking problem.

5.1 RECEDING-HORIZON REGULATION

Considering the results in Chapters 2 and 4, steady-state LQR can be looked at as a methodology for designing state-feedback regulators capable of stabilizing arbitrary linear plants while optimizing an engineering meaningful performance index. However, we have seen in Secs. 2.6 and 2.7 that two simplified variants of LQR, viz., cheap control and single-step regulation, are severely limited in their applicability. Consequently, at this stage, it appears that, for LQR design, solving either an ARE or a spectral factorization problem is mandatory in that the easier ways of feedback computation, pertaining to either cheap control or single-step regulation, are in general prevented. Now, solving an RDE over a semiinfinite horizon, or an ARE, typically entails iterations, as seen in Sec. 2.5. These, in the RDE case, are slowly converging, whereas the Kleinman algorithm for solving the ARE must be imperatively initialized from a stabilizing feedback, possibly, for fast convergence, not far from the optimal one. Further, the latter algorithm involves at each iteration step a rather high computational effort for solving the related Lyapunov equation (2.5-3). Comparable computational difficulties are associated with the spectral factorization problem, particularly in the multiple-input case.

Receding-horizon regulation (RHR) was first proposed to relax the computational shortcomings of steady-state LQR. In RHR, the current input $u(t)$ at time t and state x is obtained by determining, over an N-step horizon, the input sequence $\hat{u}_{[t,t+N)}$ optimal in a constrained LQR sense, and setting $u(t) = \hat{u}(t)$, the whole procedure being repeated at time $t+1$ to select $u(t+1)$. Accordingly, at every time, the plant is fed by the initial vector of the optimal input sequence whose subsequent $N-1$ vectors are discarded. The applied input would be optimal should the subsequent part of the optimal sequence be used as plant input at the subsequent $N-1$ steps. This is not purposely done, so RHR can be hardly considered "optimal" in any well-defined sense. Nevertheless, if the constraints are judiciously chosen, RHR can acquire attractive features. Let us consider the following as a formal statement of RHR.

Receding Horizon Regulation (RHR). Consider the time-invariant linear plant:

$$\left. \begin{array}{rcl} x(k+1) & = & \Phi x(k) + G u(k) \\ x(t) & = & x \in \mathbb{R}^n \\ y(k) & = & H x(k) \end{array} \right\} \tag{5.1-1}$$

with $u(k) \in \mathbb{R}^m$, and $y(k) \in \mathbb{R}^p$. Define the quadratic performance index over the N-step *prediction horizon* $[t+1, t+N]$:

$$J\left(x, u_{[t,t+N)}\right) = \sum_{k=t}^{t+N-1} \ell(k, x(k), u(k)) + \|x(t+N)\|_{\psi_x(N)}^2 \qquad (5.1\text{-}2)$$

$$\ell(k, x, u) = \|x\|_{\psi_x(k)}^2 + \|u\|_{\psi_u(k)}^2 \qquad (5.1\text{-}3)$$

with $\psi_x(k) = H'\psi_y(k)H$, $\psi_y(k) = \psi_y'(k) \geq 0$, $\psi_u(k) = \psi_u'(k) \geq 0$, and $\psi_x(N) = \psi_x'(N) \geq 0$. Find, whenever it exists, an optimal input sequence $\hat{u}_{[t,t+N)}$ to the plant (1) minimizing (2) while satisfying the following set of contraints:

$$\sum_{k=t}^{t+N} \left[X_i(k)x(k) + U_i(k)u(k) \right] \left\{ \begin{matrix} \leq \\ = \end{matrix} \right\} C_i \qquad (5.1\text{-}4)$$

where $i = 1, 2, \cdots, I$; $X_i(k)$ and $U_i(k)$ are matrices; C_i are vectors of compatible dimensions; and the inequality/equality sign indicates that either the first or the latter applies for the ith constraint. The RHR input to the plant is then chosen at time t to be $u(t) = \hat{u}(t)$. In case $\hat{u}_{[t,t+N)}$ can be found in an explicit open-loop form as follows:

$$\hat{u}(t+i) = f(i, x), \qquad i = 0, 1, \cdots, N-1 \qquad (5.1\text{-}5)$$

the time-invariant state-feedback given by

$$u(t) = f(0, x(t)), \qquad \forall t \in \mathbb{Z} \qquad (5.1\text{-}6)$$

is referred to as the *RHR law* relative to (1) to (4).

Example 5.1-1   Consider the problem of finding an input sequence

$$u_0^{n-1} := \begin{bmatrix} u(n-1) & \cdots & u(0) \end{bmatrix}'$$

that transfers the initial state $x(0) = x$ of (1) to a target state $x(n)$ at time $n = \dim \Phi$. Here we assume that the plant has a single input, $u(k) \in \mathbb{R}$, and is completely reachable. We have

$$x(k) = \Phi^k x + R_k u_0^{k-1} \qquad (5.1\text{-}7)$$

for $k = 0, 1, 2, \cdots$, with

$$R_k := \begin{bmatrix} G & \Phi G & \cdots & \Phi^{k-1}G \end{bmatrix} \in \mathbb{R}^{n \times k} \qquad (5.1\text{-}8)$$

Because the reachability matrix of (1)

$$R := R_n = \begin{bmatrix} G & \Phi G & \cdots & \Phi^{n-1}G \end{bmatrix} \qquad (5.1\text{-}9)$$

is nonsingular, the solution is uniquely given by

$$u_0^{n-1} = R^{-1}[x(n) - \Phi^n x] \qquad (5.1\text{-}10)$$

If terminal state $x(n)$ is constrained to equal O_n, we find

$$u_0^{n-1} = -R^{-1}\Phi^n x \qquad (5.1\text{-}11)$$

This specifies the open-loop input sequence (5) when, in (1) to (4), $N = n$, $m = 1$, (4) reduces to $x(n) = O_n$, and (2) can be any being ineffective. The RHR law (6) becomes

$$u(t) = -\underline{e}_n' R^{-1}\Phi^n x(t) \qquad (5.1\text{-}12)$$

where $\underline{e}_n$ is the nth vector of the natural basis of $\mathbb{R}^n$:

$$\underline{e}_n = \begin{bmatrix} 0 & \cdots & 0 & 1 \end{bmatrix} \in \mathbb{R}^n$$

In its general formulation, (1) to (6), RHR has to be regarded as a problem of convex quadratic programming that can be tackled with existing software tools [BB91]. In general, this is a quite formidable problem from a computational viewpoint, particularly if on-line solutions are required. Nevertheless, RHR possesses such potential favorable features to even justify a significant computational load. Among these features, we mention the capability of RHR of combining, thanks to the presence of suitable constraints, short-term behavior with long-range properties, for example, stability requirements. There are, however, RHR laws both computable with moderate efforts and yielding attractive properties to the regulated system that can be expressed in an explicit form. They will be the main subject of the remaining part of this chapter.

We warn the reader from believing that in general function $f(i,x)$ in (5) can be obtained in a closed analytic form. In fact, this happens only in specific cases, for example, Example 1. Whenever $f(i,x)$ cannot be found, one is forced to solve (1) to (5) numerically, to apply the input (6) at time t, and to repeat the whole optimization procedure (1) to (5) over the next prediction horizon $[t+2, t+N+1]$ so as to find numerically the input at time $t+1$.

Before progressing, it is convenient to point out that cheap control (Sec. 2.6) and single-step regulation (Sec. 2.7) can be embedded in the RHR formulation. For both of them, the set of constraints (4) is void, and the prediction horizon is made up by the shortest possible prediction horizon, a single step. As was remarked, the resulting regulation laws are unacceptable in many pratical cases mainly because of their unsatisfactory stabilizing properties. In this respect, a definite improvement can be obtained by enlarging the prediction horizon and/or adding suitable constraints. In order to gain some understanding on how to rationally make this selection, we introduce next a tool for analyzing the stabilizing properties of some RHR laws.

Main Points of the Section. Though RHR was first introduced to lighten the computational load of steady-state LQR, in its general formulation, it is a convex quadratic programming problem with an associated high computational burden. Nevertheless, a judicious choice of RHR design knobs can lead to regulation laws both computable with moderate efforts and yielding attractive properties to the regulated system.

5.2   RDE MONOTONICITY AND STABILIZING RHR

A convenient tool for studying the stabilizing properties of some RHR schemes is the *fake algebraic Riccati equation* (FARE) introduced in Problem 2.4-12. The FARE argument we shall mostly use is restated here.

> **FARE Argument.** Consider a stabilizable and detectable plant (Φ, G, H) and the related LQOR problem of Sec. 2.4. Let $P(k)$ be the solution of the relevant forward RDE:
>
> $$P(k+1) = \Phi'P(k)\Phi - \Phi'P(k)G\Big[\psi_u + G'P(k)G\Big]^{-1}G'P(k)\Phi + \psi_x \quad (5.2\text{-}1)$$
>
> initialized from some $P(0) = P'(0) \geq 0$. Then
>
> $$P(k+1) \leq P(k) \qquad\qquad\qquad (5.2\text{-}2)$$
>
> implies that the state-feedback-gain matrix
>
> $$F(k+1) = -\Big[\psi_u + G'P(k)G\Big]^{-1}G'P(k)\Phi \qquad\qquad (5.2\text{-}3)$$
>
> stabilizes the plant, viz., $\Phi + GF(k+1)$ is a stability matrix.

If we can associate a suitable RDE (1) to a given RHR problem whose solution satisfies (2), asymptotic stability of the regulated plant follows, whenever the state feedback has the form (3).

Other useful features of the RDE are its monotonicity properties, viz.,

$$P(k+1) \leq P(k) \quad \Rightarrow \quad P(k+2) \leq P(k+1) \qquad (5.2\text{-}4)$$
$$P(k+1) \geq P(k) \quad \Rightarrow \quad P(k+2) \geq P(k+1) \qquad (5.2\text{-}5)$$

Equation (4) tells us that if $F(k+1)$ can be proved, via the FARE argument, to be a stabilizing state feedback, $F(k+2)$ is stabilizing as well. Conversely, (5) shows that if $F(k+1)$ cannot be proved to be stabilizing via the FARE argument, the same is true for $F(k+2)$.

The preceding monotonicity properties (4) and (5) can be proved by using the next result, given in [dS89]. See also [Yaz89].

> **Fact 5.2-1.**   Let $P^1(k)$ and $P^2(k)$ denote the solutions of two forward RDEs with the same (Φ, G) and ψ_u, but possibly different initializations $P^1(0)$ and $P^2(0)$, respectively, and different weights ψ_x: ψ_x^1 and ψ_x^2, respectively. Let:
>
> $$\tilde{P}(k) := P^2(k) - P^1(k)$$
>
> $$\tilde{\psi}_x := \psi_x^2 - \psi_x^1 \qquad \tilde{\psi}_u := \psi_u + G'P^1(k)G$$
> $$\Phi(k) := \Phi - G\tilde{\psi}_u^{-1}G'P^1(k)\Phi$$

Then, $\tilde{P}(k)$ satisfies the forward RDE equation:

$$\tilde{P}(k+1) = \Phi'(k)\tilde{P}(k)\Phi(k) \tag{5.2-6}$$
$$- \Phi'(k)\tilde{P}(k)G\Big[\tilde{\psi}_u(k) + G'\tilde{P}(k)G\Big]^{-1}G'\tilde{P}(k)\Phi(k) + \tilde{\psi}_x$$

Proposition 5.2-1. *Let $P(k)$ be the nonnegative definite solution of the RDE (1). Then:*

(i) *If $P(k)$ is monotonically nonincreasing at step k, that is, $P(k+1) \le P(k)$, it is monotonically nonincreasing for all subsequent steps, that is, $P(k+i+1) \le P(k+i)$, for all $i \ge 0$.*

(ii) *If $P(k)$ is monotonically nondecreasing at step k, that is, $P(k+1) \ge P(k)$, it is monotonically nondecreasing for all subsequent steps, that is, $P(k+i+1) \ge P(k+i)$, for all $i \ge 0$.*

Proof: It is directly based on (6).

(i) Let $P^2(k) := P(k)$ and $P^1(k) := P(k+1)$. Then, $\tilde{P}(k) := P^2(k) - P^1(k) = \tilde{P}'(k) \ge 0$. Further, $\tilde{P}(k+1)$ satisfies (6) with $\tilde{\psi}_x = 0$. Then, it follows from Theorem 2.3-1 that $\tilde{P}(k+1) \ge 0$. Therefore, assertion (i) can be proved by induction.

(ii) Let $P^2(k) := P(k+1)$ and $P^1(k) := P(k)$. Then, $\tilde{P}(k) := P^2(k) - P^1(k) = \tilde{P}'(k) \ge 0$ and we can use the same argument as in (i) to prove assertion (ii).

In order to exploit Proposition 1 in the FARE argument, the easiest idea that comes to mind is to leave the constraint set (1-4) void and to choose $P(0) = \psi_x(N)$ very large, for example, $P(0) = rI_n$, with r positive and very large. In this way, one might think that a single iteration of the RDE (1) would yield $P(1) \le P(0)$. The next problem shows that this conjecture is generally false.

Problem 5.2-1 Show that in the single-input case, if $P(0) = rI_n$, $r\|G\|^2 \gg \psi_u$, and Φ is nonsingular, the RDE (1) gives

$$P(1) - P(0) \simeq r\left[\Phi'\left(I_n - \Phi^{-T}\Phi^{-1} - \frac{GG'}{\|G\|^2}\right)\Phi\right] + \psi_x \tag{5.2-7}$$

where $\Phi^{-T} := (\Phi')^{-1}$.

Next, verify that, while for Φ scalar $P(0) - P(1) + \psi_x = r > 0$, for the following second-order system

$$\Phi = \begin{bmatrix} 1 & 0 \\ a & 10 \end{bmatrix} \qquad G = \begin{bmatrix} 1 \\ 0 \end{bmatrix} \tag{5.2-8}$$

$P(0) - P(1)$ given by (7) is not nonnegative definite, irrespective of r, a, and ψ_x.

Finally, show that in the case (8),

$$\Phi + GF(1) = \Phi - G\Big[\psi_u + G'P(0)G\Big]^{-1}G'P(0)\Phi$$

is not a stability matrix.

The previous discussion indicates that in order to possibly exploit the FARE argument in RHR, we have to introduce some active constraint in (1) to (4). In the next section, it will be shown that this is the case if the terminal state $x(N)$ is constrained to O_n and N is made larger than $\dim \Phi$.

Main Points of the Section. The FARE argument provides a sufficient condition for establishing if the LQOR state-feedback-gain matrix computed via an RDE stabilizes the closed-loop system.

5.3 ZERO-TERMINAL-STATE RHR

We show that RHR yields an asymptotically stable closed-loop system, provided that the prediction horizon is long enough and a zero-terminal-state constraint is used. In this connection, we first discuss in the next example how a state-deadbeat system can be obtained via RHR for single–input completely reachable plants.

Example 5.3-1 Consider again the problem of Example 1-1 and the related zero terminal state RHR law (1-12)

$$u(t) = -\underline{e}'_n R^{-1} \Phi^n x(t) \qquad (5.3\text{-}1)$$

We show that (1) gives rise to a *state-deadbeat* closed-loop system, that is a system with a closed-loop characteristic polynomial $\bar{\chi}_{cl}(z) = z^n$. In order to establish this property, we use Ackermann's formula [Kai80]. This states that if (1-1) is a single-input completely reachable plant, the state-feedback-regulation law $u(t) = Fx(t)$ needed to get a closed-loop characteristic polynomial $\bar{\chi}_{cl}(z)$ equals

$$u(t) = -\underline{e}'_n R^{-1} \bar{\chi}_{cl}(\Phi) x(t) \qquad (5.3\text{-}2)$$

Then, that (1) yields a state-deadbeat system immediately follows from (2).

We now turn to consider the more general case of multiinput multioutput (MIMO) plants when in (1-1) to (1-6): $\psi_y(k) \equiv \psi_y = \psi'_y \geq 0$; $\psi_u(k) \equiv \psi_u = \psi'_u \geq 0$; prediction horizon length N is arbitrary; and the constraints reduce to the zero terminal state. Specifically, we shall consider the following version of (1-1) to (1-5).

Zero-Terminal-State Regulation. Consider the problem of finding, whenever it exists, an input sequence $u_{[t,t+N)}$:

$$u(t + N - k) = F(k)x(t) \qquad k = 1, \cdots, N \qquad (5.3\text{-}3)$$

to the plant (1-1), minimizing the performance index

$$\sum_{i=0}^{N-1} \left[\|y(t+i)\|_{\psi_y}^2 + \|u(t+i)\|_{\psi_u}^2 \right] \qquad (5.3\text{-}4)$$

under the zero-terminal-state constraint

$$x(t + N) = O_n \qquad (5.3\text{-}5)$$

We shall prove the following classic result on the stabilizing properties of the RHR law based on zero-terminal-state regulation.

Theorem 5.3-1. *Consider the zero terminal state regulation (3) to (5) with $\psi_u > 0$. Let the plant (1-1) be completely reachable. Then, the feedback RHR law*

$$u(t) = F(N)x(t) \qquad (5.3\text{-}6)$$

exists and stabilizes (1-1) under the following conditions.

Case (a) $\psi_y = O_{p \times p}$:

$$N \geq n \qquad (5.3\text{-}7)$$

Case (b) $\psi_y > 0$:   *The plant (1-1) is detectable, Φ is nonsingular and*

$$N \geq n + 1 \qquad (5.3\text{-}8)$$

Further, for single-input completely reachable plants, irrespective of Φ, ψ_y and ψ_u, (6) yields state–deadbeat regulation whenever

$$N = n \qquad (5.3\text{-}9)$$

The results (7) and (8) can be made sharper by replacing the plant dimension n with the reachability index ν, $\nu \leq n$, of the pair (Φ, G) [Kai80].

The state-deadbeat-regulation property has been constructively proved in Example 1. That in **Case (a)** of Theorem 1 under (8), (6) is a stabilizing regulation law, is shown in [Kle74] for an invertible Φ, and for any square Φ in [KP75] for the single-input case and in [SE88] for the multi–input case. Our interest in **Case (a)** is quite limited, the main reason being that **Case (b)** gives us some extra freedom that can be conveniently exploited for regulation design. Hence, the interested reader is directly referred to [Kle74], [KP75] and [SE88] for details on the proof of **Case (a)**. On the contrary, the proof of **Case (b)** is next reported in depth by following the approach of [BGW90], based on the FARE argument. This proof is based on an alternative form for the RDE that requires nonsingularity of Φ. The extension of **Case (b)** to possibly singular state-transition matrices will be dealt with before closing the section.

Intuitively, (5) can be implicitly embodied in the RHR formulation (1-1) to (1-6) by setting $\psi_x(N) = rI_n$ with $r \to \infty$ and (1-4) void. This amounts to using the dynamic programming formulas (2.4-9) to (2.4-15) with such a $\psi_x(N)$, or, more precisely,

$$P^{-1}(0) = O_{n \times n} \qquad (5.3\text{-}10)$$

In order to directly exploit (10), we reexpress (2.4-9) to (2.4-15) in terms of an iterative equation for $P^{-1}(k)$. To this end, it is convenient to set

$$\Pi(k+1) := P(k) - P(k)G\left[\psi_u + G'P(k)G\right]^{-1}G'P(k) \qquad (5.3\text{-}11)$$

and

$$\Omega(k) := \Pi^{-1}(k) \tag{5.3-12}$$

The relevant results are summed up in the next lemma.

Lemma 5.3-1. *Assume that the state-transition matrix Φ of the plant (1-1) be nonsingular. Then, whenever $P^{-1}(k)$ exists, we have*

$$\Omega(k+1) = P^{-1}(k) + G\psi_u^{-1}G' \tag{5.3-13}$$

Further, $\Omega(k)$, $k = 0, 1, 2, \cdots$, satisfies the following forward RDE:

$$
\begin{aligned}
\Omega(k+1) &= \Phi^{-1}\Omega(k)\Phi^{-T} - \Phi^{-1}\Omega(k)\Phi^{-T}\psi_x^{T/2} \\
&\quad \times \left[I_p + \psi_x^{1/2}\Phi^{-1}\Omega(k)\Phi^{-T}\psi_x^{T/2} \right]^{-1} \psi_x^{1/2}\Phi^{-1}\Omega(k)\Phi^{-T} \\
&\quad + G\psi_u^{-1}G'
\end{aligned}
\tag{5.3-14}
$$

Finally, provided that $\Omega(k+1)$ is nonsingular, the LQOR feedback-gain matrix $F(k+1)$, as in (2.4-10), can be expressed as follows:

$$
\begin{aligned}
F(k+1) &= -\left[\psi_u + G'P(k)G \right]^{-1} G'P(k)\Phi \\
&= -\psi_u^{-1}G'\Omega^{-1}(k+1)\Phi
\end{aligned}
\tag{5.3-15}
$$

In (14): $\Phi^{-T} := (\Phi')^{-1}$; $\psi_x^{1/2} := \psi_y^{1/2}H$; $\psi_x^{T/2} := (\psi_x^{1/2})'$; and $\psi_y^{1/2}$ is the square root of ψ_y [GVL83], $\psi_y = (\psi_y^{1/2})^2$.

To prove Lemma 1, we shall make repeated use of the following result.

Matrix Inversion Lemma *Let A, C, and $DA^{-1}B + C^{-1}$ be nonsingular matrices. Then*

$$\left(A + BCD \right)^{-1} = A^{-1} - A^{-1}B\left(DA^{-1}B + C^{-1} \right)^{-1} DA^{-1} \tag{5.3-16}$$

Proof: Let

$$
\left.
\begin{aligned}
C^{-1}x &= Du \\
y &= Bx + Au
\end{aligned}
\right\}
\tag{5.3-17}
$$

with u, x, and y vectors of suitable dimensions. Then, we have $y = (A+BCD)u$. Consider the "inverse" system of (17):

$$
\left.
\begin{aligned}
C^{-1}x &= DA^{-1}(y - Bx) \\
u &= -A^{-1}Bx + A^{-1}y
\end{aligned}
\right\}
$$

from which we get $u = A^{-1}y - A^{-1}B\left(DA^{-1}B + C^{-1} \right)^{-1} DA^{-1}y$.

Proof of Lemma 1: Using (16), we establish (13)

$$
\begin{aligned}
\Omega(k+1) &= \Pi^{-1}(k+1) \\
&= P^{-1}(k) - P^{-1}(k)[-P(k)G] \\
&\quad \times \left\{ G'P(k)P^{-1}(k)[-P(k)G] + G'P(k)G + \psi_u \right\}^{-1} G'P(k)P^{-1}(k) \\
&= P^{-1}(k) + G\psi_u^{-1}G'
\end{aligned}
$$

Next, from (11), it follows that

$$P(k) = \Phi' \Pi(k) \Phi + \psi_x$$

Hence,

$$
\begin{aligned}
W(k) &:= P^{-1}(k) & \text{(5.3-18)} \\
&= \left[\Phi' \Pi(k) \Phi + \psi_x \right]^{-1} = \Phi^{-1} \left[\Pi(k) + \Phi^{-T} \psi_x \Phi^{-1} \right]^{-1} \Phi^{-T} \\
&= \Phi^{-1} \Big\{ \Pi^{-1}(k) - \Pi^{-1}(k) \Phi^{-T} \psi_x^{T/2} \\
&\quad \times \left[I_p + \psi_x^{1/2} \Phi^{-1} \Pi^{-1}(k) \Phi^{-T} \psi_x^{T/2} \right]^{-1} \psi_x^{1/2} \Phi^{-1} \Pi^{-1}(k) \Big\} \Phi^{-T}
\end{aligned}
$$

Then, by (13), (14) follows.
Finally,

$$
\begin{aligned}
F(k+1) &= - \left[\psi_u + G' P(k) G \right]^{-1} G' P(k) \Phi \\
&= - \left[\psi_u + G' W^{-1}(k) G \right]^{-1} G' W^{-1}(k) \Phi & \text{[(18)]} \\
&= -\psi_u^{-1} \Big\{ G' - G' \left[G \psi_u^{-1} G' + W(k) \right]^{-1} G \psi_u^{-1} G' \Big\} W^{-1}(k) \Phi & \text{[(16)]} \\
&= -\psi_u^{-1} \Big\{ G' - G' \Omega^{-1}(k+1) \left[\Omega(k+1) - W(k) \right] \Big\} W^{-1}(k) \Phi & \text{[(13)]}
\end{aligned}
$$

which yields (15).

We note that (14) is the forward RDE associated with the following LQOR problem:

$$
\left.
\begin{aligned}
\xi(t+1) &= \Phi^{-T} \xi(t) + \Phi^{-T} \psi_x^{T/2} \nu(t) \\
\gamma(t) &= G' \xi(t)
\end{aligned}
\right\}
\qquad \text{(5.3-19)}
$$

$$\ell(\xi, \nu) = \| \gamma(t) \|_{\psi_u^{-1}}^2 + \| \nu(t) \|^2 \qquad \text{(5.3-20)}$$

In order to impose the zero terminal condition (5), we see from (13) that the iterations (14) must be initialized from $\Omega(0) = O_{n \times n}$. In such a case, the following lemma ensures nonsingularity of $\Omega(k)$, provided that $k \geq \dim \Phi$.

Lemma 5.3-2. *Consider the solution $\Omega(k)$ of the forward RDE (14) initialized from $\Omega(0) = O_{n \times n}$. Then, provided that (Φ, G) is a reachable pair,*

$$\det \Omega(n+i) \neq 0 , \quad \forall i \geq 0 \qquad \text{(5.3-21)}$$

with $n = \dim \Phi$.

Proof (by contradiction): First, by nonsingularity of Φ, complete reachability of (Φ, G) is equivalent to complete observability of (Φ^{-T}, G'), that is, complete observability of the dynamic system (19). Next, consider the LQOR problem (19) to (20). Assume that there is a vector ξ, $\xi \neq O_n$, such that $\xi'\Omega(k)\xi = 0$. This means (cf. (2.4-15)) that if the plant initial state is ξ, the optimal input sequence $u^0_{[0,k)}$ is such that

$$\sum_{i=0}^{k-1} \left[\|y^0(i)\|^2_{\psi_u^{-1}} + \|u^0(i)\|^2 \right] = 0$$

where $y^0(i)$ is the output of (19) at time i for $\xi(0) = \xi$ and inputs $u^0_{[0,i)}$. Then, $u^0_{[0,k)}$ and $y^0_{[0,k)}$ are both zero. For $k \geq n$, this contradicts complete observability of (19).

Proof of Theorem 3-1: Complete reachability ensures that problem (5) is solvable for any $N \geq n$. Concomitant minimization of (4) yields uniqueness of $u^0_{[0,N)}$. Further, according to (21), $F(N)$, $N \geq n$, is computable via (14) and (15). By (11), (12), and (21), we find

$$P(k) = \Phi'\Omega^{-1}(k)\Phi + \psi_x , \quad \forall k \geq n \qquad (5.3\text{-}22)$$

Further,

$$\Omega(1) \geq \Omega(0) = O_{n \times n}$$

implies, by (14) and Proposition 2-1, that

$$\Omega(k+1) \geq \Omega(k) , \quad \forall k \geq 0$$

Then, from (22), it follows that

$$P(k+1) \leq P(k) , \quad \forall k \geq n$$

By the FARE argument, we conclude that the state-feedback-gain matrix

$$F(k+1) = -\left[\psi_u + G'P(k)G \right]^{-1} G'P(k)\Phi , \quad \forall k \geq n$$

stabilizes the plant (1).

The major limitation of the stability results of Theorem 1 lies in the non-singularity assumption on the plant state-transition matrix Φ. In this respect, one could argue that this is not a real limitation, because a discrete-time plant resulting from using a zero-order-hold input and sampling uniformly in time any continuous-time *finite-dimensional* linear time-invariant system has always a non-singular state-transition matrix for suitable choices of the state. In fact, if the continuous-time system is given by $\frac{dz(\tau)}{d\tau} = Az(\tau) + Bv(\tau)$, $\tau \in \mathbb{R}$, and T_s is the sampling interval, the corresponding discrete-time plant $x(t+1) = \Phi x(t) + Gu(t)$, $t \in \mathbb{Z}$, has $\Phi = \exp(AT_s)$, for $x(t) = z(tT_s)$. The conclusion, hence, would be that, because our interests are mainly in sampled-data plants, Φ singularity entails little limitation. However, the situation is not so simple, because we are also interested in controlling sampled-data continuous-time *infinite-dimensional* linear time-invariant systems, viz., systems with *dead time* or I/O *transport delay*. In such a case, the previous argument does not hold true any longer.

Problem 5.3-1 Consider a SISO plant described by the following difference equation:

$$y(t) + a_1 y(t-1) + \cdots + a_{n_a} y(t - n_a) = b_1 u(t-1) + \cdots + b_{n_b} u(t - n_b) \qquad (5.3\text{-}23)$$

with $a_{n_a} \cdot b_{n_b} \neq 0$. Show that its state-space canonical reachability representation (2.6-8) has a nonsingular state-transition matrix if and only if $n_b \leq n_a$.

Problem 5.3-2 Consider the plant of Problem 1 but now with an extra unit I/O delay, for example its output $\gamma(t)$ is obtained by delaying $y(t)$ by one step, that is, $\gamma(t) = y(t-1)$. Show that this new plant, if $n_b = n_a$, has state-space canonical reachability representation with singular state-transition matrix.

Another situation in which we have to tackle a singular Φ arises when we want to use the RHR law (7) with a state $x(t)$ that does not coincide with $z(tT_s)$. This happens, for instance, in the practically important case where the $x(t)$ components are externally accessible variables such as past I/O pairs.

Example 5.3-2 Consider the SISO plant described by the difference equation (23). Let (cf. Problem 2.4-5)

$$s(t) := \left[\ \left(y_t^{t-n_a+1}\right)'\ \ \left(u_{t-1}^{t-n_b+1}\right)'\ \right]' \in \mathbb{R}^{n_a + n_b - 1} \qquad (5.3\text{-}24)$$

where $u_t^{t-n} := \left[\ u(t-n)\ \cdots\ u(t)\ \right]'$. Then (23) can be represented in state-space form as follows:

$$s(t+1) = \Phi s(t) + G u(t)$$
$$y(t) = H s(t)$$

with

$$\Phi := \begin{bmatrix} O_{(n_a-1)\times 1} & I_{n_a-1} & O_{(n_a-1)\times(n_b-1)} \\ -a_{n_a} \quad \cdots \quad -a_1 & b_{n_b} \quad \cdots \quad b_2 \\ O_{(n_b-2)\times(n_a+1)} & I_{n_b-2} \\ 0 \quad \cdots \quad \cdots \quad \cdots \quad 0 \end{bmatrix} \qquad (5.3\text{-}25)$$

$$G := \left[\ 0\ \cdots\ 0\ b_1\ 0\ \cdots\ 0\ 1\ \right]' = b_1 \underline{e}_{n_a} + \underline{e}_{n_a+n_b-1} \qquad (5.3\text{-}26)$$

$$H := \underline{e}'_{n_a} \qquad (5.3\text{-}27)$$

where

$$\underline{e}_i := [\underbrace{0 \cdots 0\ 1}_{i}\ 0 \cdots 0]'$$

Note that Φ is singular, though, under appropriate assumptions on (23), (Φ, G, H) is completely reachable and reconstructible.

Finally, the nonsingularity condition on Φ rules out plants described by FIR models. Because FIR models can approximate as tightly as we wish open-loop stable plants, the lack in this case of a proof of the stabilizing property for the RHR law (7) appears conceptually disturbing and restrictive for some applications. For the foregoing reasons, we shall now move on proving that the stabilizing properties of the zero-terminal-state RHR extend to the general case of a possibly singular state-transition matrix. Such an extension is obtained via two different methods of proof. These methods hinge upon two different monotonicity properties of the cost: one upon the cost monotonicity w.r.t. an increase of the prediction horizon for a fixed initial state; the other upon the cost monotonicity along the trajectories of the controlled system for a fixed prediction horizon length. While the former proof relies again on the FARE argument, the latter does not use linearity of the plant and, hence, can also cover nonlinear plants and input and state–related constraints.

Method of proof 1.   Consider again the plant (1-1) with $\Sigma = (\Phi, G, H)$, completely reachable and detectable. Let ν, $\nu \le n = \dim \Phi$, be the reachability index of Σ, viz., the smallest positive integer such that the νth order reachability matrix $\tilde{R}_\nu$ of Σ

$$\tilde{R}_\nu := \begin{bmatrix} \Phi^{\nu-1}G & \cdots & \Phi G & G \end{bmatrix}$$

has full row rank. Then, there are input sequences $u_{[0,\nu)}$ which satisfy the zero terminal state constraint

$$O_X = x(\nu) = \Phi^\nu x + \tilde{R}_\nu u^0_{\nu-1} \qquad (5.3\text{-}28)$$

for any initial state $x := x(0)$. For every $u_{[0,\nu)}$ *satisfying (28)* we write

$$J\left(x, u_{[0,\nu)} \mid x(\nu) = O_X\right) := \sum_{k=0}^{\nu-1} \ell\left(y(k), u(k)\right) \qquad (5.3\text{-}29)$$

$$\ell\left(y(k), u(k)\right) := \|y(k)\|^2_{\psi_y} + \|u(k)\|^2_{\psi_u}$$

with $\psi_y = \psi'_y > 0$ and $\psi_u = \psi'_u > 0$. Defining

$$V_\nu\left(x \mid x(\nu) = O_X\right) := \min_{u_{[0,\nu)}} J\left(x, u_{[0,\nu)} \mid x(\nu) = O_X\right) \qquad (5.3\text{-}30)$$

we show next that, irrespective of the possible singularity of Φ, we have

$$V_\nu\left(x \mid x(\nu) = O_X\right) = x'P(\nu)x \qquad (5.3\text{-}31a)$$

$$P(\nu) = P'(\nu) \ge 0 \qquad (5.3\text{-}31b)$$

Hereafter, we show how to compute $P(\nu)$ without resorting to (22). To this end, set

$$\begin{aligned} y(k) &= Hx(k) \\ &= w_1 u(k-1) + \cdots + w_k u(0) + S_k x \end{aligned}$$

where

$$w_k := H\Phi^{k-1}G$$

and

$$S_k := H\Phi^k$$

Hence,

$$y_{\nu-1}^1 = W u_{\nu-1}^0 + \Gamma x$$

where

$$W := \begin{bmatrix} w_1 & & & & \\ w_2 & w_1 & & 0 & \\ \vdots & \vdots & \ddots & & 0 \\ w_{\nu-1} & w_{\nu-2} & \cdots & w_1 & \end{bmatrix} \qquad (5.3\text{-}32)$$

and

$$\Gamma := \begin{bmatrix} S_1' & S_2' & \cdots & S_{\nu-1}' \end{bmatrix}' \qquad (5.3\text{-}33)$$

Therefore

$$
\begin{aligned}
\sum_{k=0}^{\nu-1} \ell\left(y(k), u(k)\right) &= \|y(0)\|_{\psi_y}^2 + \|y_{\nu-1}^1\|_{L_y}^2 + \|u_{\nu-1}^0\|_{L_u}^2 \\
&= \|y(0)\|_{\psi_y}^2 + \|W u_{\nu-1}^0 + \Gamma x\|_{L_y}^2 + \|u_{\nu-1}^0\|_{L_u}^2 \qquad (5.3\text{-}34)
\end{aligned}
$$

where

$$
\begin{aligned}
L_y &:= \text{block--diag} \{\psi_y\}, \quad ((\nu-1) \text{ times}) \\
L_u &:= \text{block--diag} \{\psi_u\}, \quad (\nu \text{ times})
\end{aligned}
$$

In order to minimize (29) under the constraint (28), we form the Lagrangian function [Lue69]

$$\mathcal{L} := \sum_{k=0}^{\nu-1} \ell\left(y(k), u(k)\right) + \left[\tilde{R}_\nu u_{\nu-1}^0 + \Phi^\nu x\right]' \lambda$$

where $\lambda \in \mathbb{R}^n$ is a vector of Lagrangian multipliers. The gradient of $\mathcal{L}$ w.r.t. $u_{\nu-1}^0$ vanishes for

$$\hat{u}_{\nu-1}^0 = -M^{-1} \left[W' L_y \Gamma x + \frac{1}{2} \tilde{R}_\nu' \lambda\right] \qquad (5.3\text{-}35)$$

$$M := L_u + W' L_y W$$

Premultiplying both sides of (35) by $\tilde{R}_\nu$, we get

$$
\begin{aligned}
\tilde{R}_\nu \hat{u}_{\nu-1}^0 &= -\tilde{R}_\nu M^{-1} \left[W' L_y \Gamma x + \frac{1}{2} \tilde{R}_\nu' \lambda\right] \\
&= -\Phi^\nu x \qquad [(28)] \qquad (5.3\text{-}36)
\end{aligned}
$$

Using (36) into (35), we find

$$\hat{u}_{\nu-1}^0 = -M^{-1} \left[\left(I - Q\tilde{R}_\nu M^{-1}\right) W' L_y \Gamma + Q\Phi^\nu\right] x \qquad (5.3\text{-}37)$$

$$Q := \tilde{R}'_\nu \left(\tilde{R}_\nu M^{-1} \tilde{R}'_\nu \right)^{-1}$$

Thus, $P(\nu)$ in (31) can be found by using (37) into (34). Hence, (31) is established without requiring nonsingularity of Φ.

Consider next the zero-terminal-state regulation over the interval $[0, \nu]$. Taking into account (31a), we have

$$V_{\nu+1} \left(x \mid x(\nu+1) = O_X \right) \tag{5.3-38}$$

$$= \min_{u(0)} \left\{ \|y(0)\|^2_{\psi_y} + \|u(0)\|^2_{\psi_u} + V \left(x(1) \mid x(\nu) = O_X \right) \right\}$$

$$= \min_{u(0)} \left\{ \|y(0)\|^2_{\psi_y} + \|u(0)\|^2_{\psi_u} + x'(1) P(\nu) x(1) \right\} \tag{5.3-39}$$

where

$$x(1) = \Phi x + G u(0)$$

Equation (38) is the same as a dynamic programming step in a standard LQOR problem and yields (cf. Theorem 2.4-1)

$$u(0) = - \left[\psi_u + G' P(\nu) G \right]^{-1} G' P(\nu) \Phi x \tag{5.3-40}$$

$$V_{\nu+1} \left(x \mid x(\nu+1) = O_X \right) = x' P(\nu+1) x \tag{5.3-41}$$

$$P(\nu+1) = \Phi' P(\nu) \Phi - \Phi' P(\nu) G \left(\psi_u + G' P(\nu) G \right)^{-1} G' P(\nu) \Phi + H' \psi_y H$$

Further,

$$P(\nu+1) \leq P(\nu) \tag{5.3-42}$$

In fact,

$$x' P(\nu+1) x = \min_{u_{[0,\nu]}} J \left(x, u_{[0,\nu]} \mid x(\nu+1) = O_X \right)$$

$$\leq J \left(x, \hat{u}_{[0,\nu)} \otimes O_U \mid x(\nu+1) = O_X \right)$$

$$= J \left(x, \hat{u}_{[0,\nu)} \mid x(\nu) = O_X \right) = x' P(\nu) x$$

where $\hat{u}_{[0,\nu)}$ denotes the input sequence in (37), and $\otimes$ concatenation. By RDE monotonicity (cf. Proposition 2-1), (41) yields

$$P(k+1) \leq P(k), \quad \forall k \geq \nu \tag{5.3-43}$$

Then, by the FARE Argument of Sec. 2 we conclude that the RHR law related to the zero-terminal-state regulation problem (3) to (5) yields an asymptotically stable closed–loop system whenever $N \geq \nu + 1$, irrespective of the possible singularity of Φ.

Theorem 5.3-2.   *Consider the zero-terminal-state regulation (3) to (5) with*
$\psi_y > 0$ *and* $\psi_u > 0$. *Let the plant (1-1) be completely reachable and detectable with reachability index* ν. *Then, irrespective of the possible singularity of* Φ, *the RHR law (7) relative to (3) to (5) exists unique for every* $N \geq \nu$ *and stabilizes (1-1) whenever*

$$N \geq \nu + 1 \tag{5.3-44}$$

Further for single-input completely reachable plants, irrespective of ψ_y *and* ψ_u, *(7) yields state-deadbeat regulation whenever*

$$N = n \tag{5.3-45}$$

Problem 5.3-3   Consider the plant (1-1) and the zero-terminal-state regulation problem (3) to (5). Show that the conclusions of Theorem 2, except possibly for the deadbeat result, are still valid with (8) replaced by $N \geq \max\{\nu_r + 1, \nu_{\bar{r}}\}$, provided that the plant is completely controllable, the completely reachable subsystem of the GK canonical reachability decomposition of (1-1) has reachability index ν_r, and $\nu_{\bar{r}}$ is the smallest nonnegative integer such that $(\Phi_{\bar{r}})^{\nu_{\bar{r}}} = O_{n_{\bar{r}} \times n_{\bar{r}}}$, $\Phi_{\bar{r}}$ being the state–transition matrix of the $n_{\bar{r}}$–dimensional unreachable subsystem.

Method of proof 2.   The second stability proof is based on a monotonicity property similar but yet different from the one in (43). Its interest consists of the fact that it encompasses *nonlinear plants*

$$
\begin{aligned}
x(k+1) &= \varphi\left(x(k), u(k)\right) & \text{(5.3-46a)} \\
y(k) &= \eta\left(x(k)\right) & \text{(5.3-46b)}
\end{aligned}
$$

for which O_X is an equilibrium point

$$
\begin{aligned}
O_X &= \varphi\left(O_X, O_U\right) & \text{(5.3-47a)} \\
O_Y &= \eta\left(O_X\right) & \text{(5.3-47b)}
\end{aligned}
$$

Assume that for such a plant the problem (3) to (6) is uniquely solvable with (3) and (6) replaced respectively by $u(t + N - k) = f(k, x(t))$ and $u(t) = f(N, x(t))$. We shall refer to the latter as the *zero-terminal-state RHR feedback law* with *prediction horizon* N. Under the above assumption consider for a *fixed* N the Bellman function $V(t) := V_N\left(x(t) \mid x(t+N) = O_X\right)$, the R.H.S. being defined as in (30), *along the trajectories of the controlled system*. Let $\hat{u}_{[t,t+N)}$ be the optimal input sequence for the initial state $x(t)$. We see that $\hat{u}_{[t+1,t+N)} \otimes O_U$ drives the plant state from $x(t+1) = \varphi\left(x(t), \hat{u}(t)\right)$ to O_X at time $t+N$ and hence by (47) also at time $t+1+N$. Then we have, by virtue of (47),

$$V(t) - V(t+1) \geq \|y(t)\|_{\psi_y}^2 + \|u(t)\|_{\psi_u}^2 \tag{5.3-48}$$

Therefore $\{V(t)\}_{t=0}^{\infty}$ is a monotonically nonincreasing sequence. Hence, being $V(t)$ nonnegative, as $t \to \infty$ it converges to V_∞, $0 \leq V_\infty \leq V(0)$. Consequently, summing the two sides of (48) from $t = 0$ to $t = \infty$, we get

$$\infty > V(0) - V(\infty) \geq \sum_{t=0}^{\infty} \left[\|y(t)\|_{\psi_y}^2 + \|u(t)\|_{\psi_u}^2 \right] \qquad (5.3\text{-}49)$$

This, in turn, implies for $\psi_y > 0$ and $\psi_u > 0$

$$\lim_{t \to \infty} y(t) = O_Y \qquad \text{and} \qquad \lim_{t \to \infty} u(t) = O_U \qquad (5.3\text{-}50)$$

Theorem 5.3-3. *Suppose that the zero-terminal-state regulation problem (3) to (6) with $\psi_y > 0$, $\psi_u > 0$ and (3) and (6) replaced respectively by $u(t+N-k) = f(k, x(t))$ and $u(t) = f(N, x(t))$ be uniquely solvable for the nonlinear plant (46) to (47). Then, the zero terminal state RHR law yields asymptotically vanishing I/O variables.*

Remark 5.3-1.

1. For linear controllable and detectable plants, Theorem 3 implies at once asymptotic stability of the controlled system.

2. The method of proof of Theorem 3, though simple and general, does not unveil the strict connection between zero-terminal-state RHR and LQR, nor solvability conditions, issues which are instead explicitly on focus in the constructive method of proof of Theorem 2.

3. The method of proof of Theorem 3 can be used to cover also the case of weights $\psi_y(i) > 0$ and $\psi_u(i) > 0$, $i = 0, 1, \cdots, N - 1$, in (3). In such a case the conclusions of Theorem 3 can be readily shown to hold true provided that

$$0 < \psi_y(i) \leq \psi_y(i+1) \qquad \text{and} \qquad 0 < \psi_u(i) \leq \psi_x(i+1)$$

for $i = 0, 1, \cdots, N - 2$.

4. Theorem 3 is relevant for its far–reaching consequences on the stability of the zero-terminal-state RHR applied to nonlinear plants once solvability is insured and complementary system–theoretic properties are added.

5. The reader can verify that the presence of hard constraints on input and state–dependent variables over the semi–infinite horizon $[t, \infty)$ is compatible with the method of proof of Theorem 3. This makes the results of Theorem 3 of paramount importance for practical applications where control problems with constraints are ubiquitous.

Main Points of the Section. For completely reachable and detectable linear plants, zero-terminal-state RHR yields an asymptotically stable closed–loop system whenever the prediction horizon is larger than the plant reachability index. For time-invariant nonlinear plants, the zero terminal state RHR law, whenever it exists, yields asymptotically vanishing I/O variables. The foregoing properties also hold true in the presence of hard constraints on input and state-dependent variables.

5.4 STABILIZING DYNAMIC RHR

We concentrate on SISO plants, initially with unit I/O delay, viz. the intrinsic one. Later, we shall take into account the possible presence of larger I/O delays.

Let us consider the SISO plant described by the difference equation (3-23). We shall rewrite it formally by polynomials as follows (cf. Sect. 3.4):

$$A(d)y(t) = B(d)u(t) \qquad\qquad (5.4\text{-}1)$$

where

$$A(d) = 1 + a_1 d + \cdots + a_{n_a} d^{n_a} \qquad\qquad (5.4\text{-}2)$$
$$B(d) = \qquad b_1 d + \cdots + b_{n_b} d^{n_b} \qquad\qquad (5.4\text{-}3)$$

are coprime polynomials with $a_{n_a} \cdot b_{n_b} \neq 0$ and $b_1 \neq 0$. Consider the state

$$s(t) := \left[\; (y_t^{t-n_a+1})' \quad (u_{t-1}^{t-n_b+1})' \; \right]' \qquad\qquad (5.4\text{-}4)$$

with

$$n_a := \partial A(d) \qquad \text{and} \qquad n_b := \partial B(d) \qquad\qquad (5.4\text{-}5)$$

The following lemma points out some structural properties of the state-space representation of (1) with state vector (4).

Lemma 5.4-1.   *Consider the plant (1). Let the polynomials $A(d)$ and $B(d)$ in (2) and (3) be coprime. Then, the state-space representation of (1) with state vector (4) is completely reachable and reconstructible.*

Proof: Reachability is discussed in Problem 2.4-5. Reconstructibility trivially follows from the state choice (4).

The next lemma specializes the stability results of Theorems 3-1 and 3-2 to SISO plants.

Lemma 5.4-2.   *Let the SISO plant $\Sigma = (\Phi, G, H)$ be completely reachable and detectable. Then the RHR law (3-7) relative to the zero-terminal-state regulation (3-3) to (3-5) stabilizes the plant, whenever the prediction horizon satisfies the following inequality:*

$$N \geq n_x := \dim \Sigma \qquad\qquad (5.4\text{-}6)$$

According to Lemmas 1 and 2, we can directly construct a stabilizing I/O RHR for the SISO plant (1) to (5) as follows. Find an optimal open-loop sequence $\hat{u}_{[0,N)}$ to the plant initialized from the state $s(0)$:

$$\hat{u}(N - k) = F(k)s(0) \; , \qquad k = 1, \cdots, N \qquad\qquad (5.4\text{-}7)$$

minimizing the performance index (3-4) under the zero-terminal-state constraint

$$s(N) = \left[\ \left(y_N^{N-n_a+1}\right)' \quad \left(u_{N-1}^{N-n_b+1}\right)' \ \right]' = O_{n_a+n_b-1} \qquad (5.4\text{-}8)$$

Then, the feedback-regulation law

$$u(t) = F(N)s(t) \qquad (5.4\text{-}9)$$

is the I/O RHR law of interest.

We simplify the previous formulation by referring to the extended state

$$\bar{s}(t) := \left[\ \left(y_t^{t-n+1}\right)' \quad \left(u_{t-1}^{t-n+1}\right)' \ \right]' \qquad (5.4\text{-}10)$$

where

$$n := \max(n_a, n_b) \qquad (5.4\text{-}11)$$

denotes the plant order. In fact, $\bar{s}(t)$ has the same reachability and reconstructibility properties as $s(t)$ (cf. Problem 2.4-5). Thus, referring to $\bar{s}(t)$, taking into account the implication on the summation (3-4) of the terminal-state constraint $\bar{s}(N) = O_{2n-1}$, and setting

$$T := N - n + 1 \qquad (5.4\text{-}12)$$

we can adopt the following formal statement.

Stabilizing I/O RHR (SIORHR). Consider the problem of finding, whenever it exists, an input sequence $\hat{u}_{[0,T)}$

$$\hat{u}(k) = \mathcal{F}(k)s(0) \ , \qquad k = 0, \cdots, T-1 \qquad (5.4\text{-}13)$$

to the SISO plant (1) to (5) minimizing

$$J\left(s(0), u_{[0,T)}\right) = \sum_{k=0}^{T-1} \left[\psi_y y^2(k) + \psi_u u^2(k)\right] \qquad (5.4\text{-}14)$$

under the constraints

$$u_{T+n-2}^T = O_{n-1} \qquad y_{T+n-1}^T = O_n \qquad (5.4\text{-}15)$$

Then, the dynamic feedback-regulation law

$$u(t) = \mathcal{F}(0)s(t) \qquad (5.4\text{-}16)$$

will be referred to as the *stabilizing I/O RHR (SIORHR) law* with prediction horizon T and design knobs (T, ψ_y, ψ_u).

Considering that (6), referred to state vector $\bar{s}(t)$, yields $N \geq 2n-1$ or $T \geq n$, we arrive at the following conclusion.

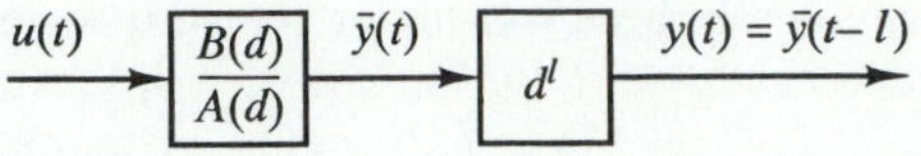

$$\begin{array}{ccc} u(t) & \boxed{\dfrac{B(d)}{A(d)}} & \bar{y}(t) \quad \boxed{d^l} \quad y(t) = \bar{y}(t-l) \end{array}$$

Figure 5.4-1: Plant with I/O transport delay ℓ.

Theorem 5.4-1. *Let polynomials $A(d)$ and $B(d)$ be coprime. Then, provided that*

$$\psi_u > 0 \qquad\qquad (5.4\text{-}17)$$

the I/O RHR law (16) stabilizes the SISO plant (1) whenever

$$T \geq n \qquad\qquad (5.4\text{-}18)$$

Further, irrespective of ψ_u, for

$$T = n \qquad\qquad (5.4\text{-}19)$$

(16) yields a state-deadbeat closed-loop system.

The next step is to extend the foregoing stabilizing properties to plants not only with the intrinsic I/O delay, but possibly exhibiting arbitrary I/O transport delays. In particular, we focus on plants described by difference equations of the form

$$\begin{aligned}
A(d)y(t) &= d^\ell B(d)u(t) & (5.4\text{-}20)\\
&= B(d)u(t-\ell)\\
&= \mathcal{B}(d)u(t)
\end{aligned}$$

where $\mathcal{B}(d) := d^\ell B(d)$; ℓ can be any nonnegative integer and denotes the plant dead time or I/O transport delay; and $A(d)$ and $B(d)$ are polynomials as in (2) and (3). Let us assume that $A(d)$ and $B(d)$ are coprime. Then, for $\ell = 0$, (20) satisfies the conditions in Theorem 1 under which stabilizing I/O RHR exists. In order to deal with a nonzero dead time, it is convenient to represent the plant (20) as in Fig. 1. Here, an intermediate variable $\bar{y}(t) = y(t + \ell)$ is indicated.

The I/O RHR (11) to (16) with $y(k)$ replaced by $\bar{y}(k) = y(k + \ell)$ yields, according to Theorem 1, a stabilizing compensator:

$$u(t) = \mathcal{F}(0)\tilde{s}(t) \qquad\qquad (5.4\text{-}21)$$

$$\begin{aligned}
\tilde{s}(t) &:= \left[\ (\bar{y}_t^{\,t-n_a+1})'\quad (u_{t-1}^{t-n_b+1})'\ \right]'\\
&= \left[\ \left(y_{t+\ell}^{t+\ell-n_a+1}\right)'\quad (u_{t-1}^{t-n_b+1})'\ \right]' & (5.4\text{-}22)
\end{aligned}$$

whenever $T \geq n$. The problem here is that (21) is anticipative. However, by noting that the system in Fig. 1 with output $\bar{y}(t)$ has a state-space description (Φ, G, H) with state $\tilde{s}(t)$, one sees that the anticipative entries of $\tilde{s}(t)$ can be expressed in terms of y^t and u^{t-1} (2) as follows, for $i = 0, 1, \cdots, \ell - 1$:

$$
\begin{aligned}
y(t + \ell - i) &= \bar{y}(t - i) \\
&= H \left[Gu(t - i - 1) + \cdots + \Phi^{\ell - i - 1} Gu(t - \ell) + \Phi^{\ell - i} \tilde{s}(t - \ell) \right]
\end{aligned}
\tag{5.4-23}
$$

Hence, we can conclude that (21) can be uniquely written as a linear combination of the components of a new state:

$$
s(t) := \left[\begin{array}{cc} \left(y_t^{t - n_a + 1} \right)' & \left(u_{t-1}^{t - \ell - n_b + 1} \right)' \end{array} \right]'
\tag{5.4-24}
$$

Therefore, in the presence of a plant dead time ℓ, (13) to (16) are modified as follows.

SIORHR ($\ell \geq 0$). Let $s(t)$ be as in (24). Consider the problem of finding, whenever it exists, an input sequence $\hat{u}_{[0,T)}$

$$
\hat{u}(k) = \mathcal{F}(k) s(0) , \qquad k = 0, \cdots, T - 1
\tag{5.4-25}
$$

to the SISO plant (20) minimizing

$$
J \left(s(0), u_{[0,T)} \right) = \sum_{k=0}^{T-1} \left[\psi_y y^2(k + \ell) + \psi_u u^2(k) \right]
\tag{5.4-26}
$$

under the constraints

$$
u_{T+n-2}^T = O_{n-1} \qquad y_{T+\ell+n-1}^{T+\ell} = O_n
\tag{5.4-27}
$$

Then, the SIORHR law is given by the dynamic feedback compensation

$$
u(t) = \mathcal{F}(0) s(t)
\tag{5.4-28}
$$

We note that (25) to (28) subsume (13) to (16). Nonetheless, according to our considerations preceding (25), the conclusions of Theorem 1 hold true in this more general case.

Theorem 5.4-2. *Consider the SISO plant (20) with dead time ℓ and $A(d)$ and $B(d)$ coprime polynomials. Then, provided that $\psi_u > 0$, the SIORHR law (28) stabilizes (20) whenever*

$$
T \geq n
\tag{5.4-29}
$$

n being the plant order. Further, irrespective of $\psi_u \geq 0$, for

$$
T = n
\tag{5.4-30}
$$

(28) yields a state-deadbeat closed-loop system.

The final comment here is that, as can be seen from (25) to (28), SIORHR approaches, as T increases, the steady-state LQOR.

Main Points of the Section. Zero-terminal-state RHR can be adapted to regulate sampled-data plants with I/O transport delays. The resulting dynamic compensator, referred to as SIORHR (stabilizing I/O receding-horizon regulator), yields stable closed-loop systems under sharp conditions. By varying its prediction horizon length, SIORHR yields different regulation laws, ranging from state-deadbeat to steady-state LQOR.

5.5   SIORHR COMPUTATIONS

We now proceed to find an algorithm for computing (4-28), which, though not the most convenient numerically, is helpful for uncovering the relationship between SIORHR and other RHR laws to be discussed next. Consider again the state (4-24):

$$s(t) := \left[\ \left(y_t^{t-n_a+1}\right)' \ \ \left(u_{t-1}^{t-\ell-n_b+1}\right)' \ \right]' \in \mathbb{R}^{n_s} \tag{5.5-1}$$

for the plant (4-20). It can be updated in the usual way:

$$\left.\begin{array}{rcl} s(t+1) & = & \Phi s(t) + G u(t) \\ y(t) & = & H s(t) \end{array}\right\} \tag{5.5-2}$$

for a suitable matrix triplet (Φ, G, H). Then, similarly to (1-7),

$$s(k) = \Phi^k s(0) + \tilde{R}_k u_{k-1}^0 \tag{5.5-3}$$

$$\tilde{R}_k := \left[\ \Phi^{k-1} G \ \ \cdots \ \ \Phi G \ \ G \ \right] \tag{5.5-4}$$

$$\begin{array}{rcl} y(k) & = & H s(k) \\ & = & w_1 u(k-1) + \cdots + w_k u(0) + S_k s(0) \end{array} \tag{5.5-5}$$

where

$$w_k := H \Phi^{k-1} G \tag{5.5-6}$$

and

$$S_k := H \Phi^k \tag{5.5-7}$$

Because $w_1 = w_2 = \cdots = w_\ell = 0$, we can also write

$$y_{\ell+T+n-1}^{\ell+1} = W u_{T+n-2}^0 + \Gamma s(0) \tag{5.5-8}$$

with n as in (4-18), W the lower triangular Toeplitz matrix having on its first column the first nonzero $T+n-1$ samples of the impulse response associated with the plant transfer function $d^\ell B(d)/A(d)$

$$W := \begin{bmatrix} w_{\ell+1} & & & \\ w_{\ell+2} & w_{\ell+1} & & 0 \\ \vdots & \vdots & \ddots & \\ w_{\ell+T+n-1} & w_{\ell+T+n-2} & \cdots & w_{\ell+1} \end{bmatrix} \qquad (5.5\text{-}9)$$

and

$$\Gamma := \begin{bmatrix} S'_{\ell+1} & S'_{\ell+2} & \cdots & S'_{\ell+T+n-1} \end{bmatrix}' \qquad (5.5\text{-}10)$$

Let W and Γ be partitioned as follows:

$$W := \begin{array}{c} T-1\{ \\ n\{ \end{array} \begin{bmatrix} W_1 & 0 \\ \\ L & W_2 \end{bmatrix} \qquad (5.5\text{-}11)$$
$$\underbrace{}_{T} \underbrace{}_{n-1}$$

$$\Gamma := \begin{array}{c} T-1\{ \\ n\{ \end{array} \begin{bmatrix} \Gamma_1 \\ \\ \Gamma_2 \end{bmatrix} \qquad (5.5\text{-}12)$$

Then, (8) can be rewritten as follows:

$$y^{\ell+1}_{\ell+T-1} = W_1 u^0_{T-1} + \Gamma_1 s(0) \qquad (5.5\text{-}13)$$

$$y^{\ell+T}_{\ell+T+n-1} = L u^0_{T-1} + W_2 u^T_{T+n-2} + \Gamma_2 s(0) \qquad (5.5\text{-}14)$$

and (4-26) becomes

$$\begin{aligned} J\left(s(0), u^0_{T-1}\right) &= \psi_y \|y^{\ell+1}_{\ell+T-1}\|^2 + \psi_u \|u^0_{T-1}\|^2 \qquad (5.5\text{-}15) \\ &= \psi_y \|W_1 u^0_{T-1} + \Gamma_1 s(0)\|^2 + \psi_u \|u^0_{T-1}\|^2 \end{aligned}$$

Further, the constraints (4-27) become

$$u^T_{T+n-2} = O_{n-1} \qquad (5.5\text{-}16)$$

$$L u^0_{T-1} + \Gamma_2 s(0) = O_n \qquad (5.5\text{-}17)$$

In conclusion, problem (4-25) to (4-27) can be solved by finding vector $u^0_{T-1} \in \mathbb{R}^T$ minimizing the Lagrangian function

$$\mathcal{L} := J\left(s(0), u^0_{T-1}\right) + \left[L u^0_{T-1} + \Gamma_2 s(0)\right]' \lambda$$

where $\lambda \in \mathbb{R}^n$ is a vector of Lagrangian multipliers. The gradient of $\mathcal{L}$ w.r.t. u^0_{T-1} vanishes for

$$u^0_{T-1} = -M^{-1}\left(\psi_y W'_1 \Gamma_1 s(0) + \frac{1}{2} L' \lambda\right) \qquad (5.5\text{-}18)$$

$$M := \psi_u I_T + \psi_y W_1' W_1 \tag{5.5-19}$$

Premultiplying both sides by L, we get

$$
\begin{aligned}
L u_{T-1}^0 &= -\psi_y L M^{-1} W_1' \Gamma_1 s(0) - \frac{1}{2} L M^{-1} L' \lambda \\
&= -\Gamma_2 s(0) \hspace{3cm} [(17)] \tag{5.5-20}
\end{aligned}
$$

Using (20) into (18), we find

$$u_{T-1}^0 = -M^{-1} \left[\psi_y \left(I_T - Q L M^{-1} \right) W_1' \Gamma_1 + Q \Gamma_2 \right] s(0) \tag{5.5-21}$$

$$Q := L' \left(L M^{-1} L' \right)^{-1} \tag{5.5-22}$$

We note that Q exists provided that matrix L in (11) has rank n. Now

$$
L = \begin{bmatrix}
w_{\ell+T} & w_{\ell+T-1} & \cdots & w_{\ell+2} & w_{\ell+1} \\
w_{\ell+T+1} & w_{\ell+T} & \cdots & w_{\ell+3} & w_{\ell+2} \\
\vdots & \vdots & \ddots & \vdots & \vdots \\
w_{\ell+T+n-1} & w_{\ell+T+n-2} & \cdots & w_{\ell+n+1} & w_{\ell+n}
\end{bmatrix} \tag{5.5-23}
$$

is obtained by reversing the column order of the Hankel matrix [Kai80] $H_{n,T}$ associated with $d^\ell B(d)/A(d)$. Then, under the same assumptions as in Theorem 4-2, rank $L = n$, provided that $T \geq n$.

Theorem 5.5-1.   *Under the same assumptions as in Theorem 4-2, the open-loop solution of (4-25) to (4-27) for $T \geq n$ is given by (21). Then, the I/O RHR law*

$$u(t) = -\underline{e}_1' M^{-1} \left[\psi_y \left(I_T - Q L M^{-1} \right) W_1' \Gamma_1 + Q \Gamma_2 \right] s(t) \tag{5.5-24}$$

stabilizes the plant whenever

$$T \geq n \tag{5.5-25}$$

Further, for $T = n$, (24) becomes

$$u(t) = -\underline{e}_1' L^{-1} \Gamma_2 s(t) \tag{5.5-26}$$

and yields a state-deadbeat closed-loop system.

Equation (24) is a formula for computing SIORHR. There are, however, various ways to carry out the involved computations. Two alternatives are considered now.

One basic ingredient of (24) is (5). In fact, the plant parameters in (5), viz., w_i, $i = 1, 2, \cdots, k$, and S_k, are needed to compute $\Gamma_1 s(t)$ and $\Gamma_2 s(t)$ in (24). In the present deterministic context, we shall refer to (5) as the *k–step ahead output evaluation* formula in that it yields $y(k)$ in terms of the state $s(0)$, once the exogenous sequence $u_{[0,k)}$ is specified. The use of a special name for (5) is justified because, as will be seen in the remaining part of this chapter, (5) plays a central role in other dynamic RHR problems. We describe two ways to compute (5).

The first, referred to as the *long-division procedure*, is based on solving w.r.t. the polynomial pair $(Q_k(d), G_k(d))$ the Diophantine equation

$$\left.\begin{array}{c} 1 = A(d)Q_k(d) + d^k G_k(d) \\ \partial Q_k(d) \le k - 1 \end{array}\right\} \qquad (5.5\text{-}27)$$

Because $A(d)$ and d^k are coprime, there is a unique solution of (27) (Appendix C). Multiplying both sides of (4-20) by $Q_k(d)$, we get

$$y(k) = Q_k(d)\mathcal{B}(d)u(k) + G_k(d)y(0) \qquad (5.5\text{-}28)$$

It is immediate to relate the polynomials in (28) with parameters w_i and S_k in (5), for example, $w_1, \cdots, w_k$ are given by the k leading coefficients of $Q_k(d)B(d)$.

Note that there exist recursive formulas for computing $(Q_{k+1}(d), G_{k+1}(d))$ given $(Q_k(d), G_k(d))$. To see this, it is enough to consider that (27) represents the kth stage of the long division of 1 by $A(d)$, with $Q_k(d) = q_0 + q_1 d + \cdots + q_{k-1}d^{k-1}$, the quotient, and $d^k G_k(d)$, the remainder. Thus, going to the next stage of the long division, we have

$$Q_{k+1}(d) = Q_k(d) + q_k d^k \qquad (5.5\text{-}29)$$

and $d^{k+1}G_{k+1}(d) = d^k G_k(d) - q_k d^k A(d)$ or

$$dG_{k+1}(d) = G_k(d) - q_k A(d) \qquad (5.5\text{-}30)$$

Setting $d = 0$ in (30), we find

$$q_k = G_k(0) \qquad (5.5\text{-}31)$$

Then, $Q_{k+1}(d)$ and $G_{k+1}(d)$ can be recursively computed via (29) to (31) initialized as follows:

$$Q_1(d) = 1 \qquad ; \qquad dG_1(d) = 1 - A(d)$$

The other way to compute (5) will be referred to as the *plant response procedure*. In order to describe it, it is convenient to denote by $\mathcal{S}\left(k, s(0), u_{k-1}^0\right)$ the plant output response at time k, from state $s(0)$ at time 0, to inputs u_{k-1}^0. Then, we have

$$w_k = \mathcal{S}\left(k, s(0) = 0, u_{k-1}^0 = \underline{e}_1\right) \qquad (5.5\text{-}32)$$

where $\underline{e}_1$ is the first vector of the natural basis of $\mathbb{R}^k$. Further, if $S_{k,r}$ denotes the rth component of row vector S_k, we have

$$S_{k,r} = \mathcal{S}\left(k, \underline{e}_r \in \mathbb{R}^{n_s}, O_k\right) \qquad (5.5\text{-}33)$$

Here $\underline{e}_r$ denotes the rth vector of the natural basis of $\mathbb{R}^{n_s}$.

We finally mention another possibility closely related to the plant response procedure. We see that the last additive term of (5) can be obtained as follows

$$\begin{aligned} p(k) \quad &:= \quad S_k s(0) \qquad\qquad (5.5\text{-}34) \\ &= \quad \mathcal{S}\left(k, s(0), O_k\right) \end{aligned}$$

Then, (24) can be computed via algebraic operations on data obtained by running the plant model as in (32) and (34) and setting

$$\Gamma_1 s(0) = \left[\ \ p(\ell+1) \quad \cdots \quad p(\ell+T-1)\ \ \right]' \qquad (5.5\text{-}35)$$

and

$$\Gamma_2 s(0) = \left[\ \ p(\ell+T) \quad \cdots \quad p(\ell+T+n-1)\ \ \right]' \qquad (5.5\text{-}36)$$

Although in the present deterministic context, when the plant is preassigned and fixed, the use of (32) and (33) appears the most convenient in that it allows us to compute the feedback in (24) once for all; when the plant is time-varying, for example, in program control or adaptive regulation, the use of (32), (34) to (36) can become preferable. In fact, the latter procedure circumvents an explicit feedback computation by providing us directly with the control variable $u(t)$ in (24).

Main Points of the Section. The SIORHR law is given by formula (24). This can be used provided that the plant parameters in the k-step-ahead output evaluation formula (5) are computed. To this end, two alternative procedures are described: the long-division procedure (27) to (31) and the plant response procedure (33) or (32) and (34) to (36).

Problem 5.5-1 Consider the problem of finding an input sequence $\hat{u}_{[0,T+n)}$, $\hat{u}(k) = \mathcal{F}(k)s(0)$, $k = 0, \cdots, T+n-1$, to the plant (4-20) minimizing

$$J\left(s(0), u_{[0,T+n)}\right) = \sum_{k=0}^{T-1} \left[\psi_y y^2(k+\ell) + \psi_u u^2(k)\right] + \lambda \sum_{k=T}^{T+n-1} \left[\psi_y y^2(k+\ell) + \psi_u u^2(k)\right]$$

for $\psi_y \geq 0$, $\psi_u > 0$, and $\lambda > 0$. Show that the limit for $\lambda \to \infty$ of the solution of such a problem coincides with (21), provided that the latter is well-defined.

5.6   GENERALIZED PREDICTIVE REGULATION

Generalized predictive regulation (GPR) is a form of dynamic compensation stemming from a RHR problem similar to the one of (4-13) to (4-16).

GPR. Define $s(t)$ as in (4-14). Consider the problem of finding, whenever it exists, the input sequence $\hat{u}_{[0,N_u)}$:

$$\hat{u}(k) = \mathcal{F}(k)s(0) \qquad k = 0, \cdots, N_u - 1 \qquad (5.6\text{-}1)$$

to the SISO plant (4-1) to (4-3) minimizing for $N_u \leq N_2$:

$$J_{\text{GPR}} = \sum_{k=N_1}^{N_2} y^2(k) + \psi_u \sum_{k=0}^{N_u-1} u^2(k) \qquad (5.6\text{-}2)$$

under the constraints

$$u_{N_2-1}^{N_u} = O_{N_2-N_u} \tag{5.6-3}$$

Then, the dynamic feedback regulation law

$$u(t) = \mathcal{F}(0)s(t) \tag{5.6-4}$$

is referred to as the *GPR law* with *prediction horizon* N_2 and *design knobs* (N_1, N_2, N_u, ψ_u).

If $N_1 = 0$, we have

$$J_{\text{GPR}} = \sum_{k=0}^{N_u-1} \left[y^2(k) + \psi_u u^2(k)\right] + \sum_{k=N_u}^{N_2} y^2(k) \tag{5.6-5}$$

Then, under (5), as $N_u \to \infty$, GPR approaches the steady-state LQOR. The stabilizing properties of the latter are then inherited by GPR as $N_u \to \infty$. This feature does not reassure us on the stability of the GPR-compensated system for small or moderate values of N_u. In fact, because the GPR computational burden greatly increases with N_u, it is mandatory, particularly in adaptive applications, to use small or moderate values of N_u. On the other hand, we already know (Sec. 2.5) that the RDE is slowly converging to its steady-state solution as its regulation horizon increases. For this reason, the claim that GPR stabilizes the plant for a "large enough" N_u, even if true, is of modest practical interest because the values of N_u for which GPR is stabilizing may be too large. In addition, such values are in general not so easily predictable, their determination typically requiring computer analysis.

A design knob choice under which, if $A(d)$ and $B(d)$ are coprime, GPR can be proved [CM89] to be stabilizing is the following:

$$N_1 = N_u \geq n , \qquad N_2 - N_1 \geq n - 1 , \qquad \psi_u \downarrow 0 \tag{5.6-6}$$

Even if limiting arguments are avoided, the difficulty here is that one has to take a "vanishingly small" value for ψ_u. For a given plant, it is not immediate to establish how small such a value must be so as to guarantee stability to the closed-loop system. In this respect, for the second-order plant:

$$A(d) = 1 - 0.7d \qquad B(d) = 0.9d - 0.6d^2$$

and GPR design knob settings:

$$N_1 = N_u = 2 \qquad \text{and} \qquad N_2 = 4$$

a computer analysis, reported in [BGW90], shows that, in order for closed-loop stability to be achieved, it is required to take $\psi_u \leq 5.2 \times 10^{-13}$.

To this inconvenience, we may add that the design knob selection (6) makes GPR close to a dynamic version of **case (a)** of Theorem 3-1, which, as remarked, is of little interest in applications.

In order to uncover GPR stabilization properties for finite prediction horizons, it is convenient to look for conditions under which GPR and SIORHR coincide. By comparing (3) with (4-15), we see that if

$$N_u = T \qquad \text{and} \qquad N_2 = T + n - 1 \tag{5.6-7}$$

GPR and SIORHR input constraints coincide. Further, if

$$N_1 = T \qquad \text{and} \qquad \psi_u = 0 \tag{5.6-8}$$

(2) becomes

$$J_{\text{GPR}} = \sum_{k=T}^{T+n-1} y^2(k) \tag{5.6-9}$$

and the constraints (3) are now

$$u_{T+n-2}^T = O_{n-1} \tag{5.6-10}$$

We already know from Theorem 4-1 that for $T = n$, there exists a unique input sequence $u_{[0,n)}$ making (9) equal to zero. Then, the GPR law related to (7) and (8) yields for $T = n$ a state-deadbeat closed-loop system, provided that polynomials $A(d)$ and $B(d)$ in the plant transfer function are coprime. We also see that the previously state-deadbeat property is retained if, keeping the other design knobs unchanged, N_2 exceeds $2n - 1$:

$$N_2 \geq 2n - 1 \tag{5.6-11}$$

Other GPR stabilizing results for open-loop stable plants are reported in [SB90].

Referring to the state-space representation with state vector $s(t)$, the GPR law can be computed via the related forward RDE, taking into account that constraints (3) amount to setting $\psi_u = \infty$ for the first $N_2 - N_u$ iterations initialized from $P(0) = H'H$. However, it is more convenient to follow an approach similar to the one adapted for SIORHR in Sec. 5. We first explicitly consider the presence of a plant I/O transport delay ℓ as in (4-20) by restating for such a case the GPR formulation.

GPR ($\ell \geq 0$). Let $s(t)$ be as follows:

$$s(t) := \left[\ \left(y_t^{t-n_a+1}\right)' \quad \left(u_{t-1}^{t-\ell-n_b+1}\right)' \ \right]' \tag{5.6-12}$$

Find, whenever it exists, an input sequence $\hat{u}_{[0,N_u)}$:

$$\hat{u}(k) = \mathcal{F}(k)s(0) \tag{5.6-13}$$

to the plant (4-20) minimizing for $N_u \leq N_2$

$$J_{\text{GPR}} = \sum_{k=N_1}^{N_2} y^2(k+\ell) + \psi_u \sum_{k=0}^{N_u-1} u^2(k) \qquad (5.6\text{-}14)$$

under the constraints

$$u_{N_2-1}^{N_u} = O_{N_2-N_u} \qquad (5.6\text{-}15)$$

Then, the dynamic feedback-regulation law

$$u(t) = \mathcal{F}(0)s(t) \qquad (5.6\text{-}16)$$

is referred to as the *GPR law* with *prediction horizon* N_2 and *design knobs* (N_1, N_2, N_u, ψ_u).

Let W be the lower triangular Toeplitz matrix defined as in (5-9) but taken here to have dimension $N_2 \times N_2$:

$$W := \begin{bmatrix} w_{\ell+1} & & & \\ w_{\ell+2} & w_{\ell+1} & & 0 \\ \vdots & \vdots & \ddots & \\ w_{\ell+N_2} & w_{\ell+N_2-1} & \cdots & w_{\ell+1} \end{bmatrix} \qquad (5.6\text{-}17)$$

Let Γ be as in (5-10) but taken here to have N_2 rows:

$$\Gamma := \begin{bmatrix} S'_{\ell+1} & S'_{\ell+2} & \cdots & S'_{\ell+N_2} \end{bmatrix}' \qquad (5.6\text{-}18)$$

with S_k as in (5-7). Let W and Γ be partitioned as follows:

$$W := \left. N_1-1 \left\{ \begin{bmatrix} /// & /// \\ W_G & /// \end{bmatrix} \right\} \right. N_2 \qquad (5.6\text{-}19)$$

$$\underbrace{}_{N_u} \underbrace{}_{N_2-N_u}$$

$$\Gamma := \left. N_1-1 \left\{ \begin{bmatrix} /// \\ \Gamma_G \end{bmatrix} \right\} \right. N_2 \qquad (5.6\text{-}20)$$

Then

$$y_{N_2}^1 = W u_{N_2-1}^0 + \Gamma s(0)$$

or

$$\begin{bmatrix} y_{N_1-1}^1 \\ y_{N_2}^{N_1} \end{bmatrix} = \begin{bmatrix} /// & /// \\ W_G & /// \end{bmatrix} \begin{bmatrix} u_{N_u-1}^0 \\ u_{N_2-1}^{N_u} \end{bmatrix} + \begin{bmatrix} /// \\ \Gamma_G \end{bmatrix} s(0)$$

Taking into account (15), we get

$$y_{N_2}^{N_1} = W_G u_{N_u-1}^0 + \Gamma_G s(0) \qquad (5.6\text{-}21)$$

Provided that

$$M_G := W_G' W_G + \psi_u I_{N_u} \qquad (5.6\text{-}22)$$

is nonsingular, it follows that

$$u_{N_u-1}^0 = -M_G^{-1} W_G' \Gamma_G s(0) \qquad (5.6\text{-}23)$$

Once the GPR design knobs are set as in (8) and $T = n$, we find that (23) equals $L^{-1}\Gamma_2 s(t)$, with L and Γ_2 as in (5-26). Therefore, our remarks on the state-deadbeat version of GPR after (7) are definitely confirmed here. The next theorem sums up some of the results so far obtained on GPR.

Theorem 5.6-1.   *Provided that matrix M_G in (22) is nonsingular, the GPR law is given by*

$$u(t) = -\underline{e}_1' M_G^{-1} W_G' \Gamma_G s(t) \qquad (5.6\text{-}24)$$

Under the design knob choice,

$$N_1 = N_u = n \; ; \qquad N_2 \geq 2n - 1 \; ; \qquad \psi_u = 0 \qquad (5.6\text{-}25)$$

and provided that $A(d)$ and $B(d)$ in (4-20) are coprime, (24) yields a state-deadbeat closed-loop system.

Problem 5.6-1   Adapt both the long-division procedure and the plant response procedure of Sec. 5 to compute the GPR law (24).

If we compare (24) with (5-24), we see that, from a computational viewpoint, GPR is less complex than SIORHR. In fact, computation of the latter requires the inversion of matrix $LM^{-1}L'$, which, in turn, is nonsingular if and only if rank $L = n$. On the contrary, GPR requires only the inversion of M_G, which is nonsingular whenever $\psi_u > 0$.

Main Points of the Section.   Like SIORHR, GPR is obtained by solving an input/output RHR problem. However, unlike SIORHR, which involves both input and output constraints, in GPR, only input constraints are considered. Consequently, GPR has a lower computational complexity than SIORHR. The price paid for it is that only few sharp results on GPR stabilizing properties are available.

Problem 5.6-2 (Connection between zero terminal state RHR and GPR for FIR plants)   Consider the SISO finite-impulse-response (FIR) plant

$$y(t) = w_1 u(t-1) + \cdots + w_n u(t-n) \qquad (w_n \neq 0)$$

with state vector

$$x(t) := \left[u_{t-1}^{t-n} \right]$$

and the following zero-terminal-state regulation problem. Find, whenever it exists, an input sequence $\hat{u}_{[0,N)}$, $\hat{u}(k) = F(k)x(0), k = 0, 1, \cdots, N-1$, to this plant minimizing

$$J\left(x(0), u_{[0,N)}\right) = \sum_{k=0}^{N-1} \left[y^2(k) + \rho u^2(k)\right] \qquad (\rho > 0)$$

subject to the constraint $x(N) = O_n$. Show that (a) this problem is solvable for $N \geq n$ and the related RHR law $u(t) = F(0)x(t)$ stabilizes the plant; and (b) the GPR problem with $N_1 = 0$, $N_2 = N - 1$, $N_u = N - n$, and $\psi_u = \rho$ is equivalent to the foregoing zero-terminal-state regulation problem.

5.7 RECEDING-HORIZON ITERATIONS

5.7.1 Modified Kleinman Iterations

In view of their control-theoretic interpretation, as depicted in Fig. 2.5-1, Kleinman iterations are closely related to RHR. In fact, in Kleinman iterations, the feedback updating from F_k to F_{k+1} is based on the solution of the RHR problem (1-1) to (1-6) with prediction horizon of infinite length:

$$J\left(x, u(\cdot)\right) = \sum_{j=0}^{\infty} \left[\|x(j)\|_{\psi_x}^2 + \|u(j)\|_{\psi_u}^2\right] \qquad (5.7\text{-}1)$$

and constraints (1-4) given as follows:

$$u(j) = F_k x(j), \qquad j = 1, 2, \cdots \qquad (5.7\text{-}2)$$

As already remarked, Kleinman iterations have a fast rate of convergence in the vicinity of the steady-state LQOR feedback but are affected by one major defect: They must be imperatively initialized from a stabilizing feedback-gain matrix. Further, they have two other negative features: Their speed of convergence may slow down if F_k is far from the steady-state LQOR feedback F_{LQ}; and at each iteration step, the computationally cumbersome Lyapunov equation (2.5-3) has to be solved. On the contrary, Riccati iterations do not suffer from such difficulties, but their speed of convergence is not generally fast. It is worth trying to combine Kleinman and Riccati-like iterations so as to possibly obtain iterations not requiring a stabilizing initial feedback and having a more uniform rate of convergence. We now discuss one such modification.

Rewrite (1) and (2) in a single equation:

$$J\left(x, u, F_k\right) = \|x\|_{\psi_x}^2 + \|u\|_{\psi_u}^2 + \sum_{j=1}^{\infty} \|x(j)\|_{\psi_x(k)}^2 \qquad (5.7\text{-}3)$$

where $x = x(0)$, $u = u(0)$, and

$$\psi_x(k) := \psi_x + F_k' \psi_u F_k \qquad (5.7\text{-}4)$$

The last term on the R.H.S. of (3) can be reorganized as follows:

$$\sum_{j=1}^{T} \|x(j)\|_{\psi_x(k)}^2 + \sum_{j=T+1}^{\infty} \|x(j)\|_{\psi_x(k)}^2 = \qquad (5.7\text{-}5)$$

$$\|x(1)\|_{\mathcal{L}_T(k)}^2 + \|x(T+1)\|_{\mathcal{L}(k)}^2$$

where $\mathcal{L}(k)$ satisfies the Lyapunov equation (cf. (2.5-3)):

$$\mathcal{L}(k) = \Phi_k' \mathcal{L}(k)\Phi_k + \psi_x(k) \qquad (5.7\text{-}6)$$

and $\mathcal{L}_T(k)$ is given by

$$\mathcal{L}_T(k) \quad := \quad \sum_{r=0}^{T-1} (\Phi_k')^r \psi_x(k)\Phi_k^r$$

$$= \quad \Phi_k' \mathcal{L}_T(k)\Phi_k + \psi_x(k) - (\Phi_k')^T \psi_x(k)\Phi_k^T \qquad (5.7\text{-}7)$$

In both (7) and (8), Φ_k denotes the closed-loop state-transition matrix

$$\Phi_k := \Phi + GF_k \qquad (5.7\text{-}8)$$

Problem 5.7-1 Prove the identity in the second line of (7).

The modification we consider consists of replacing the second additive term on the R.H.S. of (5), $\|x(T+1)\|_{\mathcal{L}(k)}^2$, by $\|x(T+1)\|_{P(k)}^2$, with $P(k)$ given by the following pseudo-Riccati iterative equation:

$$P(k) = \Phi_k' P(k-1)\Phi_k + \psi_x(k) \qquad (5.7\text{-}9)$$

initialized from an arbitrary $P(0) = P'(0) \geq 0$. In conclusion, F_{k+1} is obtained by minimizing w.r.t. u, instead of (3), the following modified cost:

$$\|x\|_{\psi_x}^2 + \|u\|_{\psi_u}^2 + \|\Phi x + Gu\|_{\Pi(k)}^2 \qquad (5.7\text{-}10)$$

with

$$\Pi(k) = \mathcal{L}_T(k) + (\Phi_k')^T P(k)\Phi_k^T \qquad (5.7\text{-}11)$$

Problem 5.7-2 Show that the symmetric nonnegative definite matrix $\Pi(k)$ in (11) satisfies the updating identity:

$$\Pi(k) \quad = \quad \Phi_k' \Pi(k-1)\Phi_k + \psi_x(k) \qquad (5.7\text{-}12)$$

$$+\Phi_k' \left[\mathcal{L}_T(k) - \mathcal{L}_T(k-1) + (\Phi_k')^T P(k-1)\Phi_k^T - (\Phi_{k-1}')^T P(k-1)\Phi_{k-1}^T \right] \Phi_k$$

To sum up, we have the following.

Modified Kleinman Iterations (MKI). Given any feedback-gain matrix F_k and any symmetric nonnegative definite matrix $P(k-1)$, compute

$$\mathcal{L}_T(k) = \sum_{r=0}^{T-1} \left(\Phi_k'\right)^r \psi_x(k)\Phi_k^r \tag{5.7-13}$$

Further, update $P(k-1)$ via (9) to find $P(k)$. Then, compute $\Pi(k)$ as in (11). The next feedback-gain matrix is then obtained as follows:

$$F_{k+1} = -\left[\psi_u + G'\Pi(k)G\right]^{-1} G'\Pi(k)\Phi \tag{5.7-14}$$

Should $\Pi(k)$ be updated via the Riccati equation

$$\Pi(k) = \Phi_k'\Pi(k-1)\Phi_k + \psi_x(k) \tag{5.7-15}$$

under stabilizability and detectability of (Φ, G, H) and $\psi_u > 0$, (14) would asymptotically yield the steady-state LQOR feedback gain. Now the true updating equation for $\Pi(k)$ is not given by (15) but, instead, by (12). The latter is the same as (15) except for an additive perturbation term. If F_k converges, this perturbation term converges to zero. Hence, under convergence, the asymptotic behavior of (9), (11), (13), and (14) coincides with that of (14) and (15).

Proposition 5.7-1.   *Let $\psi_u > 0$ and (Φ, G, H) be stabilizable and detectable. Then, the only possible convergence point of MKI is the steady-state solution of the LQR problem for the given plant and performance index (1).*

Though there is no proof of convergence, computer analysis indicates that MKI have excellent convergence properties irrespective of their initialization. In particular, if T is at least comparable with the largest time constant of the LQOR closed-loop system, MKI exhibit a rate of convergence close to that of Kleinman iterations, whenever initialization takes place from a stabilizing feedback. Further, unlike Kleinman iterations, MKI appear to have the advantage of neither requiring the solution of a Lyapunov equation at each step nor being jeopardized by an unstable initial closed-loop system.

Example 5.7-1   Consider the open-loop unstable nonminimum-phase plant $A(d)y(t) = B(d)u(t)$ with

$$A(d) = (1-\alpha d)^2 \qquad \text{and} \qquad B(d) = d - 1.999d^2 \tag{5.7-16}$$

If $\alpha \cong 1.999$, the plant has an almost unstable hidden eigenvalue. If state $x(t)$ used in MKI coincides with that of the plant canonical reachable representation, we have an almost unstable undetectable eigenvalue. If LQOR-regulation laws are computed via Riccati iterations initialized from $P(0) = O_{2\times 2}$, after the second iteration, we get single-step regulation and, hence, an unstable closed-loop system. By increasing the number of iterations, we eventually obtain a stabilizing feedback. Because of the almost undetectable unstable eigenvalue, we expect that the first stabilizing feedback is obtained after quite a

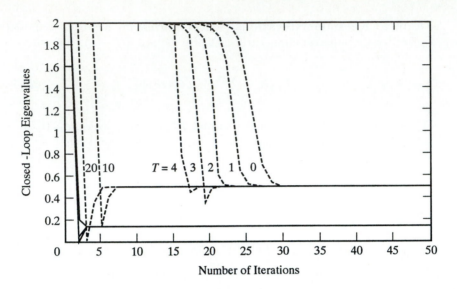

Figure 5.7-1: MKI closed-loop eigenvalues for the plant (16) with $\alpha = 2$.

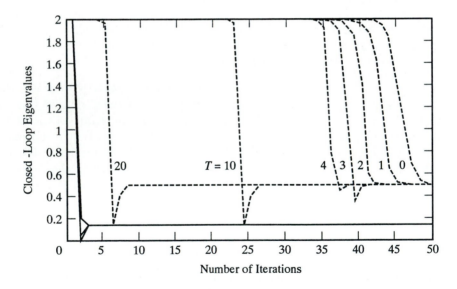

Figure 5.7-2: MKI closed-loop eigenvalues for the plant (16) with $\alpha = 1.999001$.

large number of iterations. Figures 1 and 2 show the closed-loop system eigenvalues when the plant is fed back by F_k computed via MKI. The eigenvalues are given as a function of the number of iterations k with T as a parameter. Both figures refer to $\rho = \psi_u/\psi_y = 0.1$ and MKI initialized from $P(0) = O_{2\times2}$ and $F(0) = O_{1\times2}$. Note that with $F(0) = 0_{1\times2}$ and $T = 0$, MKI and Riccati iterations yield the same feedback. Figures 1 and 2 refer to two different choices for α: $\alpha = 2$ and $\alpha = 1.999001$, respectively. They show that whereas Riccati iterations require at least 25 and 45 iterations, respectively, to yield a stabilizing feedback, MKI with $T = 20$ yield stabilizing feedback gains after 3 and 5 iterations, respectively.

5.7.2   Truncated Cost Iterations

Truncated cost iterations originate from the RHR problem with performance index

$$J(x, u(\cdot)) = \sum_{j=0}^{T} \left[\|x(j)\|_{\psi_x}^2 + \|u(j)\|_{\psi_u}^2 \right] \tag{5.7-17}$$

and constraints

$$u(j) = F_k x(j) , \qquad j = 1, 2, \cdots, T \tag{5.7-18}$$

Equations (17) and (18) can be embodied into a single equation:

$$\begin{aligned} J_T(x, u, F_k) &= \|x\|_{\psi_x}^2 + \|u\|_{\psi_u}^2 + \sum_{j=1}^{T} \|x(j)\|_{\psi_x(k)}^2 \tag{5.7-19} \\ &= \|x\|_{\psi_x}^2 + \|u\|_{\psi_u}^2 + \|x(1)\|_{\mathcal{L}_T(k)}^2 \end{aligned}$$

with $x = x(0)$, $u = u(0)$, $\psi_x(k)$ as in (4), and $\mathcal{L}_T(k)$ as in (13).

Truncated Cost Iterations (TCI). Given any feedback-gain matrix F_k, compute $\mathcal{L}_T(k)$ as in (13). The next feedback-gain matrix is then given by

$$F_{k+1} = -\left[\psi_u + G'\mathcal{L}_T(k)G \right]^{-1} G'\mathcal{L}_T(k)\Phi \tag{5.7-20}$$

As seen in (7), $\mathcal{L}_T(k)$ satisfies the identity

$$\mathcal{L}_T(k) = \Phi_k' \mathcal{L}_T(k)\Phi_k + \psi_x(k) - (\Phi_k')^T \psi_x(k)\Phi_k^T \tag{5.7-21}$$

We shall refer to (21) as the *truncated cost Lyapunov equation* because $\mathcal{L}(k)$ in (6) equals $\sum_{r=0}^{\infty}(\Phi_k')^r \psi_x(k)\Phi_k^r$ provided that Φ_k is a stability matrix, whereas $\mathcal{L}_T(k)$ equals the same sum truncated after the Tth term:

$$\mathcal{L}_T(k) = \sum_{r=0}^{T-1} (\Phi_k')^r \psi_x(k)\Phi_k^r \tag{5.7-22}$$

Table 5.7-1: TCI convergence feedback row vectors for the plant of Example 2, $\rho = 0.15$, zero initial feedback gain, and various prediction horizons T

T	1	2	3	∞
$\#i$	1	5	5	–
F	$\begin{bmatrix} 0.6727 \\ 0.4545 \end{bmatrix}'$	$\begin{bmatrix} 0.6741 \\ 0.4358 \end{bmatrix}'$	$\begin{bmatrix} 0.6752 \\ 0.4465 \end{bmatrix}'$	$F_{\text{LQ}} = \begin{bmatrix} 0.6757 \\ 0.4501 \end{bmatrix}'$

In the same way, (20) and (22) are called *truncated cost iterations (TCI)*. Although (20) and (22) do not appear amenable to convergence analysis and, consequently, sharp stabilizing results are unavailable, we make some considerations on TCI behavior. For the sake of simplicity, we consider a SISO plant $A(d)y(t) = d^\ell B(d)u(t)$ as in (4-20) with $A(d)$ and $B(d)$ coprime. We denote again ψ_u/ψ_y by ρ. If $\rho = 0$, for any $T \geq 1 + \ell$, TCI coincide with cheap control. Hence, irrespective of initial conditions, in a single iteration, TCI yield F_{CC}, the cheap control feedback. If $\rho > 0$ and small two alternative situations take place.

If the plant is minimum-phase and, hence, F_{CC} is a stabilizing feedback, we can expect that, as long as $\rho > 0$ and small so as to make $F_{\text{CC}} \cong F_{\text{LQ}}$, TCI globally converge close to such a feedback. The next example is an excerpt of extensive computer analysis on the subject confirming this conjecture.

Example 5.7-2   Consider the minimum-phase plant $A(d)y(t) = B(d)u(t)$ with

$$A(d) = 1 + d + 0.74d^2 \qquad B(d) = (1 + 0.5d)d$$

Refer the feedback row vectors to the plant canonical reachable representation. Because the plant is of minimum-phase, we expect that, for small ρ and any T, TCI globally converge to $F_{\text{LQ}} \cong F_{\text{CC}} = \begin{bmatrix} 0.74 & 0.5 \end{bmatrix}$. Table 1 shows TCI convergence feedback row vectors for $\rho = 0.15$ and zero initial feedback. The row labeled $\#i$ reports the number of iterations required to achieve convergence. Convergence is claimed at the kth iteration if k is the smallest integer for which $\|F_{k+1} - F_k\|_\infty < 10^{-5}$. Table 2 reports some computer analysis results showing that TCI are insensitive to feedback F_0, even if the latter makes the closed-loop system unstable. In Table 2, $\Phi_0 = \Phi + GF_0$ indicates the initial closed-loop transition matrix. All the results refer to $T = 4$ and $\rho = 10^{-5}$. Because of such a small value of ρ, F_{LQ} is the indistinguishable from F_{CC}.

If the plant is of nonminimum phase, more complications arise. We discuss the situation qualitatively, assuming that $\rho > 0$ is small enough so as to make $F_{\text{SS}} \cong F_{\text{CC}}$, and $F_{\text{LQ}} \cong F_{\text{SCC}}$, where F_{SS} and F_{SCC} denote the single-step regulation

Table 5.7-2: TCI convergence feedback gains for the plant of Example 2, $T = 4$, $\rho = 10^{-5}$, and various initial feedback row vectors F_0. F denotes TCI convergence feedback

F_0	$\begin{bmatrix} 0 & 0 \end{bmatrix}$	$\begin{bmatrix} 0.74 & -0.5 \end{bmatrix}$	$\begin{bmatrix} 0.74 & 5 \end{bmatrix}$	$\begin{bmatrix} 0.74 & -5 \end{bmatrix}$
Φ_0	stable	unstable	unstable	unstable
#i	5	5	6	6
F	F_{LQ}	F_{LQ}	F_{LQ}	F_{LQ}

feedback and the stabilizing cheap control feedback, respectively. For $T = 1 + \ell$, TCI yield F_{SS}. For higher values of T, TCI acquire a second possible converging point close to F_{SCC}. As T increases, the F_{SCC} domain of attraction expands, whereas the one of F_{SS} shrinks. As the size of the latter becomes smaller than the available numerical precision, global convergence close to F_{SCC} is experienced.

Example 5.7-3   Consider again the open-loop unstable nonminimum-phase plant of Example 1 with $\alpha = 2$. TCI are run for $\rho = 0.1$ and various initial feedback gains and prediction horizons T. Figure 3 exhibits TCI convergence closed-loop eigenvalues as a function of T for high- (curve (h)) and low-precision computations (curve (l)). The latter is obtained from the first by rounding off F_k beyond its third significant digit at each iteration. All TCI results of Fig. 3 are obtained for the most unfavorable initialization, viz., by choosing F_0 close to F_{SS}. Note that in the high-precision case, TCI require a prediction horizon larger than or equal to 16 to converge close to F_{LQ}. In the low-precision case, such a horizon is more than halved to 7. Figure 4 exhibits results similar to the ones in Fig. 3 pertaining to high-precision computations and the most favorable initialization $F_0 = F_{\mathrm{LQ}}$. Note that, for such an initial feedback, prediction horizons larger than or equal to 3 allow TCI to converge close to F_{LQ}.

As we see from Example 3, for a given plant, there exists a critical value of T, call it T^*, such that for all $T \geq T^*$, TCI converge close to F_{LQ}. T^* depends upon the precision of computations as well as the open-loop plant zero/pole locations. The reason is that each unstable closed-loop eigenvalue associated to $F_{\mathrm{SS}} \cong F_{\mathrm{CC}}$, viz., approximately each unstable plant zero, must give a significant contribution to J_{T^*}. We express this property by saying that each unstable plant zero must be well detectable within the critical prediction horizon T^*. For a second-order nonminimum-phase plant with $A(d) = 1 + a_1 d + a_2 d^2$ and $B(d) = b_1(1 + \beta d)d$, $|\beta| \geq 1$, detectability of the β mode within T^* increases with $|\beta^{T^*} \det \Theta|$ (cf. Problem 3).

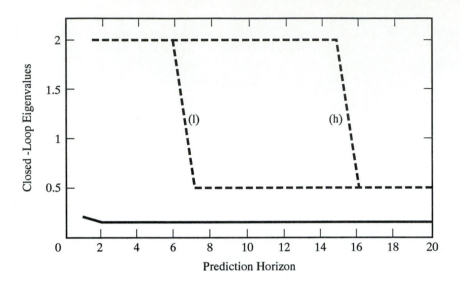

Figure 5.7-3: TCI closed-loop eigenvalues for the plant (16) with $\alpha = 2$ and $\rho = 0.1$, when: high precision (h) and low precision (l) computations are used. TCI are initialized from a feedback close to F_{SS}.

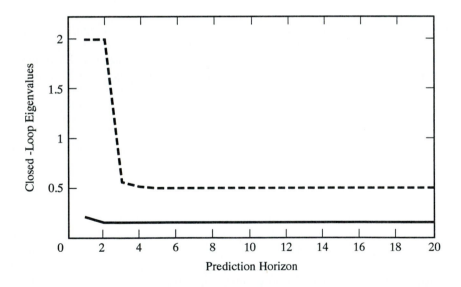

Figure 5.7-4: TCI closed-loop eigenvalues for the plant (16) with $\alpha = 2$, $\rho = 0.1$ and high precision computations. TCI are initialized from F_{LQ}.

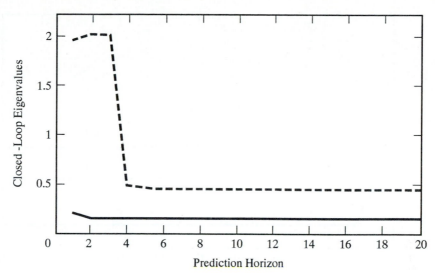

Figure 5.7-5: TCI closed-loop eigenvalues for the plant of Example 4 and $\rho = 0.1$, when high precision computations are used. TCI are initialized from a feedback close to F_{SS}.

Here $\det \Theta = b_1^2 \left(\beta^2 - a_1 \beta + a_2 \right)$ equals the determinant of the observability matrix Θ of the reachable canonical state representation of the plant. Note that an almost pole/zero cancellation yields a small value of $|\det \Theta|$ and, hence, makes detectability only possible for very large values of T. For example, for the plant of Example 3, we find $|\det \Theta| = 10^{-6}$. The closeness of the double pole in 2 to the zero in 1.999 is responsible for such a small value and the corresponding large critical prediction horizon $T^* = 14$. The next example shows that if $|\det \Theta|$ is increased to 2.74 by moving the former double pole to $2 \pm j0.7$, the critical prediction horizon T^* decreases to 4.

Example 5.7-4 Consider the open-loop unstable nonminimum-phase plant $A(d)y(t) = B(d)u(t)$ with

$$A(d) = 1 - 4d + 4.49d^2 \qquad B(d) = d - 1.999d^2$$

As in Example 3, we set $\rho = 0.1$. With reference to feedback gains pertaining to the plant canonical reachable state representation, TCI are computed for various initial feedback gains and prediction horizon T. The results are reported in Fig. 5 under the same conditions of Fig. 3, case (h).

For all the examples considered so far, the eigenvalue of the LQ-regulated system approximately equals 0.5 and, for the resulting values of T^*, it turns out that $|0.5^{T^*}| \ll 0.1$. This implies that truncation after T^* of the performance index does not remove F_{LQ} from the possible converging points of TCI. In more general situations, it is not granted that good detectability within T of open-loop unstable zeros implies that $|\lambda_M^{T-\ell}| \ll 0.1$, λ_M being the eigenvalue of the LQ-regulated

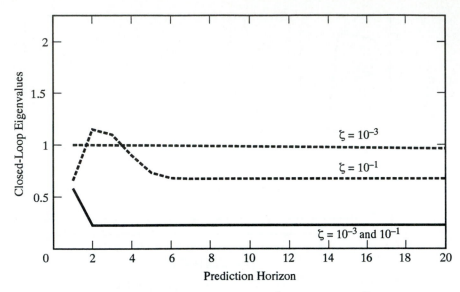

Figure 5.7-6: TCI feedback gains for $\rho = 10^{-1}$ and $\rho = 10^{-3}$, for the plant of Example 5.

system with maximum modulus. Consequently, in order to let TCI converge close to F_{LQ} from all possible initializations, T^* must be large enough so as to make open-loop unstable zeros well detectable and, at the same time, guarantee that $|\lambda_M^{T^*-\ell}| \ll 0.1$.

Example 5.7-5 Consider the open-loop unstable nonminimum-phase plant $A(d)y(t) = B(d)u(t)$ with
$$A(d) = 1 - 4d + 4.49d^2 \qquad B(d) = d - 1.01d^2$$
Here, stabilizing cheap control yields a closed-loop eigenvalue approximately equal to 0.99. For such a plant, Fig. 6 reports convergence TCI feedback gains for $\rho = 10^{-1}$ and $\rho = 10^{-3}$. We see that because the closed-loop eigenvalue λ_M is closer to one in the latter case, TCI require a much larger T to converge close to $F_{\mathrm{LQ}} \cong F_{\mathrm{SCC}}$ for $\rho = 10^{-3}$ than for $\rho = 10^{-1}$.

The foregoing qualitative considerations on TCI pertain to vanishingly small values of ρ. However, they can be extended *mutatis mutandis* to any possible $\rho > 0$ by considering that for $T = 1+\ell$, TCI yield F_{SS}, the single-step regulation feedback, and, as T increases, TCI acquire another possible convergence point close to F_{LQ}, the LQOR feedback. As T increases, the F_{SS} domain of attraction shrinks, whereas the one of F_{LQ} expands. As the size of the first becomes smaller than the available numerical precision, global convergence close to F_{LQ} is experienced.

Problem 5.7-3 Consider the SISO plant (4-20) under cheap control regulation $u(t) = F_{\mathrm{CC}}s(t) + \eta(t)$ with $s(t)$ as in (6-12), where $\eta(t)$ plays the role of an exogenous variable. Let $H_{y\eta}(d)$ and $H_{u\eta}(d)$ be the closed-loop transfer functions from $\eta(t)$ to $y(t)$ and $u(t)$, respectively. Find

$$H_{y\eta}(d) = d^{1+\ell}b_1 \qquad \text{and} \qquad H_{u\eta}(d) = b_1 \frac{dA(d)}{B(d)}$$

Conclude that for $A(d) = 1 + a_1 d + a_2 d^2$ and $B(d) = b_1(1 + \beta d)d$, $|w_{\ell+k}|$, $w_{\ell+k}$ being the sample of the impulse response associated with $H_{u\eta}(d)$, for $k \geq 2$, equals $|\beta^{k-2} \det \Theta / b_1^2|$, with $\det \Theta = b_1^2(\beta^2 - a_1\beta + a_2)$.

5.7.3 TCI Computations

There are better ways to carry out TCI than using (20) and (22). The starting point is to see how to embody constraints (18) in the i-step-ahead output-evaluation formula (5-5). This is rewritten here for a generic MIMO plant (Φ, G, H) with state vector $x(t) \in \mathbb{R}^n$:

$$y(i) = w_1 u(i-1) + \cdots + w_i u(0) + S_i x(0) \qquad (5.7\text{-}23)$$

$$w_i := H\Phi^{i-1}G \qquad S_i := H\Phi^i \qquad (5.7\text{-}24)$$

We recall that with TCI, the problem is to find, given F_k, the next feedback-gain matrix F_{k+1} in accordance with the RHR problem (17) and (18). We consider this problem for $\psi_x = H'\psi_y H$ and $y(i) = Hx(i)$. This is a simple optimization problem in that the only vector to be chosen is the first in the sequence $u_{[0,T)}$, all the remaining vectors being given by (18). We point out that, taking into account (18), (23) can be rewritten as follows:

$$y(i) = \theta_i(k)u(0) + \Gamma_i(k)x(0) \qquad i = 1, 2, \cdots, T \qquad (5.7\text{-}25)$$

Similarly,

$$u(i-1) = \mu_i(k)u(0) + \Lambda_i(k)x(0) \qquad (5.7\text{-}26)$$

with

$$\mu_1(k) = I_m \qquad \text{and} \qquad \Lambda_1(k) = O_{m \times n} \qquad (5.7\text{-}27)$$

Problem 5.7-4   Find recursive formulas in index i to express matrices $\theta_i(k)$, $\Gamma_i(k)$, $\mu_i(k)$, and $\Lambda_i(k)$ in terms of (Φ, G, H) and feedback-gain matrix F_k.

It is worth pointing out the substantial difference between (23) and (25). Unlike (23), where $u_{[1,i)}$ appears explicitly, (25) depends implicitly only on $u_{[1,i)}$ via matrices $\theta_i(k)$ and $\Gamma_i(k)$. These are in fact feedback-dependent. The same holds true for (26). In order to underline this property, we will refer to (25) as the *closed-loop i-step-ahead output-evaluation* formula, and to (26) as the *closed-loop (i − 1)-step-ahead input-evaluation* formula. Further, (25) and (26) will be referred to as output and input, respectively, *many-steps ahead evaluation* formulas whenever no specification of a particular step is desired or needed.

Once (25) and (26) are given, it is a simple matter to find the desired feedback updating formula.

Proposition 5.7-2.   *Let the closed-loop many-steps ahead output- and input-evaluation formulas be given as in (25) and (26), respectively, when feedback F_k is used. Then, the next TCI feedback is given by*

$$F_{k+1} \quad = \quad -\Xi_k^{-1} \sum_{i=1}^{T} [\theta_i'(k)\psi_y \Gamma_i(k) + \mu_i'(k)\psi_u \Lambda_i(k)] \tag{5.7-28}$$

$$\Xi_k \quad := \quad \sum_{i=1}^{T} [\theta_i'(k)\psi_y \theta_i(k) + \mu_i'(k)\psi_u \mu_i(k)] \tag{5.7-29}$$

We note that, by virtue of (27), the $m \times m$ matrix Ξ_k is nonsingular, irrespective of F_k, whenever $\psi_u > 0$.

Problem 5.7-5   Verify the TCI feedback updating formula (28).

A convenient procedure for computing matrices $\theta_i(k)$, $\Gamma_i(k)$, $\mu_i(k)$, and $\Lambda_i(k)$ in (25) and (26) is now discussed. This will be referred to as the *closed-loop system response procedure.* It is the counterpart in TCI regulation of the plant response procedure used in SIORHR and GPR. Let

$$\mathcal{S}_k^y(i, x(0), u(0)) \qquad \text{and} \qquad \mathcal{S}_k^u(i, x(0), u(0)) \tag{5.7-30}$$

denote the plant output, and input response, respectively, at time i to input $u(0)$ at time 0, from state $x(0)$ at time 0 with inputs $u_{[1,i)}$ given by the time-invariant state feedback:

$$u(j) = F_k x(j) , \qquad j = 1, \cdots, i-1$$

Let $\theta_{i,r}(k)$ denote the rth column of $\theta_i(k)$:

$$\theta_i(k) = \begin{bmatrix} \theta_{i,1}(k) & \cdots & \theta_{i,n}(k) \end{bmatrix} \tag{5.7-31}$$

Let us adopt similar notations to denote the columns of the other matrices of interest. Then, the following equalities can be drawn from (25) and (26):

$$\theta_{i,r}(k) \quad = \quad \mathcal{S}_k^y(i, O_n, \underline{e}_r \in \mathbb{R}^m) \tag{5.7-32}$$

$$\mu_{i,r}(k) \quad = \quad \mathcal{S}_k^u(i-1, O_n, \underline{e}_r \in \mathbb{R}^m) \tag{5.7-33}$$

$$\Gamma_{i,r}(k) \quad = \quad \mathcal{S}_k^y(i, \underline{e}_r \in \mathbb{R}^n, O_m) \tag{5.7-34}$$

$$\Lambda_{i,r}(k) \quad = \quad \mathcal{S}_k^u(i-1, \underline{e}_r \in \mathbb{R}^n, O_m) \tag{5.7-35}$$

where $\underline{e}_r$ denotes the rth vector of the natural basis of the space to which it belongs.

Proposition 5.7-3.   *The columns of the matrices, which, for a given feedback F_k, parameterize the closed-loop many-steps-ahead input/output evaluation formulas (25) and (26), can be obtained by running the plant fed back by F_k as indicated in (32) to (35).*

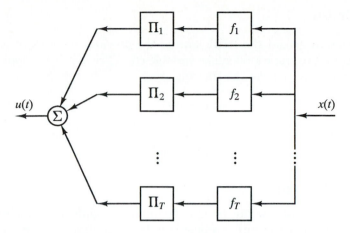

Figure 5.7-7: Realization of the TCI regulation law via a bank of T parallel feedback-gain matrices.

Problem 5.7-6 Show that, using (28) with $F := F_{k+1}$, the TCI feedback can be rewritten as follows:

$$F = \sum_{i=1}^{T} \Pi_i f_i \qquad (5.7\text{-}36)$$

with

$$\sum_{i=1}^{T} \Pi_i = I_m \qquad (5.7\text{-}37)$$

Express Π_i and f_i in terms of θ_i, Γ_i, μ_i, and Λ_i. Note that by (36), the TCI regulation law $u(t) = Fx(t)$ is realizable by a bank of T parallel feedback gains, as shown in Fig. 7.

Main Points of the Section. Modified Kleinman iterations yield at convergence the steady-state LQOR feedback. Though they can have a rate of convergence close to that of standard Kleinman iterations, they are not jeopardized by an initially unstable closed-loop system, and do not require the solution of a Lyapunov equation at each step.

Computer analysis is used to study convergence properties of TCI. This study reveals that there exists a critical prediction horizon T^* such that for all $T \geq T^*$, TCI converge close to F_{LQ}. Such a critical horizon depends on both the open-loop unstable zero/pole locations and the largest time constant of the LQ regulated system.

Under TCI setting, plant outputs and inputs can be expressed by the closed-loop many-steps-ahead evaluation formulas (25) and (26) whose parameter matrices are feedback-dependent. These closed-loop parameter matrices allow one to update the feedback gain by the simple formula (28).

5.8   TRACKING

We study how to extend the RHR laws of the previous sections so as to enable plant output $y(t)$ to track a reference variable $r(t)$. The aim is to modify the basic RH regulators so as to make the tracking error

$$\varepsilon(t) := y(t) - r(t) \qquad (5.8\text{-}1)$$

small or possibly zero as $t \to \infty$, irrespective of the initial conditions. The reader is referred to Sec. 3.5-1 for the necessary preliminaries.

5.8.1   1-DOF Trackers

Every stabilizing RHR law can be modified in a straightforward manner so as to obtain a 1-DOF tracker ensuring, under standard conditions, asymptotic tracking. Of paramount interest in applications is to guarantee asymptotic tracking for constant references as well as asymptotic rejection of constant disturbances. This can be done as follows.

Let the plant be as in (4-20). Next, the model (3.5-3) for a constant reference is given by

$$\left.\begin{array}{rcl} \Delta(d)r(t) & = & 0 \\ \Delta(d) & := & 1 - d \end{array}\right\} \qquad (5.8\text{-}2)$$

Hence, following the procedure that led us to (3.5-4), we find

$$\left.\begin{array}{rcl} A(d)\Delta(d)\varepsilon(t) & = & \mathcal{B}(d)\delta u(t) := B(d)\delta u(t - \ell) \\ \delta u(t) & := & u(t) - u(t - 1) \\ B(1) & \neq & 0 \end{array}\right\} \qquad (5.8\text{-}3)$$

This is the new plant to be used in the RHR law synthesis. In particular, the resulting control law is of the form

$$\delta u(t) = Fs(t) \qquad (5.8\text{-}4)$$

$$s(t) = \left[\ \left(\varepsilon_t^{t-n_a}\right)' \ \ \left(\delta u_{t-1}^{t-\ell-n_b+1}\right)' \ \right]' \qquad (5.8\text{-}5)$$

We note that (4) can be rewritten as a difference equation:

$$R(d)\delta u(t) = -S(d)\varepsilon(t) \qquad (5.8\text{-}6)$$

with $R(d)$ and $S(d)$ coprime polynomials in the backward-shift operator d such that $R(0) = 1$, $\partial R(d) \leq \ell + n_b - 1$, and $\partial S(d) \leq n_a$. Provided that the closed-loop system (3) and (6) is internally stable, viz., $A(d)\Delta(d)R(d) + B(d)S(d)$ is strictly Hurwitz, from Theorem 3.5-1, it follows that, thanks to the presence of the integral action in the loop, both asymptotic tracking of constant references (*set points*) and asymptotic rejection of constant disturbances are achieved.

5.8.2   2-DOF Trackers

The starting point of our discussion on 2-DOF trackers is to begin with a plant model as in (3), where the input increment $\delta u(t)$ appears as the new input variable, so as to have an integral action in the loop.

We first show how to design 2-DOF LQ trackers, and, next, how to modify the various RHR laws so as to obtain 2-DOF trackers equipped with integral action. The reason to start with LQ tracking is that the related results clearly reveal the improvement in tracking performance that can be achieved by independently using the reference variable in the control law. The next example, in which 1-DOF and 2-DOF controllers are compared, shows that such an improvement may turn out to be quite dramatic.

Example 5.8-1   Consider the discrete-time open-loop stable nonminimum-phase plant $A(d)\Delta(d)y(t) = B(d)\delta u(t)$ with

$$A(d)\Delta(d) = 1 - 1.8258d + 0.8630d^2 - 0.0376d^3 + 0.0004d^4$$
$$B(d) = 0.1669d + 0.3246d^2 - 0.6832d^3 - 0.0192d^4$$

This is obtained from the following continuous-time open-loop stable nonminimum-phase plant:

$$(s+1)\left(1 + \frac{1}{15}s\right)^2 y(\tau) = \omega^2(s-1)u(\tau - 0.2)$$

where $\tau \in \mathbb{R}$, and s denotes the time-derivative operator, viz., $sy(\tau) = dy(\tau)/d\tau$, by sampling the output every $T_s = 0.25$ second and holding the input constant and equal to $u(t)$ over the interval $[tT_s, (t+1)T_s]$. Figures 1 and 2 show the reference, a square wave, along with the plant output, when the plant input is controlled by a 1-DOF and a 2-DOF, respectively, LQ tracker. Both trackers are optimal w.r.t. the performance index

$$\sum_{k=0}^{\infty} \left\{\varepsilon^2(k) + 5\delta u^2(k)\right\} \tag{5.8-7}$$

Further, according to the pertinent results in the following theorem Theorem 1, the 2-DOF LQ tracker exploits the knowledge of the reference over 15 steps in the future.

LQ Trackers.   We study how to modify the pure LQ-regulator law so as to single out 2-DOF controllers. There are standard ways ([KS72], [AM90], and [BGW90]) to do this via the Riccati-based approach. However, our intention is to use the polynomial equation approach of Chapter 4, ultimately exploiting the results in an I/O system description framework.

We consider a MIMO plant

$$A(d)\Delta(d)y(t) = \mathcal{B}(d)\delta u(t) \tag{5.8-8}$$

with $\Delta(d) = (1-d)I_p$, $A(d)$ and $\mathcal{B}(d)$ left-coprime, and $\det \mathcal{B}(1) \neq 0$. The last two conditions are equivalent to saying that $A(d)\Delta(d)$ and $\mathcal{B}(d)$ are left-coprime.

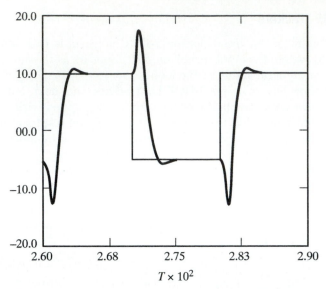

Figure 5.8-1: Reference and plant output when a 1-DOF LQ controller is used for the tracking problem of Example 1.

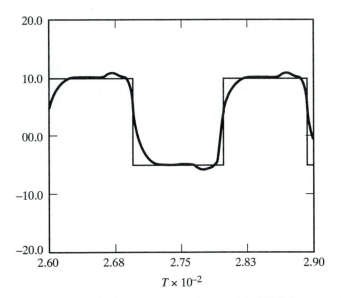

Figure 5.8-2: Reference and plant output when a 2-DOF LQ controller is used for the tracking problem of Example 1.

In (8), we adopt the usual choice of dealing with input increments $\delta u(t)$, in place of inputs $u(t)$, so as to introduce an integral action in the feedback-control system. We assume that a state vector $x(t)$, $\dim x(t) = n_x$, is chosen to represent (8) via a stabilizable and detectable state-space representation as in (4.1-1). For example, similar to (5-1), such a state vector can be made up by I/O pairs. Consider next the steady-state LQOR problem (4.1-1) to (4.1-4). By (4.4-24), its solution is of the form $X\delta u(t) = -Yx(t)$. We consider the following modified version of such a regulation law:

$$X\delta u(t) = -Yx(t) + v(t), t \in \mathbb{Z}_+ \tag{5.8-9}$$

In (9), $v(t) \in \mathbb{R}^m$ has to be chosen, so as to make the tracking error (1) small in a sense to be specified, amongst all $\mathbb{R}^m$-valued functions of $x_{[0,t]}$, $u_{[0,t)}$, and $r(\cdot) = r_{[0,\infty)}$:

$$v(t) = g\left(x_{[0,t]}, \delta u_{[0,t)}, r(\cdot)\right), t \in \mathbb{Z}_+ \tag{5.8-10}$$

It follows from (10) that (9) can be any linear or nonlinear 2-DOF controller, causal from $x(t)$ to $\delta u(t)$. On the other hand, (9) is permitted to be anticipative from $r(t)$ to $\delta u(t)$, the whole reference sequence $r(\cdot)$ being assumed to be available to the controller at any time. The problem is to choose $v(t)$ in such a way that the feedback-control system is stable and the performance index

$$J := \|\varepsilon(\cdot)\|_{\psi_y}^2 + \|\delta u(\cdot)\|_{\psi_u}^2 \tag{5.8-11}$$

is minimized. For the sake of simplicity, we stipulate temporarily that the plant state is initially zero:

$$x(0) = O_{n_x} \tag{5.8-12}$$

Further, it is assumed that

$$\|r(\cdot)\|_{\psi_y}^2 < \infty \tag{5.8-13}$$

We recall that, because (X, Y) solves the steady-state LQOR problem with performance index (11) for $r(t) \equiv O_p$, according to (4.3-9) and (4.1-12), we have

$$XA_2(d) + YB_2(d) = E(d) \tag{5.8-14}$$

$$E^*(d)E(d) = A_2^*(d)\psi_u A_2(d) + B_2^*(d)H'\psi_y HB_2(d) \tag{5.8-15}$$

with $HB_2(d)A_2^{-1}(d) = \Delta^{-1}(d)A^{-1}(d)\mathcal{B}(d)$, $A_2(d)$ and $B_2(d)$ right-coprime, and $E(d)$ assumed here to be strictly Hurwitz. Using the d-representations of the involved sequences, from (8) and (9), we find

$$\begin{aligned}
\hat{x}(d) &= \left[I_{n_x} + B_2(d)A_2^{-1}(d)X^{-1}Y\right]^{-1} B_2(d)A_2^{-1}(d)X^{-1}\hat{v}(d) \\
&= \left\{I_{n_x} + B_2(d)\left[E(d) - YB_2(d)\right]^{-1}Y\right\}^{-1} \\
&\quad \times B_2(d)A_2^{-1}(d)X^{-1}\hat{v}(d) \qquad [(14)]
\end{aligned}$$

By the matrix-inversion lemma (3-16), we get

$$\hat{x}(d) = \left[I_{n_x} - B_2(d)E^{-1}(d)Y\right]B_2(d)A_2^{-1}(d)X^{-1}\hat{v}(d)$$

which, by (14), becomes

$$
\begin{aligned}
\hat{x}(d) &= B_2(d)\left\{I - E^{-1}(d)\left[E(d) - XA_2(d)\right]\right\}A_2^{-1}(d)X^{-1}\hat{v}(d) \\
&= B_2(d)E^{-1}(d)\hat{v}(d) \qquad\qquad (5.8\text{-}16)
\end{aligned}
$$

By similar arguments, we also find

$$
\widehat{\delta u}(d) = A_2(d)E^{-1}(d)\hat{v}(d) \qquad\qquad (5.8\text{-}17)
$$

Then, using (3.1-12),

$$
\begin{aligned}
J &= \langle \hat{\varepsilon}^*(d)\psi_y\hat{\varepsilon}(d) + \widehat{\delta u}^*(d)\psi_u\widehat{\delta u}(d)\rangle \\
&= \langle \hat{y}^*(d)\psi_y\hat{y}(d) + \widehat{\delta u}^*(d)\psi_u\widehat{\delta u}(d) \\
&\quad + \hat{r}^*(d)\psi_y r(d) - \hat{y}^*(d)\psi_y\hat{r}(d) - \hat{r}^*(d)\psi_y\hat{y}(d)\rangle
\end{aligned}
$$

Now the sum of the first two terms in the latter expression equals

$$
\begin{aligned}
\langle \left[HB_2(d)E^{-1}(d)\hat{v}(d)\right]^* \psi_y HB_2(d)E^{-1}(d)\hat{v}(d) \\
+ \left[A_2(d)E^{-1}(d)\hat{v}(d)\right]^* \psi_u A_2(d)E^{-1}(d)\hat{v}(d)\rangle \\
= \langle \hat{v}^*(d)\hat{v}(d)\rangle \qquad [(15)]
\end{aligned}
$$

In conclusion, the quantity to be minimized w.r.t. $\hat{v}(d)$ becomes

$$
\left\| \hat{v}(d) - \left[E^{-*}(d)B_2^*(d)H'\psi_y\hat{r}(d)\right]_+ \right\|^2
$$

where, as in (4.1-24), $[\ell(d)]_+$ denotes the causal part of the two-sided sequence $\ell(d)$. Hence, the optimal choice turns out to be

$$
\hat{v}(d) = \left[E^{-*}(d)B_2^*(d)H'\psi_y\hat{r}(d)\right]_+ \qquad\qquad (5.8\text{-}18)
$$

or, in terms of the impulse-response matrix of the strictly causal and stable transfer matrix $HB_2(d)E^{-1}(d)$,

$$
HB_2(d)E^{-1}(d) = \sum_{i=1}^{\infty} h'(i)d^i \qquad\qquad (5.8\text{-}19)
$$

$$
v(t) = \sum_{i=1}^{\infty} h(i)\psi_y r(t+i), \forall t \in \mathbb{Z}_+ \qquad\qquad (5.8\text{-}20)
$$

We see that the optimal feedforward term $v(t)$ in the 2-DOF LQ tracker (2) turns out to be a linear combination of *future* samples of the reference to be tracked. By recalling (4.2-2), (18) can be equivalently written as

$$
\hat{v}(d) = \left[\bar{E}^{-1}(d)\bar{B}_2(d)H'\psi_y\hat{r}(d)\right]_+ \qquad\qquad (5.8\text{-}21)
$$

The reader is warned not to believe that, thanks to (21), (9) can be reorganized as follows:

$$\bar{E}(d)X\delta u(t) = -\bar{E}(d)Yx(t) + \bar{B}_2(d)H'\psi_y r(t) \qquad (5.8\text{-}22)$$

In fact, it is easily seen that, with $\bar{E}(d)$ anti-Hurwitz, (22) yields an unstable closed-loop system. We insist on underlining that the correct interpretation of (21) is to provide a *command* or *feedforward* input increment in terms of a linear combination of future reference samples.

Example 5.8-2 Consider a SISO nonminimum-phase plant with $HB_2(d) = d(1 - bd)$, $|b| > 1$, and the performance index (11) with $\psi_u = 0$ and $\psi_y = 1$. Then, for $r(t) \equiv 0$, stabilizing cheap control results. Hence,

$$E(d) = k\left(1 - b^{-1}d\right), \qquad\qquad k := \left[\frac{1 + b^2}{1 + b^{-2}}\right]^{1/2} \qquad (5.8\text{-}23)$$

The optimal feedforward input increment (18) equals

$$\hat{v}(d) = \left[k^{-1}\frac{\left(1 - bd^{-1}\right)d^{-1}}{1 - b^{-1}d^{-1}}\hat{r}(d)\right]_+ \qquad (5.8\text{-}24)$$

or for $t \in \mathbb{Z}_+$

$$v(t) = k^{-1}\left[r(t+1) + (1 - b^2)\sum_{j=1}^{\infty} b^{-j}r(t+j+1)\right] \qquad (5.8\text{-}25)$$

Problem 5.8-1 Consider the same tracking problem as in Example 2 with the exception that $|b| < 1$, viz., the plant is here minimum-phase. Show that, instead of (25), here we get $v(t) = r(t+1)$.

Problem 1 and Example 2 point out the different amount of information on the reference required for computing the feedforward input $v(t)$ in the minimum-phase and the nonminimum-phase plant case, respectively. Whereas in the first case only, $r(t+1)$ is needed at time t, the latter case requires the knowledge of the whole reference future, viz., $r_{[t+1,\infty)}$. Hence, we can expect that exploitation of the future of the reference can yield significant tracking-performance improvements in the nonminimum-phase plant case. For instance, for the tracking problem of Example 1, it can be found that setting the reference future equal to the current reference value, the 2-DOF LQ tracking performance deteriorates from that in Fig. 2 to approximately the one in Fig. 1.

Problem 5.8-2 Show that for a square plant ($m = p$) the 2-DOF controller (9) and (18) yields an offset-free closed-loop system, provided that no unstable hidden modes are present. (*Hint*: Use (16) and (15).)

Despite that (18) is obtained under the limitative assumption (12), it is reassuring that the 2-DOF LQ control law has the form (9). In fact, the latter shows that if $r(t) \equiv O_p$, the controller acts as the steady-state LQOR, and, hence, counteracts nonzero initial states in an optimal way. With some extra algebraic effort, it can be proved that (18) still holds true for the generic situation of nonzero initial state. Equation (18) will be extended in Sec. 7.5 to a stochastic setting. From now on, we shall refer to (9) and (18) as the 2-DOF LQ control law.

Theorem 5.8-1. **(2-DOF LQ Control)** *The 2-DOF control law minimizing (11) for a finite energy reference and a plant with any initial state is given by*

$$X\delta u(t) = -Yx(t) + v(t) \tag{5.8-26}$$

where X and Y are the constant matrices in (14) solving the pure underlying LQOR problem and $v(t)$ is the command *or feedforward input*

$$
\begin{aligned}
\hat{v}(d) &= [E^{-*}(d)B_2^*(d)H'\psi_y\hat{r}(d)]_+ \\
&= [\bar{E}^{-1}(d)\bar{B}_2(d)H'\psi_y\hat{r}(d)]_+
\end{aligned}
\tag{5.8-27}
$$

Provided that $E(d)$ is strictly Hurwitz, the plant is square, and the modified plant with input $\delta u(t)$ and output $y(t)$ is free of unstable hidden modes, the 2-DOF LQ controller yields an offset-free closed-loop system and, thanks to its integral action, asymptotic rejection of constant disturbances.

Theorem 1 can be used so as to single out a 2-DOF controller based on MKI. The same holds true for **TCC** (*truncated cost control*), the 2-DOF tracking extension of TCI. Whereas in the first the underlying pure regulation problem coincides with LQOR, this is still essentially true in the second case provided that the prediction horizon is taken large enough. Nevertheless, both MKI and TCI yield the LQOR law $\delta u(t) = Fx(t)$, where clearly $F = -X^{-1}Y$. The problem here is to determine the 2-DOF control law

$$\delta u = Fx(t) + X^{-1}v(t) \tag{5.8-28}$$

by computing $X^{-1}v(t)$ directly from the feedback-gain matrix and knowledge of the plant, without the need of solving the spectral factorization problem (15). To this end, rewrite (14) as follows:

$$Q(d) := X^{-1}E(d) = A_2(d) - FB_2(d) \tag{5.8-29}$$

Then, because $A_2(1) = O_p$ and, by Theorem 1, the closed-loop system is offset-free, we get

$$X^{-1}\hat{v}(d) = X^{-1}Q^{-*}(d)(X')^{-1}B_2^*(d)H'\psi_y\hat{r}(d) \tag{5.8-30}$$

$$X^{-1} = -FB_2(1)\left[HB_2(1)\right]^{-1}\varphi_y^{-1} \tag{5.8-31}$$

where $\varphi_y'\varphi_y = \psi_y$.

Problem 5.8-3   Verify (30) and (31).

Equations (29) to (31) allow us to compute the 2-DOF control law (28) without the need of solving the spectral factorization problem (15), by only using the plant model and feedback F.

SIORHC.   The *stabilizing I/O receding-horizon controller* (SIORHC) is the 2-DOF tracking extension of SIORHR. SIORHC is obtained by modifying (4-25) to (4-28) as follows. Given

$$s(t) := \left[\ \left(y_t^{t-n_a}\right)' \quad \left(\delta u_{t-1}^{t-\ell-n_b+1}\right)' \ \right]' \tag{5.8-32}$$

consider the problem of finding, whenever it exists, an input sequence $\widehat{\delta u}_{[t,t+T)}$ to the SISO plant

$$A(d)\Delta(d)y(t) = B(d)\delta u(t) \tag{5.8-33}$$

minimizing

$$J\left(s(t), \delta u_{[t,t+T)}\right) = \sum_{k=t}^{t+T-1} \left[\psi_y \varepsilon^2(k+\ell) + \psi_u \delta u^2(k)\right] \tag{5.8-34a}$$

under the constraints

$$\delta u_{t+T+n-2}^{t+T} = O_{n-1} \ , \qquad\qquad y_{t+T+\ell+n-1}^{t+T+\ell} = \underline{r}(t+T+\ell) \tag{5.8-34b}$$

with

$$n := \max\left\{n_a + 1, n_b\right\} \tag{5.8-35}$$

and

$$\underline{r}(k) := \left. \begin{bmatrix} r(k) \\ \vdots \\ r(k) \end{bmatrix} \right\} n \tag{5.8-36}$$

Then, the plant input at time t given by SIORHC equals

$$\delta u(t) = \widehat{\delta u}(t) \tag{5.8-37}$$

The solution of problem (32) to (36) can be obtained *mutatis mutandis* as in Sec. 5 in the following form:

$$\begin{aligned} \widehat{\delta u}_{t+T-1}^{t} \ &= \ -M^{-1}\Big\{\psi_y \left(I_T - QLM^{-1}\right) W_1' \left[\Gamma_1 s(t) - r_{t+\ell+T-1}^{t+\ell+1}\right] \\ &\quad + Q\left[\Gamma_2 s(t) - \underline{r}(t+\ell+T)\right]\Big\} \end{aligned} \tag{5.8-38}$$

with M, Q, L, Γ_1, and Γ_2 as in (5-21). Hence, the SIORHC law equals

$$\delta u(t) = \underline{e}_1' \widehat{\delta u}_{t+T-1}^{t} \tag{5.8-39}$$

which, in turn, can be also written in polynomial form as follows:

$$R(d)\delta u(t) = -S(d)y(t) + Z^*(d)r(t+\ell) \qquad (5.8\text{-}40)$$

In (40), $R(d)$ and $S(d)$ are polynomials similar to the ones in (6) and

$$Z^*(d) = z_1 d^{-1} + \cdots + z_T d^{-T} \qquad (5.8\text{-}41)$$

Problem 5.8-4 Verify that (38) solves the problem (32) to (36).

We note that the stabilizing properties of (39) or (40) can be directly deduced from the ones of SIORHR in Theorem 5-1, taking into account that the initial plant $A(d)$ polynomial has now become $A(d)\Delta(d)$. Further, we show hereafter that (33) controlled by SIORHC is offset-free whenever the closed-loop system is stable. Using (33) and (40), we find the following equation for the controlled system:

$$[A(d)\Delta(d)R(d) + B(d)S(d)] \begin{bmatrix} y(t) \\ \delta u(t) \end{bmatrix} = \begin{bmatrix} B(d)Z^*(d) \\ A(d)\Delta(d)Z^*(d) \end{bmatrix} r(t+\ell) \qquad (5.8\text{-}42)$$

Hence, provided that $A(d)\Delta(d)R(d) + B(d)S(d)$ is strictly Hurwitz, if $r(t) \equiv r$, we have $\bar{y} := \lim_{t\to\infty} y(t) = [Z^*(1)/S(1)]r$ and $\bar{\delta u} := \lim_{t\to\infty} \delta u(t) = 0$. In order to prove that $S(1) = Z^*(1)$, comparing (38) and (40), we show that every row of Γ_1 and Γ_2 has its first $n_a + 1$ entries that sum up to 1. To see this, we consider an equation similar to (5-27) in the present context:

$$\left. \begin{array}{r} 1 = A(d)\Delta(d)Q_k(d) + d^k G_k(d) \\ \partial Q_k(d) \le k-1 \end{array} \right\} \qquad (5.8\text{-}43)$$

For $d = 1$, we get

$$G_k(1) = 1 \qquad (5.8\text{-}44)$$

Hence, the coefficients of polynomials $G_k(d)$ sum up to 1. Finally, the desired property follows, because, by (5-8) and (5-28), the first n entries of rows of Γ_1 and Γ_2 coincide with the coefficients of $G_k(d)$.

The foregoing results are summed up in the following theorem.

Theorem 5.8-2. (SIORHC) *Under the same assumptions as in Theorem 4-2 with $A(d)$ replaced by $A(d)\Delta(d)$, the SIORHC law*

$$\begin{aligned} \delta u(t) \quad = \quad &-\underline{e}_1' M^{-1} \Big\{ \psi_y \left(I_T - QLM^{-1} \right) W_1' \left[\Gamma_1 s(t) - r_{t+\ell+T-1}^{t+\ell+1} \right] \\ &+ Q \left[\Gamma_2 s(t) - \underline{r}(t+\ell+T) \right] \Big\} \end{aligned} \qquad (5.8\text{-}45)$$

where $s(t)$ is as in (32), inherits all the stabilizing properties of SIORHR. Further, whenever stabilizing, SIORHC yields an offset-free closed-loop system and, thanks to its integral action, asymptotic rejection of constant disturbances.

GPC. The *generalized predictive control (GPC)* is the 2-DOF tracking extension of GPR. GPC is obtained by modifying (6-12) to (6-16) as follows. Given $s(t)$ as in (32), find an input sequence $\widehat{\delta u}_{[t,t+N_u]}$ to the plant (33) minimizing, for $N_u \leq N_2$,

$$J_{\text{GPC}} = \sum_{k=t+N_1}^{t+N_2} \varepsilon^2(k+\ell) + \psi_u \sum_{k=t}^{t+N_u-1} \delta u^2(k) \qquad (5.8\text{-}46)$$

under the constraints

$$\delta u_{t+N_2-1}^{t+N_u} = O_{N_2-N_1} \qquad (5.8\text{-}47)$$

$$r_{t+\ell+N_2}^{t+\ell+N_u} = \underline{r}(t+\ell+N_u) := \left.\begin{bmatrix} r(t+\ell+N_u) \\ \vdots \\ r(t+\ell+N_u) \end{bmatrix}\right\} N_2 - N_u + 1 \qquad (5.8\text{-}48)$$

Then, the GPC plant input at time t is given by

$$\delta u(t) = \widehat{\delta u}(t) \qquad (5.8\text{-}49)$$

A remark on (48) is in order. It is seen that (48) is a constraint on the reference to be tracked. Consequently, (48) should not be included in the GPC formulation but, on the contrary, it should be fulfilled by the reference itself. On the other hand, to hold for all t, (48) implies, for $N_u < N_2$, that the reference is constant: a contradiction with our goal to use 2-DOF controllers to get high-performance tracking with general reference sequences. The correct interpretation for (48) is that, whatever the reference future behavior, the controller *pretends* that the reference is constant from time $t + \ell + N_u$ throughout $t + \ell + N_2$. This assumption, consistent with the input constraints (47), will be referred to as the *reference consistency constraint*. Taken into account (34b), we note that such a constraint is embedded in SIORHC formulation. As the next example shows, the reference consistency constraint is important for ensuring good tracking performance.

Example 5.8-3 Consider the SISO plant $A(d)\Delta(d)y(t) = B(d)\delta u(t)$ with

$$A(d) = 1 + 0.9d - 0.5d^2 \qquad\qquad B(d) = d + 1.01d^2$$

SIORHC, with $T = 3$, and GPC, with $N_1 = N_u = 3$, $N_2 = 5$, and $\rho = 0$, both yield a state-deadbeat controller whose tracking performance, when the reference consistency condition is satisfied, is shown in Fig. 3. Figure 4 shows that if the reference consistency condition is violated, the tracking performance becomes unacceptable.

Following developments similar to (6-17) to (6-23), we find

$$\widehat{\delta u}_{t+N_u-1}^{t} = -M_G^{-1}W_G'\left[\Gamma_G s(t) - \bar{r}\left(t+\ell, N_1, N_u\right)\right] \qquad (5.8\text{-}50)$$

$$\bar{r}(t+\ell, N_1, N_u) := \left.\begin{bmatrix} r_{t+\ell+N_u-1}^{t+\ell+N_1} \\ \underline{r}(t+\ell+N_u) \end{bmatrix}\right\} N_2 - N_1 + 1 \qquad (5.8\text{-}51)$$

By the same arguments as in (43) and (44), it follows that every row of Γ_G has its first $n_a + 1$ entries that sum up to 1. Hence, as with SIORHC, GPC yields zero offset.

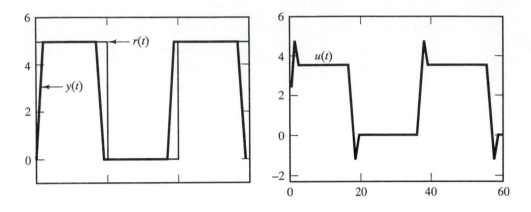

Figure 5.8-3: Deadbeat tracking for the plant of Example 3 controlled by SIORHC (or GPC) when the reference consistency condition is satisfied. $T = 3$ is used for SIORHC ($N_1 = N_u = 3$, $N_2 = 5$ and $\psi_u = 0$ for GPC).

Problem 5.8-5   Verify that (50) solves problem (46) to (48).

Theorem 5.8-3.   (GPC) *Under the same assumption as in Theorem 6-1 with $A(d)$ replaced by $A(d)\Delta(d)$, the GPC law given by*

$$\delta u(t) = -\underline{e}_1' M_G^{-1} W_G' \left[\Gamma_G s(t) - \bar{r}\left(t + \ell, N_1, N_u\right) \right] \qquad (5.8\text{-}52)$$

inherits the state-deadbeat property of GPR. Further, whenever stabilizing, GPC yields an offset-free closed-loop system and, thanks to its integral action, asymptotic rejection of constant disturbances.

5.8.3   Reference Management and Predictive Control

In many cases, it is convenient to distinguish between the reference sequence $r(\cdot)$ used in the control laws and the desired plant output $w(\cdot)$. An example is to let $r(\cdot)$ be a filtered version of $w(\cdot)$.

$$r(t) = \frac{\epsilon}{1 - (1 - \epsilon)d} w(t) \qquad (5.8\text{-}53)$$

or

$$H(d)r(t) = w(t) \qquad (5.8\text{-}54)$$

with $H(d) = \epsilon^{-1}[1 - (1 - \epsilon)d]$, $H(1) = 1$, and ϵ such that $0 < \epsilon < 1$ and small for low-pass filtering $w(t)$, for example, $\epsilon = 0.25$. For 2-DOF trackers based on RHR, another possibility to make smooth the transition from the current output $y(t)$ to a desired constant set point w is to let

$$\left.\begin{array}{rcl} r(t) & = & y(t) \\ r(t+i) & = & (1 - \epsilon)r(t + i - 1) + \epsilon w \end{array}\right\} \qquad (5.8\text{-}55)$$

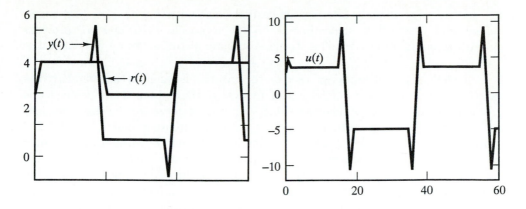

Figure 5.8-4: Tracking performance for the plant of Example 3 controlled by GPC ($N_1 = N_u = 3$, $N_2 = 5$ and $\psi_u = 0$) when the reference consistency condition is violated, *viz.* the time-varying sequence $r(t + N_u + i)$, $i = 1, \cdots, N_2 - N_u$, is used in calculating $u(t)$.

In designing 2-DOF trackers, a different approach consists of leaving $w(\cdot)$ unaltered, and, instead, filtering $y(\cdot)$ and $u(\cdot)$. Specifically, the performance index (11) is changed in

$$J = \|y_H(\cdot) - w(\cdot)\|^2_{\psi_y} + \|\delta u_H(\cdot)\|^2_{\psi_u} \qquad (5.8\text{-}56)$$

where $y_H(\cdot)$ and $u_H(\cdot)$ are filtered versions of $y(\cdot)$ and $u(\cdot)$, respectively:

$$y_H(t) = H(d)y(t) \qquad\qquad u_H(t) = H(d)u(t) \qquad (5.8\text{-}57)$$

with $H(d)$ a strictly Hurwitz polynomial such that $H(1) = 1$. Note that, taking into account that the plant can also be represented as

$$A(d)\Delta(d)y_H(t) = \mathcal{B}(d)\delta u_H(t)$$

we now find for the 2-DOF LQ control law, instead of (26),

$$X\delta u_H(t) = -Y x_H(t) + v(t)$$

with $v(t)$ given again by (27), $\hat{v}(d) = [E^{-*}(d)B_2^*(d)H'\psi_y \hat{w}(d)]_+$, and $x_H(t) = H(d)x(t)$. Consequently,

$$X\delta u(t) = -Y x(t) + \frac{v(t)}{H(d)}$$

Hence, we conclude that filtering $y(t)$ and $u(t)$ as in (56) and (57) has the effect of leaving everything unchanged with the exception of changing $w(t)$ into $r(t) = w(t)/H(d)$. In other terms, in the present deterministic context, (56) and (57) are equivalent to directly filtering the desired plant output $w(t)$ as in (54) to get reference $r(t)$. As will be shown in Sec. 7.5, the approach of (56) and (57) provides us with some additional benefits should stochastic disturbances be present in the loop.

1-DOF or 2-DOF tracking based on the receding-horizon method is referred to as *receding-horizon control (RHC)*. This reduces to RHR when the reference to be tracked is identically zero. The name *multistep predictive control* or *long-range predictive control* has been customarily used in the literature to designate 2-DOF RHC whereby the control action is selected taking into account the future evolution over a multistep prediction horizon of both the plant state and the reference as provided by "predictive" dynamic models. From the results of this section, particularly Theorem 1, it follows that 2-DOF LQ control satisfies the prior characterization and hence will be regarded as a long-range predictive controller as well. For the sake of brevity, unless needed to avoid possible confusion, from now on, we shall often omit the attribute "multistep" or "long range" and simply refer to the preceding class as *predictive control*.

A peculiar and important feature in predictive control is that the future evolution of the reference can be designed in real time (cf. (55)). This can be done, taking into account the current value of the plant output $y(t)$ and the desired set point w, so as to ensure that the plant input $u(t)$ be within admissible bounds and, hence, avoid saturation phenomena. However, this mode of operation, whereby $r(t + i)$ is made dependent on $y(t)$ as in (55), introduces an extra feedback loop that must be proved not to destabilize the closed-loop system.

Main Points of the Section. 1-DOF and 2-DOF controllers can be designed by suitably modifying the basic RHR laws so as to ensure asymptotic rejection of constant disturbances and zero offset. In constrast with 1-DOF controllers, in 2-DOF controllers, the reference to be tracked is processed, independently of the plant output, which is fed back, by a feedforward filter so as to enhance the tracking performance of the controlled system. Predictive control is a 2-DOF RHC whereby the feedforward action depends on the future reference evolution, which, in turn, can be selected on-line so as to avoid saturation phenomena and fulfill state-related constraints.

NOTES AND REFERENCES

At the beginning of the seventies, two papers ([Kle70] and [Kle74]), proposed a simple method to stabilize time-invariant linear plants. This method was later adopted in [Tho75] by using the concept of a receding horizon. [KP75] and [KP78] extended RHR to stabilize time-varying linear plants. See also [KBK83]. It took longer than 15 years [CM93] to extend the stabilizing property of zero-terminal-state RHR to the case of a possibly singular state-transition matrix as in Theorem 3-2.

In a different direction, [CS82], [Sha79], and [TSS77] considered nonlinear state-dependent RHR for time-invariant linear plants so as to speed up the response to large regulation errors. Extensions of RHR to stabilize continuous-time nonlinear systems were reported in [MM90a], [MM90b], [MM91a], and [MM91b]. For RHR with possibly nonquadratic costs and discrete-time nonlinear systems, see [KG85], [KG86], and [KG88].

The concept of predictive control, wherein on-line reference design takes place, first appeared in [Mar76a] and [Mar76b]. Subsequent approaches to predictive control, particularly from the standpoint of industrial process control, were referred to as *model predictive control* [RRTP78] and *dynamic matrix control* [CR80]. See also the survey [GPM89].

SIORHR and SIORHC were introduced in [MLZ90], [MZ92], and, independently, [CS91]. GPC was introduced in [CMT85] and discussed in [CMT87a], [CMT87b], and [CM89]. For continuous-time GPC, see [DG91] and [DG92]. MKI and TCI are related to the self-tuning algorithms first reported in [ML89], and [Mos83], respectively. For other contributions, see also [dKvC85], [LM85], [Pet84], [RT90], [Soe92], [TC88], and [Yds84].

PART
2

STATE ESTIMATION, SYSTEM IDENTIFICATION, AND LQ AND PREDICTIVE STOCHASTIC CONTROL

CHAPTER 6

RECURSIVE STATE FILTERING AND SYSTEM IDENTIFICATION

This chapter addresses problems of state and system parameter estimation. Our interest is mainly directed to solutions usable in real time, viz., recursive estimation algorithms. They are introduced by initially considering a simple but pervasive estimation problem consisting of solving a system of linear equations in an unknown time-invariant vector. At each time step, a number of new equations becomes available and the estimate is required to be recursively updated.

In Sec. 1, this problem is formulated as an indirect-sensing measurement problem and solved via an orthogonalization method. When the unknown coincides with the state of a stochastic linear dynamic system, we get a Kalman filtering problem of which in Sec. 2, we derive in detail the Riccati-based solution and its duality with the LQR problem of Chapter 2. We also briefly describe the polynomial equations for the steady-state Kalman filter as well as how the so-called innovations representation originates. In Sec. 3, we consider several system identification algorithms. In this respect, to break the ice, a simple deterministic system parameter estimation algorithm is derived by direct use of the indirect-sensing measurement problem solution. This simple algorithm is next modified so as to take into account several impairments due to disturbances. In this way, RLS, RELS, RML, and the stochastic gradient algorithm are introduced and related to algorithms derived systematically via the prediction-error method. In Sec. 4, convergence analysis of the foregoing algorithms is considered.

6.1   INDIRECT-SENSING MEASUREMENT PROBLEMS

We begin with addressing problems of filtering and system parameter estimation within a common framework. The distinct elementary ingredients are an unknown w, a set of time-indexed stimuli ρ_t by which w can be probed, and a set of accessible reactions m_t of w to ρ_t. The cause-and-effect correspondence between ρ_t and m_t via the unknown w is linear and of the simplest nontrivial form. Loosely stated, the problem is to find, at every t, an approximation to the unknown w based on the knowledge of present and past stimuli and corresponding reactions. This simple scheme encompasses seemingly different applications, ranging from problems of system parameter estimation to Kalman filtering.

In order to accommodate different applications in a single mathematical framework, an abstract setting has to be used. Let $\mathcal{H}$ be a vector space, not necessarily of finite dimension, equipped with an inner product $\langle \cdot, \cdot \rangle$. We say that an ordered pair

$$(\rho, m) \in \mathcal{H} \times \mathbb{R} \tag{6.1-1}$$

is an *indirect-sensing linear measurement (ISLM)*, or simply a measurement, on an unknown vector $w \in \mathcal{H}$ if m equals the value taken on by the inner product $\langle w, \rho \rangle$:

$$m = \langle w, \rho \rangle \tag{6.1-2}$$

In such a case, we call m the *outcome* and ρ the measurement *representer*.[1] It is assumed that a sequence of integer-indexed measurements

$$(\rho_k, m_k) \ , \qquad k \in \mathbb{Z}_1 := \{1, 2, \cdots\} \tag{6.1-3}$$

is available. Let $\underline{r} := \{1, 2, \cdots, r\}$. The sequence $\mathcal{E}^r$, made up of the initial r measurements,

$$\mathcal{E}^r := \Big\{ (\rho_k, m_k) \ , \ k \in \underline{r} \Big\} \tag{6.1-4}$$

will be referred to as the *experiment* up to the integer r. We say that a vector $v \in \mathcal{H}$ *interpolates* $\mathcal{E}^r$ if $\langle v, \rho_k \rangle = m_k, \ k \in \underline{r}$.

The *ISLM problem* is to find a recursive formula for

$$\hat{w}_{|r} := \text{the minimum-norm vector in } \mathcal{H} \text{ interpolating } \mathcal{E}^r \tag{6.1-5}$$

The vector $\hat{w}_{|r}$ will be hereafter referred to as the ISLM *estimate* of w based on $\mathcal{E}^r$. The norm alluded to in (5) is the one induced by $\langle \cdot, \cdot \rangle$, viz., $\|w\| := +\sqrt{\langle w, w \rangle}$.

Figure 1 depicts the geometry of the problem that has been set up. Though the vector w is unknown, given ρ_k and $m_k = \langle w, \rho_k \rangle, \ k \in \underline{r}$, we can find $\hat{w}_{|r}$ as follows. Let $[\rho^r]$ be the linear manifold generated by $\rho^r := \{\rho_k\}_{k=1}^r$:

$$[\rho^r] := \text{Span} \{\rho_k\}_{k=1}^r$$

---

[1] We recall that if $\mathcal{H}$ is a Hilbert space, every continuous linear functional on $\mathcal{H}$ has the form $\langle \cdot, \rho \rangle$, and ρ is called the functional representer [Lue69]. This accounts for the adopted terminology.

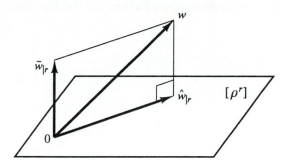

Figure 6.1-1: The ISLM estimate as given by an orthogonal projection.

Then $\hat{w}_{|r}$ equals the orthogonal projection in $\mathcal{H}$ of the unknown w onto $[\rho^r]$. According to the orthogonal projection theorem [Lue69], this can be found by using the following conditions:

(a) $\hat{w}_{|r} \in [\rho^r]$, that is, $\hat{w}_{|r} = \sum_{k=1}^{r} \alpha_k \rho_k$, $\alpha_k \in \mathbb{R}$

(b) $\tilde{w}_{|r} := w - \hat{w}_{|r} \perp [\rho^r]$

Combining (a) and (b), we get

$$\sum_{k=1}^{r} \alpha_k \langle \rho_k, \rho_i \rangle = \langle w, \rho_i \rangle = m_i , \qquad i \in \underline{r} \qquad (6.1\text{-}6)$$

Note that $\hat{w}_{|r}$ minimizes the norm of $\tilde{w}_{|r} := w - \hat{w}_{|r}$ among all vectors belonging to $[\rho^r]$:

$$\hat{w}_{|r} = \arg \min_{v \in [\rho^r]} \left\{ \|w - v\|^2 \right\} \qquad (6.1\text{-}7)$$

Then the ISLM estimate $\hat{w}_{|r}$ is the same as the minimum-norm error estimate of w based linearly on ρ^r.

Equation (6) is a system of r linear equations in r unknowns $\{\alpha_k\}_{k=1}^r$. These equations are known as the *normal equations*. A system of normal equations can be set up and solved in order to find the minimum-norm solution to an underdetermined system of linear equations, viz., a system where $\dim[\rho^r] < \dim w$. The direct solution of (6) is impractical for two main reasons: first, the number of the equations grows with r, and, second, the solution at integer r does not explicitly use the one at the integer $r - 1$. The recursive solution that we intend to find circumvents these difficulties.

The following examples show that the simple abstract framework that has been set up encompasses seemingly different applications of interest.

Example 6.1-1 (Deterministic System Parameter Estimation)
Consider the SISO system

$$\left. \begin{array}{l} A(d)y(k) = B(d)u(k) \\ A(d) := 1 + a_1 d + \cdots + a_{n_a} d^{n_a} \\ B(d) := \qquad b_1 d + \cdots + b_{n_b} d^{n_b} \end{array} \right\} \qquad (6.1\text{-}8)$$

$k = 1, 2, \cdots, t$. This can be rewritten as

$$y(k) = \varphi'(k-1)\theta \qquad (6.1\text{-}9)$$

$$\varphi(k-1) := \left[\begin{array}{cc} \left(-y_{k-n_a}^{k-1}\right)' & \left(u_{k-n_b}^{k-1}\right)' \end{array} \right]' \in \mathbb{R}^{n_\theta} \qquad (6.1\text{-}10)$$

$$\theta := \left[\begin{array}{ccccc} a_1 & \cdots & a_{n_a} & b_1 & \cdots & b_{n_b} \end{array} \right]' \in \mathbb{R}^{n_\theta} \qquad (6.1\text{-}11)$$

with $n_\theta := n_a + n_b$. If $\mathcal{H}$ equals the Euclidean space of vectors in $\mathbb{R}^{n_\theta}$ with inner product $\langle w, \rho_k \rangle = \rho_k' w$ and we set

$$w := \theta \qquad\qquad \rho_k := \varphi(k-1) \qquad\qquad m_k := y(k) \qquad (6.1\text{-}12)$$

the ISLM problem amounts to finding a recursive formula for updating the minimum-norm system parameter vector θ interpolating the I/O data up to time t.

Example 6.1-2 (Linear MMSE Estimation)
Consider (cf. Appendix D) a real-valued random variable v defined in an underlying probability space $(\Omega, \mathcal{F}, \mathbb{P})$ of elementary events $\omega \in \Omega$, with a σ-algebra $\mathcal{F}$ of subsets of Ω, and probability measure $\mathbb{P}$. We remind that a random variable v is an $\mathcal{F}$-measurable function $v : \Omega \rightarrow \mathbb{R}$. In particular, we are interested in the set of all random variables with a finite second moment:

$$\mathcal{E}\left\{v^2\right\} := \int_\Omega v^2(\omega)\, \mathbb{P}(d\omega) < \infty$$

where $\mathcal{E}$ denotes expectation. This set can be made a vector space over the real field via the usual operations of pointwise sum of functions and multiplication of functions by real numbers. Further,

$$\langle u, v \rangle := \mathcal{E}\{uv\} \qquad (6.1\text{-}13)$$

satisfies all the axioms of an inner product [Lue69]. This vector space equipped with the inner product (13) will be denoted by $L_2(\Omega, \mathcal{F}, \mathbb{P})$. Setting $\mathcal{H} = L_2(\Omega, \mathcal{F}, \mathbb{P})$ in the ISLM problem, the outcome (2) becomes the cross-correlation $m = \mathcal{E}\{w\rho\}$, and the experiment (4) consists of r ordered pairs made up by the random variables ρ_k and the reals $m_k = \mathcal{E}\{w\rho_k\}$. Here the ISLM estimate equals

$$\hat{w}_{|r} = \arg \min_{v \in [\rho^r]} \left\{ \mathcal{E}\left\{ (w-v)^2 \right\} \right\} \qquad (6.1\text{-}14)$$

If w and ρ_k have both zero mean, $\mathcal{E}\{w\} = \mathcal{E}\{\rho_k\} = 0$, $\hat{w}_{|r}$ coincides with the linear *minimum mean-square-error (MMSE)* estimate of w based on the observations ρ^r.

For some applications, we need to consider a generalization of the foregoing setting. There are, in fact, cases in which w and ρ_k have a number of components in $\mathcal{H}$. Then, in general,

$$w = \left[\begin{array}{ccc} w^1 & \cdots & w^n \end{array} \right]' \in \mathcal{H}^n \qquad (6.1\text{-}15)$$

$$\rho_k = \left[\begin{array}{ccc} \rho_k^1 & \cdots & \rho_k^p \end{array} \right]' \in \mathcal{H}^p \qquad (6.1\text{-}16)$$

(2) takes the form of an outer-product matrix:

$$m = \{\langle w, \rho \rangle\} := \begin{bmatrix} \langle w^1, \rho^1 \rangle & \cdots & \langle w^1, \rho^p \rangle \\ \vdots & & \vdots \\ \langle w^n, \rho^1 \rangle & \cdots & \langle w^n, \rho^p \rangle \end{bmatrix} \in \mathbb{R}^{n \times p} \qquad (6.1\text{-}17)$$

and the minimum-norm ISLM problem is then to find a recursive formula for

$$\hat{w}_{|r} := \begin{bmatrix} \hat{w}^1_{|r} & \cdots & \hat{w}^n_{|r} \end{bmatrix}' \qquad (6.1\text{-}18)$$

where, for each $i \in \underline{n}$,

$$\hat{w}^i_{|r} := \text{the minimum-norm vector in } \mathcal{H} \text{ interpolating} \\ \mathcal{E}^r = \left\{ \left(\rho_k, \{\langle w^i, \rho_k \rangle\} \right), k \in \underline{r} \right\} \qquad (6.1\text{-}19)$$

A remark here is in order. In general, $\hat{w}^i_{|r}$ cannot be found by just considering the "reduced" experiment

$$\left\{ \left(\rho^j_k, \langle w^i, \rho^j_k \rangle \right), j \in \underline{p}, \ k \in \underline{r} \right\}$$

and disregarding the remaining experiments that pertain to the other $n-1$ vectors listed in (15). In fact, because the ρ_k's may depend on w, and consequently $\hat{w}^i_{|r}$ and $\hat{w}^s_{|r}$, $i \neq s$, may turn out to be interdependent, it is not possible to solve the problem of finding $\hat{w}^i_{|r}$ separately from that of $\hat{w}^s_{|r}$. The Kalman filtering problem studied in the next section is an important example of such a situation.

Problem 6.1-1   Consider the outer-product matrix $\{\langle \cdot, \cdot \rangle\}$ defined in (17). Show that

(a) $\{\langle u, v \rangle\} = \{\langle v, u \rangle\}'$. $u \in \mathcal{H}^m$, $v \in \mathcal{H}^n$.

(b) $\{\langle Mu, v \rangle\} = M \{\langle u, v \rangle\}$ for every matrix $M \in \mathbb{R}^{p \times m}$.
Consequently, $\{\langle u, Mv \rangle\} = \{\langle u, v \rangle\} M'$, for every matrix $M \in \mathbb{R}^{p \times n}$.

The geometric interpretation of Fig. 1 carries over to the general case (15) to (19). To this end, it is enough to set

$$[\rho^r] := \text{Span} \{\rho_k, k \in \underline{r}\} = \text{Span} \left\{ \rho^j_k, j \in \underline{p}, k \in \underline{r} \right\} \qquad (6.1\text{-}20)$$

Thus, $\hat{w}^i_{|r}$ coincides with the orthogonal projection in $\mathcal{H}$ of the unknown w^i onto $[\rho^r]$:

$$\hat{w}^i_{|r} = \text{Projec} \left[w^i \mid \rho^r \right] \qquad (6.1\text{-}21\text{a})$$

For the sake of brevity, setting $\hat{w}_{|r} = \begin{bmatrix} \hat{w}^1_{|r} & \cdots & \hat{w}^n_{|r} \end{bmatrix}'$ instead of (21a), we shall simply write

$$\hat{w}_{|r} = \text{Projec}\left[w \mid \rho^r\right] \qquad\qquad (6.1\text{-}21b)$$

Further, $\hat{w}^i_{|r}$ is uniquely specified by two requirements:

(a) $\qquad\quad \hat{w}^i_{|r} \in [\rho^r] \;, \qquad i \in \underline{n} \qquad\qquad\qquad (6.1\text{-}22a)$

(b) $\qquad\quad \{\langle \tilde{w}_{|r}, \rho_k \rangle\} = O_{n\times p} \;, \qquad k \in \underline{r} \qquad\qquad (6.1\text{-}22b)$

where

$$\tilde{w}_{|r} := w - \hat{w}_{|r} \qquad\qquad (6.1\text{-}23)$$

Problem 6.1-2 Let $u \in \mathcal{H}^m$ and $v \in \mathcal{H}^n$. Let $\text{Projec}\,[u \mid v]$ denote the componentwise orthogonal projection in $\mathcal{H}$ of u onto $\text{Span}\{v\}$. Then show that

$$\text{Projec}\,[u \mid v] = \{\langle u, v \rangle\}\{\langle v, v \rangle\}^{-1} v \qquad\qquad (6.1\text{-}24)$$

6.1.1 Solution by Innovations

Let us now construct from the representers $\{\rho_k, k \in \mathbb{Z}_1\}$ an orthonormal sequence $\{\nu_k, k \in \mathbb{Z}_1\}$ in $\mathcal{H}^p$ by the *Gram–Schmidt orthogonalization* procedure [Lue69]. Here by orthonormality, we mean

$$\{\langle \nu_r, \nu_k \rangle\} = I_p \delta_{r,k} \qquad \forall r, k \in \mathbb{Z}_1 \qquad\qquad (6.1\text{-}25)$$

where $\delta_{r,k}$ denotes the Kronecker symbol. Accordingly,

$$e_r := \rho_r - \sum_{k=1}^{r-1} \{\langle \rho_r, \nu_k \rangle\} \nu_k \qquad\qquad (6.1\text{-}26)$$

$$\nu_r := \begin{cases} L_r^{-1/2} e_r &, \quad L_r \text{ nonsingular} \\ O_{\mathcal{H}_p} &, \quad \text{otherwise} \end{cases} \qquad\qquad (6.1\text{-}27)$$

where L_r equals the symmetric positive definite matrix

$$L_r := \{\langle e_r, e_r \rangle\} \in \mathbb{R}^{p\times p} \qquad\qquad (6.1\text{-}28)$$

$L^{T/2} := \left(L^{1/2}\right)'$ and $L_r^{1/2}$ any $p \times p$ matrix such that $L_r^{1/2} L_r^{T/2} = L_r$. Eq. (26) can be rewritten as

$$e_r = \rho_r - \hat{\rho}_{r|r-1} \;, \qquad \hat{\rho}^j_{r|r-1} := \text{Projec}\left[\rho^j_r \mid \rho^{r-1}\right] \qquad (6.1\text{-}29)$$

so every e_r is obtained by subtracting from ρ_r its ISLM one-step-ahead prediction, that is, its ISLM estimate based on the experiment

$$\{(\rho_k, m_k = \{\langle \rho_r, \rho_k \rangle\}) \;, \quad k \in \underline{r-1}\}$$

up to the immediate past. For this reason, hereafter $\{e_r, r \in \mathbb{Z}_1\}$ will be called the sequence of *innovations* of $\{\rho_r, r \in \mathbb{Z}_1\}$ and $\{\nu_r, r \in \mathbb{Z}_1\}$ that of the *normalized innovations*.

The introduction of the innovations leads to easily finding an updating equation for the orthogonal projector

$$S_r(\cdot) := \mathrm{Projec}\,[\,\cdot\mid\rho^r]:\mathcal{H}\to[\rho^r] \qquad (6.1\text{-}30)$$

which will be referred to as the *estimator* at step r associated to the given experiment. Indeed, because $[\nu_r] := \mathrm{Span}\{\nu_r\}$ is the orthogonal complement of $[\rho^{r-1}]$ in $[\rho^r]$, we have that

$$\left[\,\rho^r\,\right] = \left[\,\rho^{r-1}\,\right]\oplus\left[\,\nu_r\,\right] \qquad (6.1\text{-}31)$$

Therefore, setting $S_0(\cdot) := O_{\mathcal{H}}$, we have

$$\begin{aligned}
S_r(\cdot) &= S_{r-1}(\cdot) + \{\langle\cdot,\nu_r\rangle\}\,\nu_r \qquad &(6.1\text{-}32)\\
&= S_{r-1}(\cdot) + \{\langle\cdot,e_r\rangle\}\,L_r^{-1}e_r
\end{aligned}$$

The *n-order extension* of $S_r(\cdot)$ is defined as follows:

$$\begin{aligned}
S_r^n &: \left[\begin{array}{ccc} v^1 & \cdots & v^n\end{array}\right]' \mapsto \left[\begin{array}{ccc} S_r(v^1) & \cdots & S_r(v^n)\end{array}\right]' \qquad &(6.1\text{-}33)\\
&: \mathcal{H}^n \to [\rho^r]^n
\end{aligned}$$

The *innovator* $\mathcal{I}_r(\cdot)$ at step r is defined as the orthogonal projector mapping $\mathcal{H}$ onto the orthogonal complement $[\rho^{r-1}]^\perp$ of $[\rho^{r-1}]$:

$$\mathcal{I}_r(\cdot) = I(\cdot) - S_{r-1}(\cdot):\mathcal{H}\to[\rho^{r-1}]^\perp \qquad (6.1\text{-}34)$$

where $I(\cdot)$ is the identity transformation in $\mathcal{H}$. Its n-order extension is defined as follows:

$$\mathcal{I}_r^n : \left[\begin{array}{ccc} v^1 & \cdots & v^n\end{array}\right]' \mapsto \left[\begin{array}{ccc} \mathcal{I}_r(v^1) & \cdots & \mathcal{I}_r(v^n)\end{array}\right]' \qquad (6.1\text{-}35)$$

The following identity justifies the terminology used for $\mathcal{I}_r(\cdot)$:

$$\begin{aligned}
e_r &= \rho_r - \hat{\rho}_{r|r-1} = \left[I_p - S_{r-1}^p\right](\rho_r)\\
&= \mathcal{I}_r^p(\rho_r) \qquad &(6.1\text{-}36)
\end{aligned}$$

Proposition 6.1-1.   *The innovator and the estimator, respectively, satisfy the following recursions:*

$$\begin{aligned}
\mathcal{I}_r^n(\cdot) &= \mathcal{I}_{r-1}^n(\cdot) - \left\{\langle\cdot,\mathcal{I}_{r-1}^p(\rho_{r-1})\rangle\right\}L_{r-1}^{-1}\mathcal{I}_{r-1}^p(\rho_{r-1}) \qquad &(6.1\text{-}37)\\
S_r^n(\cdot) &= S_{r-1}^n(\cdot) + \left\{\langle\cdot,\mathcal{I}_r^p(\rho_r)\rangle\right\}L_r^{-1}\mathcal{I}_r^p(\rho_r) \qquad &(6.1\text{-}38)
\end{aligned}$$

where $\mathcal{I}_1(\cdot) = I(\cdot)$, $S_0(\cdot) = O_{\mathcal{H}}$, *and*

$$L_r = \left\{\langle\rho_r,\mathcal{I}_r^p(\rho_r)\rangle\right\} \qquad (6.1\text{-}39)$$

Proof: Equations (37) and (38) follows at once from (32) to (36). Equation (39) is proved by using self-adjointness and idempotency of $\mathcal{I}_r(\cdot)$. In fact, the latter being an orthogonal projector is idempotent, that is, $\mathcal{I}_r^2(\cdot) = \mathcal{I}_r(\cdot)$, and self-adjoint, that is, $\langle \mathcal{I}_r(u), v \rangle = \langle u, \mathcal{I}_r(v) \rangle$, for all $u, v \in \mathcal{H}$. Hence,

$$
\begin{aligned}
L_r &= \{\langle \mathcal{I}_r^p(\rho_r), e_r \rangle\} && [(36)] \\
&= \{\langle \rho_r, \mathcal{I}_r^p(e_r) \rangle\} && \text{[self-adjointness]} \\
&= \{\langle \rho_r, (\mathcal{I}_r^p)^2(\rho_r) \rangle\} && [(36)] \\
&= \{\langle \rho_r, \mathcal{I}_r^p(\rho_r) \rangle\} && \text{[idempotency]}
\end{aligned}
$$

The desired recursive formula for $\hat{w}_{|r}$ can now be obtained by simply applying (38) to the unknown w. Indeed,

$$
\begin{aligned}
\hat{w}_{|r} = S_r^n(w) &= S_{r-1}^n(w) + \{\langle w, \mathcal{I}_r^p(\rho_r) \rangle\} L_r^{-1} \mathcal{I}_r^p(\rho_r) \\
&= \hat{w}_{|r-1} + \{\langle w, \mathcal{I}_r^p(\rho_r) \rangle\} L_r^{-1} \mathcal{I}_r^p(\rho_r) \qquad (6.1\text{-}40)
\end{aligned}
$$

Further, self-adjointness of $\mathcal{I}$ yields

$$
\begin{aligned}
\{\langle w, \mathcal{I}_r^p(\rho_r) \rangle\} &= \{\langle \mathcal{I}_r^n(w), \rho_r \rangle\} && (6.1\text{-}41) \\
&= \{\langle w - \hat{w}_{|r-1}, \rho_r \rangle\} && [(34)] \\
&= \tilde{m}_{r|r-1}
\end{aligned}
$$

where

$$
\tilde{m}_{r|r-1} := m_r - \hat{m}_{r|r-1} \qquad (6.1\text{-}42)
$$

is the one-step-ahead prediction error on the measurement outcome at step r and

$$
\hat{m}_{r|r-1} := \{\langle \hat{w}_{|r-1}, \rho_r \rangle\} \qquad (6.1\text{-}43)
$$

Theorem 6.1-1.   *The ISLM estimate $\hat{w}_{|r}$ of $w \in \mathcal{H}^n$ based on the experiment $\mathcal{E}^r$ satisfies the following recursion:*

$$
\hat{w}_{|r} = \hat{w}_{|r-1} + \tilde{m}_{r|r-1} L_r^{-1} \mathcal{I}_r^p(\rho_r) \qquad (6.1\text{-}44)
$$

where $\mathcal{I}_t(\cdot)$ is the innovator satisfying (37), L_r and $\tilde{m}_{r|r-1}$ are given by (39), and (42), respectively, and $\hat{w}_{|0} = O_{\mathcal{H}^n}$.

The results of Theorem 1 are schematically depicted in Fig. 2.

We consider next the matrix $\{\langle \tilde{w}_{|r}, \tilde{w}_{|r} \rangle\}$ that quantifies the estimation errors at the step r:

$$
\begin{aligned}
M_{r+1} &:= \{\langle \tilde{w}_{|r}, \tilde{w}_{|r} \rangle\} = \{\langle \mathcal{I}_{r+1}^n(w), \mathcal{I}_{r+1}^n(w) \rangle\} && [(34)] \\
&= \{\langle \mathcal{I}_{r+1}^n(w), w \rangle\} && \text{[self-adjointness and idempotency]} \\
&= M_r - \{\langle w, \mathcal{I}_r^p(\rho_r) \rangle\} L_r^{-1} \{\langle \mathcal{I}_r^p(\rho_r), w \rangle\} && (6.1\text{-}45)
\end{aligned}
$$

with $M_1 = \{\langle w, w \rangle\}$. In linear algebra, M_r is known [Lue69] as the Gram matrix of the vectors collected in $\tilde{w}_{|r}$. Its recursive formula (45) is reminiscent of a Riccati

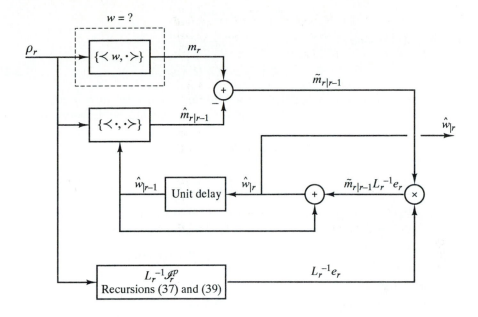

Figure 6.1-2: Block diagram view of algorithm (37)–(44) for computing recursively the ISLM estimate $\hat{w}_{|r}$.

difference equation. In fact, as will be shortly shown, it becomes a Riccati equation once suitable auxiliary assumptions on the structure of the problem at hand are made.

Main Points of the Section. On-line linear MMSE estimation of a random variable from time-indexed observations as well as deterministic system parameter estimation from I/O data can be both seen as particular versions of the ISLM problem. The latter consists of finding recursively the minimum-norm solution to an underdetermined system of linear equations, the number of equations in the system growing by n at each time step.

The ISLM problem can be recursively solved via the Gram–Schmidt orthogonalization procedure by constructing the innovations of the measurement representers.

6.2 KALMAN FILTERING

6.2.1 The Kalman Filter

We apply (1-44) to the linear MMSE estimation setting of Example 1-2 generalized to the random vector case. In such a case,

$$\tilde{m}_{r|r-1} \;=\; \mathcal{E}\{\tilde{w}_{|r-1}\rho'_r\} \qquad (6.2\text{-}1)$$

$$L_r = \mathcal{E}\{\rho_r e_r'\} \tag{6.2-2}$$

Therefore, (1-44) yields

$$\hat{w}_{|r} = \hat{w}_{|r-1} + \mathcal{E}\left\{\tilde{w}_{|r-1}\,\rho_r'\right\}\left(\mathcal{E}\left\{\rho_r e_r'\right\}\right)^{-1} e_r \tag{6.2-3}$$

This is an updating formula for the linear MMSE estimate yet at a quite unstructured level. We next assume that the following relationship holds between w and ρ_r:

$$\rho_r := z(t) = H(t)w + \zeta(t) \tag{6.2-4}$$

where

$$t := r + t_0 - 1 \in \left\{t_0, t_0 + 1, \cdots\right\} \tag{6.2-5}$$

and where $H(t)$ is a $(p \times n)$ matrix with real entries, and $\zeta(t)$ a vector-valued random sequence with zero mean, white, with covariance matrix $\Psi_\zeta(t)$:

$$\mathcal{E}\{\zeta(t)\} = O_p \qquad \mathcal{E}\{\zeta(t)\zeta'(\tau)\} = \Psi_\zeta(t)\delta_{t,\tau} \tag{6.2-6}$$

and uncorrelated with w:

$$\mathcal{E}\{w\zeta'(t)\} = O_{n \times p} \tag{6.2-7}$$

Then

$$\begin{aligned} e(t) &:= & e_r &= \rho_r - \hat{\rho}_{r|r-1} \\ &&&= z(t) - H(t)\hat{w}(t-1) \\ &&&= H(t)\tilde{w}(t-1) + \zeta(t) \end{aligned} \tag{6.2-8}$$

where

$$\hat{w}(t) := \hat{w}_{|t} \qquad \text{and} \qquad \tilde{w}(t) := w - \hat{w}(t) \tag{6.2-9}$$

Further,

$$L(t) := L_r = H(t)M(t)H'(t) + \Psi_\zeta(t) \tag{6.2-10}$$

$$\mathcal{E}\{\tilde{w}(t-1)z'(t)\} = M(t)H'(t)$$

where $M(t) := \mathcal{E}\{\tilde{w}(t-1)\tilde{w}'(t-1)\}$ takes here the form of the following *Riccati difference equation (RDE)*:

$$\begin{aligned} M(t+1) & \\ &= M(t) - \{\langle w, \mathcal{I}_r^n\,(\rho_r)\rangle\}\,L_r^{-1}\,\{\langle \mathcal{I}_r^n\,(\rho_r), w\rangle\} & [(1\text{-}45)] \\ &= M(t) - M(t)H'(t)\Big[H(t)M(t)H'(t) + \Psi_\zeta(t)\Big]^{-1}H(t)M(t) \tag{6.2-11} \end{aligned}$$

with $M(t_0) = \mathcal{E}\{ww'\}$. Hence, (3) becomes

$$\begin{aligned} \hat{w}(t) &= \hat{w}(t-1) \tag{6.2-12} \\ &\quad + M(t)H'(t)\Big[H(t)M(t)H'(t) + \Psi_\zeta(t)\Big]^{-1}\Big[z(t) - H(t)\hat{w}(t-1)\Big] \end{aligned}$$

This result can be extended in a straightforward way to cover the linear MMSE estimation problem of the state

$$w := x(t) \tag{6.2-13}$$

of a dynamic system evolving in accordance with the following equation:

$$x(t+1) = \Phi(t)x(t) + \xi(t) \tag{6.2-14}$$

where $t = t_0, t_0 + 1, \cdots$; $\Phi(t) \in \mathbb{R}^{n \times n}$; and $\xi(t)$ is a vector-valued random sequence with zero mean, white, with covariance matrix $\Psi_\xi(t)$:

$$\mathcal{E}\{\xi(t)\} = O_p \qquad \mathcal{E}\{\xi(t)\xi'(\tau)\} = \Psi_\xi(t)\delta_{t,\tau} \tag{6.2-15}$$

and uncorrelated with $\zeta(\tau)$:

$$\mathcal{E}\{\xi(t)\zeta(\tau)\} = 0 , \quad \forall t, \tau \geq t_0 \tag{6.2-16}$$

Finally, the initial state $x(t_0)$ is a random vector with zero mean and uncorrelated both with $\xi(t)$ and $\zeta(t)$:

$$\mathcal{E}\{x(t_0)\} = 0 \qquad \mathcal{E}\{x(t_0)\xi'(t)\} = 0 \qquad \mathcal{E}\{x(t_0)\zeta'(t)\} = 0 \tag{6.2-17}$$

Theorem 6.2-1. (**Kalman Filter I**) *Consider the linear MMSE estimate $x(t \mid t)$ of the state vector $x(t)$ of the dynamic system (14) to (17) based on the observations z^t:*

$$z^t := \left\{ z(\tau) \right\}_{\tau=t_0}^{t}$$

$$z(\tau) = H(\tau)x(\tau) + \zeta(\tau) \tag{6.2-18}$$

with ζ satisfying (6) and (7). Then, provided that the matrix $L(\tau)$ in (10) is non-singular $\forall \tau = t_0, \cdots, t$, $x(t \mid t)$ is given in recursive form by the estimate update:

$$x(t \mid t) = x(t \mid t-1) + \tilde{K}(t)e(t) \tag{6.2-19}$$

and the time-prediction update:

$$x(t+1 \mid t) = \Phi(t)x(t \mid t) \tag{6.2-20}$$

$$x(t_0 \mid t_0 - 1) = O_n \tag{6.2-21}$$

where

$$\tilde{K}(t) := \Pi(t)H'(t)L^{-1}(t) \tag{6.2-22a}$$

$$K(t) := \Phi(t)\Pi(t)H'(t)L^{-1}(t) \tag{6.2-22b}$$

is the Kalman gain,

$$e(t) = z(t) - H(t)x(t \mid t-1) \tag{6.2-23}$$

is the innovation of the observation process $z(t)$,

$$L(t) = H(t)\Pi(t)H'(t) + \Psi_\zeta(t) \qquad (6.2\text{-}24)$$

$\Pi(t)$ *equals the covariance matrix of the one–step–ahead state-prediction error* $\tilde{x}(t \mid t-1) := x(t) - x(t \mid t-1)$:

$$\Pi(t) = \mathcal{E}\{\tilde{x}(t \mid t-1)\tilde{x}'(t \mid t-1)\} \qquad (6.2\text{-}25)$$

and satisfies the following RDE:

$$
\begin{aligned}
\Pi(t+1) &= \Phi(t)\Pi(t)\Phi'(t) \\
&\quad - \Phi(t)\Pi(t)H'(t)L^{-1}(t)H(t)\Pi(t)\Phi'(t) + \Psi_\xi(t) \qquad (6.2\text{-}26) \\
&= \Phi(t)\Pi(t)\Phi'(t) - K(t)L(t)K'(t) + \Psi_\xi(t) \qquad (6.2\text{-}27) \\
&= [\Phi(t) - K(t)H(t)]' \, \Pi(t) \, [\Phi(t) - K(t)H(t)] \\
&\quad + K(t)\Psi_\zeta(t)K'(t) + \Psi_\xi(t) \qquad (6.2\text{-}28)
\end{aligned}
$$

with $\Pi(t_0) = \mathcal{E}\{\tilde{x}(t_0 \mid t_0-1)\tilde{x}'(t_0 \mid t_0-1)\}$, *the a priori covariance of* $x(t_0)$. *Further, the linear MMSE one-step-ahead prediction of the state is given by*

$$x(t+1 \mid t) = \Phi(t)x(t \mid t-1) + K(t)e(t) \qquad (6.2\text{-}29)$$

Proof: Using (13) in (12), we get

$$
\begin{aligned}
x(t \mid t) &= x(t \mid t-1) + M(t \mid t)H'(t)L^{-1}(t)[z(t) - H(t)x(t \mid t-1)] \\
L(t) &= H(t)M(t \mid t)H'(t) + \Psi_\zeta(t) \\
M(t \mid t) &:= \mathcal{E}\{\tilde{x}(t \mid t-1)\tilde{x}'(t \mid t-1)\}
\end{aligned}
$$

where

$$\tilde{x}(t \mid \tau) := x(t) - x(t \mid \tau)$$

Then (19) and (22) are proven if we can show that $\Pi(t) := M(t \mid t)$ satisfies (26). Now, by (16), (20) holds, and $\tilde{x}(t+1 \mid t) = \Phi(t)\tilde{x}(t \mid t) + \xi(t)$. Consequently,

$$
\begin{aligned}
M(t+1 \mid t+1) &= \Phi(t)M(t \mid t+1)\Phi'(t) + \Psi_\xi(t) \qquad (6.2\text{-}30) \\
M(t \mid t+1) &= \mathcal{E}\{\tilde{x}(t \mid t)\tilde{x}'(t \mid t)\}
\end{aligned}
$$

Further, (11) yields

$$M(t \mid t+1) = M(t \mid t) - M(t \mid t)H'(t)L^{-1}(t)H(t)M(t \mid t) \qquad (6.2\text{-}31)$$

Finally, setting

$$\Pi(t) := M(t \mid t) \qquad (6.2\text{-}32)$$

and combining (30) and (31), we get (26). Equation (29) is obtained by substituting (19) into (20).

We now intend to extend the Kalman filter to cover the case of nonzero means. To this end, consider Example 1-2, where now

$$\left.\begin{array}{ll} \mathcal{E}\{w\} = \bar{w} & \underline{w} := w - \bar{w} \\ \mathcal{E}\{\rho_k\} = \bar{\rho}_k & \underline{\rho}_k := \rho_k - \bar{\rho}_k \end{array}\right\} \tag{6.2-33}$$

Here $\underline{w}$ and $\underline{\rho}_k$ are *centered* random vectors, viz., zero-mean random vectors. Let, similarly to (1-20),

$$\begin{aligned} [\bar{\rho}^r] &:= \quad \text{Span}\left\{\underline{\rho}_k, \bar{\rho}_k, k \in \underline{r}\right\} \\ &= \quad \text{Span}\left\{\rho_k, \bar{\rho}_k, k \in \underline{r}\right\} \\ &= \quad \text{Span}\left\{\rho_k, \mathbb{1}, k \in \underline{r}\right\} \end{aligned} \tag{6.2-34}$$

where $\mathbb{1}$ denotes the random variable equal to one. Then, we refer to

$$\hat{w}_{|r} = \arg\min_{v \in [\bar{\rho}^r]} \left\{\mathcal{E}\left[(w-v)^2\right]\right\} \tag{6.2-35}$$

as the *affine MMSE estimate* of w based on ρ^r or the *linear MMSE estimate* of w based on $\{\rho^r, \mathbb{1}\}$.

Problem 6.2-1 Show that $\hat{w}_{|r}$ in (35) equals

$$\hat{w}_{|r} = \bar{w} + \text{Projec}\left[\underline{w} \mid \underline{\rho}^r\right] \tag{6.2-36}$$

that is, the sum of its a priori mean with the linear MMSE of the centered random vector $\underline{w}$ based on the centered observations $\underline{\rho}^r$.

We use the foregoing result so as to find the affine MMSE estimate of the state $x(t)$ of the system

$$\left.\begin{array}{ll} x(t+1) &= \quad \Phi(t)x(t) + G_u(t)u(t) + \xi(t) \\ z(t) &= \quad H(t)x(t) + \zeta(t) \end{array}\right\} \tag{6.2-37}$$

based on z^t, the observations up to time t,

$$z^t := \left\{ z(t), \quad z(t-1), \quad \cdots, \quad z(t_0) \right\}$$

It is assumed that

$$\mathcal{E}\{x(t_0)\} = x_0 \tag{6.2-38}$$

$$\mathcal{E}\{u(t)\} = \bar{u}(t) \in \mathbb{R}^m \tag{6.2-39}$$

Then, setting $\bar{x}(t) := \mathcal{E}\{x(t)\}$, we have from (37)

$$\left.\begin{array}{ll} \bar{x}(t+1) &= \quad \Phi(t)\bar{x}(t) + G_u(t)\bar{u}(t) \\ \bar{x}(t_0) &= \quad x_0 \end{array}\right\} \tag{6.2-40}$$

Further, letting $\underline{x}(t) := x(t) - \bar{x}(t)$, $\underline{u}(t) := u(t) - \bar{u}(t)$, and $\underline{z}(t) := z(t) - H(t)\bar{x}(t)$, we find

$$\left.\begin{array}{rcl} \underline{x}(t+1) & = & \Phi(t)\underline{x}(t) + G_u(t)\underline{u}(t) + \xi(t) \\ \underline{z}(t) & = & H(t)\underline{x}(t) + \zeta(t) \\ \mathcal{E}\{\underline{x}(t_0)\} & = & 0 , \qquad \mathcal{E}\{\underline{x}(t_0)\underline{x}'(t_0)\} = \Pi(t_0) \end{array}\right\} \qquad (6.2\text{-}41)$$

Before proceeding any further, let us comment on the decomposition $x(t) = \bar{x}(t) + \underline{x}(t)$. First, note that (40) is a predictable system in that $\bar{x}(t)$ can be precomputed. On the opposite, (41) is unpredictable owing to the uncertainty on its initial state $\underline{x}(t_0)$, and the presence of the inaccessible disturbance ξ.

Let us assume that $\underline{u}(t)$ equals a vector-valued linear function of $\underline{z}(\tau)$ up to time t, viz.,

$$\underline{u}(t) = \underline{f}\left(t, \underline{z}^t\right) \qquad (6.2\text{-}42)$$

The linear MMSE estimate $\underline{x}(t \mid t)$ of $\underline{x}(t)$ based on $\underline{z}^t$ equals

$$\left.\begin{array}{rcl} \underline{x}(t \mid t) & = & \underline{x}(t \mid t-1) \\ & & + \Pi(t)H'(t)L^{-1}(t)\left[\underline{z}(t) - H(t)\underline{x}(t \mid t-1)\right] \\ \underline{x}(t+1 \mid t) & = & \Phi(t)\underline{x}(t \mid t) + G_u(t)\underline{u}(t) \end{array}\right\} \qquad (6.2\text{-}43)$$

If we add $\bar{x}(t)$ to the first of (43), and $\bar{x}(t+1)$ to the second of (43), and set

$$\left.\begin{array}{rcl} x(t \mid t) & := & \underline{x}(t \mid t) + \bar{x}(t) \\ x(t+1 \mid t) & := & \underline{x}(t+1 \mid t) + \bar{x}(t+1) \end{array}\right\} \qquad (6.2\text{-}44)$$

we find, according to (36), the desired result.

Theorem 6.2-2. **(Kalman Filter II)** *Consider the state vector $x(t)$ of the dynamic system (37), where all the centered random vectors satisfy (6) to (7) and (14) to (17), and its linear MMSE estimate $x(t \mid t)$ based on the observations $\{z^t, \mathbf{I\!I}\}$. Let (38) hold and $u(t)$ be linear in $\{z^t, \mathbf{I\!I}\}$:*

$$\left.\begin{array}{l} u(t) = f\left(t, z^t, \mathbf{I\!I}\right) \\ f(t, \cdot, \cdot) \quad \text{linear} \end{array}\right\} \qquad (6.2\text{-}45)$$

Then, under the same assumptions as in Theorem 1, $x(t \mid t)$ satisfies the estimate update:

$$x(t \mid t) = x(t \mid t-1) + \tilde{K}(t)e(t) \qquad (6.2\text{-}46)$$

and the time prediction update:

$$x(t+1 \mid t) = \Phi(t)x(t \mid t) + G_u(t)u(t) \qquad (6.2\text{-}47)$$

where $x(t_0 \mid t_0 - 1) = \mathcal{E}\{x(t_0)\}$ and (22) to (28) hold true. Furthermore,

$$x(t+1 \mid t) = \Phi(t)x(t \mid t-1) + G_u(t)u(t) + K(t)e(t) \qquad (6.2\text{-}48)$$

Proof: It suffices to note that

$$\underline{z}(t) - H(t)\underline{x}(t \mid t-1) = z(t) - H(t)\left[\bar{x}(t) + \underline{x}(t \mid t-1)\right] = z(t) - H(t)x(t \mid t-1)$$

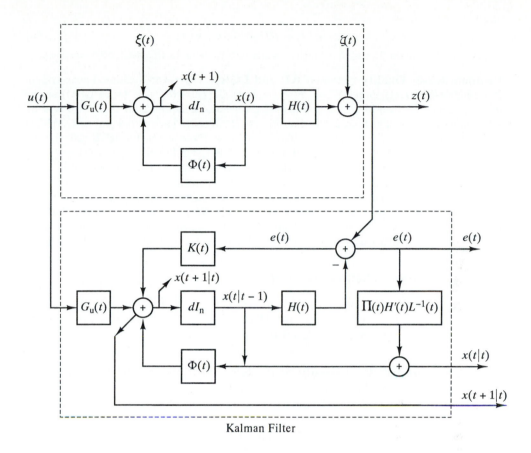

Kalman Filter

Figure 6.2-1: Illustration of the Kalman filter.

Terminology. For the sake of simplicity, from now on we shall refer to the affine MMSE estimate (35) as the *linear* MMSE estimate $\hat{w}_{|r}$ of w based on ρ^r by adhering to the convention to include in ρ^r the random variable $\rho_0 := \mathbb{I}$.

Equations (46) and (47), or (48), along with (22) to (28), are called the *Kalman filter (KF)*. Figure 1 shows a diagrammatic representation of the KF. The KF outputs indicated in Fig. 1 are the innovations process $e(t)$, the state *filtered estimate* $x(t \mid t)$, and the one-step-ahead *prediction estimate* $x(t + 1 \mid t)$. The first, which in view of (23) is a by-product of $x(t \mid t - 1)$, is indicated to point out that the KF can be also considered as an innovations generator.

Problem 6.2-2   Show that for every integer $k \in \mathbb{Z}_+$ the linear MMSE k-step-ahead prediction of $x(t + k)$ based on $\{z^t, \mathbb{I}\}$ is given by

$$x(t + k \mid t) = \Phi(t + k, t)x(t \mid t) + \varphi(t + k, t, O_x, u_{[t,t+k)}) \qquad (6.2\text{-}49)$$

and

$$z(t+k \mid t) = H(t+k)x(t+k \mid t) \tag{6.2-50}$$

where $x(t \mid t)$ is as in Theorem 2 and the same notations as in Problem 2.2-1 are used.

Problem 6.2-3 (Duality between KF and LQR) Given the dynamic linear system $\Sigma = (\Phi(t), G(t), H(t))$, define $\Sigma^* = (\Phi^*(t) := \Phi'(t), G^*(t) := H'(t), H^*(t) := G'(t))$ as the *dual system* of Σ. Consider next the LQOR problem of Chapter 2 for $M(t) \equiv 0$ and its solution as given by Theorem 2.3-1. Consider $\Psi_\xi(t)$ in (15). Factorize $\Psi_\xi(t)$ as $G(t)G'(t)$ with $G(t)$ a full column-rank matrix. Show that, if $\Psi_\zeta(t) = \psi_u(t)$, the RDE (26) for the KF and the system Σ is the same as the one of (2.3-3) for LQOR with $\psi_y(t) = I$, and the plant Σ^*, the only difference being that whereas the first is updated forward in time, the latter is a backward difference equation. Show that under the prior duality conditions the state-transition matrix $\Phi^*(t) + G^*(t)F(t)$ of the closed-loop LQOR system is the transposed of the state-transition matrix $\Phi(t) - K(t)H(t)$ of the KF provided that the two RDEs (2.3-3) and (26), respectively, are iterated, backward and forward, by the same number of steps starting from a common initial nonnegative definite matrix $\Pi_0 = \mathcal{P}(T)$.

6.2.2   Steady-State Kalman Filtering

Consider the time invariant KF problem, viz.: $\Phi(t) = \Phi$; $\Psi_\xi(t) = \Psi_\xi = GG'$, with G of full column rank; $H(t) = H$; and $\Psi_\zeta(t) = \Psi_\zeta$. We can use duality between KF and LQR to adapt Theorem 2.4-5 to the present context.

Theorem 6.2-3.   (Steady-State KF) *Consider the time-invariant KF problem and the related matrix sequence $\{\Pi(t)\}_{t=0}^{\infty}$ generated via the Riccati iterations (26) to (28) initialized from any $\Pi(0) = \Pi'(0) \geq 0$. Assume that $\Psi_\zeta = \Psi'_\zeta > 0$. Then there exists*

$$\Pi = \lim_{t \to \infty} \Pi(t) \tag{6.2-51}$$

such that

$$
\begin{aligned}
\Pi &= \lim_{t \to \infty} \mathcal{E}\left\{\tilde{x}(t \mid t-1)\tilde{x}'(t \mid t-1)\right\} \tag{6.2-52} \\
&= \lim_{t \to \infty} \min_{v \in \left[z^{t-1}, \mathbf{\Pi}\right]} \mathcal{E}\left\{[x(t) - v][x(t) - v]'\right\}
\end{aligned}
$$

Further the limiting filter

$$
\begin{aligned}
x(t \mid t) &= x(t \mid t-1) + \Pi H' L^{-1} e(t) \tag{6.2-53} \\
x(t+1 \mid t) &= \Phi x(t \mid t) + G_u(t)u(t) \tag{6.2-54}
\end{aligned}
$$

with

$$L = H\Pi H' + \Psi_\zeta \tag{6.2-55}$$

or

$$
\begin{aligned}
x(t+1 \mid t) &= \Phi x(t \mid t-1) + G_u(t)u(t) + Ke(t) \tag{6.2-56} \\
e(t) &= z(t) - Hx(t \mid t-1) \tag{6.2-57}
\end{aligned}
$$

with K the steady-state Kalman gain

$$K = \Phi \Pi H' L^{-1} \qquad (6.2\text{-}58)$$

is asymptotically stable, that is, $\Phi - KH$ is a stability matrix, if and only if the system (Φ, G, H) generating the observations $z(t)$ is stabilizable and detectable. Further, under such conditions, the matrix Π in (51) coincides with the unique symmetric nonnegative definite solution of the following algebraic Riccati equation (ARE):

$$\Pi = \Phi \Pi \Phi' - \Phi \Pi H' \left(H \Pi H' + \Psi_\zeta\right)^{-1} H \Pi \Phi' + \Psi_\xi \qquad (6.2\text{-}59)$$

Terminology. We shall refer to the conditions of Theorem 3 along with the properties of stabilizability and detectability of (Φ, G, H) as the *standard case* of steady-state KF.

6.2.3   Correlated Disturbances

We again consider the system

$$\left. \begin{array}{rcl} x(t+1) & = & \Phi(t)x(t) + G_u(t)u(t) + \xi(t) \\ z(t) & = & H(t)x(t) + \zeta(t) \end{array} \right\} \qquad (6.2\text{-}60)$$

along with the usual assumptions. Instead of (16), it is assumed hereafter that

$$\mathcal{E}\{\xi(t)\zeta'(\tau)\} = S(t)\delta_{t,\tau} \qquad (6.2\text{-}61)$$

with $S(t) \in \mathbb{R}^{n \times p}$ possibly a nonzero matrix. In order to extend Theorem 1 to the present case, it is convenient to introduce the linear MMSE estimate $\check{\xi}(t)$ of $\xi(t)$ based on ζ^t:

$$\begin{array}{rcll} \check{\xi}(t) & := & \text{Projec}[\xi(t) \mid \zeta^t] & (6.2\text{-}62\text{a}) \\ & = & \text{Projec}[\xi(t) \mid \zeta(t)] & [(61)] \\ & = & \mathcal{E}\{\xi(t)\zeta'(t)\}\{\mathcal{E}\{\zeta(t)\zeta'(t)\}\}^{-1}\zeta(t) & [(1\text{-}24)] \\ & = & S(t)\Psi_\zeta^{-1}(t)\zeta(t) & \end{array}$$

Set

$$\tilde{\xi}(t) := \xi(t) - \check{\xi}(t) \qquad (6.2\text{-}62\text{b})$$

Problem 6.2-4   Show that $\tilde{\xi}(t)$ is uncorrelated with $\zeta(\tau)$:

$$\mathcal{E}\{\tilde{\xi}(t)\zeta(\tau)\} = O_{n \times p}, \quad \forall t, \tau \geq t_0 \qquad (6.2\text{-}62\text{c})$$

and white with covariance matrix

$$\Psi_{\tilde{\xi}}(t) = \Psi_\xi(t) - S(t)\Psi_\zeta(t)S'(t) \qquad (6.2\text{-}62\text{d})$$

By using (62), the first of (60) can be rewritten as follows:

$$
\begin{aligned}
x(t+1) &= \Phi(t)x(t) + G_u(t)u(t) + \check{\xi}(t) + \tilde{\xi}(t) \qquad\qquad\qquad (6.2\text{-}63) \\
&= \Phi(t)x(t) + G_u(t)u(t) + S(t)\Psi_\zeta^{-1}(t)\Big[z(t) - H(t)x(t)\Big] + \tilde{\xi}(t) \\
&= \Big[\Phi(t) - S(t)\Psi_\zeta^{-1}(t)H(t)\Big]x(t) \\
&\quad + G_u(t)u(t) + S(t)\Psi_\zeta^{-1}(t)z(t) + \tilde{\xi}(t)
\end{aligned}
$$

This system is now in the standard form (40) because $\tilde{\xi}(t)$ and $\zeta(t)$ are both zero mean, white, and mutually uncorrelated, and the extra input $S(t)\Psi_\zeta^{-1}(t)z(t) \in$ Span $\{z^t\}$. Thus, Theorem 2 can be used to get an estimate update identical with (46) and the following time-prediction update:

$$
\begin{aligned}
x(t+1\mid t) &= \Big[\Phi(t) - S(t)\Psi_\zeta^{-1}(t)H(t)\Big]x(t\mid t) \qquad\qquad (6.2\text{-}64\text{a}) \\
&\quad + G_u(t)u(t) + S(t)\Psi_\zeta^{-1}(t)z(t)
\end{aligned}
$$

Problem 6.2-5 Prove that the state prediction-error covariance $\Pi(t)$, as defined as in (24), in the present case satisfies the recursion

$$
\begin{aligned}
\Pi(t+1) &= \Phi(t)\Pi(t)\Phi'(t) \qquad\qquad\qquad\qquad\qquad (6.2\text{-}64\text{b}) \\
&\quad - \Big[\Phi(t)\Pi(t)H'(t) + S(t)\Big]L^{-1}(t)\Big[\Phi(t)\Pi(t)H'(t) + S(t)\Big]' + \Psi_\xi(t)
\end{aligned}
$$

Problem 6.2-6 Prove that the recursive equation for $x(t\mid t-1)$ can be rearranged as follows:

$$
x(t+1\mid t) = \Phi(t)x(t\mid t-1) + G_u(t)u(t) + K(t)[z(t) - H(t)x(t\mid t-1)] \quad (6.2\text{-}64\text{c})
$$

where $K(t)$ is the Kalman gain:

$$
K(t) = \Big[\Phi(t)\Pi(t)H'(t) + S(t)\Big]L^{-1}(t) \qquad\qquad\qquad (6.2\text{-}64\text{d})
$$

Problem 6.2-7 Consider the time-invariant linear system

$$
\left.
\begin{aligned}
x(t+1) &= \Phi x(t) + G_u u(t) + G\zeta(t) \\
z(t) &= Hx(t) + \zeta(t)
\end{aligned}
\right\}
$$

where the possibly non-Gaussian process ζ satisfies the following martingale difference properties (*cf.* Appendix D.5 and Sec. 7.2 for the definition of $\{\mathcal{F}_k\}$):

$$
\mathcal{E}\left\{\zeta(t)\mid \mathcal{F}_{t-1}\right\} = O_p \quad \text{a.s.}
$$

$$
\infty > \mathcal{E}\left\{\zeta(t)\zeta'(t)\mid \mathcal{F}_{t-1}\right\} = \Psi_\zeta > 0 \quad \text{a.s.}
$$

and

$$
\mathcal{E}\left\{\zeta(t_0)x'(t_0)\right\} = O_{p\times m}
$$

Let $\Phi - GH$ be a stability matrix and $u(t)$ satisfy (45). Let $x(t\mid k) := \mathcal{E}\{x(t)\mid z^k\}$. Then, show that $\lim_{t\to\infty} \Pi(t) = 0$, as $t \to \infty$, $x(t\mid t) \to x(t\mid t-1)$, and $x(t\mid t-1)$ satisfies the recursions

$$
\left.
\begin{aligned}
x(t+1\mid t) &= \Phi x(t\mid t-1) + G_u u(t) + G\zeta(t) \\
\zeta(t) &= z(t) - Hx(t\mid t-1)
\end{aligned}
\right\}
$$

6.2.4   Distributional Interpretation of the Kalman Filter

It is known [Cai88] that if u and v are two jointly Gaussian random vectors with zero mean, the orthogonal projection $\text{Projec}[u \mid [v]]$ coincides with the conditional expectation $\mathcal{E}\{u \mid v\}$, which, in turn, yields the unconstrained, viz., linear or nonlinear, MMSE estimate of u based on v. Then, it follows that if $x(t_0)$, $\xi(\tau)$, and $\zeta(\tau)$ are jointly Gaussian, $x(t \mid t-1)$ and $\Pi(t)$ generated by the KF coincide with the conditional expectation $\mathcal{E}\left\{x(t) \mid z^{t-1}\right\}$ and conditional covariance $\text{Cov}\left(x(t) \mid z^{t-1}\right)$, respectively. Further, because under the stated assumptions, the conditional probability distribution $P\left(x(t) \mid z^{t-1}\right)$ of $x(t)$ given z^{t-1} is Gaussian, we have

$$P\left(x(t) \mid z^{t-1}\right) = N\left(x(t \mid t-1), \Pi(t)\right) \qquad (6.2\text{-}65)$$

where $N(\bar{x}, \Pi)$ denotes the Gaussian or normal probability distribution with mean $\bar{x}$ and covariance Π, viz., assuming Π nonsingular and hence considering the probability density function $n(\hat{x}, \Pi)$ corresponding to $N(\hat{x}, \Pi)$,

$$n(\hat{x}, \Pi) = \left[(2\pi)^n \det \Pi\right]^{-1/2} \exp\left[-\frac{1}{2}\|\hat{x}\|^2_{\Pi^{-1}}\right]$$

The operation of the KF under such hypotheses is referred to as the *distributional version*, or *interpretation, of the Kalman filter.*

An interesting and useful extension of the distributional KF is the *conditionally Gaussian Kalman filter* wherein all the deterministic quantities at time t in the filter derivation are known once a realization of z^t is given. For example, (46) to (48) are still valid in case $u(t) = f(t, z^t)$ with $f(t, \cdot)$ possibly nonlinear. This is of interest in problems where input u is generated by a causal feedback.

Fact 6.2-1.   (Kalman Filter III) *Suppose that in the dynamic system (37) $x(t_0)$, ξ, and ζ are jointly Gaussian. Let all the centered random vectors satisfy (6) to (7) and (14) to (17). Assume that*

$$u(t) = f\left(t, z^t\right)$$

with $f(t, \cdot)$ possibly nonlinear. Then, the conditional expectation $\mathcal{E}\{x(t) \mid z^{t-1}\}$ of state $x(t)$ given z^{t-1} coincides with vector $x(t \mid t-1)$ generated by the KF equations of Theorem 2 or their extension (64).

Recall (Appendix D) that the conditional mean $\mathcal{E}\{x(t) \mid z^{t-1}\}$ is the MMSE estimator of $x(t)$ based on z^{t-1} amongst all possible linear and nonlinear estimators of $x(t)$.

6.2.5   Innovations Representation

As shown in Fig. 1, one of the KF outputs is the innovations process e. In this respect, the KF plays the role of a whitening filter, its input process z being trans-

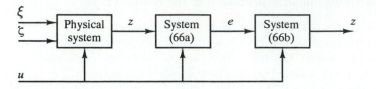

Figure 6.2-2: Illustration of the KF as an innovations generator. The third system recovers z from its innovations e.

formed into the white innovations process e:

$$
\begin{aligned}
x(t+1\mid t) &= \begin{aligned}[t] &[\Phi(t) - K(t)H(t)]\,x(t\mid t-1) \\ &+ G_u(t)u(t) + K(t)z(t) \end{aligned} \\
e(t) &= -H(t)x(t\mid t-1) + z(t)
\end{aligned}
\left.\vphantom{\begin{aligned}x\\+\\e\end{aligned}}\right\} \qquad (6.2\text{-}66a)
$$

On the other hand, as was noticed in [Kai68], the observation process z can be recovered via the "inverse" of (66a):

$$
\begin{aligned}
x(t+1\mid t) &= \Phi(t)x(t\mid t-1) + G_u(t)u(t) + K(t)e(t) \\
z(t) &= H(t)x(t\mid t-1) + e(t)
\end{aligned}
\left.\vphantom{\begin{aligned}x\\z\end{aligned}}\right\} \qquad (6.2\text{-}66b)
$$

The situation is depicted in Fig. 2, where it is also shown that z is generated by the dynamic system (37), labelled "Physical System." Equation (66b) is called the *innovations representation* of process z.

In the time-invariant case, under the validity conditions of Theorem 3 and in steady state, we can compute the transfer matrices associated with (66a) and (66b). We find

$$
\begin{aligned}
& \begin{bmatrix} H_{eu}(d) & H_{ez}(d) \end{bmatrix} \\
&= -H\Big(I_n - d(\Phi - KH)\Big)^{-1} \begin{bmatrix} dG_u & dK \end{bmatrix} + \begin{bmatrix} O_{p\times m} & I_p \end{bmatrix} \qquad (6.2\text{-}67a) \\
&= C^{-1}(d) \begin{bmatrix} B(d) & A(d) \end{bmatrix} \qquad (6.2\text{-}67b)
\end{aligned}
$$

$$
\begin{aligned}
& \begin{bmatrix} H_{zu}(d) & H_{ze}(d) \end{bmatrix} \\
&= H\Big(I_n - d\Phi\Big)^{-1} \begin{bmatrix} dG_u & dK \end{bmatrix} + \begin{bmatrix} O_{p\times m} & I_p \end{bmatrix} \qquad (6.2\text{-}67c) \\
&= A^{-1}(d) \begin{bmatrix} -B(d) & C(d) \end{bmatrix} \qquad (6.2\text{-}67d)
\end{aligned}
$$

Problem 6.2-8 By using the matrix-inversion lemma (5.3-21), verify that in (67a) and (67c), $H_{ze}^{-1}(d) = H_{ez}(d)$. Further, check that $H_{eu}(d) = -H_{ez}(d)H_{zu}(d)$, and $H_{zu}(d) = -H_{ze}(d)H_{eu}(d)$.

In (67b), $C^{-1}(d)\begin{bmatrix} B(d) & A(d) \end{bmatrix}$ denotes a left-coprime MFD of the transfer matrix in (67a). Then, it follows from Fact 3.1-1 that

$$\det C(d) \mid \det C(0) \cdot \chi_{\Phi_K}(d) \tag{6.2-68a}$$

$$\Phi_K := \Phi - HK \tag{6.2-68b}$$

In the standard case of steady-state KF, the state-transition matrix Φ_K is asymptotically stable. Hence, the $p \times p$ polynomial matrix $C(d)$ is strictly Hurwitz.

The discussion is summarized in the following theorem.

Theorem 6.2-4. *The KF (66a) causally transforms the observation process z into its innovations process e. This transformation admits a causal inverse (66b), called* innovations representation *of z. In the standard case of steady-state KF, (66a) becomes time-invariant and asymptotically stable, yielding a stationary innovations process with covariance $H'\Pi H + \Psi_\zeta$.*

Finally, the $p \times p$ polynomial matrix $C(d)$ in the I/O innovations representation obtained from (67d)

$$A(d)z(t) = B(d)u(t) + C(d)e(t) \tag{6.2-69a}$$

satisfies (68) and hence, in the standard case, $C(d)$ is strictly Hurwitz.

In the statistics and engineering literature, the innovation representation (69) is called an *ARMAX (autoregressive moving average with exogenous inputs)* or a *CARMA (controlled ARMA)* model. ARMAX models have become widely known and exploited in time-series analysis, econometrics, and engineering ([Åst70] and [BJ76]).

The word *exogenous* has been adopted in time-series analysis and econometrics to describe any influence that originates outside the system. In control theory, however, the process u appearing in an ARMAX system is a control input that may be a function of past values of y and u. For this reason, in such cases, an ARMAX system is more appropriately referred to as a CARMA model.

Whenever $C(d) = I_p$ in (9a), the resulting representation

$$A(d)z(t) = B(d)u(t) + e(t) \tag{6.2-69b}$$

is called an *ARX (autoregressive with exogenous inputs)* or a *CAR (controlled AR)* model.

6.2.6 Solution via Polynomial Equations

The polynomial equation approach of Chapter 4 can be adapted *mutatis mutandis* to solve the steady-state KF problem. We give here the relevant polynomial equations without a detailed derivation. The reason is that the results that follow consist of the direct dual equations of the ones obtained in Chapter 4. The interested reader is referred to [CM92b] for a thorough discussion of the topic.

We consider the following time-invariant version of (37):

$$
\left.
\begin{array}{rl}
x(t+1) &= \Phi x(t) + G_u(t)u(t) + G\nu(t) \\
z(t) &= Hx(t) + \zeta(t)
\end{array}
\right\} \tag{6.2-70a}
$$

where ν and ζ are zero-mean mutually uncorrelated processes with constant covariance matrices

$$
\mathcal{E}\{\nu(t)\nu'(\tau)\} = \Psi_\nu \delta_{t,\tau} \qquad \mathcal{E}\{\zeta(t)\zeta'(\tau)\} = \Psi_\zeta \delta_{t,\tau} \tag{6.2-70b}
$$

where $\Psi_\nu = \Psi_\nu' > 0$, and $\Psi_\zeta = \Psi_\zeta' \geq 0$. Define the following polynomial matrices:

$$
A(d) := I_n - d\Phi \qquad\qquad B(d) := dH \tag{6.2-70c}
$$

Find a left-coprime MDF $A_1^{-1}(d)B_1(d)$ of $B(d)A^{-1}(d)$:

$$
A_1^{-1}(d)B_1(d) = B(d)A^{-1}(d) \tag{6.2-70d}
$$

Note that $B(d)A^{-1}(d)$ is a MFD of the transfer matrix from $\xi(t) := G\nu(t)$ and $z(t)$. Next, find a $p \times p$ Hurwitz polynomial matrix $C(d)$ solving the following *left-spectral factorization problem*:

$$
C(d)C^*(d) = A_1(d)\Psi_\zeta A^*(d) + B_1(d)\Psi_\xi B_1^*(d) \tag{6.2-71a}
$$

with $\Psi_\xi := G\Psi_\nu G'$. Let

$$
q := \max\{\partial A_1(d), \partial B_1(d)\} \tag{6.2-71b}
$$

where $\partial A(d)$ denotes the degree of polynomial matrix $A(d)$. Define

$$
\bar{C}(d) := d^q C^*(d) ; \quad \bar{A}_1(d) := d^q A_1^*(d) ; \quad \bar{B}_1(d) := d^q B_1^*(d) \tag{6.2-71c}
$$

Let the greatest common right divisors of $A(d)$ and $B(d)$ be strictly Hurwitz, that is, (cf. B.1-5) the pair (Φ, H) detectable. Then, from the dual of Lemma 4.4-1, it follows that there is a unique solution $(X, Y, Z(d))$ of the following system of bilateral Diophantine equations:

$$
Y\bar{E}(d) + A(d)Z(d) = \Psi_\xi \bar{B}_1(d) \tag{6.2-72a}
$$
$$
X\bar{E}(d) - B(d)Z(d) = \Psi_\zeta \bar{A}_1(d) \tag{6.2-72b}
$$

with $\partial Z(d) < \partial \bar{E}(d)$.

Problem 6.2-9 Show that a triplet $(X, Y, Z(d))$ is a solution of (72a) and (72b) if and only if it solves (72a) and

$$
A_1(d)X + B_1(d)Y = C(d) \tag{6.2-72c}
$$

The dual of Theorem 4.4-1 gives the polynomial solution of the steady-state KF problem.

Theorem 6.2-5. *Let (Φ, H) be a detectable pair, or equivalently, $A(d) := I - d\Phi$ and $B(d) := dH$ have strictly Hurwitz gcrd's. Let $(X, Y, Z(d))$ be the minimum degree solution w.r.t. $Z(d)$ of the bilateral Diophantine equations (72a) and (72b) [or (72a) and (72c)]. Then, the constant matrix*

$$K = YX^{-1} \qquad\qquad (6.2\text{-}73)$$

makes $\Phi_K = \Phi - HK$ a stability matrix if and only if the spectral factor $C(d)$ in (71a) is strictly Hurwitz. In such a case, (73) yields the Kalman gain of the steady-state KF. Further, if (Φ, H) is reconstructible, the d-characteristic polynomial $\chi_{\Phi_K}(d)$ equals

$$\chi_{\Phi_K}(d) = \frac{\det C(d)}{\det C(0)} \qquad\qquad (6.2\text{-}74)$$

If Ψ_ζ is positive definite, $C(d)$ is strictly Hurwitz if (Φ, G, H) is stabilizable and detectable.

Finally, if (Φ, H) is an observable pair, the matrix pair (X, Y) in (73) is the constant solution of the unilateral Diophantine equation (72c).

The results dual to the ones of Sec. 4.5 that give the relationship between the polynomial and the Riccati-based solution of Theorem 3 follow:

$$\begin{aligned}
\Pi &= \Phi\Pi\Phi' - YY' + \Psi_\xi & (6.2\text{-}75a) \\
YY' &= \Phi\Pi H' \left(\Psi_\zeta + H\Pi H'\right)^{-1} H\Pi\Phi' & (6.2\text{-}75b) \\
XX' &= \Psi_\zeta + H\Pi H' & (6.2\text{-}75c) \\
YX' &= \Phi\Pi H' & (6.2\text{-}75d) \\
Z(d) &= \Pi\bar{B}_1(d) & (6.2\text{-}75e)
\end{aligned}$$

Main Points of the Section. The Kalman filter of Theorem 2 or its extension (64) for correlated disturbances gives the recursive linear MMSE estimate of the state of a stochastic linear dynamic system based on noisy output observations. Under the Gaussian regime, the Kalman filter yields the conditional distribution of the state given the observations.

State-space and I/O innovations representations, such as ARMAX and ARX processes, result from Kalman filtering theory.

6.3   SYSTEM PARAMETER ESTIMATION

The theory of optimal filtering and control assumes the availability of a mathematical model capable of adequately describing the behavior of the system under consideration. Such models can be obtained from the physical laws governing the system or by some form of data analysis. The latter approach, referred to as *system identification*, is appropriate when the system is highly complex or imprecisely understood, but the behavior of the relevant I/O variables can be adequately described by simple models. System identification should not be necessarily seen as an alternative to physical modeling in that it can be used to refine an incomplete model derived via the latter approach.

The system identification methodology involves a number of steps like the following:

1. Selection of a *model set* from which a model that adequately fits the experimental data has to be chosen.

2. *Experiment design* whereby the inputs to the unknown system are chosen and the measurements to be taken planned.

3. *Model selection* from the experimental data.

4. *Model validation* where the selected model is accepted or rejected on the grounds of its adequacy to some specific task such as prediction or control system design.

For these various aspects of system identification, we refer the reader to related specific standard textbooks, for example, [Lju87] and [SS89].

In this section, we limit our considerations to models consisting of linear time-invariant dynamic systems parameterized by a vector with real components. We focus the attention on how to suitably choose one model fitting the experimental data. This aspect of the system identification methodology is usually referred to as *parameter estimation*. This terminology is somewhat misleading in that it suggests the existence of a "true" parameter by which the system can be exactly represented in the model set. In fact, because in practice this is never achieved, the aim of system identification merely consists in the selection of a model whose response is capable of adequately approximating that of the unknown underlying system.

Also with parameter estimation algorithms, our choice has been quite selective. In fact, the main emphasis is on algorithms that admit a prediction-error formulation. In particular, no description is given here of instrumental variables methods for which we refer the reader to standard textbooks.

We now consider various recursive algorithms for estimating the parameters of a time-invariant linear dynamic system model from I/O data.

6.3.1   Linear-Regression Algorithms

We start by assuming that the system with inputs $u(k) \in \mathbb{R}$ and outputs $y(k) \in \mathbb{R}$ is exactly represented by the difference equation:

$$A(d)y(k) = B(d)u(k) \tag{6.3-1a}$$

$$A(d) = 1 + a_1 d + \cdots + a_{n_a} d^{n_a} \tag{6.3-1b}$$

$$B(d) = \qquad b_1 d + \cdots + b_{n_b} d^{n_b} \tag{6.3-1c}$$

where n_a and n_b are assigned, and the $n_\theta := n_a + n_b$ parameters a_i and b_i are unknown reals. We shall rewrite (1a) in the form

$$y(k) = \varphi'(k-1)\theta \tag{6.3-2a}$$

$$\varphi(k-1) \quad := \quad \left[\ (-y_{k-n_a}^{k-1})' \quad (u_{k-n_b}^{k-1})' \ \right]' \quad \in \mathbb{R}^{n_\theta} \tag{6.3-2b}$$

$$\theta \quad := \quad \left[\ a_1 \quad \cdots \quad a_{n_a} \quad b_1 \quad \cdots \quad b_{n_b} \ \right]' \quad \in \mathbb{R}^{n_\theta} \tag{6.3-2c}$$

The problem is to estimate θ from the knowledge of $y(k)$ and $\varphi(k-1)$, $k \in \mathbb{Z}_1$. We begin by applying Theorem 1-1 so as to find the ISLM estimate of θ.

As in Example 1-1, we set

$$w := \theta \qquad \rho_k := \varphi(k-1) \qquad m_k := y(k) \tag{6.3-3}$$

$$\mathcal{H} \quad : \quad \left\{ \begin{array}{l} \text{Euclidean vector space of dimension } n_\theta \\ \text{with inner product} \langle w, \rho_k \rangle = \rho_k' w \end{array} \right\}$$

Here the ISLM problem of Sec. 1 amounts to finding a recursive formula for updating the minimum-norm system parameter vector θ interpolating the I/O data up to time t.

Because here $\dim \mathcal{H} = n_\theta$, the t–innovator (1-34) consists of an $n_\theta \times n_\theta$ symmetric nonnegative definite matrix:

$$P(t-1) := \mathcal{I}_t = I_{n_\theta} - S_{t-1} \tag{6.3-4}$$

$P(t-1)$ can be computed recursively via (1-37) as follows. First, note that, because of (1-39),

$$L_t = \varphi'(t-1)P(t-1)\varphi(t-1) \tag{6.3-5}$$

Then, setting

$$\theta(t) := \hat{\theta}_{|t} \tag{6.3-6}$$

the following algorithm follows at once from (1-44).

Orthogonalized Projection Algorithm

$$\theta(t+1) \quad = \quad \left\{ \begin{array}{ll} \theta(t) & , \ \text{if } L_{t+1} = 0 \\ \theta(t) + \frac{P(t)\varphi(t)}{\varphi'(t)P(t)\varphi(t)} \left[y(t+1) - \varphi'(t)\theta(t) \right] & , \ \text{otherwise} \end{array} \right. \tag{6.3-7a}$$

$$P(t+1) \quad = \quad \left\{ \begin{array}{ll} P(t) & , \ \text{if } L_{t+1} = 0 \\ P(t) - \frac{P(t)\varphi(t)\varphi'(t)P(t)}{\varphi'(t)P(t)\varphi(t)} & , \ \text{otherwise} \end{array} \right. \tag{6.3-7b}$$

with $\theta(0) = O_{n_\theta}$, and $P(0) = I_{n_\theta}$. Note that $L_{t+1} = 0$ is equivalent to the condition $\varphi(t) \in \text{Span}\{\varphi(k)\}_{k=0}^{t-1}$.

The name of the algorithm (7) is justified because $\theta(t)$ given by (7) equals the orthogonal projection of the unknown θ onto $\text{Span}\{\varphi(k)\}_{k=0}^{t-1}$. The condition on L_{t+1} has to be used because $\varphi(t)$ can be linearly dependent on $\{\varphi(k)\}_{k=0}^{t-1}$. In any case, after n_θ output observations corresponding to n_θ linearly independent vectors $\varphi(k)$, the orthogonalized projection algorithm converges to the true θ vector in the ideal deterministic case under consideration. As will be seen soon, in order to face the nonideal case in which (7) become impractical, the algorithm can be modified in various ways, for example, recursive least squares. These modified algorithms are usually started up by assigning arbitrary initial values to $\theta(0)$. This makes the estimates $\theta(t)$ dependent on the chosen initialization. A correction to this procedure is to exploit the innovative initial portion of the orthogonalized projection algorithm so as to get a data-based initial guess on the unknown θ to start up the modified recursions.

By adhering to a terminology borrowed from statistics, we shall refer to $\varphi(t)$ and $y(t+1)$ as the *regressor* and the *regressand*, respectively.

Example 6.3-1 (FIR Estimation by PRBS)

Consider the system (1) with $n_a = 0$, viz., $y(t) = B(d)u(t)$. This is usually called a finite impulse-response (FIR) system. Note that, because $n_a = 0$, here $\varphi(k-1) = u_{k-n_b}^{k-1}$, $\theta = \begin{bmatrix} b_1 & b_2 & \cdots & b_{n_b} \end{bmatrix}'$, and hence $n_\theta = n_b$. We assume that input signal $u(t)$ is a specific probing signal made up of a periodic pseudorandom binary sequence (PRBS) ([PW71] and [SS89]). This signal has amplitude $+V$ and $-V$ and a period of L steps, with L, called the length of the sequence, taking on the values $L = 2^i - 1$, $i = 2, 3, \cdots$. It is assumed that $n_b \leq L$, that is, that the system memory does not exceed the sequence period. This assumption is only made for the sake of simplicity, being inessential in view of the results in [FN74]. If the $y(t)$ samples are used in (7) after the test input has been applied to the system for at least L steps, thanks to the characterizing property of the PRBS autocorrelation function, we have

$$\langle \rho_k, \rho_\tau \rangle = \varphi'(\tau - 1)\varphi(k-1) = \begin{cases} LV^2, & k = \tau + iL \\ -V^2, & \text{elsewhere} \end{cases} \tag{6.3-8}$$

Instead of using (7) directly, we can exploit the PRBS autocorrelation-function property to rewrite (7) in the following simplified form [Mos75]:

$$\theta(t) = \theta(t-1) + \frac{\alpha_{t+1} e_t}{(L+1)V^2} \varepsilon_t \tag{6.3-9a}$$

$$e_t = \varphi(t-1) - \varphi(t-2) + \alpha_t e_{t-1} \tag{6.3-9b}$$

$$\varepsilon_t = y(t) - y(t-1) + \alpha_t \varepsilon_{t-1} \tag{6.3-9c}$$

$$\alpha_t = \frac{L-t+3}{L-t+2} \tag{6.3-9d}$$

$t = 1, 2, \cdots, L$; $e_0 = \varphi(-2) = O_{n_b}$; and $\varepsilon_0 = y(0) = 0$.

Assume that the system simply delays the input by 16 steps. Use (9) with a PRBS of length $L = 31$ and also assume $n_b = 31$. Three estimates of $\theta(t)$, $t = 15$, 30, and 31, are shown in Fig. 1. Note that $\theta(31)$ is an exact reproduction of the impulse response of the system because, as a result of (8), $\{\varphi(k-1)\}_{k=1}^{31}$ is a set of linearly independent vectors in $\mathbb{R}^{31}$.

The previous discussion concerns an ideal deterministic situation. In a more realistic case, system output $y(t)$ is affected by a noise or disturbance $n(t)$:

$$y(t) = \varphi'(t-1)\theta + n(t)$$

Provided that $\mathcal{E}\{n(t)\} = 0$ and $\mathcal{E}\{n(t)n(t+k)\} = 0$, $k \geq 31$, this situation can be tackled by performing successively a number of separate estimates $\theta(31)$ based on 31 I/O pairs and averaging them.

A simplification is to set $P(t) = I_{n_\theta}$ in the orthogonalized projection algorithm. The resulting algorithm is called the projection algorithm.

Projection Algorithm

$$\theta(t+1) = \begin{cases} \theta(t) & , \quad \|\varphi(t)\| = 0 \\ \theta(t) + \frac{\varphi(t)}{\|\varphi(t)\|^2}\left[y(t+1) - \varphi'(t)\theta(t)\right] & , \quad \text{otherwise} \end{cases} \qquad (6.3\text{-}10)$$

It follows from (2) that

$$\frac{\varphi(t)}{\|\varphi(t)\|^2}\left[y(t+1) - \varphi'(t)\theta(t)\right] = \frac{\varphi(t)\varphi'(t)}{\|\varphi(t)\|^2}\tilde{\theta}(t)$$

$$= \left\langle \tilde{\theta}(t), \frac{\varphi(t)}{\|\varphi(t)\|} \right\rangle \frac{\varphi(t)}{\|\varphi(t)\|}$$

$$\tilde{\theta}(t) := \theta - \theta(t)$$

The estimate $\theta(t)$ of θ is then updated by summing to $\theta(t)$ the orthogonal projection of the estimation error $\tilde{\theta}(t)$ onto $\varphi(t)/\|\varphi(t)\|$. Figure 2 illustrates how the algorithm works assuming that $\theta \in \mathbb{R}^2$ and $\theta(0) = O_2$. We see that, despite the fact that $\varphi(0)$ and $\varphi(1)$ are linearly independent, $\theta(2) \neq \theta$. Note that whereas the orthogonalized projection algorithm yields $\theta(2) = \theta$, provided that $\text{Span}\{\varphi(0), \varphi(1)\} = \mathbb{R}^2$, this holds true with the projection algorithm if and only if it also happens that $\varphi(1) \perp \varphi(0)$.

An alternative to (10) that avoids the need of checking $\|\varphi(t)\|$ for zero is the following slightly modified form of the algorithm. In some filtering literature [Joh88], this algorithm is also known as the *normalized least-mean-squares* algorithm [WS85].

Modified Projection Algorithm

$$\theta(t+1) = \theta(t) + \frac{a\varphi(t)}{c + \|\varphi(t)\|^2}\left[y(t+1) - \varphi'(t)\theta(t)\right] \qquad (6.3\text{-}11)$$

with $\theta(0)$ given, and $c > 0$, $0 < a < 2$.

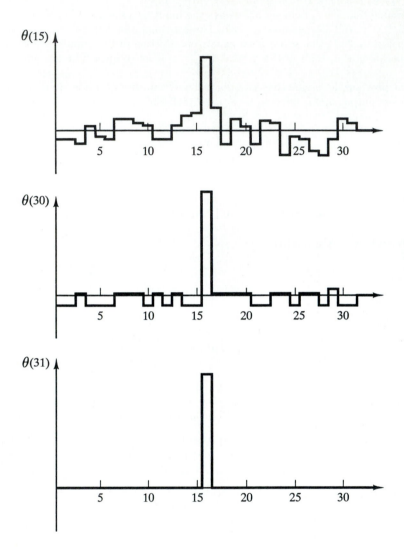

Figure 6.3-1: Orthogonalized projection algorithm estimate of the impulse response of a 16 steps delay system when the input is a PRBS of period 31.

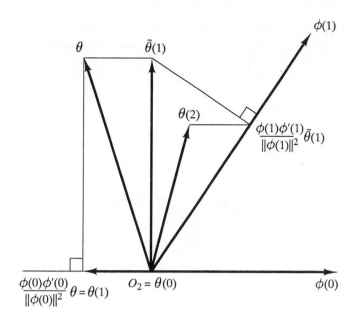

Figure 6.3-2: Geometric interpretation of the projection algorithm.

Example 6.3-2 [LM76]. Figure 3 reports results of simulations of recursive estimation of the impulse response θ of a six-pole Butterworth filter with a cutoff frequency of 0.8 kHz and a sampling rate equal to 2.4 kHz, using as a probing signal the same PRBS as in Example 1. Two estimation algorithms are considered: the orthogonalized projection algorithm (9) (solid lines), and the modified projection algorithm (11) (dashed lines). As indicated in Example 1, for both algorithms, the observations $y(t)$ were affected by an additive stationary zero-mean white Gaussian noise with variance σ_n^2. SNR denotes the signal-to-noise ratio in dB given by

$$\text{SNR} = 10 \log_{10} \left(\frac{\sum_1^{31} \|\varphi(k)\theta\|^2}{31 \cdot \sigma_n^2} \right).$$

Fig. 3 shows the experimental resulting mean-square error in estimating θ. The estimation based on (9) was obtained by carrying out N separate estimates of θ, each based on $L = 31$ input/output pairs and then averaging the corresponding estimates. In Fig. 3, the abscissa is N for algorithm (9), whereas N_i is the overall number of recursions t for algorithm (11). The various curves for a given SNR correspond to different noise sequences. Note that algorithm (9) yields a lower residual error $\|\tilde{\theta}(t)\|^2$ than (11). Further, if SNR decreases by 20 dB, $\|\tilde{\theta}(t)\|^2$ decreases by the same amount for algorithm (9). Note that this is not true for algorithm (11).

As can be seen from the previous discussion, in the deterministic ideal case where no disturbances are present and hence (2) holds true, the orthogonalized projection algorithm and the projection or modified projection algorithm have a

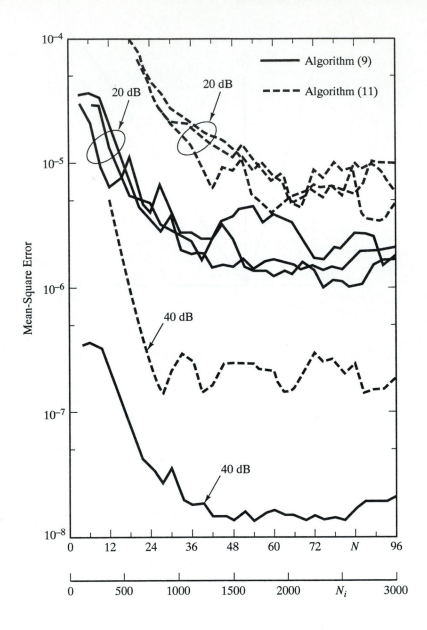

Figure 6.3-3: Recursive estimation of the impulse response θ of the 6-pole Butterworth filter of Example 2.

comparable rate of convergence provided that initially regressors $\varphi(k)$ are almost mutually orthogonal. Thanks to (8), this happens with PRBSs of period L large enough. In the general case, however, the orthogonalized projection algorithm exhibits a much faster convergence than the projection or modified projection algorithm. It is therefore important to suitably modify the orthogonalized projection algorithm so as to retain its favorable convergence properties in nonideal cases where (2) does not hold true exactly any longer. For instance, the need of checking L_t for zero at each step of the orthogonalized projection algorithm can be avoided by modifying (7) as follows:

$$\theta(t+1) \;=\; \theta(t) + \frac{P(t)\varphi(t)}{c + \varphi'(t)P(t)\varphi(t)}\,[y(t+1) - \varphi'(t)\theta(t)] \qquad (6.3\text{-}12\text{a})$$

$$P(t+1) \;=\; P(t) - \frac{P(t)\varphi(t)\varphi'(t)P(t)}{c + \varphi'(t)P(t)\varphi(t)} \qquad (6.3\text{-}12\text{b})$$

where $c > 0$. When $c = 1$, (12) becomes the well-known recursive least-squares algorithm whose origin, as discussed in [You84], can be traced back to Gauss [Gau63], who used the least-squares technique for calculating orbits of planets.

Recursive Least-Squares (RLS) Algorithm

$$\theta(t+1) \;=\; \theta(t) + \frac{P(t)\varphi(t)}{1 + \varphi'(t)P(t)\varphi(t)}\,[y(t+1) - \varphi'(t)\theta(t)] \qquad (6.3\text{-}13\text{a})$$

$$=\; \theta(t) + P(t+1)\varphi(t)\,[y(t+1) - \varphi'(t)\theta(t)] \qquad (6.3\text{-}13\text{b})$$

$$P(t+1) \;=\; P(t) - \frac{P(t)\varphi(t)\varphi'(t)P(t)}{1 + \varphi'(t)P(t)\varphi(t)} \qquad (6.3\text{-}13\text{c})$$

$$=\; P(t) - K(t)\,[1 + \varphi'(t)P(t)\varphi(t)]\,K'(t) \qquad (6.3\text{-}13\text{d})$$

$$=\; [I - K(t)\varphi'(t)]'\,P(t)\,[I - K(t)\varphi'(t)] + K(t)K'(t) \qquad (6.3\text{-}13\text{e})$$

with

$$K(t) = \frac{P(t)\varphi(t)}{1 + \varphi'(t)P(t)\varphi(t)} \qquad (6.3\text{-}13\text{f})$$

with $\theta(0)$ given and $P(0)$ any symmetric and positive definite matrix.

Problem 6.3-1 Use the matrix-inversion lemma (5.3-16) to show that the inverse $P^{-1}(t)$ of $P(t)$ satisfying (13c) fulfills the following recursion:

$$P^{-1}(t+1) = P^{-1}(t) + \varphi(t)\varphi'(t) \qquad (6.3\text{-}13\text{g})$$

Proposition 6.3-1.   (RLS and Normal Equations)   *Let* $\{\theta(k)\}_{k=0}^{t}$ *be given by the RLS algorithm (13). Then,* $\theta(t)$ *satisfies the normal equations*

$$\left[\sum_{k=0}^{t-1}\varphi(k)\varphi'(k)\right]\theta(t) = \sum_{k=0}^{t-1}\varphi(k)y(k+1) + P^{-1}(0)\,[\theta(0) - \theta(t)] \qquad (6.3\text{-}14)$$

and, hence minimizes the criterion

$$J_t(\theta) = \frac{1}{2} \left\{ \sum_{k=0}^{t-1} [y(k+1) - \varphi'(k)\theta]^2 + \|\theta - \theta(0)\|_{P^{-1}(0)}^2 \right\} \qquad (6.3\text{-}15)$$

Proof: Premultiply both sides of (12b) by $P^{-1}(t+1)$ to get

$$
\begin{aligned}
P^{-1}(t+1)\theta(t+1) &= P^{-1}(t+1)\theta(t) + \varphi(t)\left[y(t+1) - \varphi'(t)\theta(t)\right] \\
&= P^{-1}(t)\theta(t) + \varphi(t)y(t+1) \qquad\qquad [(13\mathrm{g})] \\
&= P^{-1}(0)\theta(0) + \sum_{k=0}^{t} \varphi(k)y(k+1)
\end{aligned}
$$

Further, by (13g),

$$P^{-1}(t+1) = P^{-1}(0) + \sum_{k=0}^{t} \varphi(k)\varphi'(k)$$

which, substituted in the L.H.S. of the prevision equation, yields (14). Note that, because by the assumed initialization $P^{-1}(0) > 0$, the system of normal equation (14) always has a unique solution $\theta(t)$. Furthermore, it is a simple matter to check that $\theta(t)$ satisfying (14) minimizes $J_t(\theta)$.

In order to give a geometric interpretation to the RLS algorithm, let us consider the normal equations (14), assuming that $P^{-1}(0)$ is small enough so as to make $P^{-1}(0)[\theta(0) - \theta(t)]$ negligible w.r.t. the other terms:

$$\left[\sum_{k=0}^{t-1} \varphi(k)\varphi'(k) \right] \theta(t) = \sum_{k=0}^{t-1} \varphi(k)y(k+1)$$

Now set

$$
\begin{aligned}
Y &:= y_t^1 \in \mathbb{R}^t \\
\rho_i' &:= \left[\ \varphi_i(0) \quad \cdots \quad \varphi_i(t-1)\ \right], \quad i \in \underline{n}_\theta \\
\theta(t) &:= \left[\ \theta_1(t) \quad \cdots \quad \theta_{n_\theta}(t)\ \right]'
\end{aligned}
$$

where $\underline{n}_\theta := \{1, 2, \cdots, n_\theta\}$. With such new notations, the previous set of equations can be rewritten as follows:

$$\sum_{j=1}^{n_\theta} \theta_j(t)\langle \rho_j, \rho_i \rangle = \langle Y, \rho_i \rangle, \quad i \in \underline{n}_\theta$$

Comparing this equation with (1-6), we obtain Fig. 4, which is the analog of Fig. 1-1 to the present case. In Fig. 4, $\hat{Y}$ denotes the orthogonal projection of Y onto the subspace $[\rho^{n_\theta}]$ in $\mathbb{R}^t$ generated by $\{\rho_i, i \in \underline{n}_\theta\}$ and $\tilde{Y} := Y - \hat{Y}$. According to condition (a) before (1-6), vector $\theta(t)$ is then such that

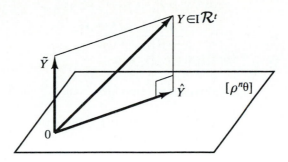

Figure 6.3-4: Geometrical illustration of the Least Squares solution.

$$\hat{Y} = \sum_{j=1}^{n_\theta} \theta_j(t)\rho_j$$

$$= \sum_{j=1}^{n_\theta} \theta_j(t) \left[\varphi_j(0) \quad \cdots \quad \varphi_j(t-1) \right]'$$

$$= \begin{bmatrix} \varphi'(0) \\ \vdots \\ \varphi'(t-1) \end{bmatrix} \theta(t)$$

If, instead of (2a), we consider as equations to be "solved" w.r.t. θ:

$$y(k) = \varphi'(k-1)\theta + n(k) , \quad k \in \underline{t}$$

where $n(k)$ represents an unknown equation error, for $P^{-1}(0)$ small enough and $n_\theta \le \dim[\rho^t]$, the RLS find a "solution" $\theta(t)$ to the previous *hyperdetermined* *system* of equations that minimizes $\sum_{k=0}^{t-1} [y(k+1) - \varphi'(k)\theta]^2$.

Problem 6.3-2 (RLS with Data Weighting)   Consider the criterion

$$\bar{J}_t(\theta) = S_t(\theta) + \frac{1}{2}\|\theta - \theta(0)\|_{P^{-1}(0)}^2 \qquad (6.3\text{-}16\text{a})$$

$$S_t(\theta) = \frac{1}{2}\sum_{k=1}^{t} c(k) \left[y(k) - \varphi'(k-1)\theta\right]^2 \qquad (6.3\text{-}16\text{b})$$

where $c(k)$ are nonnegative weighting coefficients, and $P^{-1}(0) > 0$. Show that vector θ minimizing $\bar{J}_t(\theta)$ is given by the following sequential algorithm called *RLS with data weighting*:

$$\theta(t+1) = \theta(t) + \frac{c(t)P(t)\varphi(t)}{1 + c(t)\varphi'(t)P(t)\varphi(t)} \left[y(t+1) - \varphi'(t)\theta(t)\right]$$

$$= \theta(t) + c(t)P(t+1)\varphi(t)\left[y(t+1) - \varphi'(t)\theta(t)\right] \qquad (6.3\text{-}17\text{a})$$

$$P(t+1) = P(t) - \frac{c(t)P(t)\varphi(t)\varphi'(t)P(t)}{1 + c(t)\varphi'(t)P(t)\varphi(t)} \qquad (6.3\text{-}17\text{b})$$

$$P^{-1}(t+1) = P^{-1}(t) + c(t)\varphi(t)\varphi'(t) \qquad (6.3\text{-}17\text{c})$$

Note that the orthogonalized projection algorithm (7) can be recovered from (17) by setting

$$c(k) = \begin{cases} 0, & \text{for } \varphi'(k)P(k)\varphi(k) = 0 \\ \infty, & \text{otherwise} \end{cases} \qquad (6.3\text{-}18)$$

This choice is in accordance with the use in algorithm (7) of the quantity $L_{t+1} = \varphi'(t)P(t)\varphi(t)$ as an indicator of the new information contained in $\varphi(t)$.

RLS and Kalman Filtering.   We now give a Kalman filter interpretation to RLS. Consider the following stochastic model for an unknown possibly time-varying system parameter vector $x(t)$:

$$\left.\begin{aligned} x(t+1) &= x(t) + \xi(t) \\ y(t) &= \varphi'(t-1)x(t) + \zeta(t) \end{aligned}\right\} \qquad (6.3\text{-}19)$$

Let (19) satisfy the same conditions as (2-37). Applying Theorem 2-1, we find

$$x(t+1 \mid t) = x(t \mid t-1) + K(t)\left[y(t) - \varphi'(t-1)x(t \mid t-1)\right] \qquad (6.3\text{-}20\text{a})$$

$$K(t) = \frac{\Pi(t)\varphi(t-1)}{\Psi_\zeta(t) + \varphi'(t-1)\Pi(t)\varphi(t-1)} \qquad (6.3\text{-}20\text{b})$$

$$\Pi(t+1) = \Pi(t) - \frac{\Pi(t)\varphi(t-1)\varphi'(t-1)\Pi(t)}{\Psi_\zeta(t) + \varphi'(t-1)\Pi(t)\varphi(t-1)} + \Psi_\xi(t) \qquad (6.3\text{-}20\text{c})$$

Case 1 ($\Psi_\xi(t) \equiv O_{n_\theta \times n_\theta}$, $\Psi_\zeta(t) \equiv \psi_\zeta$). In such a case, the first of (19) becomes $x(t+1) = x(t) = \theta$. Furthermore, setting $\theta(t) := x(t+1 \mid t)$ and $P(t-1) := \Pi(t)/\psi_\zeta$, the equations in (20) become the same as the RLS algorithm (13).

Case 2 ($\Psi_\xi(t)$ arbitrary, $\Psi_\zeta \equiv \psi_\zeta$). Setting $\theta(t) := x(t+1 \mid t)$, $P(t-1) := \Pi(t)/\psi_\zeta$, and $Q(t) := \Psi_\xi(t)/\psi_\zeta$, the equations in (19) become the *RLS with covariance modification*:

$$\theta(t+1) = \theta(t) + \frac{P(t)\varphi(t)}{1 + \varphi'(t)P(t)\varphi(t)} \qquad (6.3\text{-}21\text{a})$$
$$\times \left[y(t+1) - \varphi'(t)\theta(t)\right]$$

$$P(t+1) = P(t) - \frac{P(t)\varphi(t)\varphi'(t)P(t)}{1 + \varphi'(t)P(t)\varphi(t)} + Q(t+1) \qquad (6.3\text{-}21\text{b})$$

Case 3 ($\Psi_\xi(t) = O_{n_\theta \times n_\theta}$, $\Psi_\zeta(t+1) = \lambda(t)\Psi_\zeta(t)$, $0 < \lambda(t) \leq 1$). Setting again $\theta(t) := x(t+1 \mid t)$, $P(t-1) := \Pi(t)/\Psi_\zeta(t)$, the equations in (20) become the *exponentially weighted RLS:*

$$\theta(t+1) \;=\; \theta(t) + \frac{P(t)\varphi(t)}{1+\varphi'(t)P(t)\varphi(t)} \tag{6.3-22a}$$
$$\times\,[y(t+1)-\varphi'(t)\theta(t)]$$
$$=\; \theta(t) + \lambda(t+1)P(t+1)\varphi(t)\,[y(t+1)-\varphi'(t)\theta(t)]$$

$$P(t+1) \;=\; \frac{1}{\lambda(t+1)}\left[P(t)-\frac{P(t)\varphi(t)\varphi'(t)P(t)}{1+\varphi'(t)P(t)\varphi(t)}\right] \tag{6.3-22b}$$

$$P^{-1}(t+1) \;=\; \lambda(t+1)\left[P^{-1}(t)+\varphi(t)\varphi'(t)\right] \tag{6.3-22c}$$

The foregoing Kalman filter solutions give valuable insights into the RLS algorithm:

- If, as in Case 1, the regredend and regressor are related by

$$y(t) = \varphi'(t-1)\theta + \zeta(t) \tag{6.3-23}$$

for example,

$$A(d)y(t) = B(d)u(t) + \zeta(t)$$

with $A(d)$, $B(d)$, and $\varphi(t-1)$ as in (1) and (2), and ζ zero-mean, white, and Gaussian, the distributional interpretation of the Kalman filter tells us that the conditional probability distribution of θ given y^t equals

$$P\left(\theta \mid y^t\right) = N\left(\theta(t), \Psi_\zeta P(t)\right) \tag{6.3-24}$$

where $\theta(t)$ and $P(t)$ are generated by the RLS (13) initialized from $\theta(0) = \mathcal{E}\{\theta(0)\}$ and $P(0) = \mathcal{E}\{\tilde{\theta}(0)\tilde{\theta}'(0)\}$. In words, this means that $\theta(0)$ is what we guess the parameter vector to be before the data are acquired, and $P(0)$ is larger the lower is our confidence in this guess.

- The comparison between (20) with $\Psi_\xi(t) \equiv O_{n_\theta \times n_\theta}$ and (17) leads one to conclude that RLS with data weighting are the same as the Kalman filter provided that $c(k) = \Psi_\zeta^{-1}(k)$. In words, the larger the output noise at a given time, the smaller the weight in (16).

- Case 2 corresponds to a time-varying system parameter vector consisting of a process with uncorrelated increments (or a random walk). The related solution (21) tells us that, in order to take into account these time variations, to compute $P(t+1)$, the symmetric nonnegative definite matrix $Q(t+1)$ must be added to the L.H.S. of the updating equation (13c) of the standard RLS. This suggests that matrix $P(t)$, and hence the updating gain $P(t)\varphi(t)[1+\varphi'(t)P(t)\varphi(t)]^{-1}$, is prevented from becoming too small.

The next problem indicates why algorithm (22) of the previous Case 3 is referred to as the exponentially weighted RLS.

Problem 6.3-3 (Exponentially Weighted RLS)   Consider again criterion (16) with (16b) now modified so as to make the weighting coefficient dependent on t in an exponential fashion:

$$S_t(\theta) = \frac{1}{2} \sum_{k=1}^{t} c(t,k) \left[y(k) - \varphi'(k-1)\theta\right]^2 \qquad (6.3\text{-}25a)$$

$$\begin{cases} c(t,t) &= 1 \\ c(t,k) &= \lambda(t-1)c(t-1,k) \\ &= \prod_{i=k}^{t-1} \lambda(i) \end{cases} \qquad (6.3\text{-}25b)$$

with

$$0 < \lambda(i) \leq 1 \qquad (6.3\text{-}25c)$$

In particular, if $\lambda(i) \equiv \lambda$, we have

$$c(t,k) = \lambda^{t-k} \qquad (6.3\text{-}25d)$$

Show that from (25b), it follows that

$$S_{t+1}(\theta) = \lambda(t)S_t(\theta) + \frac{1}{2} \cdot \left[y(t+1) - \varphi'(t)\theta\right]^2 \qquad (6.3\text{-}25e)$$

viz. the data in $S_t(\theta)$ are discounted in $S_{t+1}(\theta)$ by the factor $\lambda(t)$. Prove that the vector θ minimizing (16) with $S_t(\theta)$ as in (25a) is given by the recursive algorithm (26).

6.3.2   Pseudolinear Regression Algorithms

Recursive Extended Least Squares (RELS(PR)) (A Priori Prediction Errors).   This algorithm originates from the problem of recursively fitting an ARMAX model:

$$A(d)y(t) = B(d)u(t) + C(d)e(t)$$

to the system I/O data sequence. The previous ARMAX model can be also represented as

$$y(t) = \varphi_e'(t-1)\theta + e(t)$$

$$\varphi_e(t-1) = \begin{bmatrix} -y(t-1) \\ \vdots \\ -y(t-n_a) \\ u(t-1) \\ \vdots \\ u(t-n_b) \\ e(t-1) \\ \vdots \\ e(t-n_c) \end{bmatrix}$$

$$\theta = \begin{bmatrix} a_1 & \cdots & a_{n_a} & b_1 & \cdots & b_{n_b} & c_1 & \cdots & c_{n_c} \end{bmatrix} \qquad (6.3\text{-}26a)$$

If $\varphi_e(t-1)$ were available, the RLS algorithm could be used to recursively estimate θ. In reality, all of $\varphi_e(t-1)$ is known except for its last n_c components. In RELS(PR), such components are replaced by using the *a priori* prediction errors:

$$\varepsilon(k) := y(k) - \varphi'(k-1)\theta(k-1) \qquad (6.3\text{-}26\text{b})$$

where $\varphi(t-1)$ is given here by the pseudo-regressor:

$$\varphi(t-1) = \begin{bmatrix} -y(t-1) \\ \vdots \\ -y(t-n_a) \\ u(t-1) \\ \vdots \\ u(t-n_b) \\ \varepsilon(t-1) \\ \vdots \\ \varepsilon(t-n_c) \end{bmatrix} \qquad (6.3\text{-}26\text{c})$$

and, for the rest, the RELS(PR) is the same as the RLS (13):

$$\theta(t+1) = \theta(t) + P(t+1)\varphi(t)\varepsilon(k+1) \qquad (6.3\text{-}26\text{d})$$

$$P(t+1) = P(t) - \frac{P(t)\varphi(t)\varphi'(t)P(t)}{1+\varphi'(t)P(t)\varphi(t)} \qquad (6.3\text{-}26\text{e})$$

Recursive Extended Least Squares (RELS(PO)) (A Posteriori Prediction Errors). This method is the same as RELS(PR) with the only exception that here, instead of the a priori prediction errors $\varepsilon(k)$, the *a posteriori* prediction errors

$$\bar{\varepsilon}(k) := y(k) - \varphi'(k-1)\theta(k) \qquad (6.3\text{-}27\text{a})$$

are used in the pseudo-regression vector (26c). Hence, (26c) is replaced here with

$$\varphi(t-1) = \begin{bmatrix} -y(t-1) \\ \vdots \\ -y(t-n_a) \\ u(t-1) \\ \vdots \\ u(t-n_b) \\ \bar{\varepsilon}(t-1) \\ \vdots \\ \bar{\varepsilon}(t-n_c) \end{bmatrix} \qquad (6.3\text{-}27\text{b})$$

The RELS(PR) and RELS(PO) methods will be both simply referred to as the RELS method whenever no further distinction is required. The RELS method was first proposed in [ÅBW65], [May65], [Pan68], and [You68].

The use of a posteriori prediction errors in RELS was introduced by [You74] and turned out ([Che81] and [Sol79]) to be instrumental for avoiding parameter estimate *monitoring*, viz., the projection of parameter estimates into a stability region so as to ensure the stability of the recursive scheme ([Han76] and [Lju77a]).

In accordance with the terminology that we have already adopted, the RELS regressors in (26c) and (27b) are often called *pseudo–regression* vectors, and the RELS are sometimes referred to as the *pseudo-linear regression algorithm*, so as to point out the intrinsic nonlinearity in θ of the algorithm, where $\varphi(t-1)$ is dependent on previous estimates via (26b) or (27b). Another name for RELS is the *approximate maximum-likelihood* algorithm, this choice being justified in that RELS can be regarded as a simplification of the next algorithm.

Recursive Maximum Likelihood (RML). This algorithm, similarly to RELS, aims at recursively fitting an ARMAX model to the system I/O data sequence. The RML algorithm is given as follows:

$$\theta(t+1) \;=\; \theta(t) + P(t+1)\psi(t)\varepsilon(t+1) \qquad (6.3\text{-}28a)$$

$$P(t+1) \;=\; P(t) - \frac{P(t)\psi(t)\psi'(t)P(t)}{1 + \psi'(t)P(t)\psi(t)} \qquad (6.3\text{-}28b)$$

with $\varepsilon(t)$ as in (26b). In order to define $\psi(t)$, let

$$C(t,d) := 1 + c_1(t)d + \cdots + c_{n_c}(t)d^{n_c} \qquad (6.3\text{-}28c)$$

where the $c_i(t)$'s, $i = 1, \cdots, n_c$ are the last n_c components of $\theta(t)$:

$$\theta(t) = \left[\begin{array}{ccccccccc} -a_1(t) & \cdots & -a_{n_a}(t) & b_1(t) & \cdots & b_{n_b}(t) & c_1(t) & \cdots & c_{n_c}(t) \end{array} \right]'$$
$$(6.3\text{-}28d)$$

Then $\psi(t)$ is obtained by filtering the pseudo-regressor $\varphi(t)$ in (26c) as follows:

$$C(t,d)\psi(t) = \varphi(t) \qquad (6.3\text{-}28e)$$

or

$$\psi(t) = \varphi(t) - c_1(t)\psi(t-1) - \cdots - c_{n_c}(t)\psi(t-n_c) \qquad (6.3\text{-}28f)$$

This means that

$$
\psi(t-1) = \begin{bmatrix} -y_f(t-1) \\ \vdots \\ -y_f(t-n_a) \\ u_f(t-1) \\ \vdots \\ u_f(t-n_b) \\ \varepsilon_f(t-1) \\ \vdots \\ \varepsilon_f(t-n_c) \end{bmatrix} \tag{6.3-28g}
$$

where

$$
C(t,d)y_f(t) = y(t) \tag{6.3-28h}
$$

and similarly for $u_f(t)$ and $\varepsilon_f(t)$. It must be emphasized that the "exact" construction of $\psi(t-1)$ requires the use of the n_c "fixed" filters $C(t-i,d)$, $i = 1, \cdots, n_c$, for example, $C(t-i,d)y_f(t-i) = y(t-i)$, and hence storage of the related n_c^2 parameters.

The previous algorithm can be properly called the RML with *a priori* prediction errors. The RML with *a posteriori* prediction errors is instead obtained by substituting the definition of $\psi(t-1)$ in (28g) with

$$
\psi(t-1) = \begin{bmatrix} -y_f(t-1) \\ \vdots \\ -y_f(t-n_a) \\ u_f(t-1) \\ \vdots \\ u_f(t-n_b) \\ \bar{\varepsilon}_f(t-1) \\ \vdots \\ \bar{\varepsilon}_f(t-n_c) \end{bmatrix} \tag{6.3-29a}
$$

with $\bar{\varepsilon}_f(t)$ the following filtered a posteriori prediction error:

$$
\begin{aligned}
C(t,d)\bar{\varepsilon}_f(t) &= \bar{\varepsilon}(t) \\
&= y(t) - \varphi'(t-1)\theta(t)
\end{aligned} \tag{6.3-29b}
$$

Stochastic Gradient Algorithms.   They resemble either the RLS or the RELS but have the simplifying feature that matrix $P(t+1)$ in either (13b) or (26d) is replaced by $a/\operatorname{Tr} P^{-1}(t+1)$:

$$
\theta(t+1) = \theta(t) + \frac{a\varphi(t)}{q(t+1)}\varepsilon(t+1) \ , \qquad a > 0
$$

$$\begin{aligned}
\varepsilon(t+1) &= y(t+1) - \varphi'(t)\theta(t) \\
q(t+1) &= q(t) + \|\varphi(t)\|^2
\end{aligned}$$

where $\theta(0) \in \mathbb{R}^{n_\theta}$ and $q(0) > 0$. According to the model that has to be fitted to the experimental data, vector $\varphi(t)$ is as in (26) for an ARX model or, alternatively, as in (26c) or (27b) for an ARMAX model. Stochastic gradient algorithms can be also considered as extensions of the modified projection algorithm (11).

6.3.3   Parameter Estimation for MIMO Systems

In the previous part of this section, we have taken the system to be SISO to simplify the notation. We now indicate how parameter estimation algorithms can be extended to the MIMO case. This extension is straightforward and the reader should have no difficulty in constructing the appropriate MIMO versions of the previous algorithms.

We base our discussion on a MIMO-system ARMAX model $A(d)y(t) = B(d)u(t) + C(d)e(t)$ with $A(d) = I_p + A_1 d + \cdots + A_{n_a} d^{n_a}$, $B(d) = B_1 d + \cdots + B_{n_b} d^{n_b}$, and $C(d) = I_p + C_1 d + \cdots + C_{n_c} d^{n_c}$. We have

$$\begin{aligned}
y^i(t) &= -a_1^i y(t-1) - \cdots - a_{n_a}^i y(t - n_a) \\
&\quad + b_1^i u(t-1) + \cdots + b_{n_b}^i u(t - n_b) \\
&\quad + c_1^i e(t-1) + \cdots + c_{n_c}^i e(t - n_c) + e^i(t) \\
&= \varphi'_e(t-1)\theta^i + e^i(t) \tag{6.3-30a}
\end{aligned}$$

where the following notations are used:

$$\begin{aligned}
y^i(t) \quad &: \quad i\text{th component of } y(t) \in \mathbb{R}^p, \ i = 1, \cdots, p \\
a_j^i \quad &: \quad i\text{th row of } A_j, \ j = 1, \cdots, n_a \\
b_j^i \quad &: \quad i\text{th row of } B_j, \ j = 1, \cdots, n_b \\
c_j^i \quad &: \quad i\text{th row of } C_j, \ j = 1, \cdots, n_c
\end{aligned}$$

$$\varphi_e(t-1) := \begin{bmatrix} -y(t-1) \\ \vdots \\ -y(t-n_a) \\ u(t-1) \\ \vdots \\ u(t-n_b) \\ e(t-1) \\ \vdots \\ e(t-n_c) \end{bmatrix} \tag{6.3-30b}$$

$$\theta^i := \begin{bmatrix} a_1^i & \cdots & a_{n_a}^i & b_1^i & \cdots & b_{n_b}^i & c_1^i & \cdots & c_{n_c}^i \end{bmatrix}' \tag{6.3-30c}$$

Then, we can write

$$y(t) = \phi'(t-1)\theta + e(t) \tag{6.3-30d}$$

$$\phi'_e(t-1) := \left.\begin{bmatrix} \varphi'_e(t-1) & & & \\ & \varphi'_e(t-1) & & \\ & & \ddots & \\ & & & \varphi'_e(t-1) \end{bmatrix}\right\} p \tag{6.3-30e}$$

$$\underbrace{\phantom{\begin{bmatrix} \varphi'_e(t-1) & & & \\ & \varphi'_e(t-1) & & \\ & & \ddots & \\ & & & \varphi'_e(t-1) \end{bmatrix}}}_{n_\theta}$$

$$\theta := \begin{bmatrix} (\theta^1)' & (\theta^2)' & \cdots & (\theta^p)' \end{bmatrix}' \in \mathbb{R}^{n_\theta}$$

We indicate the related RELS algorithm with a priori prediction errors:

$$\theta(t+1) = \theta(t) + P(t+1)\phi(t)\varepsilon(t+1) \tag{6.3-31a}$$

$$P(t+1) = P(t) - P(t)\phi(t)\left[I_p + \phi'(t)P(t)\phi(t)\right]^{-1}\phi(t) \tag{6.3-31b}$$

$$\varphi(t-1) = \begin{bmatrix} -y'(t-1)- \\ \vdots \\ -y'(t-n_a) \\ u'(t-1) \\ \vdots \\ u'(t-n_b) \\ \varepsilon'(t-1) \\ \vdots \\ \varepsilon'(t-n_c) \end{bmatrix} \tag{6.3-31c}$$

$$\phi'(t-1) := \left.\begin{bmatrix} \varphi'(t-1) & & & \\ & \varphi'(t-1) & & \\ & & \ddots & \\ & & & \varphi'(t-1) \end{bmatrix}\right\} p \tag{6.3-31d}$$

$$\underbrace{\phantom{\begin{bmatrix} \varphi'(t-1) & & & \\ & \varphi'(t-1) & & \\ & & \ddots & \\ & & & \varphi'(t-1) \end{bmatrix}}}_{n_\theta}$$

$$\varepsilon(t+1) = y(t+1) - \phi'(t)\theta(t) \tag{6.3-31e}$$

6.3.4 The Minimum Prediction-Error Method

So far we have discussed in a quite informal way a number of system parameter estimation algorithms. Nevertheless, our initial thrust, the orthogonalized projection algorithm, was justified by considering (2) an *exact deterministic system model* relating the unknown parameter vector θ to the I/O data. Our departure from an exact deterministic modeling assumption was undertaken by adopting sensible modifications to the initial deterministic algorithm. The resulting modified algorithms, basically variants of the RLS algorithm, were next reinterpreted as solutions of Kalman filtering problems for *exact stochastic system models*. A conclusion to the previous considerations is that so far our basic underlying presumption has been the availability of an exact, either deterministic or stochastic, system model.

The basic idea of the *minimum prediction-error (MPE)* method is to fit a prediction model, parameterized by a vector θ, to the recorded I/O data. The parameter θ selected by the method is then the one for which the prediction errors are minimized in some sense. In this way, the search for a true parameterized model is abandoned, and what is sought instead is the best parameterized predictor in a given class. Consequently, the MPE method focuses attention on the approximation of the observed data through models of limited or reduced complexity. Our main goal is to introduce the MPE method and show that the majority of the estimation algorithms discussed so far can be derived within the MPE framework.

The kind of approximation that is sought in the MPE method is motivated by the fact that in many applications, the system model is used for prediction. This is often inherently the case for control system synthesis. Most systems are stochastic, viz., the output at time t cannot be exactly determined from I/O data at time $t-1$. We have already touched upon the topic in Problem 2-2 for stochastic state-space descriptions and Kalman filtering. Here we denote by $\hat{y}(t \mid t - 1; \theta)$ the (one-step-ahead) *prediction* of the system output $y(t)$. $\hat{y}(t \mid t - 1; \theta)$ depends on both the I/O data up to time $t - 1$ and model parameter vector θ. The rule according to which the prediction is computed is called the *predictor* and

$$\varepsilon(t, \theta) := y(t) - \hat{y}(t \mid t - 1; \theta) \qquad (6.3\text{-}32\text{a})$$

is the *prediction error*. It is therefore appealing to determine θ by minimizing the cost:

$$J_M(\theta) = \frac{1}{M} \sum_{t=1}^{M} \|\varepsilon(t, \theta)\|_Q^2 \qquad (6.3\text{-}32\text{b})$$

with $Q = Q' > 0$. In the discussion following Proposition 1, we have seen that for $P^{-1}(0)$ small enough and M large, the RLS algorithm tends to minimize (32b) when

$$\hat{y}(t \mid t - 1; \theta) = \varphi'(t - 1)\theta \qquad (6.3\text{-}33)$$

with $\varphi(t - 1)$ known at time t and hence independent on θ.

A model parameter vector θ obtained by minimizing of (32b) is called a *minimum prediction-error (MPE)* estimate.

Example 6.3-3 (Prediction for an ARMAX Model)
Consider the ARMAX model (2-69)

$$A(d)y(t) = B(d)u(t) + C(d)e(t) \qquad (6.3\text{-}34\text{a})$$

or

$$y(t) = [I_p - A(d)] y(t) + B(d)u(t) + [C(d) - I_p] e(t) + e(t)$$

Because the innovation $e(t)$ at time t can be computed in terms of y^t, u^{t-1} via

$$e(t) = C^{-1}(d)A(d)y(t) - C^{-1}(d)B(d)u(t) \qquad (6.3\text{-}34b)$$

a reasonable choice is to set

$$
\begin{aligned}
\hat{y}(t \mid t-1; \theta) &= [I_p - A(d)] y(t) + B(d)u(t) + [C(d) - I_p] e(t) \\
&= C^{-1}(d) \left\{ [C(d) - A(d)] y(t) + B(d)u(t) \right\} \qquad [(34b)]
\end{aligned}
$$

or

$$C(d)\hat{y}(t \mid t-1; \theta) = [C(d) - A(d)] y(t) + B(d)u(t) \qquad (6.3\text{-}34c)$$

From (32a), it then follows that

$$C(d)\varepsilon(t, \theta) = A(d)y(t) - B(d)u(t) \qquad (6.3\text{-}34d)$$

where θ is the vector collecting all the free entries of matrices $A(d)$, $B(d)$, and $C(d)$ that parameterize the ARMAX model (34a). Notice that $C(d)$ is not completely free, because it is required, according to Theorem 2-4, to be strictly Hurwitz. Hence, from (34d) and (34a), it follows that $\varepsilon(t, \theta) = e(t)$ under the choice (34c) for $\hat{y}(t \mid t-1; \theta)$. It is a simple exercise to see that (34c) yields the MMSE prediction of $y(t)$ based on y^{t-1}, u^{t-1} provided that (34a) is a correct model for the I/O data. In fact, if we let $\bar{y}(t)$ be any function of y^{t-1}, u^{t-1}, the minimum of

$$
\begin{aligned}
\mathcal{E}\left\{ \|y(t) - \bar{y}(t)\|_Q^2 \right\} &= \mathcal{E}\left\{ \|\hat{y}(t \mid t-1; \theta) + e(t) - \bar{y}(t)\|_Q^2 \right\} \\
&= \mathcal{E}\left\{ \|\hat{y}(t \mid t-1; \theta) - \bar{y}(t)\|_Q^2 \right\} + \mathcal{E}\left\{ \|e(t)\|_Q^2 \right\}
\end{aligned}
$$

is attained at $\bar{y}(t) = \hat{y}(t \mid t-1; \theta)$.

It is to be remarked that (34c) does not allow $\hat{y}(t \mid t-1; \theta)$ to be expressed in terms of a finite numbers of past I/O pairs. This only happens when $C(d) = I_p$ and hence the ARMAX model (34a) collapses to the ARX model $A(d)y(t) = B(d)u(t) + e(t)$. As seen after (32b), in the latter case, the MPE estimate is given as $M \to \infty$ by the RLS estimate, initialized by a small $P^{-1}(0)$.

Figure 5, where the "Process" indicates the real system with input $u(t)$ and output $y(t)$, provides an illustration of the MPE method. To be specific, we have indicated in (32b) only one possible form of the cost to be minimized. Another possible choice, motivated by maximum-likelihood estimation [SS89], is $J_M = M^{-1} \det \left[\sum_{t=1}^{M} \varepsilon(t, \theta)\varepsilon'(t, \theta) \right]$.

In the special case where $\varepsilon(t, \theta)$ depends linearly on θ, the minimization of $J_M(\theta)$ can be carried out analytically. This is the case of linear regression that can be solved via off-line least squares or the RLS algorithm. In most cases, the minimization must be performed by using a numerical search routine. In this regard, a commonly used tool is the *Newton–Raphson* iterations [Lue69]:

$$\theta^{(k+1)} = \theta^{(k)} - \left[J_M^{(2)}\left(\theta^{(k)}\right) \right]^{-1} J_M^{(1)}\left(\theta^{(k)}\right) \qquad (6.3\text{-}35a)$$

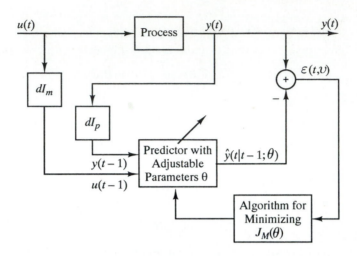

Figure 6.3-5: Block diagram of the MPE estimation method.

where $\theta^{(k)}$ denotes the kth iteration in the search, $J_M^{(1)}\left(\theta^{(k)}\right)$ the gradient of $J_M(\theta)$ w.r.t. θ evaluated at $\theta^{(k)}$:

$$J_M^{(1)}\left(\theta^{(k)}\right) := \left.\frac{\partial J_M(\theta)}{\partial\theta}\right|_{\theta=\theta^{(k)}} \in \mathbb{R}^{n_\theta}$$

and $J_M^{(2)}\left(\theta^{(k)}\right)$ the Hessian matrix of the second derivatives of $J_M(\theta)$ evaluated at $\theta^{(k)}$:

$$J_M^{(2)}\left(\theta^{(k)}\right) := \left.\frac{\partial^2 J_M(\theta)}{\partial\theta^2}\right|_{\theta=\theta^{(k)}}$$

Referring to (32b) and, for the sake of simplicity, to the single-output case, we find for $Q=1$

$$J_M^{(1)}(\theta) = -\frac{2}{M}\sum_{t=1}^{M}\psi(t,\theta)\varepsilon(t,\theta) \tag{6.3-35b}$$

$$\psi(t,\theta) = -\frac{\partial\varepsilon(t,\theta)}{\partial\theta} := -\left[\begin{array}{ccc}\frac{\partial\varepsilon}{\partial\theta_1} & \cdots & \frac{\partial\varepsilon}{\partial\theta_{n_\theta}}\end{array}\right]' \tag{6.3-35c}$$

$$J_M^{(2)}(\theta) = \frac{2}{M}\sum_{t=1}^{M}[\psi(t,\theta)\psi'(t,\theta) - H(t,\theta)\varepsilon(t,\theta)] \tag{6.3-35d}$$

$$H(t,\theta) = -\left\{\frac{\partial^2\varepsilon(t,\theta)}{\partial\theta_i\partial\theta_j}\right\} \tag{6.3-35e}$$

$$= - \begin{bmatrix} \frac{\partial^2 \varepsilon}{\partial \theta_1^2} & \frac{\partial^2 \varepsilon}{\partial \theta_1 \partial \theta_2} & \cdots & \frac{\partial^2 \varepsilon}{\partial \theta_1 \partial \theta_{n_\theta}} \\ \vdots & \vdots & \ddots & \vdots \\ \frac{\partial^2 \varepsilon}{\partial \theta_{n_\theta} \partial \theta_1} & \frac{\partial^2 \varepsilon}{\partial \theta_{n_\theta} \partial \theta_2} & \cdots & \frac{\partial^2 \varepsilon}{\partial \theta_{n_\theta}^2} \end{bmatrix}$$

Suppose that the real system is exactly described by the adopted model for $\theta = \theta_0$ in the sense that

$$y(t) = \hat{y}(t \mid t - 1; \theta_0) + e(t) \tag{6.3-36}$$

where θ_0 denotes the true parameter vector. Then $\varepsilon(t, \theta_0) = e(t)$, with $\{e(t)\}$ as in (2-69). Note that the entries of $H(t, \theta)$ only depend on y^{t-1}, u^{t-1}. Hence, under stationariety and the usual ergodicity conditions [Cai88], the second term on the R.H.S. of (35d) vanishes for $\theta = \theta_0$. If we extrapolate such a conclusion for any θ, (35a) becomes

$$\theta^{(k+1)} = \theta^{(k)} + \left[\sum_{t=1}^{M} \psi\left(t, \theta^{(k)}\right) \psi'\left(t, \theta^{(k)}\right) \right]^{-1} \left[\sum_{t=1}^{M} \psi\left(t, \theta^{(k)}\right) \varepsilon\left(t, \theta^{(k)}\right) \right] \tag{6.3-37}$$

These are called the *Gauss–Newton* iterations.

Problem 6.3-4 (Least Squares as a MPE Method)   Consider the linear-regression model (33). Show that (37) yields a system of normal equations that for every k gives an off-line or batch least-squares estimate of θ.

Example 6.3-4 (Gauss–Newton Iterations for the ARMAX Model)
Consider again Example 3 for a SISO ARMAX model. First, differentiate (34d) w.r.t. a_i to get

$$C(d)\frac{\partial \varepsilon(t, \theta)}{\partial a_i} = y(t - i) \qquad i = 1, \cdots, n_a \tag{6.3-38a}$$

Next, differentation of (34d) w.r.t. b_i gives

$$C(d)\frac{\partial \varepsilon(t, \theta)}{\partial b_i} = -u(t - i) \qquad i = 1, \cdots, n_b \tag{6.3-38b}$$

Similarly, differentiate (34d) w.r.t. c_i to get

$$\varepsilon(t - i, \theta) + C(d)\frac{\partial \varepsilon(t, \theta)}{\partial c_i} = 0 \qquad i = 1, 2, \cdots, n_c \tag{6.3-38c}$$

If we set

$$\theta := \begin{bmatrix} a_1 & \cdots & a_{n_a} & b_1 & \cdots & b_{n_b} & c_1 & \cdots & c_{n_c} \end{bmatrix}' \tag{6.3-38d}$$

we find for (35c)

$$\psi(t,\theta) = \frac{1}{C(d)}\begin{bmatrix} -y(t-1) \\ \vdots \\ -y(t-n_a) \\ u(t-1) \\ \vdots \\ u(t-n_b) \\ \varepsilon(t-1,\theta) \\ \vdots \\ \varepsilon(t-n_c,\theta) \end{bmatrix} = \begin{bmatrix} -y_f(t-1) \\ \vdots \\ -y_f(t-n_a) \\ u_f(t-1) \\ \vdots \\ u_f(t-n_b) \\ \varepsilon_f(t-1,\theta) \\ \vdots \\ \varepsilon_f(t-n_c,\theta) \end{bmatrix} \qquad (6.3\text{-}38e)$$

With $y_f(t)$ as in (28h). The foregoing expression should be compared with (28g) used in the RML algorithm (28). It elucidates the operations that must be carried out at each iteration of (37).

The Gauss–Newton iterations yield properly an off-line or batch estimate of θ. However, these iterations can be suitably modified by using further simplifications so as to provide recursive algorithms. For instance, by recursively minimizing at time t the following exponentially weighted cost

$$J_t(\theta) = \sum_{k=1}^{t} \lambda^{t-k}\|\varepsilon(k,\theta)\|_Q^2 \qquad (6.3\text{-}39a)$$

the following *recursive minimum-prediction error (RMPE)* algorithm can be obtained [SS89]:

$$\theta(t+1) = \theta(t) + \lambda P(t+1)\psi(t)Q\varepsilon(t+1) \qquad (6.3\text{-}39b)$$

$$P(t+1) = \frac{1}{\lambda}\Big\{P(t) - P(t)\psi(t)\big[Q^{-1} + \psi'(t)P(t)\psi(t)\big]^{-1}$$
$$\psi'(t)P(t)\Big\} \qquad (6.3\text{-}39c)$$

$$\varepsilon(t) := \varepsilon(t,\theta(t-1)) \qquad (6.3\text{-}39d)$$

$$\psi(t) := \frac{\partial\varepsilon(t,\theta)}{\partial\theta}\Big|_{\theta(t-1)} = \begin{bmatrix} \frac{\partial\varepsilon_1}{\partial\theta_1} & \cdots & \frac{\partial\varepsilon_p}{\partial\theta_1} \\ \vdots & \ddots & \vdots \\ \frac{\partial\varepsilon_1}{\partial\theta_{n_\theta}} & \cdots & \frac{\partial\varepsilon_p}{\partial\theta_{n_\theta}} \end{bmatrix}_{\theta(t-1)} \qquad (6.3\text{-}39e)$$

The RMPE algorithm (39) can be applied to different prediction models. It is simple to see that for a linear-regression model, it gives the RLS with exponential forgetting factor λ (*cf.* (22)). In light of the results of Example 4, particularly (38c), it is not surprising that (39) applied to a SISO ARMAX model yields, once suitable simplifications are made, the RML algorithm (28) with forgetting factor λ.

Problem 6.3-5 (*RMPE and RML Algorithms*) Consider the ARMAX model of Example 3 and the related results of Example 4. Find the simplifications that are needed to make the RMPE algorithm for $\lambda = 1$ coincident with the RML algorithm (28).

6.3.5   Tracking and Covariance Management

There are several issues that must be taken into account in the practical use of the recursive estimation algorithms introduced in this section. Though we discuss one of them with reference to the RLS algorithm, it is common to the other recursive algorithms as well.

An important reason for using recursive estimation in practice is that the system can be time-varying, and its variations have to be tracked. The adoption of the exponentially weighted RLS (22) related to the minimization of (16a) with

$$S_t(\theta) = \frac{1}{2} \sum_{k=1}^{t} \lambda^{t-k} \left[y(k) - \varphi'(k-1)\theta \right]^2 \qquad (6.3\text{-}40\text{a})$$

where $\lambda \in (0, 1)$ seems to be a quite natural choice. In this case, λ is called the *forgetting factor*. Because $\lambda^k = e^{k \ln \lambda} \cong e^{-k(1-\lambda)}$, the measurements that are older than $1/(1-\lambda)$ are included in the criterion with weights smaller than $e^{-1} \approx 36\%$ of the most recent measurement. Therefore, we can associate to λ a *data memory* M:

$$M = \frac{1}{1-\lambda} \qquad (6.3\text{-}40\text{b})$$

Roughly, M indicates the number of past measurements that the current estimate is effectively based on. Typical choices for λ are in the range between 0.98 ($M = 50$) and 0.995 ($M = 200$).

Problem 6.3-6   Consider the exponentially weighted RLS (22) with $\lambda(t + 1) \equiv \lambda$, $\lambda \in (0, 1)$. Suppose that the regressor sequence $\{\varphi(k)\}_{k=0}^{t-1}$ lies in a hyperplane of dimension lower than n_θ. Show that as $t \to \infty$, $P^{-1}(t)$ becomes singular, and hence $P(t)$ diverges, irrespective of $P^{-1}(0)$.

As the foregoing problem suggests, a potential difficulty with exponentially weighted RLS is the so-called *covariance wind-up* phenomenon. If the regressor vectors bring little or no information, viz., according to the comment after (18), $P(k)\varphi(k) \approx 0$, it follows from (22) that $P(k+1) \approx P(k)/\lambda$. The forgetting therefore has the effect of increasing the size of $P(k)$ from one recursion to the next. Then, if no information enters the estimator over a long period, the division by λ at every step causes $P(k)$ to become very large, leading to erratic behavior of the estimates and possibly numerical overflow.

According to the preceding, exponentially weighted RLS must be carefully used. The main idea is to ensure that $P(k)$ stays bounded. In particular, whenever possible, a *dither* signal should be added to the system input so as to prevent the algorithm from incurring the covariance wind-up phenomenon. Another possibility is to equip RLS with a *covariance resetting* logic fix according to which $P(k)$ is reset to a given positive definite matrix, whenever its value computed via (22b) becomes too large. A useful procedure, viz., the *dead-zone* fix, is to stop the updating of the parameter vector and the covariance matrix when $P(k)\varphi(k)$ and/or $\varepsilon(k)$ are sufficiently small. We next focus on specific mechanisms for preventing covariance wind-up, such as directional forgetting and constant-trace RLS.

Directional Forgetting RLS. In exponentially weighted RLS, the covariance wind-up phenomenon is caused by the fact that at each updating step, the normalized information matrix $P^{-1}(t)$ is reduced by the multiplicative factor λ in all directions in $\mathbb{R}^{n_\theta}$, except along the direction of the incoming regressor $\varphi(t)$ where matrix $\lambda\varphi(t)\varphi'(t)$ is added to $\lambda P^{-1}(t)$. In directional forgetting ([Hag83], [KK84], and [Kul87]), the idea is to modify $P^{-1}(t)$ only along the direction of the incoming regressor according to the formula

$$P^{-1}(t+1) = P^{-1}(t) + \eta(t)\varphi(t)\varphi'(t) \qquad (6.3\text{-}41a)$$

This should be compared with (22c). In (41a), $\eta(t)$ is a real number to be suitably chosen under the constraint that $P^{-1}(t+1) > 0$ provided that $P^{-1}(t) > 0$

Problem 6.3-7 Let $P^{-1} = P^{-T} > 0$ and $\hat{P}^{-1} = P^{-1} + \eta\varphi\varphi'$, with $\eta \in \mathbb{R}$. Show that $\hat{P}^{-1} > 0$ if and only if

$$-\frac{1}{\varphi'P\varphi} < \eta \qquad (6.3\text{-}41b)$$

(*Hint*: Prove that (41b) implies $\hat{P} > 0$. Next, show that if (41b) is not true, vectors $x = P\varphi$ make $x'\hat{P}^{-1}x \le 0$.)

In [KK84], the following choice for $\eta(t)$ is derived via a Bayesian argument

$$\eta(t) = \lambda - \frac{1-\lambda}{\varphi'(t)P(t)\varphi(t)} \qquad (6.3\text{-}41c)$$

Here λ plays a role similar to that of a fixed forgetting factor. Note that $\eta(t)$ as defined before satisfies the inequality (41b). The RLS with directional forgetting (41c) update the θ estimate as in (13a) with

$$P(t+1) = P(t) - \frac{P(t)\varphi(t)\varphi'(t)P(t)}{\eta^{-1}(t) + \varphi'(t)P(t)\varphi(t)} \qquad (6.3\text{-}41d)$$

the latter being obtained from (41a) via the matrix-inversion lemma (5.3-16).

Constant-Trace RLS. A constant covariance trace algorithm can be built out of the RLS with covariance modification (21) by simply choosing $Q(t+1)$ so as to make $\mathrm{Tr}\, P(t+1) = \mathrm{Tr}\, P(t) = \mathrm{Tr}\, P(0)$, viz.,

$$\mathrm{Tr}\, Q(t+1) = \frac{\varphi'(t)P^2(t)\varphi(t)}{1 + \varphi'(t)P(t)\varphi(t)} \qquad (6.3\text{-}42a)$$

One possible choice for $Q(t+1)$ is then to set

$$Q(t+1) = \frac{\varphi'(t)P^2(t)\varphi(t)}{n_\theta[1 + \varphi'(t)P(t)\varphi(t)]}I_{n_\theta} \qquad (6.3\text{-}42b)$$

An alternative to the foregoing is to start with the exponentially weighted RLS and choose the time-varying forgetting factor $\lambda(t+1)$ so as to make the covariance trace constant, viz.,

$$\lambda(t+1) = 1 - \frac{1}{\operatorname{Tr} P(0)} \frac{\varphi'(t)P^2(t)\varphi(t)}{1 + \varphi'(t)P(t)\varphi(t)} \tag{6.3-43}$$

Note that though we have described directional forgetting RLS and constant-trace RLS as algorithms for coping with the covariance wind-up phenomenon, they are also suitable for estimating time-varying parameters. A similar remark can be applied to other estimation methods for covariance management such as the ones based on covariance resetting [GS84] and the variable forgetting factor of [FKY81].

6.3.6 Numerically Robust Recursions

The recursive identification algorithms, as they have been given so far, are known not to be numerically robust. In particular, the RLS algorithm (13) hinges on (13c) to (13d), which is seen to be a Riccati equation. By Problem 2-3, this is the dual of the Riccati equation relevant for the LQOR problem. Then, from the discussion in Sec. 2.5, it follows that (13e) is the numerically robustified form of the Riccati recursions for RLS. The use of (13e) in the RLS algorithm yields in most circumstances completely satisfactory results. Nonetheless, more robust numerical implementations are obtained by factorizing the "covariance matrix" $P(t)$ in terms of a "square root" matrix $S(t)$, viz., $P(t) = S(t)S'(t)$. The RLS algorithm can be implemented by updating $S(t)$ in each recursion. This is roughly equivalent to computing $P(t)$ in double precision and ensures that $P(t)$ remains positive definite. If the rounding errors are significant, implementations based on factorization methods yield definitely superior results than the ones achievable with the standard RLS algorithm (13).

We next describe RLS recursions based on the $U - D$ factorized form:

$$P(t) = U(t)D(t)U'(t)$$

where $U(t)$ is an $n_\theta \times n_\theta$ upper triangular matrix with unit diagonal elements, and $D(t)$ a diagonal matrix of dimension n_θ. The recursions are obtained by slightly modifying the U–D covariance factorization in [Bie77] so as to consider the exponentially weighted RLS (22).

U–D Recursions for Exponentially Weighted RLS.   Let

$$P(t-1) = UDU' \qquad \text{and} \qquad \theta(t-1) = \theta \tag{6.3-44}$$

$$U = \begin{bmatrix} u_1 & \cdots & u_{n_\theta} \end{bmatrix} \qquad\qquad D = \operatorname{diag}\{\delta_i, i \in \underline{n}_\theta\}$$

Denote by

$$y = y(t) \qquad \text{and} \qquad \varphi = \varphi(t-1) \tag{6.3-45}$$

the regressand and the regressor, respectively, at time t. Then

$$P(t) = \hat{U}\hat{D}\hat{U}' \qquad \text{and} \qquad \theta(t) = \theta + K(y - \varphi'\theta) \qquad (6.3\text{-}46)$$

$$\hat{U} = \begin{bmatrix} \hat{u}_1 & \cdots & \hat{u}_{n_\theta} \end{bmatrix} \qquad \hat{D} = \text{diag}\left\{\hat{\delta}_i, i \in \underline{n}_\theta\right\}$$

are generated as follows.

Step 1. Compute vectors f and v:

$$\left.\begin{array}{l} f = \begin{bmatrix} f_1 & \cdots & f_{n_\theta} \end{bmatrix} = U'\varphi \\ v = \begin{bmatrix} v_1 & \cdots & v_{n_\theta} \end{bmatrix} = Df \end{array}\right\} \qquad (6.3\text{-}47)$$

Step 2. Set

$$\hat{\delta}_1 = \frac{\delta_1}{\alpha_1\lambda} \qquad \alpha_1 = 1 + v_1 f_1 \qquad K_2 = \begin{bmatrix} v_1 & O_{1\times(n_\theta-1)} \end{bmatrix}' \qquad (6.3\text{-}48)$$

Step 3. For $i = 2, \cdots, n_\theta$ recursively cycle through (49) to (53):

$$\alpha_i = \alpha_{i-1} + v_i f_i \qquad (6.3\text{-}49)$$

$$\hat{\delta}_i = \frac{\delta_i \alpha_{i-1}}{\alpha_i \lambda} \qquad (6.3\text{-}50)$$

$$\mu_i = -\frac{f_i}{\alpha_{i-1}} \qquad (6.3\text{-}51)$$

$$\hat{u}_i = u_i + \mu_i K_i \qquad (6.3\text{-}52)$$

$$K_{i+1} = K_i + v_i u_i \qquad (6.3\text{-}53)$$

Step 4. Compute

$$K = \frac{K_{n_\theta+1}}{\alpha_{n_\theta}} \qquad (6.3\text{-}54)$$

Main Points of the Section. Several parameter estimation algorithms have been introduced for recursively identifying dynamic linear I/O models, such as FIR, ARX, and ARMAX models. These algorithms can be classified as either linear or pseudo-linear-regression algorithms, according to the independence or dependence of the regressor on the estimated vector. The majority of the algorithms considered admit a minimum prediction-error formulation and hence can be seen as tools to fit the observed data by models of limited and reduced complexity. The need of discounting old data suggests adopting suitable provisions aimed at ensuring boundedness of the $P(t)$ matrix. These include fixes like dither signals, covariance resetting, dead zones, directional forgetting, constant trace, and combinations thereof. In order to enhance numerical robustness, the recursions have to be carried out via factorization methods, for example, the $U - D$ estimate-covariance updating algorithm.

6.4 CONVERGENCE OF RECURSIVE IDENTIFICATION ALGORITHMS

In this section, we give an account of the convergence properties of the recursive estimation algorithms introduced in Sec. 3. The discussion is carried out in detail only for the RLS algorithm, the results for the other algorithms being briefly sketched. The main idea is to describe some convergence analysis tools applicable to a great deal of recursive stochastic algorithms, in connection with the most frequently used identification method in adaptive control applications, viz., the RLS algorithm. We consider the RLS algorithm first in a deterministic and next in a stochastic setting. Finally, we state convergence results for some pseudo-linear-regression algorithms.

We point out from the outset that in order to prove convergence to the "true" system parameter vector θ, some strong assumptions have to be made. In particular:

1. The system model (for example, FIR, ARX, and ARMAX) and its order must be exactly known.

2. The inputs must be persistently exciting in a sense to be clarified.

3. Mean-square boundedness is required in the RELS(PO) convergence proof.

We point out that such properties cannot be a priori guaranteed in adaptive control schemes whereby the analysis must be carried out without relying on the convergence of the identifier.

There have been three major approaches to the analysis of recursive identification algorithms:

1. **Ordinary Differential Equation (ODE) Analysis.** This method consists of associating a system of ordinary differential equations to a recursive algorithm in such a way that the asymptotic behavior of the latter is described by the state evolution of the first. We do not introduce the ODE method here and postpone its description and use in subsequent chapters dealing with adaptive control.

2. **Analysis via Stochastic Lyapunov Functions.** The analysis is carried out by the direct construction of a positive supermartingale so as to exploit appropriate martingale convergence theorems. A positive supermartingale (or a stochastic function closely related to it) can be seen as the stochastic analogue of a Lyapunov function of deterministic stability theory. The Lyapunov function methods are the ones mainly used throughout this section for both deterministic and stochastic convergence analysis of RLS. The reader can thus usefully compare the two developments to find out similarities and differences in the two cases.

3. **Direct Analysis.** In some cases, the method of analysis does not follow any of the two prior approaches and is specifically tailored to the algorithm under consideration. See, for instance, the convergence proof of RLS originally obtained by [LW82] and further elaborated in [Cai88].

6.4.1 RLS Deterministic Convergence

Throughout this section, we assume that (3-1) and (3-2) hold true. This means that there are no modeling errors, the I/O data are noise-free, and there is a true system parameter vector $\theta \in \mathbb{R}^{n_\theta}$ to be determined. Setting

$$R(t) := \frac{1}{t} \sum_{k=0}^{t-1} \varphi(k)\varphi'(k) \qquad (6.4\text{-}1a)$$

the system of normal equations (3-14) yields for the RLS estimate

$$
\begin{aligned}
\theta(t) &= \left[R(t) + \frac{1}{t} P^{-1}(0) \right]^{-1} \left[\frac{1}{t} \sum_{k=0}^{t-1} \varphi(k)y(k+1) + \frac{1}{t} P^{-1}(0)\theta(0) \right] \\
&= \left[R(t) + \frac{1}{t} P^{-1}(0) \right]^{-1} \left[R(t)\theta + \frac{1}{t} P^{-1}(0)\theta(0) \right] \qquad [(3\text{-}2)] \qquad (6.4\text{-}1b)
\end{aligned}
$$

Then we see that if $R(t)$ converges to a bounded nonsingular matrix as $t \to \infty$, $\theta(t)$ converges to the true system parameter vector θ. Whereas nonsingularity of $R(t)$ for large t is unavoidable for establishing convergence and, as we shall soon see, is related to the notion of a sufficiently "exciting" regressor, boundedness of $R(t)$ as $t \to \infty$ entails stability of the system to be identified. We consider next a different tool for RLS analysis that does not require system stability.

Setting

$$\tilde{\theta}(t) := \theta(t) - \theta \qquad (6.4\text{-}2a)$$

we can write

$$
\begin{aligned}
\varepsilon(t) &:= y(t) - \varphi'(t-1)\theta(t-1) \\
&= -\varphi'(t-1)\tilde{\theta}(t-1) \qquad (6.4\text{-}2b)
\end{aligned}
$$

Subtracting θ from both sides of (3-13a, 3-13b), we get

$$
\begin{aligned}
\tilde{\theta}(t+1) &= \tilde{\theta}(t) - \frac{P(t)\varphi(t)\varphi'(t)\tilde{\theta}(t)}{1 + \varphi'(t)P(t)\varphi(t)} \qquad (6.4\text{-}2c) \\
&= \tilde{\theta}(t) - P(t+1)\varphi(t)\varphi'(t)\tilde{\theta}(t) \qquad (6.4\text{-}2d) \\
&= \left[I_{n_\theta} - P(t+1)\varphi(t)\varphi'(t) \right] \tilde{\theta}(t) \qquad (6.4\text{-}2e) \\
&= P(t+1)P^{-1}(t)\tilde{\theta}(t) \qquad [(3\text{-}13g)] \qquad (6.4\text{-}2f)
\end{aligned}
$$

Because the RLS estimation error $\tilde{\theta}(t)$ satisfies difference equation (2), $\tilde{\theta}(t)$ converges to O_{n_θ} for any $\tilde{\theta}(0)$ and $P(0)$ provided that (2) and (3-13g) is an asymptotically stable system. In order to find out conditions under which this is guaranteed, we use a Lyapunov function argument [SL91]. We first exhibit the existence of a Lyapunov function $V(\tilde{\theta}(t))$ for (2) and (3-13g), viz., a nonnegative function of $\tilde{\theta}(t)$ that is nonincreasing along the trajectories of (2) and (3-13g). Next, we find sufficient conditions under which convergence of $V(\tilde{\theta}(t))$ implies convergence of $\tilde{\theta}(t)$ to O_{n_θ}.

Theorem 6.4-1. **(RLS Convergence)** *Let the y and φ sequences be as in (3-1) and (3-2). Then the nonnegative function*

$$V(t) := \tilde{\theta}'(t)P^{-1}(t)\tilde{\theta}(t) \qquad (6.4\text{-}3)$$

is nonincreasing along the trajectories of (2) and (3-13g). Further, provided that

$$\lim_{t\to\infty} \lambda_{\min}[P^{-1}(t)] = \infty \qquad (6.4\text{-}4)$$

the RLS estimate $\theta(t)$ converges to θ as $t \to \infty$, for all $\theta(0)$ and $P(0) = P'(0) > 0$.

Proof: By (2f), (3) can be rewritten as follows:

$$V(t) = \tilde{\theta}'(t)P^{-1}(t-1)\tilde{\theta}(t-1)$$

Consequently,

$$
\begin{aligned}
V(t) - V(t-1) &= \left[\tilde{\theta}(t) - \tilde{\theta}(t-1)\right]' P^{-1}(t-1)\tilde{\theta}(t-1) \\
&= -\frac{\tilde{\theta}'(t-1)\varphi(t-1)\varphi'(t-1)\tilde{\theta}(t-1)}{1 + \varphi'(t-1)P(t-1)\varphi(t-1)} \qquad [(2c)] \\
&= -\frac{\varepsilon^2(t)}{1 + \varphi'(t-1)P(t-1)\varphi(t-1)} \qquad [(2b)] \quad (6.4\text{-}5)
\end{aligned}
$$

Hence, $V(t)$ is nonincreasing. Being also nonnegative, $V(t)$ converges to a bounded limit as $t \to \infty$. Therefore,

$$M > \lim_{t\to\infty} \tilde{\theta}'(t)P^{-1}(t)\tilde{\theta}(t) > \lim_{t\to\infty} \left\{ \lambda_{\min}[P^{-1}(t)] \cdot \|\tilde{\theta}(t)\|^2 \right\}$$

for some $M > 0$. Hence, if (4) is fulfilled, $\tilde{\theta}(t) \to O_{n_\theta}$, for any $\theta(0)$, and $P(0) = P'(0) > 0$.

Condition (4) is guaranteed provided that

$$\lim_{t\to\infty} \left\{ \lambda_{\min} \left[\sum_{k=0}^{t-1} \varphi(k)\varphi'(k) \right] \right\} = \infty \qquad (6.4\text{-}6)$$

In order to relate (6) to the system input sequence, we begin with considering a FIR system whereby

$$
\begin{aligned}
y(t) &= B(d)u(t) \qquad &(6.4\text{-}7a) \\
&= \varphi'(t-1)\theta &
\end{aligned}
$$

$$
\begin{aligned}
\varphi'(t-1) &= \left[\begin{array}{ccc} u(t-1) & \cdots & u(t-n_b) \end{array} \right] \qquad &(6.4\text{-}7b) \\
\theta' &= \left[\begin{array}{ccc} b_1 & \cdots & b_{n_b} \end{array} \right] \qquad &(6.4\text{-}7c)
\end{aligned}
$$

We say that input signal $\{u(t)\}$ is *persistently exciting* of order n if

$$\rho_1 I_n \geq \lim_{N \to \infty} \frac{1}{N} \sum_{t=1}^{N} \begin{bmatrix} u(t-1) \\ \vdots \\ u(t-n) \end{bmatrix} \begin{bmatrix} u(t-1) & \cdots & u(t-n) \end{bmatrix} \geq \rho_2 I_n \qquad (6.4\text{-}8)$$

for some $\rho_1 \geq \rho_2 > 0$. For $n \geq n_b$, this condition implies (6) and, hence, RLS convergence according to Theorem 1.

Problem 6.4-1 (RLS Rate of Convergence)   Prove that for the deterministic FIR system (7) $\|\tilde{\theta}(t)\|^2$ converges at least at the rate $1/t$ provided that system input is persistently exciting of order n_b. However, as can be verified using (2) via a scalar example where $\varphi(t-1) \equiv 1$, $\theta(t)$ converges at the rate $1/t$.

It can be shown [GS84] that a stationary input sequence whose spectrum is nonzero at n points or more is persistently exciting of order n. In particular, this happens to be true for an input of the form $u(t) = \sum_{i=1}^{s} v_i \sin(\omega_i t + \alpha_i)$, $\omega_i \in (0, \pi)$, $\omega_i \neq \omega_j$, $v_i \neq 0$, and $s \geq n/2$.

For the general deterministic recurrent system (1), RLS convergence properties are similar to the ones valid for the FIR system (7). In particular, the following result applies.

Fact 6.4-1.   [GS84].   *Let $y(\cdot)$ and $\varphi(\cdot)$ sequences be as in (3-1) and (3-2). Then, if $A(d)$ is strictly Hurwitz, the RLS estimate $\theta(t)$ converges to the true parameter vector θ provided that:*

- *input $\{u(t)\}$ is a stationary sequence whose spectral distribution is nonzero at $n_a + n_b$ points or more*

- *polynomials $A(d)$ and $B(d)$ are coprime*

A similar convergence result [GS84] is available if system (3-1) is unstable and its input $u(t)$ is given by a nonnecessarily stabilizing but piecewise constant-feedback component $F\varphi(t-1)$ plus an exogenous signal $v(t) = \sum_{i=1}^{s} v_i \sin(\omega_i t + \alpha_i)$, $\omega_i \in (0, \pi)$, $\omega_i \neq \omega_j$, $v_i \neq 0$, and $s \geq 4n$. This convergence result is subject again to the condition that $A(d)$ and $B(d)$ are coprime polynomials. We point out the importance of the latter condition. In fact, it is basically an *identifiability* condition in that it makes the representation (3-1) well-defined on the grounds of the I/O system behavior. Further, in light of Problem 2.4-5, the previous condition is equivalent to the reachability of state $\varphi(t)$ of system (3-1) (cf. Lemma 5.4-1). It is intuitively clear that reachability of $\varphi(t)$ is a key property that has to be satisfied in order that (6) be possibly achieved via a persistently exciting input signal.

A deterministic convergence analysis for the exponentially weighted RLS is reported in [JJBA82], where it is shown that, under persistent excitation, this algorithm, unlike the $\lambda = 1$ case, for $\lambda \in (0, 1)$, is exponentially convergent (see also [Joh88]). Exponential convergence is important in that it implies tracking capability for slowly varying parameters [AJ83]. However, as we have seen at the end of Sec. 3, other problems arise when $\lambda < 1$ with the exponentially weighted RLS algorithm whenever persistent excitation conditions are not satisfied.

For a deterministic convergence analysis of directional forgetting RLS, see [BBC90a]. A constant-trace normalized version of RLS is analyzed under deterministic conditions in [LG85]. This analysis is reported in Sec. 8.6, where the algorithm is used in adaptive control schemes. For conditions that guarantee convergence of the projection algorithm in the deterministic case, the reader is referred to [GS84].

6.4.2   RLS Stochastic Convergence

We first consider the RLS algorithm under the limitative assumption that y and u are finite-variance or square integrable strictly stationary ergodic processes. Hence [Cai88],

$$\lim_{N \to \infty} \frac{1}{N} \sum_{k=1}^{N} \varphi(k-1)\varphi'(k-1) = \mathcal{E}\{\varphi(t)\varphi'(t)\} = \Psi_\varphi \quad \text{a.s.} \qquad (6.4\text{-}9a)$$

where

$$\varphi(t-1) := \begin{bmatrix} y'(t-1) & \cdots & y'(t-n_a) & u'(t-1) & \cdots & u'(t-n_b) \end{bmatrix}' \qquad (6.4\text{-}9b)$$

Further, let

$$\Psi_\varphi > 0 \qquad (6.4\text{-}9c)$$

We make no assumption on how y and u are related. In particular, the underlying system whose input and output variables are the u and y processes, respectively, need not be linear or exactly described by an ARX model with $n_a = \partial A(d)$ and $n_b = \partial B(d)$.

Consider next the orthogonal projection $\hat{\mathcal{E}}\{y(t) \mid \varphi(t-1)\}$ of $y(t)$ onto $[\varphi(t-1)]$, the subspace of $L_2(\Omega, \mathcal{F}, \mathbb{P})$ (cf. Example 1-2) spanned by the random vector $\varphi(t-1)$. It results (cf. Problem 1-2)

$$\hat{\mathcal{E}}\{y(t) \mid \varphi(t-1)\} = \mathcal{E}\{y(t)\varphi'(t-1)\} \Psi_\varphi^{-1}\varphi(t-1) \qquad (6.4\text{-}10a)$$

$$= \overset{\circ}{\theta}{}' \varphi(t-1)$$

where

$$\overset{\circ}{\theta} := \Psi_\varphi^{-1}\mathcal{E}\{\varphi(t-1)y'(t)\} \in \mathbb{R}^{n_\theta} \qquad (6.4\text{-}10b)$$

Note that

$$\mathcal{E}\left\{ \left[y(t) - \overset{\circ}{\theta}{}' \varphi(t-1) \right] \varphi'(t-1) \right\} = 0 \qquad (6.4\text{-}10c)$$

The following theorem relates the RLS estimate to the previous vector $\overset{\circ}{\theta}$.

Theorem 6.4-2.   (RLS Consistency in the Ergodic Case) *Let y and u be finite-variance strictly stationary ergodic processes, and, consequently, (9a) be fulfilled. In addition, let (9c) hold. Then,*

(i) *The vector $\overset{\circ}{\theta}$ given by (10b) is the unique vector that parameterizes the orthogonal projection of $y(t)$ onto $[\varphi(t-1)]$ according to (10a) or (10c).*

(ii) *For each $t \in \mathbb{Z}_1$, there is a unique solution $\theta(t)$, the RLS estimate of $\overset{\circ}{\theta}$, to the normal equations (3-14).*

(iii) *The RLS estimator is strongly consistent, viz., $\theta(t)$ converges to $\overset{\circ}{\theta}$ a.s. as $t \to \infty$.*

Proof: For (i) and (ii), see Sec. 1 and Proposition 3-1, respectively. Setting

$$R(t) := \frac{1}{t} \sum_{k=0}^{t-1} \varphi(k)\varphi'(k)$$

from (3-14), we get by ergodicity

$$\begin{aligned}
\lim_{t\to\infty} \theta(t) &= \lim_{t\to\infty} \left\{ \left[R(t) + \frac{P^{-1}(0)}{t} \right]^{-1} \left[\frac{1}{t} \sum_{k=0}^{t-1} \varphi(k)y'(k+1) + \frac{1}{t}P^{-1}(0)\theta(0) \right] \right\} \\
&= \left[\lim_{t\to\infty} R(t) \right]^{-1} \left[\lim_{t\to\infty} \left(\frac{1}{t} \sum_{k=0}^{t-1} \varphi(k)y'(k+1) \right) \right] \\
&= \Psi_\varphi^{-1} \mathcal{E} \left\{ \varphi(k)y'(k+1) \right\} = \overset{\circ}{\theta} \qquad \text{a.s.} \qquad (6.4\text{-}11)
\end{aligned}$$

The relevance of Theorem 2 is that it tells us that, under ergodicity, the RLS-based one-step output predictor $\hat{y}(t \mid t-1; \theta(t)) := \theta'(t)\varphi(t-1)$ converges a.s. to $\hat{y}(t \mid t-1; \overset{\circ}{\theta}) = \overset{\circ}{\theta}{}' \varphi(t-1)$, the MMSE estimator of $y(t)$ based on y^{t-1}, u^{t-1}, among all estimators of the form $\theta'\varphi(t-1)$. This result consolidates a similar observation made after (3-31b).

In the last line of the foregoing proof, we have used the ergodicity property (9a) and (9c). Comparing this with (8), we see that (9c) can be interpreted as a persistency of excitation condition for the present ergodic situation. When (9c) is satisfied, φ is said to be a *persistently exciting regressor*. Under these circumstances, if $\varphi(k)$ is as in (7b), u is said to be *persistently exciting of order n_b*.

Suppose now that the data-generating system is as in (3-1) and (3-2) with $A(d)$ strictly Hurwitz and u ergodic. Then, $\varphi(k)$ as in (3-2b) is a persistently exciting regressor vector if $A(d)$ and $B(d)$ are coprime and u is a persistently exciting input of order $n_a + n_b$ [SS89]. Let the data-generating system be given by a perturbed version of the difference equation (3-1):

$$A(d)y(t) = B(d)u(t) + v(t) \qquad\qquad (6.4\text{-}12)$$

In (12), $v(t)$ represents the "disturbance" or the "equation error." It is assumed that u and v are ergodic, $\mathcal{E}\{u(t)v(\tau)\} = 0$, for all t and τ, and $A(d)$ is strictly Hurwitz. Then, $\varphi(k-1)$ as in (3-2b) is a persistently exciting regressor, provided that u is persistently exciting of order n_b, and v is persistently exciting of order n_a [SS89]. Note that the latter condition is always fulfilled if $v(t) = H(d)e(t)$ with $H(d)$ a rational transfer function and $e(t)$ white.

Rewrite (12) as follows:

$$y(t) = \theta'\varphi(t-1) + v(t) \qquad (6.4\text{-}13\text{a})$$

Consequently, (10b) becomes

$$\overset{\circ}{\theta} = \theta + \Psi_\varphi^{-1}\mathcal{E}\{\varphi(t-1)v'(t)\} \qquad (6.4\text{-}13\text{b})$$

Hence, under the stated assumptions, the RLS estimator converges a.s. to the "true" parameter vector θ, in which case we say that the RLS estimator is *asymptotically unbiased*, if and only if the equation error $v(t)$ is uncorrelated with the regressor $\varphi(t-1)$:

$$\mathcal{E}\{\varphi(t-1)v'(t)\} = 0 \qquad (6.4\text{-}13\text{c})$$

Problem 6.4-2 Assume that the data-generating system is given by the ARMAX model

$$A(d)y(t) = B(d)u(t) + C(d)e(t)$$

with $\partial A(d) = n_a$, $\partial B(d) = n_b$, and $\partial C(d) \geq 1$. Consider the RLS estimator with regressor $\varphi(k-1) = \begin{bmatrix} -y(k-1) & \cdots & -y(k-n_a) & u(k-1) & \cdots & u(k-n_b) \end{bmatrix}'$. Assume that $A(d)$ is strictly Hurwitz and u and e finite-variance ergodic processes with u persistently exciting of order n_b. Prove that such an estimator of $\theta = \begin{bmatrix} a_1 & \cdots & a_{n_a} & b_1 & \cdots & b_{n_b} \end{bmatrix}'$ is asymptotically unbiased if $\mathcal{E}\{u(t)e(\tau)\} = 0$, for all t and τ, provided that $n_a = 0$. On the opposite, show that (13c), and hence the previous property, does not hold true if $n_a > 0$ and/or $\mathcal{E}\{u(t)e(\tau)\} = 0$ only for all $t < \tau$.

Problem 2 points out that in general, the RLS estimator is *asymptotically biased*, viz., it is not consistent with the "true" θ vector. Just to mention a few relevant cases, such a difficulty is met, even when φ is a persistently exciting regressor, under the following circumstances:

- $n_a > 0$, viz., the data-generating system is not FIR, and the equation error $\{v(t)\}$ in (12) is not a white process

- n_a and/or n_b are chosen too small, and hence $v(t)$, depending on past I/O pairs, is correlated with $\varphi(t-1)$

Another important situation that prevents RLS from being consistent is the loss of regressor persistency of excitation that typically takes place when the input $u(t)$ is solely generated by a dynamic feedback from output $y(t)$.

We now turn on to analyze the RLS algorithm in the stochastic case under no ergodicity assumption. As anticipated in the beginning of this section, to this end, we follow the stochastic Lyapunov function (or stochastic stability) approach. We limit our analysis to the RLS algorithm, taken here as a representative of other identification algorithms, such as pseudo-linear-regression algorithms, for which, nonetheless, we shall indicate the conclusions achievable via a similar convergence analysis. The reader is referred to Appendix D for the necessary results on martingale convergence properties that will be used in the remaining part of this section.

We assume that the data-generating system is given by the SISO ARX model:

$$A(d)y(k) = B(d)u(k) + e(k) \qquad (6.4\text{-}14\text{a})$$

with $A(d)$ and $B(d)$ as in (3-1), and $e(k)$ the equation error, or, equivalently,

$$y(k) = \varphi'(k-1)\theta + e(k) \qquad (6.4\text{-}14\text{b})$$

with $\varphi(k-1)$ and θ as in (3-2). The stochastic assumptions are as follows. The process $\{\varphi(0), z(1), z(2), \cdots\}$, $z(k) := \begin{bmatrix} y(k) & u(k) \end{bmatrix}'$, is defined on an underlying probability space $(\Omega, \mathcal{F}, \mathbb{P})$, and we define $\mathcal{F}_0$ to be the σ-field generated by $\{\varphi(0)\}$. Further, for all $t \in \mathbb{Z}_1$, $\mathcal{F}_t$ denotes the σ-field generated by $\{\varphi(0), z(1), \cdots, z(t)\}$ or, equivalently, $\{\varphi(0), \varphi(1), \cdots, \varphi(t)\}$. Consequently, $\mathcal{F}_0 \subset \mathcal{F}_t \subset \mathcal{F}_{t+1}$, $t \in \mathbb{Z}_1$. The following independence and variance assumptions are adopted on process e:

$$\mathcal{E}\left\{e(t) \mid \mathcal{F}_{t-1}\right\} = 0 \qquad \text{a.s.} \qquad (6.4\text{-}14\text{c})$$

$$\mathcal{E}\left\{e^2(t) \mid \mathcal{F}_{t-1}\right\} = \sigma^2 \qquad \text{a.s.} \qquad (6.4\text{-}14\text{d})$$

for every $t \in \mathbb{Z}_1$. Note that by the smoothing properties of conditional expectations, (14c) and (14d) imply that $\{e(t)\}$ is zero-mean and white.

Theorem 6.4-3.   (RLS Strong Consistency) *Consider the RLS algorithm (3-13) applied to the data generated by the ARX system (14). Then, provided that:*

(i) persistent excitation

$$\lim_{t \to \infty} \lambda_{\min}\left[P^{-1}(t)\right] = \infty \qquad (6.4\text{-}15\text{a})$$

(ii) order condition

$$\limsup_{t \to \infty} \frac{\lambda_{\max}\left[P^{-1}(t)\right]}{\lambda_{\min}\left[P^{-1}(t)\right]} < \infty \qquad (6.4\text{-}15\text{b})$$

the RLS estimate is strongly convergent to θ, that is,

$$\lim_{t \to \infty} \theta(t) = \theta \qquad \text{a.s.} \qquad (6.4\text{-}16)$$

Problem 6.4-3 Prove that, if $0 < \rho_1 \leq \rho_2 < \infty$, (15a) and (15b) are implied by the following persistent excitation condition

$$\rho_1 I_{n_\theta} < \lim_{t \to \infty} \frac{1}{t} \sum_{k=1}^{t} \varphi(k-1)\varphi'(k-1) < \rho_2 I_{n_\theta} \qquad \text{a.s.} \qquad (6.4\text{-}17)$$

but not vice versa. In particular, note that asymptotic boundedness of $t^{-1} \sum_{k=1}^{t} \varphi(k - 1)\varphi'(k - 1)$ is a stability condition whereby input $u(t)$, possibly determined through feedback from $\{y(k), k \leq t\}$, stabilizes system (14).

Proof of Theorem 6.4-3: As in (2a), let $\tilde{\theta}(t) := \theta(t) - \theta$. Next, as in (3), define $V(t) := \tilde{\theta}'(t)P^{-1}(t)\tilde{\theta}(t)$. Denoting $\text{Tr}[P^{-1}(t)]$ by $r(t)$, from (13g), it follows that

$$r(t) = r(t - 1) + \|\varphi(t - 1)\|^2 \tag{6.4-18}$$

with $r(0) = \text{Tr}[P^{-1}(0)] > 0$. The proof is based on the following two key results:

$$\lim_{t \to \infty} \frac{V(t)}{r(t)} < \infty \qquad \text{a.s.} \tag{6.4-19}$$

$$\sum_{t=1}^{\infty} \frac{\|\varphi(t-1)\|^2}{r(t-1)} \frac{V(t)}{r(t)} < \infty \qquad \text{a.s.} \tag{6.4-20}$$

We first show how (19) and (20) can be used to prove the theorem and, next, we derive them via a stochastic Lyapunov function argument.

Equations (19) and (20) imply that

$$\lim_{t \to \infty} \frac{V(t)}{r(t)} = 0 \qquad \text{a.s.} \tag{6.4-21}$$

In fact, (20) can be rewritten as

$$\sum_{t=1}^{\infty} \frac{V(t)}{r(t)} \frac{r(t) - r(t - 1)}{r(t - 1)} < \infty \qquad \text{a.s.} \tag{6.4-22a}$$

Now, we show by contradiction that

$$\sum_{t=1}^{\infty} \frac{r(t) - r(t - 1)}{r(t - 1)} = \infty \tag{6.4-22b}$$

On the contrary, suppose that the foregoing is finite. Because (15a) implies that $\lim_{t \to \infty} r(t) = \infty$, Kronecker's lemma as given by Result D.6-1 yields

$$\lim_{t \to \infty} \frac{1}{r(t - 1)} \sum_{k=1}^{t} [r(k) - r(k - 1)] = 0$$

Hence, because $r(t - 1) \leq r(t)$,

$$\begin{aligned} 0 &= \lim_{t \to \infty} \frac{1}{r(t)} \sum_{k=1}^{t} [r(k) - r(k - 1)] \\ &= \lim_{t \to \infty} \left[1 - \frac{r(0)}{r(t)} \right] \end{aligned}$$

This contradicts $\lim_{t \to \infty} r(t) = \infty$. Therefore, (22b) holds. Then, (21) follows from (19), (22a), and (22b). Now from the definition of $V(t)$,

$$\frac{V(t)}{r(t)} \geq \frac{\lambda_{\min}\left[P^{-1}(t)\right] \|\tilde{\theta}(t)\|^2}{r(t)}$$

$$\geq \frac{\lambda_{\min}\left[P^{-1}(t)\right]}{n_\theta \lambda_{\max}\left[P^{-1}(t)\right]} \|\tilde{\theta}(t)\|^2$$

From (21) and the foregoing, using (15b), we have

$$\lim_{t \to \infty} \|\tilde{\theta}(t)\|^2 = 0 \qquad \text{a.s.}$$

and (16) follows.

We now proceed to prove (19) and (20). We do it in two steps.

(a) **Calculation of $V(t)$.** From (3-13) and (14), we get

$$\tilde{\theta}(t) - P(t-1)\varphi(t-1)\eta(t) = \tilde{\theta}(t-1) \qquad (6.4\text{-}23)$$

where $\eta(t)$ denotes the a posteriori error:

$$\begin{aligned} \eta(t) &= y(t) - \varphi'(t-1)\theta(t) \\ &= -\varphi'(t-1)\tilde{\theta}(t) + e(t) \\ &= \frac{\varepsilon(t)}{1 + \varphi'(t-1)P(t-1)\varphi(t-1)} \end{aligned}$$

and $\varepsilon(t)$ the a priori error $\varepsilon(t) := y(t) - \varphi'(t-1)\theta(t-1)$. Setting $b(t) := -\varphi'(t-1)\tilde{\theta}(t)$, from (23), we find

$$\begin{aligned} V(t-1) &= \tilde{\theta}'(t)P^{-1}(t-1)\tilde{\theta}(t) + 2b(t)\eta(t) + \varphi'(t-1)P(t-1)\varphi(t-1)\eta^2(t) \\ &= V(t) - b^2(t) + 2b(t)\eta(t) + \varphi'(t-1)P(t-1)\varphi(t-1)\eta^2(t) \qquad [(13\text{g})] \end{aligned}$$

and recalling that $\eta(t) = b(t) + e(t)$,

$$V(t) = V(t-1) - b^2(t) - 2b(t)e(t) - \varphi'(t-1)P(t-1)\varphi(t-1)\eta^2(t)$$

Taking conditional expectations w.r.t. the σ-field, $\mathcal{F}_{t-1}$ gives

$$\begin{aligned} \mathcal{E}\left\{V(t) \mid \mathcal{F}_{t-1}\right\} &= V(t-1) - \mathcal{E}\left\{b^2(t) \mid \mathcal{F}_{-1}\right\} + 2\varphi'(t-1)P(t)\varphi(t-1)\sigma^2 \\ &\quad - \mathcal{E}\left\{\varphi'(t-1)P(t-1)\varphi(t-1)\eta^2(t) \mid \mathcal{F}_{t-1}\right\} \qquad (6.4\text{-}24) \end{aligned}$$

Equation (24) is obtained by using the following properties: $\theta(t) \in \mathcal{F}_t$, where this notation indicates that $\theta(t)$ is $\mathcal{F}_t$-measurable; $P(t) \in \mathcal{F}_{t-1}$ by (13g); and (cf. Problem 4, which follows) $\mathcal{E}\left\{b(t)e(t) \mid \mathcal{F}_{t-1}\right\} = -\varphi'(t-1)P(t)\varphi(t-1)\sigma^2$.

(b) **Construction of a Stochastic Lyapunov Function.** Define

$$X(t) := \frac{V(t)}{r(t)} + \frac{\sum_{k=1}^{t} b^2(k)}{r(t-1)} + \sum_{k=1}^{t}\left[\frac{\varphi'(k-1)P(k-1)\varphi(k-1)}{r(k-1)}\eta^2(k)\right]$$

$$+ \sum_{k=1}^{t}\left[\frac{\|\varphi(k-1)\|^2}{r(k-1)}\frac{V(k)}{r(k)}\right] \qquad (6.4\text{-}25)$$

By using (24), we show that $X(t)$ is a stochastic Lyapunov function, in that it is a positive process and

$$\mathcal{E}\left\{X(t) \mid \mathcal{F}_{t-1}\right\} \leq X(t-1) + 2\frac{\varphi'(t-1)P(t)\varphi(t-1)}{r(t-1)} \qquad \text{a.s.} \qquad (6.4\text{-}26)$$

Since (cf. Problem 5, which follows)

$$\sum_{k=1}^{\infty} \frac{\varphi'(k-1)P(k)\varphi(k-1)}{r(k)} < \infty \qquad \text{a.s.} \qquad (6.4\text{-}27)$$

by virtue of (26), we can apply the martingale convergence theorem (Theorem D.6-1) to conclude that $\{x(t), t \in \mathbb{Z}\}$ converges a.s. to a finite random variable:

$$\lim_{t \to \infty} X(t) = X < \infty \qquad \text{a.s.} \qquad (6.4\text{-}28)$$

In particular, because all the additive terms in (25) are nonnegative, (19) and (20) follow at once.

To prove (26), we take conditional expectations w.r.t. the σ-field $\mathcal{F}_{t-1}$ of every term in (25):

$$
\begin{aligned}
\mathcal{E}\left\{X(t) \mid \mathcal{F}_{t-1}\right\} \quad = \quad & \frac{\mathcal{E}\left\{V(t) \mid \mathcal{F}_{t-1}\right\}}{r(t)} + \frac{\mathcal{E}\left\{b^2(t) \mid \mathcal{F}_{t-1}\right\}}{R(t-1)} + \sum_{k=1}^{t-1} \frac{b^2(k)}{r(k-1)} \\
& + \frac{\mathcal{E}\left\{\varphi'(t-1)P(t-1)\varphi(t-1)\eta^2(t) \mid \mathcal{F}_{t-1}\right\}}{r(t-1)} \\
& + \sum_{k=1}^{t-1} \frac{\varphi'(k-1)P(k-1)\varphi(k-1)}{r(k-1)}\eta^2(k) \\
& + \frac{\|\varphi(t-1)\|^2}{r(t-1)} \frac{\mathcal{E}\left\{V(t) \mid \mathcal{F}_{t-1}\right\}}{r(t)} + \sum_{k=1}^{t-1} \frac{\|\varphi(k-1)\|^2}{r(k-1)} \frac{V(k)}{r(k)}
\end{aligned}
$$

Because

$$\frac{1}{r(t)}\left[1 + \frac{\|\varphi(t-1)\|^2}{r(t-1)}\right] = \frac{1}{r(t-1)}$$

using (24), we get

$$
\begin{aligned}
\mathcal{E}\left\{x(t) \mid \mathcal{F}_{t-1}\right\} \quad = \quad & \frac{V(t-1)}{r(t-1)} + \sum_{k=1}^{t-1} \frac{\varphi'(k-1)P(k-1)\varphi(k-1)}{r(k-1)}\eta^2(k) \\
& + \sum_{k=1}^{t-1} \frac{\|\varphi(k-1)\|^2}{r(k-1)} \frac{V(k)}{r(k)} + 2\frac{\varphi'(t-1)P(t)\varphi(t-1)}{r(t-1)}\sigma^2
\end{aligned}
$$

Hence, (26) holds with the equality sign.

Problem 6.4-4 Consider the RLS algorithm (3-13) applied to the data generated by the ARX system (14). Let $b(t) := -\varphi'(t-1)\tilde{\theta}(t)$. Show that

$$\mathcal{E}\left\{b(t)e(t) \mid \mathcal{F}_{t-1}\right\} = \varphi'(t-1)P(t)\varphi(t-1)\sigma^2$$

Problem 6.4-5   Prove the existence of the bounded limit in (27). *Hint:* Show that the nonnegative partial sum

$$\sum_{k=1}^{N} \frac{\varphi'(k-1)P(k)\varphi(k-1)}{r(k-1)}$$

is dominated by the monotonic nonincreasing sequence

$$\sum_{k=1}^{N} \mathrm{Tr}\,[P(k-1) - P(k)] = \mathrm{Tr}\,P(0) - \mathrm{Tr}\,P(N+1)\,.$$

To estimate the parameter vector θ of (14), it is instructive to consider instead of the RLS algorithm (3-13), the off-line least-squares algorithm

$$\theta(t) \quad = \quad R^{-1}(t)\sum_{k=1}^{t}\varphi(k-1)y(k) \qquad\qquad (6.4\text{-}29\text{a})$$

$$R(t) \quad := \quad \sum_{k=1}^{t}\varphi(k-1)\varphi'(k-1) \qquad\qquad (6.4\text{-}29\text{b})$$

which can be seen to minimize the criterion (3-15) for $P^{-1}(0) = 0$. For such an algorithm, we can prove that $\lim_{t\to\infty}\theta(t) = \theta$ a.s. provided that

$$\lim_{t\to\infty}\lambda_{\min}[R(t)] = \infty \qquad \text{a.s.} \qquad\qquad (6.4\text{-}30\text{a})$$

$$\frac{R(t)}{\mathrm{Tr}[R(t)]} \geq \rho I_{n_\theta} > 0 \qquad \text{a.s.} \qquad\qquad (6.4\text{-}30\text{b})$$

Note that $P^{-1}(t)$ reduces to $R(t)$ under the initialization $P^{-1}(0) = 0$. Hence, (30a) is a persistent excitation condition, whereas (30b) is similar to (15b). It is easy to see that the equations of (30) are implied by (17) but not vice versa. The strong consistency proof of (29) under (30) can be carried out [KV86] via a martingale convergence theorem, similarly, but in a somewhat more direct fashion, to the proof of Theorem 3.

In [LW82], it was proved via a direct analysis that RLS strong consistency is still guaranteed if conditions (15) can be relaxed as follows:

$$\lim_{t\to\infty}\frac{\lambda_{\min}\left[P^{-1}(t)\right]}{\log\lambda_{\max}\left[P^{-1}(t)\right]} = \infty \qquad \text{a.s.} \qquad\qquad (6.4\text{-}31)$$

Further, (31) follows from (30a), and

$$\lim_{t\to\infty}\frac{\lambda_{\min}\left[R(t)\right]}{\log\lambda_{\max}\left[R(t)\right]} = \infty \qquad \text{a.s.} \qquad\qquad (6.4\text{-}32)$$

According to [LW82], (30a) and (32) make up in some sense the weakest possible condition for establishing RLS strong convergence for possibly unstable systems and feedback-control systems with white-noise disturbances.

6.4.3 RELS Convergence Results

We consider the RELS(PO) algorithm (3-26b), (3-26d) to (3-27b) under the assumption that the data satisfy the ARMAX model:

$$A(d)y(t) = B(d)u(t) + C(d)e(t) \tag{6.4-33a}$$

or

$$y(t) = \varphi_e'(t-1)\theta + e(t) \tag{6.4-33b}$$

where $\varphi_e(t-1)$ is the "true" parameter vector θ as in (3-26a). The stochastic assumptions are as follows. All the involved processes, as well as $\varphi_e(0)$, are defined on an underlying probability space $(\Omega, \mathcal{F}, \mathbb{P})$. $\mathcal{F}_0$ is defined to be the σ-field generated by $\{\varphi_e(0)\}$. Further, for all $t \in \mathbb{Z}_1$, $\mathcal{F}_t$ denotes the σ-field generated by $\{\varphi_e(0), z(1), \cdots, z(t)\}$, $z(k) := \begin{bmatrix} y(k) & u(k) \end{bmatrix}'$, or, equivalently, $\{\varphi_e(0), \varphi_e(1), \cdots, \varphi_e(t)\}$. Consequently, $\mathcal{F} \subset \mathcal{F}_t \subset \mathcal{F}_{t+1}$, $t \in \mathbb{Z}_1$. We have also for the process e:

$$\mathcal{E}\left\{e(t) \mid \mathcal{F}_{t-1}\right\} = 0 \qquad \text{a.s.} \tag{6.4-33c}$$

$$\mathcal{E}\left\{e^2(t) \mid \mathcal{F}_{t-1}\right\} = \sigma^2 \qquad \text{a.s.} \tag{6.4-33d}$$

$$\limsup \frac{1}{N} \sum_{k=1}^{N} e^2(k) < \infty \qquad \text{a.s.} \tag{6.4-33e}$$

In order to state the desired result, we need an extra definition. Given a $p \times p$ matrix $H(d)$ of rational functions with real coefficients, we say that $H(d)$ is *positive real (PR)* if

$$H(e^{i\omega}) + H'(e^{-i\omega}) \geq 0, \qquad \omega \in [0, 2\pi) \tag{6.4-34}$$

$H(e^{i\omega})$ is said to be *strictly positive real (SPR)* if the foregoing is a strict inequality. Note that for $p = 1$, (34) becomes $\text{Re}[H(e^{i\omega})] \geq 0$, where Re denotes the "real part."

> **Result 6.4-1. (Strong Consistency of RELS(PO))** *Consider the RELS(PO) algorithm (3-26b) and (3-26d) to (3-27b) applied to the data generated by the ARMAX system (33). Assume further that the following conditions hold:*
>
> *(i) (Stability condition)* $\det[C(d)]$ *is a strictly Hurwitz polynomial.*
>
> *(ii) (Positive real condition)* $\frac{1}{C(d)} - \frac{1}{2}$ *is SPR.*
>
> *(iii) (Persistent excitation) The sample mean limit of the outer products of the process φ_e exists a.s. with*
>
> $$\rho_1 I_{n_\theta} < \lim_{N \to \infty} \frac{1}{N} \sum_{k=1}^{N} \varphi_e(k-1)\varphi_e'(k-1) < \rho_2 I_{n_\theta} \tag{6.4-35}$$
>
> *with $0 < \rho_1 \leq \rho_2 < \infty$.*

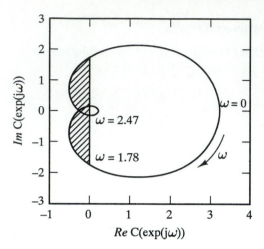

Figure 6.4-1: Polar diagram of $C(e^{i\omega})$ with $C(d)$ as in (37).

Then, the RELS(PO) estimate is strongly convergent to θ, that is,

$$\lim_{t \to \infty} \theta(t) = \theta \qquad a.s.$$

Problem 6.4-6   Let $C(d)$ be a polynomial. Show that for $\omega \in [0, 2\pi)$,

$$\left| C\left(e^{i\omega}\right) - 1 \right| < 1 \Longleftrightarrow \left[\frac{1}{C(d)} - \frac{1}{2} \right] \text{ is SPR} \Longrightarrow C(d) \text{ is SPR} \qquad (6.4\text{-}36)$$

Note that (36) indicates that the SPR condition in Result 1 amounts to assuming that (33a) is not too far from the ARX model $A(d)y(t) = B(d)u(t) + e(t)$ with $A(d)$ and $B(d)$ as in (33a).

Example 6.4-1
Consider the ARMAX model (33a) with

$$\left. \begin{aligned} A(d) &= 1 + d + 0.9d^2 \\ B(d) &= 0 \\ C(d) &= 1 + 1.5d + 0.75d^2 \end{aligned} \right\} \qquad (6.4\text{-}37)$$

Figure 1 depicts the polar diagram of $C(e^{i\omega})$. We see that $C(d)$ is not SPR and this, in turn, implies that $\frac{1}{C(d)} - \frac{1}{2}$ is not SPR. If the RELS(PO) algorithm with no input data in the pseudoregressor (27b) is applied to the data generated by the ARMA model (37), we get the results in Fig. 2. This shows the time evolution of the four components of $\theta(t) = \begin{bmatrix} a_1(t) & a_2(t) & c_1(t) & c_2(t) \end{bmatrix}'$. We see from Fig. 2 that the algorithm attempts to reach the true values. However, convergence is not achieved in that when the estimates come close to the optimal ones, they are pushed away and keep on bouncing below the true values of the parameters.

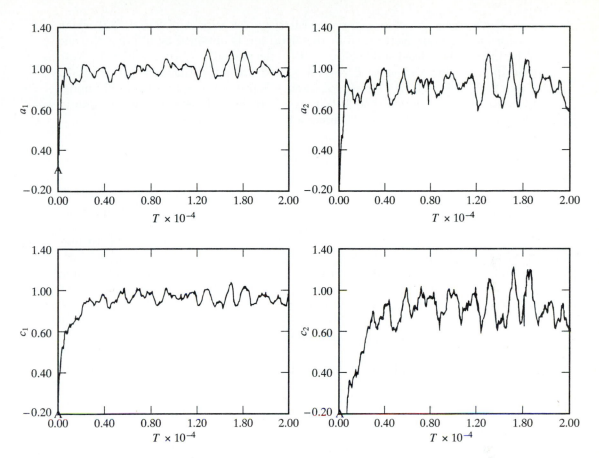

Figure 6.4-2: Time evolution of the four RELS estimated parameters when the data are generated by the ARMA model (37).

For a proof of Result 1, the reader is referred to [Cai88], pp. 556–565. This proof follows similar lines as the ones of Theorem 3 with extra complications arising from the presence here of the $C(d)$ innovations polynomial. As for the RLS algorithm, a direct approach was presented in [LW86] that allows one to replace the persistent excitation condition (35) by the weaker condition (31) provided that $P(t)$ is as in (3-26e) with $\varphi(t)$ as in (3-27b). Equation (31) is in turn implied by

$$\lim_{t\to\infty} \lambda_{\min}\left[R_e(t)\right] = \infty \qquad \text{a.s.}$$

and

$$\lim_{t\to\infty} \frac{\lambda_{\min}\left[R_e(t)\right]}{\log \lambda_{\max}\left[R_e(t)\right]} = \infty \qquad \text{a.s.}$$

where $R_e(t) := \sum_{k=1}^{t} \varphi_e(k-1)\varphi'_e(k-1)$.

For a discussion of the strong consistency of a variant of RELS(PO) where the pseudo-regressor vector used in the algorithm is obtained by filtering the one of RELS(PO) by a fixed stable filter $1/D(d)$, the reader is referred to [GS84]. Note that this identification method resembles, and has its justification in, the RML algorithm (3.28). Though strong consistency is proved for such a case only under the restrictive assumption that $A(d)$ is strictly Hurwitz, it satisfies our intuition to see that the SPR condition of Theorem 4 is modified as follows:

$$\left\{\frac{D(d)}{C(d)} - \frac{1}{2}\right\} \text{ is SPR}$$

Even if $\frac{1}{C(d)} - \frac{1}{2}$ is not SPR, the previous condition can be satisfied by choosing $D(d)$ close to $C(d)$ provided that the latter can be guessed with a good approximation.

Main Points of the Section. The Lyapunov function method and its stochastic extension based on martingale convergence theorems can be used to prove deterministic convergence and stochastic strong consistency of the RLS algorithm, respectively. The crucial conditions that must be satisfied to this end are the *model-matching condition*, viz., that the true data-generating system belongs to the model set parameterized by the vector to be identified, and the inputs satisfy appropriate *persistent excitation conditions*.

Strong convergence of pseudo-linear-regression algorithms, for example, RELS (PO), requires strong additional assumptions, such as *mean-square boundedness* of the involved signal and satisfaction of a *strict positive real condition*.

Although convergence analysis of recursive identification algorithms is highly instructive to understand their potential, it falls short for adaptive control where the previously mentioned conditions are usually not satisfied or cannot be guaranteed.

NOTES AND REFERENCES

Prediction problems of stationary time series were independently and simultaneously considered by Kolmogorov [Kol41] and Wiener [Wie49] for the discrete-time and the continuous-time parameter case, respectively. The first used Wold's idea [Wol38] of representing time series in terms of innovations. Later ([WM57] and [WM58]), Wiener used the Hilbert-space framework of Kolmogorov for addressing the problem for the multivariate stationary discrete-time processes. In 1960, Kalman [Kal60b] presented the first recursive solution to the nonstationary prediction problem for discrete-time processes represented by stochastic linear state-space models. The solution for the analogous problem with a continuous time parameter was given in [KB61]. An informative survey of the development of the subject is found in [Kai74]; see also [Kai76]. The literature on Kalman filtering is now immense, for example, [AM79], [Gel74], [Jaz70], [May79], [Med69], [Sol88], and [Won70]. For the delicate issues of robustified implementations of the Kalman filter via matrix factorizations, see [Bie77].

The problem of parameter estimation, and the associated topics of biasedness, consistency, efficiency, maximum-likelihood estimators, are well covered in books of statistics, for example, [Cra46], [KS79], and [Rao73]. In [GP77], [Lju87], and [SS89], these concepts are applied in the identification of linear systems. See also [BJ76], [Cai88], [CG91], [Che85], [Eyk74], [GS84], [HD88], [Joh88], [KR76], [Lan90], [LS83], [Men73], [MG90], [ML76], [Nor87], [TAG81], and [UR87]. For a Bayesian approach, see [Pet81]. The use of prediction models in stochastic modeling and the interpretation of the RLS and RML algorithms as prediction-error methods have been emphasized in [Cai76] and [Lju78].

The RELS method was first proposed by a number of authors: [ÅBW65], [May65], [Pan68], and [You68]. The RML was derived in [Söd73]. There is a vast literature on how to implement recursive identification algorithms via robust numerical methods ([Bie77] and [LH74]). See also [KHB$^+$85] and [Pet86].

CHAPTER 7

LQ AND PREDICTIVE STOCHASTIC CONTROL

The purpose of this chapter is to extend LQ and predictive receding-horizon control to a stochastic setting. In Secs. 1 and 2, we consider the LQ-regulation problem for stochastic linear dynamic plants when the plant state is either completely or only partially accessible to the controller. Stochastic dynamic programming is used to yield the optimal solution via the so-called certainty-equivalence principle. In Sec. 3, two distinct steady-state regulation problems for CARMA plants are considered. The first consists of a single-step regulation problem based on a performance index given by a conditional expectation. The second adopts the criterion of minimizing the unconditional expectation of a quadratic cost. Both problems are tackled via the stochastic variant of the polynomial equation approach introduced in Chapter 4. Section 4 discusses some monotonic performance properties of steady-state LQ stochastic regulation. Section 5 deals with 2-DOF tracking and servo problems. The relationship between LQ stochastic control and $\mathcal{H}_\infty$ control is pointed out in Sec. 6. Finally, Sec. 7 extends SIORHR and SIORHC, two predictive receding-horizon controllers introduced in Chapter 5, to steady-state regulation and control of CARMA and CARIMA plants.

7.1 LQ STOCHASTIC REGULATION: COMPLETE STATE INFORMATION

The time evolution of state $x(k)$ of the plant to be regulated is represented here as follows:

$$x(k + 1) = \Phi(k)x(k) + G(k)u(k) + \xi(k) \qquad (7.1\text{-}1a)$$

where $k \in [t_0, T)$, $x(k) \in \mathbb{R}^n$, $u(k) \in \mathbb{R}^m$, $\xi(k) \in \mathbb{R}^n$, $u(k)$ is the manipulated input, and $\xi(k)$ an inaccessible disturbance. The initial state $x(t_0)$ and the processes ξ

and u are defined in the underlying probability space $(\Omega, \mathcal{F}, \mathbb{P})$. We consider a nondecreasing family of sub-σ-fields $\{\mathcal{F}_k\}_{k=t_0}^{T-1}$, $\mathcal{F}_{t_0} \subset \cdots \subset \mathcal{F}_k \subset \mathcal{F}_{k+1} \subset \mathcal{F}$, such that $x(t_0) \in \mathcal{F}_{t_0}$, $\xi(k) \in \mathcal{F}_k$. Here we use the shorthand notation $v \in \mathcal{F}$ to state that v is $\mathcal{F}$-measurable. Note that if we let $\mathcal{F}_k$ be the σ-field generated by $\{x(t_0), \xi(t_0), \cdots, \xi(k)\}$

$$\mathcal{F}_k := \sigma\left\{x(t_0), \xi(t_0), \cdots, \xi(k)\right\}, \qquad \mathcal{F}_{t_0-1} := \{\emptyset, \Omega\}$$

then $\{\mathcal{F}_k\}_{k=t_0}^{t-1}$ has the stipulated properties. Further, disturbance ξ has the martingale difference property

$$\left.\begin{array}{ll} \mathcal{E}\left\{\xi(k) \mid \mathcal{F}_{k-1}\right\} = O_n & \text{a.s.} \\ \mathcal{E}\left\{\xi(k)\xi'(k) \mid \mathcal{F}_{k-1}\right\} = \Psi_\xi(k) < \infty & \text{a.s.} \end{array}\right\} \qquad (7.1\text{-}1b)$$

and

$$\mathcal{E}\left\{\xi(t_0)x'(t_0)\right\} = O_{n \times n} \qquad (7.1\text{-}1c)$$

Note that no Gaussian assumption is used here and that (1b) implies that ξ is zero-mean and white.

We next elucidate the nature of the process u. In the present complete state information case, $u(k)$ is allowed to be measurable w.r.t. the σ-field generated by $\mathcal{I}_k$:

$$u(k) \in \sigma\left\{\mathcal{I}_k\right\} \qquad (7.1\text{-}2)$$

where $\mathcal{I}_k := \left\{x^k, u^{k-1}\right\}$, and $x^k := \{x(i)\}_{i=t_0}^{k}$. In words, $u(k)$ can be computed as a function of the realizations of $\mathcal{I}_k$. Equation (2) specifies the *admissible regulation strategy*. Note that the strategy (2) is *nonanticipative* or *causal*, in that $u(k)$ can be computed in terms of past realizations of u, and present and past realizations of x.

We consider the following quadratic performance index:

$$\mathcal{E}\left\{J\left(t_0, x(t_0), u_{[t_0, T)}\right)\right\} = \mathcal{E}\left\{\sum_{k=t_0}^{T} \ell\left(k, x(k), u(k)\right)\right\} \qquad (7.1\text{-}3a)$$

$$\ell\left(k, x(k), u(k)\right) := \|x(k)\|_{\psi_x(k)}^2 + 2u'(k)M(k)x(k) + \|u(k)\|_{\psi_u(k)}^2 \geq 0 \qquad (7.1\text{-}3b)$$

$$\ell\left(T, x(T), u(T)\right) := \|x(T)\|_{\psi_x(T)}^2 \geq 0 \qquad (7.1\text{-}3c)$$

For the properties of matrices $\psi_x(k)$, $\psi_u(k)$, and $M(k)$, the reader is referred to Sec. 2.1. We wish to consider the following problem.

LQ Stochastic (LQS) Regulator with Complete State Information. Consider the stochastic linear plant (1) and the quadratic performance index (3). Find an input sequence $u_{[t_0, T)}^0$ to the plant that minimizes the performance index among all the admissible regulation strategies (2).

We tackle the problem via *stochastic dynamic programming* ([Ber76] and [BS78]). This is the extension to a stochastic setting of the dynamic programming technique discussed in Sec. 2.2.

For $t \in [t_0, T]$, we introduce the *Bellman function:*

$$
V(t, x(t)) \quad := \quad \min_{u_{[t,T)}} \mathcal{E} \left\{ \sum_{k=t}^{T} \ell(k, x(k), u(k)) \mid \mathcal{I}_t \right\}
$$

$$
= \quad \min_{u_{[t,T)}} \mathcal{E} \left\{ \sum_{k=t}^{T} \ell(k, x(k), u(k)) \mid x(t) \right\} \qquad (7.1\text{-}4)
$$

where the last equality follows because, with the plant governed by the Markovian stochastic difference equation (1a), the conditional probability distribution of future plant variables, given $\mathcal{I}_t$, depends on $x(t)$ only. This consideration leads us to conclude that the optimal input $u_{[t_0,T)}^0$ does indeed satisfy the following admissible regulation strategy:

$$
u(k) \in \sigma\{x(k)\} \qquad (7.1\text{-}5)
$$

That is, the optimal regulation law is in a state-feedback form. For any $t_1 \in [t, T)$, we can write

$$
V(t, x(t)) \quad = \quad \min_{u_{[t,t_1)}} \mathcal{E} \left\{ \sum_{k=t}^{t_1-1} \ell(k, x(k), u(k)) \right.
$$

$$
+ \quad \min_{u_{[t_1,T)}} \mathcal{E} \left\{ \sum_{k=t_1}^{T} \ell(k, x(k), u(k)) \mid x(t_1) \right\} \left. \mid x(t) \right\}
$$

$$
= \quad \min_{u_{[t,t_1)}} \mathcal{E} \left\{ \sum_{k=t}^{t_1-1} \ell(k, x(k), u(k)) + V(t_1, x(t_1)) \mid x(t) \right\} \qquad (7.1\text{-}6)
$$

The last equality follows because, by the smoothing properties of conditional expectations (cf. Appendix D.3),

$$
\min_{u_{[t_1,T)}} \mathcal{E} \left\{ \sum_{k=t_1}^{T} \ell(k, x(k), u(k)) \mid x(t) \right\}
$$

$$
= \min_{u_{[t_1,T)}} \mathcal{E} \left\{ \mathcal{E} \left\{ \sum_{k=t_1}^{T} \ell(k, x(k), u(k)) \mid x(t_1), x(t) \right\} \mid x(t) \right\}
$$

$$
= \min_{u_{[t,T)}} \mathcal{E} \left\{ V(t_1, x(t_1)) \mid x(t) \right\}
$$

Setting $t_1 = t + 1$ in (6), we get the *stochastic Bellman equation:*

$$
V(t, x(t)) = \min_{u(t)} \left\{ \ell(t, x(t), u(t)) + \mathcal{E} \left\{ V(t+1, x(t+1)) \mid x(t) \right\} \right\} \qquad (7.1\text{-}7)
$$

with terminal condition

$$V(T, x(T)) = \ell(T, x(T), u(T)) = \|x(T)\|_{\psi_x(T)}^2 \qquad (7.1\text{-}8)$$

The last two equations correspond to (2.2-8) and (2.2-9) in the deterministic setting of Sec. 2.2. The functional equation (7) can be used as follows. For $t = T - 1$, it yields

$$
\begin{aligned}
V(T-1, x(T-1)) \;=\; &\min_{u(T-1)} \Big\{ \ell(T-1, x(T-1), u(T-1)) \\
&+ \mathcal{E}\Big\{ \|\Phi(T-1)x(T-1) + G(T-1)u(T-1) \\
&+ \xi(T-1)\|_{\psi_x(T)}^2 \;\Big|\; x(T-1) \Big\} \Big\}
\end{aligned}
$$

This can be solved w.r.t. $u(T-1)$, giving the optimal input at time $T-1$ in state-feedback form:

$$u^0(T-1) = u^0(T-1, x(T-1))$$

and hence determines $V(T-1, x(T-1))$. By iterating backward the previous procedure, we can determine the optimal control law in state-feedback form:

$$u^0(k) = u^0(k, x(k)) \;, \qquad k \in [t_0, T)$$

and $V(k, x(k))$. The next theorem verifies that the previous procedure solves the LQS regulation problem with complete state information.

Theorem 7.1-1. *Suppose that $\{V(t, x)\}_{t=t_0}^{T}$ satisfies the stochastic Bellman equation (7) with terminal condition (8). Suppose that the minimum in (7) be attained at*

$$\hat{u}(t) = \hat{u}(t, x) \qquad t \in [t_0, T)$$

Then $\hat{u}_{[t_0, T)}$ minimizes the cost $\mathcal{E}\left\{ J\left(t, x(t_0), u_{[t_0, T)}\right)\right\}$ over the class of all state-feedback inputs. Further, the minimum cost equals $\mathcal{E}\left\{V(t_0, x(t_0))\right\}$.

Proof: Let $u(t, x(t))$ be an arbitrary feedback input and $x(t)$ the process generated by (1) with $u(t) = u(t, x(t))$. We have

$$V(t_0, x(t_0)) - V(T, x(T)) = \sum_{t=t_0}^{T-1} [V(t, x(t)) - V(t+1, x(t+1))] \qquad (7.1\text{-}9)$$

By the smoothing properties of conditional expectations, we obtain

$$
\begin{aligned}
\mathcal{E}\left\{V(t, x(t)) - V(t+1, x(t+1))\right\} & \\
&= \mathcal{E}\left\{ \mathcal{E}\left\{ V(t, x(t)) - V(t+1, x(t+1)) \;\Big|\; x(t) \right\} \right\} \\
&= \mathcal{E}\left\{ V(t, x(t)) - \mathcal{E}\left\{ V(t+1, x(t+1)) \;\Big|\; x(t) \right\} \right\} \\
&\leq \mathcal{E}\left\{ \ell(t, x(t), u(t)) \right\} \qquad\qquad [(7)] \quad (7.1\text{-}10)
\end{aligned}
$$

Taking the expectation of both sides of (9), we get

$$
\mathcal{E}\left\{V\left(t_0, x(t_0)\right) - V\left(T, x(T)\right)\right\} = \mathcal{E}\left\{\sum_{t=t_0}^{T-1}\left[V(t, x(t)) - V(t+1, x(t+1))\right]\right\}
$$

$$
\leq \mathcal{E}\left\{\sum_{t=t_0}^{T-1}\ell(t, x(t), u(t))\right\} \qquad [(10)] \quad (7.1\text{-}11)
$$

Hence, from (8), it follows that

$$
\mathcal{E}\left\{V\left(t_0, x(t_0)\right)\right\} \leq \mathcal{E}\left\{J\left(t_0, x(t_0), u_{[t_0,T)}\right)\right\} \qquad (7.1\text{-}12)
$$

Conversely, the same argument holds with equality instead of inequality in (11) when $u(t) = \hat{u}(t)$. Consequently,

$$
\mathcal{E}\left\{V\left(t_0, x(t_0)\right)\right\} = \mathcal{E}\left\{J\left(t_0, x(t_0), \hat{u}_{[t_0,T)}\right)\right\} \qquad (7.1\text{-}13)
$$

From (12) and (13), it follows that $\hat{u}_{[t_0,T)}$ is optimal and that the minimum cost equals $\mathcal{E}\{V(t_0, x(t_0))\}$.

The solution of the stochastic Bellman equation (7) is related in a simple way to that of its deterministic counterpart (2.3-7). In fact, in the present stochastic case, we have

$$
V(t, x) = x'\mathcal{P}(t)x + v(t) \qquad (7.1\text{-}14)
$$

where $\mathcal{P}(T) = \psi_x(T)$, and $v(T) = 0$. By assuming (14) to be true, the induction argument, as used in the deterministic case of Theorem 2.3-1, shows that

$$
V(t-1, x) = x'\mathcal{P}(t-1)x + v(t) + \mathrm{Tr}\left[\mathcal{P}(t)\Psi_\xi(t-1)\right]
$$

where $\mathcal{P}(t)$ is given by the Riccati backward iterations (2.3-3) to (2.3-6). Further, because $v(T) = 0$, working backward from $t = T$, we find that

$$
v(t) = \sum_{k=t}^{T-1} \mathrm{Tr}\left[\mathcal{P}(k+1)\Psi_\xi(k)\right] \qquad (7.1\text{-}15)
$$

We observe that $v(t)$ is not affected by $u_{[t_0,T)}$. Hence, the optimal inputs obtained by (7) are given by (2.3-1) as if the plant were deterministic, that is, $\xi(t) \equiv O_n$. Summing up, we have the following result.

Theorem 7.1-2. (LQS Regulator with Complete State Information) *Among all the admissible strategies (2), the optimal input for the LQS regulator with complete state information is given by the linear state-feedback regulation law*

$$
u(t) = F(t)x(t) \qquad t \in [t_0, T) \qquad (7.1\text{-}16)
$$

In (16), the optimal feedback-gain matrix $F(t)$ is the same as in the deterministic case ($\xi(t) \equiv O_n$) of Theorem 2.3-1 and given by (2.3-2) in terms of the solution

$\mathcal{P}(t+1)$ of the Riccati backward difference equation (2.3-3) to (2.3-6). Further, the minimum cost incurred over the regulation horizon $[t, T]$ for the optimal input sequence $u_{[t,T)}$, conditional to the initial plant state $x(t)$, is given by

$$V(t, x(t)) = \min_{u_{[t,T)}} \mathcal{E}\left\{\sum_{k=t}^{T} \ell(k, x(k), u(k)) \mid x(t)\right\}$$

$$= x'(t)\mathcal{P}(t)x(t) + \sum_{k=t}^{T-1} \operatorname{Tr}\left[\mathcal{P}(k+1)\Psi_\xi(k)\right] \qquad (7.1\text{-}17)$$

Problem 7.1-1 By using the induction argument, prove (17) and (16).

Problem 7.1-2 Taking into account (17), show that the minimum achievable cost over $[t, T)$ equals

$$\min_{u_{[t,T)}} \mathcal{E}\left\{J\left(t, x(t), u_{[t,T)}\right)\right\} \qquad (7.1\text{-}18)$$

$$= \mathcal{E}\left\{V(t, x(t))\right\}$$

$$= \|\mathcal{E}\{x(t)\}\|_{\mathcal{P}(t)}^2 + \operatorname{Tr}\left[\mathcal{P}(t)\operatorname{Cov}(x(t))\right] + \sum_{k=t}^{T-1} \operatorname{Tr}\left[\mathcal{P}(k+1)\Psi_\xi(k)\right]$$

(*Hint*: Use Lemma D.2-1 of Appendix D.)

Notice that in (18), the first two terms depend on the distribution of the initial state, and the third is due to the disturbance ξ forcing the plant (1a).

Main Points of the Section. For any horizon of finite length and possibly non-Gaussian disturbances, the LQS-regulation problem with complete state information is solved by a linear time-varying state-feedback regulation law that is the same as if the plant (1a) were deterministic, that is, $\xi(k) \equiv O_n$.

Problem 7.1-3 Consider the plant given by the SISO CAR model (cf. (6.2-69b)):

$$A(d)y(k) = B(d)u(k) + e(k) \qquad (7.1\text{-}19)$$

with polynomials $A(d)$ and $B(d)$ as in (6.1-8). Let $s(k)$ be the vector

$$s(k) := \left[\ \left(y_k^{k-n_a+1}\right)'\ \ \left(u_{k-1}^{k-n_b+1}\right)'\ \right] \in \mathbb{R}^{n_a+n_b-1} \qquad (7.1\text{-}20)$$

Then (cf. Example 5.4-1) (19) can be written in state-space form as follows:

$$\left.\begin{array}{rcl} s(k+1) & = & \Phi s(k) + G_u u(k) + Ge(k+1) \\ y(k) & = & Hs(k) \end{array}\right\} \qquad (7.1\text{-}21)$$

where (Φ, G_u, H) are defined in Example 5.4-1, and $G = \underline{e}_{n_a}$. For

$$\xi(k) := e(k+1)$$

and
$$\mathcal{F}_k := \sigma\left\{x(t_0), \xi(t_0), \cdots, \xi(k)\right\} \qquad \mathcal{F}_{t_0-1} := \{\emptyset, \Omega\}$$
assume that (1b) and (1c) are satisfied. Consider the cost

$$\mathcal{E}\left\{J\left(t_0, s(t_0), u_{[t_0,t)}\right)\right\} = \mathcal{E}\left\{\sum_{k=t_0}^{T-1}\left[y^2(k) + \rho u^2(k)\right]\right\} \qquad (7.1\text{-}22)$$

$\rho \geq 0$, and the admissible regulation strategy

$$u(k) \in \sigma\left\{s(t_0), y^k, u^{k-1}\right\} \qquad (7.1\text{-}23)$$

Show that the problem of finding, among all the admissible inputs (23), the ones minimizing (22) for the plant (19) is an LQS-regulation problem with complete state information. Further, specify suitable conditions on $A(d)$ and $B(d)$ that guarantee the existence of the limiting control law $u(t) = Fs(t)$ as $T \to \infty$. Compare the conclusion with those of Problem 2.4-5.

7.2   LQ STOCHASTIC REGULATION: PARTIAL STATE INFORMATION

7.2.1   LQG Regulation

We shall refer to the *plant* as the combination of system (1-1a) to be regulated along with a *state-sensing* device that makes available at every time k an observation $z(k)$ of linear combinations of state components x corrupted by a sensor noise $\zeta(k)$:

$$\left.\begin{array}{rcl} x(k+1) & = & \Phi(k)x(k) + G(k)u(k) + \xi(k) \\ z(k) & = & H(k)x(k) + \zeta(k) \end{array}\right\} \qquad (7.2\text{-}1a)$$

with x, u, ξ, and ζ defined in the probability space $(\Omega, \mathcal{F}, \mathbb{P})$,

$$x(k), \xi(k) \in \mathbb{R}^n, \quad u(k) \in \mathbb{R}^m, \quad z(k), \zeta(k) \in \mathbb{R}^p$$

and all matrices of compatible dimensions. Let

$$\nu(k) := \left[\begin{array}{cc} \xi'(k) & \zeta'(k) \end{array}\right]'$$

Define the family of sub-σ-fields $\{\mathcal{F}_k\}_{k=t_0}^{T-1}$ as follows:

$$\mathcal{F}_k := \sigma\left\{x(t_0), \nu(t_0), \cdots \nu(k)\right\} \qquad \mathcal{F}_{t_0-1} := \{\emptyset, \Omega\}$$

and assume that ν has the martingale difference property:

$$\left.\begin{array}{l} \mathcal{E}\left\{\nu(k) \mid \mathcal{F}_{k-1}\right\} = O_{n+p} \qquad \text{a.s.} \\[2mm] \mathcal{E}\left\{\nu(k)\nu'(k) \mid \mathcal{F}_{k-1}\right\} = \Psi_\nu(k) = \left[\begin{array}{cc} \Psi_\xi(k) & O_{n\times p} \\ O_{p\times n} & \Psi_\zeta(k) \end{array}\right] \qquad \text{a.s.} \end{array}\right\} \qquad (7.2\text{-}1b)$$

$$\mathcal{E}\left\{\nu(t_0)x'(t_0)\right\} = O_{(n+p)\times n} \qquad (7.2\text{-}1c)$$

and

$$x(t_0) \quad \text{and} \quad \{\nu(k)\}_{k=t_0}^{T-1} \quad \text{are jointly Gaussian-distributed} \qquad (7.2\text{-}1\text{d})$$

Here, for the sake of simplicity, we have taken the cross-covariance between $\xi(k)$ and $\zeta(k)$ to be zero at any instant k.

In the present partial state information case, the *admissible regulation strategy* allows $u(k)$ to be measurable w.r.t. the σ-field generated by $\{z^k, u^{k-1}\}$, $z^k := \{z(i)\}_{i=t_0}^{k}$:

$$u(k) \in \sigma\left\{z^k, u^{k-1}\right\} = \sigma\left\{z^k\right\} \qquad (7.2\text{-}2)$$

Note that by (1-1a) $\sigma\left\{z^k\right\} \subset \mathcal{F}_k$.

Following the lines of the previous section, we take the performance index to be $\mathcal{E}\left\{J\left(t_0, x(t_0), u_{[t_0,T)}\right)\right\}$ as in (1-3) and consider the following problem.

LQ Gaussian (LQG) Regulator. Consider the linear Gaussian plant (1) and the quadratic performance index (1-3). Find an input sequence $u_{[t_0,T)}^0$ to the plant that minimizes the performance index among all the admissible regulation strategies (2).

By the smoothing properties of conditional expectations, we can rewrite the performance index as follows:

$$\mathcal{E}\left\{J\left(t_0, x(t_0), u_{[t_0,T)}\right)\right\} = \mathcal{E}\left\{\sum_{k=t_0}^{T} \mathcal{E}\left\{\ell(k, x(k), u(k)) \mid z^k\right\}\right\} \qquad (7.2\text{-}3)$$

Further, recalling (1-3b), (1-3c), and Lemma D.2-1 of Appendix D,

$$\mathcal{E}\left\{\ell(k, x(k), u(k)) \mid z^k\right\}$$
$$= \ell\left(k, x(k \mid k), u(k)\right) + \mathrm{Tr}\left[\psi_x(k) \, \mathrm{Cov}\left(x(k) \mid z^k\right)\right] \qquad (7.2\text{-}4)$$

Where $x(k \mid k)$ denotes the conditional expectation

$$x(k \mid k) = \mathcal{E}\left\{x(k) \mid z^k\right\}$$

Thanks to the Gaussian assumption (1d), by virtue of Fact 6.2-1, $x(k \mid k)$ is given by the following Kalman filter formulas (6.2-46) and (6.2-47):

$$
\begin{aligned}
x(k \mid k) &= x(k \mid k-1) + \tilde{K}(k)e(k) & [(6.2\text{-}46)] \quad (7.2\text{-}5)\\
x(k+1 \mid k) &= \Phi(k)x(k \mid k) + G(k)u(k) & (7.2\text{-}6)\\
\tilde{K}(k) &= \Pi(k)H'(k)\left[H(k)\Pi(k)H'(k) + \Psi_\zeta(t)\right]^{-1} & [(6.2\text{-}22\text{a})] \quad (7.2\text{-}7)
\end{aligned}
$$

with $e(k) = z(k) - H(k)x(k \mid k-1)$ and $\Pi(k)$, the state prediction-error covariance, satisfying the forward Riccati (filter) recursion (6.2-26). The previous filtering equations are to be initialized from

$$x(t_0 \mid t_0 - 1) = \mathcal{E}\{x(t_0)\} \qquad \text{and} \qquad \Pi(t_0) = \text{Cov}(x(t_0))$$

Further, recalling (6.2-31), we have

$$
\begin{aligned}
\Pi(k \mid k) \quad &:= \quad \text{Cov}(x(k) \mid z^k) &(7.2\text{-}8)\\
&= \quad \Pi(k) - \Pi(k)H'(k)\Big[H(k)\Pi(k)H'(k) + \Psi_\zeta(t)\Big]^{-1} H(k)\Pi(k)
\end{aligned}
$$

By taking into account the foregoing considerations, (3) becomes

$$
\begin{aligned}
\mathcal{E}\left\{ J\left(t_0, x(t_0), u_{[t_0,T)}\right)\right\} \\
= \mathcal{E}\left\{ \textstyle\sum_{k=t_0}^{T} \Big[\ell(k, x(k \mid k), u(k)) + \text{Tr}\Big[\psi_x(k)\Pi(k \mid k)\Big]\Big]\right\} (7.2\text{-}9)
\end{aligned}
$$

Now $\Pi(k \mid k)$ can be precomputed, as it is only dependent on $\text{Cov}(x(t_0))$ and the $\psi_x(k)$'s are given weighting matrices. Thus, minimizing (9) w.r.t. $u_{[t_0,T)}$ under the admissible regulation strategy (2) is the same as minimizing

$$\mathcal{E}\left\{ \sum_{k=t_0}^{T} \ell\left(k, x(k \mid k), u(k)\right)\right\}$$

w.r.t. $u_{[t_0,T)}$ for the plant

$$x(k+1 \mid k+1) = \Phi(k)x(k \mid k) + G(k)u(k) + \tilde{K}(k+1)e(k+1)$$

with complete state information. In particular, note that (1-1b) and (1-1c) are satisfied for $\xi(k)$ changed into $e(k+1)$. The theorem that follows is an immediate consequence of these considerations.

Theorem 7.2-1.   (LQG Regulation) *The optimal LQG-regulation law under partial state information, viz., fulfilling the admissible regulation strategy (2), is given by the filtered-state-feedback law*

$$u(t) = F(t)x(t \mid t) \qquad t \in [t_0, T) \qquad (7.2\text{-}10)$$

In (10): $F(t)$ denotes the optimal feedback-gain matrix that is the same as in the deterministic LQR case ($\xi(t) \equiv O_p$) of Theorem 2.3-1 and given by (2.3-2) in terms of the solution $\mathcal{P}(t+1)$ of the backward (regulation) Riccati difference equation (2.3-3) to (2.3-6); and $x(t \mid t) = \mathcal{E}\{x(t) \mid z^t\}$ is the Kalman filtered state provided by (5) to (7) whose optimal gain matrix $\tilde{K}(t)$ is given by (7) in terms of the solution $\Pi(t)$ of the forward (filtering) Riccati equation (6.2-26). Further, the minimum cost

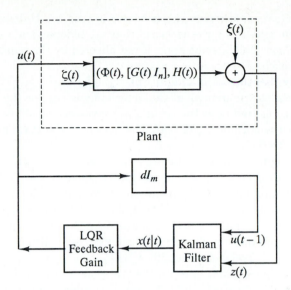

Figure 7.2-1: The LQG regulator.

incurred over the regulation horizon $[t, T]$ for the optimal input sequence $u_{[t,T)}$ is given by

$$
\min_{u_{[t,T)}} J\left(t, x(t), u_{[t,T)}\right) = \|\mathcal{E}\{x(t)\}\|_{\mathcal{P}(t)}^2 + \mathrm{Tr}\left[\mathcal{P}(t)\Pi(t \mid t)\right]
$$

$$
+ \sum_{k=t}^{T-1} \mathrm{Tr}\left[\mathcal{P}(k+1)\Psi_\xi(k)\right]
$$

$$
+ \sum_{k=t}^{T} \mathrm{Tr}\left[\psi_x(k)\Pi(k \mid k)\right] \qquad (7.2\text{-}11)
$$

with $\Pi(k \mid k)$ as in (8).

Problem 7.2-1 By using (9) and (1-17), prove (11).

It is instructive to compare (11) with (1-18). W.r.t. (1-18), the only extra term in (11) is its last summation. This depends on the posterior covariance matrices $\Pi(k \mid k) = \mathrm{Cov}(x(k) \mid z^k)$ that are nonzero in the partial state information case.

The LQG-regulator solution is depicted in Fig. 1. It is to be pointed out that the computations involved refer *separately* to the regulation and state filtering problems, respectively. In fact, computation of $F(t)$ involves the "cost" parameters $\psi_x(k)$, $M(k)$, $\psi_u(k)$, $\psi_x(T)$ but not the "noise" parameters $\Psi_\xi(k)$, $\Psi_\zeta(k)$, and

$\mathcal{E}\{x(t_0)\}$ and $\text{Cov}(x(t_0))$, whereas the converse is true for Kalman filter design. The regulator then *separates* into two parts that are independent: a *filtering stage* and a *regulation stage*. The filtering stage is not affected by the regulation objective and conversely. This is the *separation principle*.

The property that the optimal input takes the form $u(t) = F(t)x(t \mid t)$, where $F(t)$ is the feedback-gain matrix as in the deterministic case or the complete state information case, is referred to as the *certainty-equivalence principle*. This, in other words, states that the optimal LQG regulator acts as if the state filtered estimate $x(t \mid t)$ were equal to the true state $x(t)$.

It is interesting to pause so as identify the reason responsible for the validity of the certainty-equivalence principle in the LQG-regulation problem. In a general stochastic regulation problem with partial state information, input sequence $u_{[t_0,t)}$ has the so-called *dual effect* [Fel65], in that it affects both variables to be regulated and the posterior distribution of the plant state given the observations. Under such circumstances, the optimal regulator may exhibit a separation property but not satisfy the certainty-equivalence principle, ([BST74] and [Wit71]). In this connection, consider a nonlinear plant composed by a system governed by the equation $x(k+1) = \Phi(k)x(k) + G(k)u(k) + \xi(k)$, and a possibly nonlinear sensing device giving observations $z(k) = h(k, x(k), \zeta(k))$. For the sake of simplicity, assume that ξ and ζ have the same properties as in (1). Then, in [BST74], it is shown that the optimal regulator minimizing (3) fulfills the certainty-equivalence principle if and only if the posterior covariance of the plant state $x(t)$ given the observations z^t is the same as if $u_{[t_0,t)}$ were the zero sequence, viz., the input has no dual effect on the second moments of the posterior distribution. In this respect, we note that in the LQG-regulator solution, the posterior uncertainty on the true state $x(t)$, as measured by $\text{Cov}(x(t) \mid z^t)$ in (8), is not affected by $u_{[t_0,t)}$. In fact, $\text{Cov}(x(t) \mid z^t)$ can be precomputed because it is unaffected by the realization of z^t and the specific input sequence $u_{[t_0,t)}$. This means that the input has no dual effect and, hence, the certainty-equivalence principle applies.

7.2.2   Linear Non-Gaussian Plants

Consider the linear stochastic plant (1a). Again we write

$$\nu(k) := \begin{bmatrix} \xi'(k) & \zeta'(k) \end{bmatrix}' \qquad (7.2\text{-}12\text{a})$$

In contrast with (1d), here we assume that the involved random vectors are not Gaussian. Nonetheless, their means and covariances are as before:

$$\mathcal{E}\{\nu(k)\} = O_{n+p}$$

$$\left.\mathcal{E}\{\nu(k)\nu'(i)\} = \Psi_\nu(k)\delta_{k,i} = \begin{bmatrix} \Psi_\xi(k) & O_{n\times p} \\ O_{p\times n} & \Psi_\zeta(k) \end{bmatrix} \delta_{k,i}\right\} \qquad (7.2\text{-}12\text{b})$$

$$\mathcal{E}\{\nu(k)x'(t_0)\} = O_{(n+p)\times n} \qquad (7.2\text{-}12\text{c})$$

In such a case, Theorem 6.2-2 states that the Kalman filtered state $x(t \mid t)$ is only the *linear* MMSE estimate of $x(t)$ based on z^t and not the conditional mean $\mathcal{E}\{x(t) \mid z^t\}$ as in the Gaussian case. Further, $\mathrm{Cov}(\varepsilon(t))$, $\varepsilon(t) := x(t) - x(t \mid t)$, is still given by (8) also in the non-Gaussian case. Finally, by Lemma D.2-1, $\mathcal{E}\{\|x(k)\|^2_{\psi_x(k)}\}$ depends only on the mean and $\mathrm{Cov}(x(t))$. These observations lead to the following result.

Result 7.2-1.   *Consider the linear stochastic plant (1a), (12a) to (12c). Assume that the involved random variables are possibly non-Gaussian and the cost is quadratic as in (1-3). Then, the optimal input sequence $u^0_{[t_0,T)}$ to the plant (1a) that minimizes (1-3) among all the admissible* linear *regulation strategies*

$$\left. \begin{array}{r} u(t) = f\left(t, z^t, u^{t-1}\right) \\ f(t, \cdot, \cdot) \ linear \end{array} \right\} \qquad (7.2\text{-}13)$$

is given by the linear *feedback law (10) with $F(t)$ and $x(t \mid t)$ computed as indicated in Theorem 1 as if the involved random vectors were jointly Gaussian-distributed.*

Problem 7.2-2   Prove the conclusions of Result 1.

7.2.3   Steady-State LQG Regulation

Up to now, the results obtained on LQ stochastic regulation have required no stationariety assumptions on the involved processes nor time-invariant system matrices. However, as with the deterministic LQR problem of Chapter 2, an interesting topic is the analysis of the limiting behavior of the LQG regulator as the regulation horizon goes to infinity.

Consider the time-invariant linear Gaussian plant:

$$\left. \begin{array}{rcl} x(k+1) & = & \Phi x(k) + G_u u(k) + \xi(k) \\ y(k) & = & H x(k) \\ z(k) & = & H_z x(k) + \zeta(k) \end{array} \right\} \qquad (7.2\text{-}14a)$$

with

$$\nu(k) := \left[\ \xi'(k) \quad \zeta'(k) \ \right]'$$

a finite-variance wide-sense stationary Gaussian process satisfying (2-1) with

$$\Psi_\xi(k) \equiv \Psi_\xi = GG' \qquad\qquad \Psi_\zeta(k) \equiv \Psi_\zeta > 0 \qquad (7.2\text{-}14b)$$

where G has full column rank. Setting $N := T - t_0 \in \mathbb{Z}_1$, consider also the performance index

$$\mathcal{E}\left\{ \frac{1}{N} \left[\sum_{k=t_0}^{T-1} \ell\left(y(k), u(k)\right) + \|x(T)\|^2_{\psi_x(T)} \right] \right\} \qquad (7.2\text{-}15a)$$

$$\ell(y, u) := \|y\|^2_{\psi_y} + \|u\|^2_{\psi_u} \tag{7.2-15b}$$

$$\psi_y = \psi'_y > 0 \qquad \psi_u = \psi'_u > 0 \qquad \psi_x(T) = \psi'_x(T) \geq 0 \tag{7.2-15c}$$

along with an admissible regulation strategy given by (2). We see that in (14a), $y(k)$ represents an output vector to be regulated and $z(k)$ is the sensor output or observation at time k. We know from Theorem 1 that the solution of the LQG-regulation problem for every $N \in \mathbb{Z}_1$ satisfies the certainty-equivalence principle. We are now interested in establishing the limiting solution as $N \to \infty$ and $t_0 \to -\infty$. In the problem that we have just set up, there are two system triples, viz., (Φ, G_u, H) and (Φ, G, H_z). They concern the regulation and the state-filtering problem, respectively. On the other hand, we know that both problems admit limiting asymptotically stable solutions provided that

$$\left. \begin{array}{l} (\Phi, G_u, H) \\ (\Phi, G, H_z) \end{array} \right\} \text{ are both stabilizable and detectable} \tag{7.2-16}$$

Further, by the separation principle, the two limiting processes are seen to be non-interacting. These considerations make it plausible for the foregoing problem to have the following conclusions.

Result 7.2-2. (Steady–State LQG Regulation) *Consider the time-invariant linear Gaussian plant (14) and the quadratic performance index (15) with time-invariant weights. Let (16) hold. Then, as $N \to \infty$ and $t_0 \to -\infty$, the LQG-regulation law, optimal among all the admissible regulation strategies (2), equals*

$$u(t) = Fx(t \mid t) \tag{7.2-17}$$

Here F is the constant-feedback-gain matrix solving the time-invariant deterministic LQOR problem ($\xi(t) \equiv O_n$ and $\zeta(t) \equiv O_p$) as in Theorem 2.4-5:

$$F = -\left(\psi_u + G'_u P G_u\right)^{-1} G'_u P \Phi \tag{7.2-18a}$$

where $P = P' \geq 0$ satisfies the regulation ARE:

$$P = \Phi' P \Phi - \Phi' P G_u \left(\psi_u + G'_u P G_u\right)^{-1} G'_u P \Phi + H' \psi_y H \tag{7.2-18b}$$

Further, $x(t \mid t)$ is generated by the steady-state Kalman filter, as in Theorem 6.2-3:

$$\begin{aligned} x(t \mid t) &= x(t \mid t - 1) + \tilde{K} e(t) & \tag{7.2-19a} \\ x(t + 1 \mid t) &= \Phi x(t \mid t) + G_u u(t) & \tag{7.2-19b} \\ e(t) &= z(t) - H_z x(t \mid t - 1) & \tag{7.2-19c} \\ \tilde{K} &= \Pi H'_z \left(H_z \Pi H'_z + \Psi_\zeta\right)^{-1} & \tag{7.2-19d} \end{aligned}$$

where $\Pi = \Pi' \geq 0$ satisfies the filtering ARE:

$$\Pi = \Phi \Pi \Phi' - \Phi \Pi H'_z \left(H_z \Pi H'_z + \Psi_\zeta\right)^{-1} H_z \Pi \Phi' + GG' \tag{7.2-19e}$$

Equations (17) to (19) make the closed-loop system asymptotically stable. Hence, under the stated conditions, as $N \to \infty$ and $t_0 \to -\infty$, all the involved processes become stationary and (17) minimizes the stochastic steady-state cost:

$$\mathrm{Tr}\,[\psi_y \Psi_y] + \mathrm{Tr}\,[\psi_u \Psi_u] \tag{7.2-20a}$$

where

$$\Psi_y = \lim_{\substack{N \to \infty \\ t_0 \to -\infty}} \mathcal{E}\{y(k)y'(k)\} \qquad \Psi_u = \lim_{\substack{N \to \infty \\ t_0 \to -\infty}} \mathcal{E}\{u(k)u'(k)\} \tag{7.2-20b}$$

Problem 7.2-3   Let $\tilde{x}(t) := x(t) - x(t \mid t-1)$. Show that the control law (17) to (19) gives the closed-loop system:

$$\begin{bmatrix} x(t+1) \\ \tilde{x}(t+1) \end{bmatrix} = \begin{bmatrix} \Phi + G_u F & -G_u F(I_n - \tilde{K}H_z) \\ O & \Phi - KH_z \end{bmatrix} \begin{bmatrix} x(t) \\ \tilde{x}(t) \end{bmatrix}$$

$$+ \begin{bmatrix} I_n & G_u F \tilde{K} \\ I_n & -K \end{bmatrix} \begin{bmatrix} \xi(t) \\ \zeta(t) \end{bmatrix} \tag{7.2-21}$$

with $K = \Phi \tilde{K}$. Hence, by recalling (4.4-25) and (6.2-68), provided that (Φ, G_u, H_z) is controllable and reconstructible, the d-characteristic polynomial χ_{LQG} of the LQG-regulated system equals

$$\chi_{\mathrm{LQG}}(d) = \frac{\det E(d)}{\det E(0)} \times \frac{\det C(d)}{\det C(0)} \tag{7.2-22}$$

Problem 7.2-4 (*Linear Possibly Non-Gaussian Plants in Innovations Form*)   Consider the linear time-invariant plant (14a) with $\xi(k) = G\zeta(k)$, possibly non-Gaussian, satisfying (1-1b) to (1-1c). Let

$$\Phi - GH_z \quad \text{be a stability matrix}$$

and

$$(\Phi, G_u, H) \quad \text{be stabilizable and detectable}$$

Exploit the result in Problem 6.2-7 to show that as $t \to \infty$, the LQ stochastic regulator minimizing (1-3) among all the admissible regulation strategies (2) is again given by (17) to (19) with

$$x(t \mid t) = \Phi x(t-1 \mid t-1) + G_u u(t-1) + G\,[z(t-1) - H_z x(t-1 \mid t-1)]$$

Main Points of the Section.   The solution of the LQG-regulation problem for partial state information is given by the state-filtered feedback law $F(t)x(t \mid t)$ where $x(t \mid t) = \mathcal{E}\{x(t) \mid z^t\}$ is generated via Kalman filtering, and $F(t)$ is the same as in deterministic LQ regulation. The steady-state LQG regulator is obtained by cascading the deterministic steady-state LQ regulator with the steady-state Kalman filter generating $x(t \mid t)$.

7.3   STEADY-STATE REGULATION OF CARMA PLANTS:
SOLUTION VIA POLYNOMIAL EQUATIONS

In this section, we consider various stochastic regulation problems for the CARMA plant:

$$A(d)y(k) = B(d)u(k) + C(d)e(k) \qquad (7.3\text{-}1)$$

where e, the innovations process of y, will be assumed to be zero-mean wide-sense stationary white with $\mathcal{E}\{e(k)e'(k)\} = \Psi_e > 0$. Other additional requirements on e will be specified whenever needed. In connection with (1), we assume that

- $A^{-1}(d) \begin{bmatrix} B(d) & C(d) \end{bmatrix}$ is an irreducible left MFD   (7.3-2a)

- $C(d)$ is Hurwitz   (7.3-2b)

- the gcld's of $A(d)$ and $B(d)$ are strictly Hurwitz   (7.3-2c)

We point out that, by Theorem 6.2-4, (2b) entails no limitation and (2c) is a necessary condition (cf. Problem 3.2-3) for the existence of a linear compensator, acting on the manipulated input u only, capable of making the resulting feedback system internally stable.

7.3.1   Single-Step Stochastic Regulation

Here we consider the CARMA plant (1) to (2) along with the following additional assumptions on e:

$$\mathcal{E}\left\{e(k+1) \mid e^k\right\} = O_p \qquad \text{a.s.} \qquad (7.3\text{-}3a)$$

$$\mathcal{E}\left\{e(k+1)e'(k+1) \mid e^k\right\} = \Psi_e > 0 \qquad \text{a.s.} \qquad (7.3\text{-}3b)$$

As we shall see, the martingale difference properties (3) are sufficient for tackling in full generality the regulation problem we are going to set up. In particular, no Gaussian assumption will be required. Further, we assume that $u(k)$ has to satisfy the *admissible regulation strategy*

$$u(k) \in \sigma\left\{y^k, u^{k-1}\right\} = \sigma\left\{y^k\right\} \qquad (7.3\text{-}4)$$

Further, the *performance index* is given by the conditional expectation

$$\mathcal{C} = \mathcal{E}\left\{\|y(t+\tau)\|^2_{\psi_y} + \|u(t)\|^2_{\psi_u} \mid y^t\right\} \qquad (7.3\text{-}5)$$

In (5), the integer $\tau \in \mathbb{Z}_1$ equals the plant I/O delay:

$$\tau := \operatorname{ord} B(d) \geq 1$$

We consider the following problem, which is the extension of the deterministic single-step regulation of Sec. 2.7 to the present stochastic setting.

Single-Step Stochastic Regulation. Consider the CARMA plant (1) to (3) and the performance index given by the single-step conditional expectation (5). Find an optimal regulation law for the plant that makes the resulting feedback system internally stable and in stochastic steady state minimizes the performance index among all the admissible regulation strategies (4).

It is convenient to explicitly exhibit in (1) the delay τ by introducing the polynomial matrix $B_0(d)$ such that

$$d^\tau B_0(d) = B(d) \tag{7.3-6a}$$

$$B_0(d) = B_\tau + B_{\tau+1}d + \cdots + B_{\partial B}d^{\partial B - \tau} \tag{7.3-6b}$$

Consequently, (1) becomes

$$A(d)y(k) = B_0(d)u(k - \tau) + C(d)e(k) \tag{7.3-6c}$$

In order to tackle the regulation problem stated before, we first consider the following prediction problem.

MMSE τ-Step-Ahead Prediction. Consider the CARMA plant (1) to (3) whose input u satisfies (4). Find a finite-variance vector

$$\hat{y}(k + \tau \mid k) \in \sigma\left\{y^k\right\}$$

such that, for every $R = R' > 0$, in stochastic steady state,

$$\mathcal{E}\left\{\|y(k + \tau) - \hat{y}(k + \tau \mid k)\|_R^2\right\} \le \mathcal{E}\left\{\|y(k + \tau) - \bar{y}(k)\|_R^2\right\} \tag{7.3-7}$$

among all finite-variance vectors $\bar{y}(k) \in \sigma\left\{y^k\right\}$.

$\hat{y}(k + \tau \mid k)$ is called the *MMSE τ-step-ahead prediction* of $y(k + \tau)$.

Let $(Q_\tau(d), G_\tau(d))$ be the minimum-degree solution w.r.t. $Q_\tau(d)$ of the Diophantine equation

$$\left.\begin{array}{r} C(d) = A(d)Q_\tau(d) + d^\tau G_\tau(d) \\ \partial Q_\tau(d) \le \tau - 1 \end{array}\right\} \tag{7.3-8}$$

Then, (6c) can be rewritten as

$$\begin{aligned} y(k + \tau) &= Q_\tau(d)C^{-1}(d)B_0(d)u(k) + A^{-1}(d)G_\tau(d)C^{-1}(d)A(d)y(k) \\ &\quad + Q_\tau(d)e(k + \tau) \end{aligned} \tag{7.3-9}$$

and the next theorem follows.

Theorem 7.3-1. **(MMSE τ-Step-Ahead Prediction)** *Consider the CARMA plant (1) to (3) whose input u satisfies (4). Let both u and y mean-square bounded. Then, provided that the transfer matrix*

$$H_{\hat{y}|u,y}(d) = \left[\; Q_\tau(d)C^{-1}(d)B_0(d) \quad A^{-1}(d)G_\tau(d)C^{-1}(d)A(d) \; \right]$$

is stable, the MMSE τ-step-ahead prediction of y equals

$$
\begin{aligned}
\hat{y}(k+\tau \mid k) &= y(k+\tau) - Q_\tau(d)e(k+\tau) &&\text{(7.3-10)} \\
&= Q_\tau(d)C^{-1}(d)B_0(d)u(k) + A^{-1}(d)G_\tau(d)C^{-1}(d)A(d)y(k)
\end{aligned}
$$

where the polynomial matrix pair $(Q_\tau(d), G_\tau(d))$ is the minimum-degree solution of the Diophantine equation (8). Further, the MMSE prediction error $\tilde{y}(k+\tau \mid k)$ is given by the moving average

$$\tilde{y}(k+\tau \mid k) := y(k+\tau) - \hat{y}(k+\tau \mid k) = Q_\tau(d)e(k+\tau) \qquad \text{(7.3-11)}$$

Proof: Let $y_0(k)$ be given by the R.H.S. of (10). Then

$$
\begin{aligned}
\mathcal{E}\left\{ \|y(k+\tau) - \bar{y}(k)\|_R^2 \right\} &= \mathcal{E}\left\{ \|y_0(k) - \bar{y}(k) + Q_\tau(d)e(k+\tau)\|_R^2 \right\} \\
&= \mathcal{E}\left\{ \|y_0(k) - \bar{y}(k)\|_R^2 \right\} + \mathrm{Tr}\left[R\langle Q_\tau(d)\Psi_e Q_\tau^*(d)\rangle\right]
\end{aligned}
$$

Hence, $\hat{y}(k+\tau \mid k) = y_0(k)$. The last equality in the preceding equation follows because by the smoothing properties of conditional expectations,

$$
\begin{aligned}
\mathcal{E}&\left\{ \|y_0(k) - \bar{y}(k) + Q_\tau(d)e(k+\tau)\|_R^2 \right\} \\
&= \mathcal{E}\left\{ \mathcal{E}\left\{ \|y_0(k) - \bar{y}(k) + Q_\tau(d)e(k+\tau)\|_R^2 \mid y^k \right\} \right\} \\
&= \mathcal{E}\left\{ \|y_0(k) - \bar{y}(k)\|_R^2 + \mathrm{Tr}\left[RE\left\{ f(k+\tau)f'(k+\tau) \mid y^k \right\}\right] \right\} \quad \text{[Lemma D.2-1]}
\end{aligned}
$$

where $f(k+\tau) := Q_\tau(d)e(k+\tau) = \sum_{i=1}^{\tau-1} Q_i e(k+\tau-i)$ if $Q_\tau(d) = \sum_{i=1}^{\tau-1} Q_i d^i$. The conditional expectation inside the trace operator equals

$$
\begin{aligned}
\sum_{i,j=1}^{\tau-1} Q_i \mathcal{E}\left\{ e(k+\tau-i)e'(k+\tau-j) \mid y^k \right\} Q_j' &= \sum_{i=1}^{\tau-1} Q_i \Psi_e Q_i' \\
&= \langle Q_\tau(d)\Psi_e Q_\tau^*(d) \rangle
\end{aligned}
$$

Remark 7.3-1. $C(d)$ strictly Hurwitz guarantees stability of transfer matrix $H_{\hat{y}|u,y}(d)$. For $H_{\hat{y}|u,y}(d)$ stability, however, a necessary and sufficient condition is that the denominator matrices of the irreducible $MFDs$ of $H_{\hat{y}|u,y}(d)$ be strictly Hurwitz. Should $C(d)$ be Hurwitz but not strictly Hurwitz, stability may still be obtained if all the roots of $C(d)$ on the unit circle are canceled in every entry of $H_{\hat{y}|u,y}(d)$. Hence, in such a case, stability can be only concluded after computing $H_{\hat{y}|u,y}(d)$.

We now consider the minimization of (5) w.r.t. $u(t) \in \sigma\{y^t\}$. We find

$$
\begin{aligned}
\mathcal{C} &= \mathcal{E}\left\{ \|\hat{y}(t+\tau \mid t) + \tilde{y}(t+\tau \mid t)\|_{\psi_y}^2 + \|u(t)\|_{\psi_u}^2 \mid y^t \right\} \\
&= \|\hat{y}(t+\tau \mid t)\|_{\psi_y}^2 + \|u(t)\|_{\psi_u}^2 + \operatorname{Tr}\left[\psi_y \langle Q_\tau(d)\Psi_e Q_\tau^*(d) \rangle \right] \qquad [(11)]
\end{aligned}
$$

Equating to zero the first derivatives of $\mathcal{C}$ w.r.t. the components of $u(t)$ and setting

$$
\begin{aligned}
L &:= \quad Q_\tau(0)C^{-1}(0)B_0(0) & (7.3\text{-}12) \\
&= \quad A^{-1}(0)B_0(0) & [(8)]
\end{aligned}
$$

we find

$$
L'\psi_y \hat{y}(t+\tau \mid t) + \psi_u u(t) = O_m \qquad (7.3\text{-}13)
$$

or provided that

$$
\psi_u + L'\psi_y L \qquad \text{is nonsingular} \qquad (7.3\text{-}14)
$$

$$
u(t) = -\left(\psi_u + L'\psi_y L\right)^{-1} L'\psi_y \left[\hat{y}(t+\tau \mid t) - Lu(t)\right] \qquad (7.3\text{-}15)
$$

It is instructive to specialize (15) to the SISO case where w.l.o.g., we can assume

$$
\psi_y = 1; \qquad \psi_u = \rho \geq 0; \qquad A(0) = C(0) = 1 \qquad (7.3\text{-}16)
$$

Further,

$$
B_0(0) = b_\tau \qquad [(6b)]
$$

Then, the single-step stochastic regulation law becomes

$$
\left[\rho C(d) + b_\tau Q_\tau(d) B_0(d) \right] u(t) = -b_\tau G_\tau(d) y(t) \qquad (7.3\text{-}17)
$$

Problem 7.3-1   Let polynomials $\rho C(d) + b_\tau Q_\tau(d)B_0(d)$ and $G_\tau(d)$ be coprime. Show that the d-characteristic polynomial $\chi_{cl}(d)$ of the closed-loop system (1) and (17) equals

$$
\chi_{cl}(d) = \gamma \left[\frac{\rho}{b_\tau} A(d) + B_0(d) \right] C(d) \qquad (7.3\text{-}18)
$$

where $\gamma = b_\tau/(\rho + b_\tau^2)$. Compare this with (2.7-13) to conclude that $C(d)$ plays the role of the characteristic polynomial of the implicit Kalman filter embedded in the dynamic compensator (17).

Remark 7.3-2.   From (18), it follows that in the generic case, closed-loop stability requires $C(d)$ to be strictly Hurwitz. Should $C(d)$ be Hurwitz but not strictly Hurwitz, a comment similar to the one of Remark 1 applies here. Further, existence of steady-state wide-sense stationariety of the involved processes, as in the formulation of the single-step stochastic regulation problem, implicitly requires that (15) yields an internally stable closed-loop system.

Theorem 7.3-2. **(Single-Step Stochastic Regulation)** *Consider the CARMA plant (1) to (3), the single-step performance index (5), and the admissible regulation strategy (4). Let (14) be satisfied. Then, provided that (15) makes the closed–loop system internally stable, the optimal* single-step stochastic regulator *is given by (15), which in the SISO case simplifies as in (17). In the latter case, the d-characteristic polynomial of the closed-loop system generically equals (18).*

Regulator (15) or (17) is also referred to as the *generalized minimum-variance regulator*. By setting $\psi_u = O_{m \times m}$ or $\rho = 0$, the regulator (15) or (17) becomes the so-called *minimum-variance* regulator, which has to be intended as the regulator attempting to minimize the trace of the plant *output* covariance.

As was to be expected in view of Sec. 2.7 the single-step stochastic regulator suffers (cf. Problem 1) by the same limitations as in the deterministic case. In particular, for $\rho = 0$, it is inapplicable to nonminimum-phase plants, and may not be capable of stabilizing nonminimum-phase open-loop unstable plants irrespective of ρ. As seen from (18), in the stochastic case $C(d)$, representing the dynamics of the implicit Kalman filter, is a factor of the resulting closed-loop characteristic polynomial and, hence, affects the robustness properties of the compensated system.

Problem 7.3-2 (Single-Step Stochastic Servo) Consider the CARMA plant (1) to (3), the performance index

$$\mathcal{C} = \mathcal{E} \left\{ \|y(t+\tau) - r(t+\tau)\|^2_{\psi_y} + \|u(t)\|^2_{\psi_u} \ \Big| \ y^t, r^{t+\tau} \right\}$$

and the admissible control strategy

$$u(t) \in \sigma \left\{ y^t, r^{t+\tau} \right\}$$

with r, the reference to be tracked, a wide-sense stationary process independent on the innovations e. The adopted control strategy amounts to assuming that the controller at time t has full knowledge of the reference realization up to time $t + \tau$. Find how (15) must be modified to solve the above single-step stochastic servo problem, provided that the resulting control law yields an internally stable feedback system.

Problem 7.3-3 (Minimum-Variance Regulator) Consider the regulation law obtained from (17) for $\rho = 0$ and $\tau = 1$. Show that the resulting regulation law is the same as that obtained from (1) by forcing y to equal e.

7.3.2   Steady-State LQ Stochastic Linear Regulation

Hereafter, instead of the conditional expectation (5), we consider the minimization in stochastic steady-state of a *performance index* consisting of the following unconditional expectation:

$$\mathcal{C} = \mathcal{E} \left\{ \|y(k)\|^2_{\psi_y} + \|u(k)\|^2_{\psi_u} \right\} \qquad (7.3\text{-}19)$$

According to Result 2-2, minimization of (19) is achieved by steady-state LQG regulation, whose Riccati-based solution is given by (2-17) to (2-19). The aim here is to solve the problem for CARMA plants via the polynomial equation approach, so as to obtain the stochastic extension of the results of Chapter 4. We shall proceed as follows. We tackle the problem by first finding the optimal *linear* regulator and, next, showing that this is also optimal among all *nonlinear* feedback compensators. This result does not require the plant to be Gaussian, it is only sufficient that the innovations process e satisfy the martingale difference properties (3).

Consider the *plant* (1) to (2) along an *admissible regulation strategy* that restricts the plant input to be given by a causal *linear* compensator with transfer matrix $\mathcal{K}(d)$:

$$u(t) = -\mathcal{K}(d)y(t) \qquad (7.3\text{-}20)$$

making the resulting feedback system internally stable. The strategy (20), besides linearity of the regulator, is also restrictive in that plant output y to be regulated coincides with the variable at the compensator input. In this respect, more general system configurations have been considered in [CM91], [HKŠ92], and [HŠK91] at the expense of greater algebraic complications. For the sake of simplicity, we shall refrain from discussing these extensions by restricting ourselves to the following problem.

Steady-State LQ Stochastic Linear (LQSL) Regulator. Consider the CARMA plant (1) to (2) and the quadratic performance index (19). Find, whenever it exists, a linear feedback compensator (20) that makes the closed-loop system internally stable and minimizes (19).

According to Theorem 3.2-2, the previous stability requirement is equivalent to state that $\mathcal{K}(d)$ is factorizable in terms of the ratio of two stable transfer matrices $M_2(d)$ and $N_2(d)$, or $M_1(d)$ and $N_1(d)$,

$$
\begin{aligned}
\mathcal{K}(d) &= N_2(d)M_2^{-1}(d) & (7.3\text{-}21\text{a}) \\
&= M_1^{-1}(d)N_1(d) & (7.3\text{-}21\text{b})
\end{aligned}
$$

satisfying the identities

$$
\begin{aligned}
I_p &= A_1(d)M_2(d) + B_1(d)N_2(d) & (7.3\text{-}22\text{a}) \\
I_m &= M_1(d)A_2(d) + N_1(d)B_2(d) & (7.3\text{-}22\text{b})
\end{aligned}
$$

In order to minimize (19), it is first convenient to introduce some additional material on the description of wide-sense stationary processes and the transformations produced on their second-order properties when they are filtered by time-invariant linear systems. Let v be a vector-valued stochastic process with finite variance, that is, $\mathcal{E}\{\|v(t)\|^2\} < \infty$, for all $t \in \mathbb{Z}$. By assuming v to be wide-sense stationary, the two-sided sequence of its covariance matrices $K_v := \{K_v(k)\}_{k=-\infty}^{\infty}$:

$$
\begin{aligned}
K_v(k) &:= \mathcal{E}\{v(t+k)v'(t)\} & (7.3\text{-}23\text{a}) \\
&= K_v'(-k) & (7.3\text{-}23\text{b})
\end{aligned}
$$

is called the *covariance function* of v. Last equality easily follows from wide-sense stationariety of v:

$$\mathcal{E}\{v(t+k)v'(t)\} = \mathcal{E}\{v(t)v'(t-k)\} = [\mathcal{E}\{v(t-k)v'(t)\}]'$$

The d-representation $\Psi_v(d)$ (cf. Chapter 3) of K_v:

$$\Psi_v(d) \quad := \quad \sum_{k=-\infty}^{\infty} K_v(k)d^k \qquad\qquad (7.3\text{-}24a)$$

$$= \quad \Psi_v'\left(d^{-1}\right) = \Psi_v^*(d) \qquad\qquad (7.3\text{-}24b)$$

is called the *spectral density function* of v. Equation (24b) shows that for d taking values on the complex unit circle, $d = e^{i\theta}$, $\theta \in [0, 2\pi)$, Ψ_v is Hermitian symmetric:

$$\Psi_v\left(e^{i\theta}\right) = \Psi_v'\left(e^{-i\theta}\right) \qquad\qquad (7.3\text{-}24c)$$

Note that

$$\mathcal{E}\left\{\|v(t)\|_Q^2\right\} \quad = \quad \mathrm{Tr}\,[QK_v(0)]$$

$$= \quad \mathrm{Tr}\,[Q\langle\Psi_v(d)\rangle] \qquad\qquad (7.3\text{-}25)$$

where, as in (3.1-12), the symbol $\langle\ \rangle$ denotes extraction of the 0-power term. Equation (25) is the counterpart of (3.1-12) in the present stochastic setting.

Problem 7.3-4   Consider a linear time-invariant system with transfer matrix $H(d)$. Let its input u and output y be wide-sense stationary stochastic processes with spectral density functions $\Psi_u(d)$ and $\Psi_y(d)$, respectively. Then, show that

$$\Psi_y(d) = H(d)\Psi_u(d)H^*(d) \qquad\qquad (7.3\text{-}26)$$

By dealing with vector-valued wide-sense stationary processes, the notion of cross-covariance between two processes is already embedded in (23). In fact, let a process z be partitioned into two separate processes u and v, $z(t) = \begin{bmatrix} u'(t) & v'(t) \end{bmatrix}'$. Then, we have

$$K_z(k) = \begin{bmatrix} K_u(k) & K_{uv}(k) \\ K_{vu}(k) & K_v(k) \end{bmatrix}$$

where

$$K_{uv}(k) \quad := \quad \mathcal{E}\{u(t+k)v'(t)\}$$

$$= \quad K_{vu}'(-k)$$

is called the *cross-covariance function* of u and v. Similarly to (24), we define the *cross-spectral density function* of u and v:

$$\Psi_{uv}(d) \quad := \quad \sum_{k=-\infty}^{\infty} K_{uv}(k)d^k$$

$$= \quad \Psi_{vu}'(d^{-1}) = \Psi_{vu}^*(d)$$

Problem 7.3-5 Consider two wide-sense stationary processes u and v with cross-spectral density $\Psi_{uv}(d)$. Let y and z be other two wide-sense stationary processes related to u and v as follows:

$$y(t) = H_{yu}(d)u(d) \qquad z(t) = H_{zv}(d)v(t)$$

where $H_{yu}(d)$ and $H_{zv}(d)$ are rational transfer matrices. Then show that

$$\Psi_{yz}(d) = H_{yu}(d)\Psi_{uv}(d)H_{zv}^*(d)$$

Problem 7.3-6 Let u and v be wide-sense stationary processes with cross-spectral density $\Psi_{uv}(d)$. Show that

$$\mathcal{E}\left\{v'(t)Qu(t)\right\} = \operatorname{Tr}\left[Q\langle\Psi_{uv}(d)\rangle\right]$$

Further, let $H(d)$ be a rational transfer function such that $z(t) = H(d)u(t)$, $\dim z(t) = \dim v(t)$, is wide-sense stationary. Show that

$$\mathcal{E}\left\{v'(t)[H(d)u(t)]\right\} = \mathcal{E}\left\{[H^*(d)v(t)]'u(t)\right\}$$

We next express $y(t)$ and $u(t)$ in terms of the exogenous input e for the closed-loop system (1) and (20).

Let $A_1^{-1}(d)B_1(d)$ and $B_2(d)A_2^{-1}(d)$ be irreducible left and right MFDs, respectively, of $A^{-1}(d)B(d)$:

$$
\begin{align}
A^{-1}(d)B(d) &= A_1^{-1}(d)B_1(d) && \text{(7.3-27a)}\\
&= B_2(d)A_2^{-1}(d) && \text{(7.3-27b)}
\end{align}
$$

then, for the regulated system (1) and (20), we find

$$
\begin{align}
y(t) &= -A_1^{-1}(d)B_1(d)N_2(d)M_2^{-1}(d)y(t) + A^{-1}(d)C(d)e(t)\\
&= \left[I_p + A_1^{-1}(d)B_1(d)N_2(d)M_2^{-1}(d)\right]^{-1}A^{-1}(d)C(d)e(t)\\
&= M_2(d)\left[A_1(d)M_2(d) + B_1(d)N_2(d)\right]^{-1}A_1(d)A^{-1}(d)C(d)e(t)\\
&= M_2(d)A_1(d)A^{-1}(d)C(d)e(t) && \text{[(22a)]}\\
&= \left[I_p - B_2(d)N_1(d)\right]A^{-1}(d)C(d)e(t) && \text{[(3.2-23a)]} \quad \text{(7.3-28)}
\end{align}
$$

Further,

$$
\begin{align}
u(t) &= -N_2(d)M_2^{-1}(d)y(t)\\
&= -N_2(d)A_1(d)A^{-1}(d)C(d)e(t) && \text{[(28)]}\\
&= -A_2(d)N_1(d)A^{-1}(d)C(d)e(t) && \text{[(3.2-30a)]} \quad \text{(7.3-29)}
\end{align}
$$

By using (25), (19) can be rewritten as follows:

$$\mathcal{C} = \operatorname{Tr}\langle\psi_y\Psi_y(d)\rangle + \operatorname{Tr}\langle\psi_u\Psi_u(d)\rangle \qquad \text{(7.3-30)}$$

where, by (28), (29), and (26),

$$
\begin{align}
\Psi_y(d) &= \left[I_p - B_2(d)N_1(d)\right]A^{-1}(d)D(d)D^*(d)A^{-*}(d)\left[I_p - N_1^*(d)B_2^*(d)\right]\\
\Psi_u(d) &= A_2(d)N_1(d)A^{-1}(d)D(d)D^*(d)A^{-*}(d)N_1^*(d)A_2^*(d)
\end{align}
$$

In the previous equations, $A^{-*}(d) := [A^{-1}(d)]^*$, and $D(d)$ is a $p \times p$ Hurwitz polynomial matrix such that

$$D(d)D^*(d) = C(d)\Psi_e C^*(d) \tag{7.3-31}$$

Problem 7.3-7 Consider the plant

$$A(d)y(t) = B(d)u(t) + L(d)\nu(t) \tag{7.3-32}$$

with $L(d)$ possibly non-Hurwitz and rectangular, the cost (19), and the admissible regulation strategy

$$u(t) = -\mathcal{K}(d)z(t) \tag{7.3-33}$$

where

$$z(t) = y(t) + \zeta(t) \tag{7.3-34}$$

In the previous equations, ν and ζ are two mutually uncorrelated zero-mean wide-sense stationary white processes with $K_\nu(k) = \Psi_\nu \delta_{k,0}$ and $K_\zeta(k) = \Psi_\zeta \delta_{k,0}$. Show that the earlier steady-state LQ stochastic regulation problem is equivalent to the one for the CARMA plant

$$A(d)z(t) = B(d)u(t) + D(d)e(t) \tag{7.3-35}$$

the admissible regulation strategy (33), and the cost

$$\mathcal{C} = \mathcal{E}\left\{ \|z(t)\|^2_{\psi_y} + \|u(t)\|^2_{\psi_u} \right\} \tag{7.3-36}$$

provided that e is zero-mean wide-sense stationary and white with an identity covariance matrix, and $D(d)$ is a $p \times p$ Hurwitz polynomial matrix solving the following left spectral factorization problem

$$D(d)D^*(d) = L(d)\Psi_\nu L^*(d) + A(d)\Psi_\zeta A^*(d) \tag{7.3-37}$$

D(d) exists if and only if

$$\mathrm{rank} \begin{bmatrix} L(d)\Psi_\nu & A(d)\Psi_\zeta \end{bmatrix} = p = \dim z \tag{7.3-38}$$

Using the expressions for $\Psi_y(d)$ and $\Psi_u(d)$, we find

$$\begin{aligned}
\mathcal{C} = \ & \mathrm{Tr}\langle D^*(d)A^{-*}(d)\Big[N_1^*(d)E^*(d)E(d)N_1(d) \\
& - \psi_y B_2(d)N_1(d) - N_1^*(d)B_2^*(d)\psi_y + \psi_y \Big] A^{-1}(d)D(d)\rangle
\end{aligned}$$

where $E(d)$ is an $m \times m$ Hurwitz polynomial matrix solving the right spectral factorization problem (cf. (4.1-12)):

$$E^*(d)E(d) = A_2^*(d)\psi_u A_2(d) + B_2^*(d)\psi_y B_2(d) \tag{7.3-39}$$

$E(d)$ exists if and only if

$$\mathrm{rank} \begin{bmatrix} \psi_u A_2(d) \\ \psi_y B_2(d) \end{bmatrix} = m := \dim u \tag{7.3-40}$$

$\mathcal{C}$ can be reorganized as follows:

$$\mathcal{C} = \mathcal{C}_1 + \mathcal{C}_2 \tag{7.3-41a}$$

with

$$\mathcal{C}_1 := \text{Tr}\langle \mathcal{L}^*(d)\mathcal{L}(d)\rangle \tag{7.3-41b}$$

$$\mathcal{L}(d) := \left[E^{-*}(d)B_2^*(d)\psi_y - E(d)N_1(d) \right] A^{-1}(d)D(d) \tag{7.3-41c}$$

$$\mathcal{C}_2 := \text{Tr}\langle D^*(d)A^{-*}(d)\left[\psi_y - \psi_y B_2(d)E^{-1}(d)E^{-*}(d)B_2^*(d)\psi_y\right] A^{-1}(d)D(d)\rangle \tag{7.3-41d}$$

Note that $\mathcal{C}_2$ is not affected by the choice of $\mathcal{K}(d)$. Thus, the problem amounts to finding $N_1(d)$ minimizing (41b). This equals the square of the ℓ_2 norm (cf. (3.1-12)) of matrix sequence $\mathcal{L}(d)$. In turn, $\mathcal{L}(d)$ has two additive components. One, $E(d)N_1(d)A^{-1}(d)D(d)$ is causal. The other, which results from premultiplying the causal sequence $\psi_y A^{-1}(d)D(d)$ by $E^{-*}(d)B_2^*(d)$, is a two-sided matrix sequence. The situation is similar to the one met in the deterministic LQ-regulation problem: cf. (4.1-20) to (4.1-23). We then follow the same solution method as in Sec. 4.2.

Causal–Anticausal Decomposition.   Let

$$q := \max\{\partial A_2(d), \partial B_2(d), \partial E(d)\} \tag{7.3-42a}$$

$$\bar{A}_2(d) := d^q A_2^*(d); \qquad \bar{B}_2(d) := d^q B_2^*(d); \qquad \bar{E}(d) := d^q E^*(d) \tag{7.3-42b}$$

Then, (39) can be rewritten as

$$\bar{E}(d)E(d) = \bar{A}_2(d)\psi_u A_2(d) + \bar{B}_2(d)\psi_y B_2(d) \tag{7.3-43}$$

Likewise, the first additive term on the R.H.S. of (41c) can be rewritten as

$$\tilde{\mathcal{L}}(d) := \bar{E}^{-1}(d)\bar{B}_2(d)\psi_y A^{-1}(d)D(d) \tag{7.3-44}$$

Suppose now that we can find a pair of polynomial matrices $Y(d)$ and $Z(d)$ fulfilling the following bilateral Diophantine equation:

$$\bar{E}(d)Y(d) + Z(d)A_3(d) = \bar{B}_2(d)\psi_y D_2(d) \tag{7.3-45a}$$

with the degree constraint

$$\partial Z(d) < \partial \bar{E}(d) = q \tag{7.3-45b}$$

The last equality follows because, with $E(d)$ Hurwitz, $E(0)$ is nonsingular. In (45a), we have denoted by $A_3(d)D_2^{-1}(d)$ an irreducible right MFD of $A(d)D^{-1}(d)$:

$$D^{-1}(d)A(d) = A_3(d)D_2^{-1}(d) \tag{7.3-46}$$

Using (45a) in (44), we find

$$\tilde{\mathcal{L}}(d) = \tilde{\mathcal{L}}_+(d) + \tilde{\mathcal{L}}_-(d)$$

where

$$\begin{aligned}
\tilde{\mathcal{L}}_+(d) &:= Y(d)A_3^{-1}(d) \\
\tilde{\mathcal{L}}_-(d) &:= \bar{E}^{-1}(d)Z(d)
\end{aligned} \qquad (7.3\text{-}47)$$

are respectively a causal and a strictly anticausal and possibly ℓ_2 sequence (cf. (4.2-10)). In conclusion, provided that we can find a pair $(Y(d), Z(d))$ solving (45), $\mathcal{L}(d)$ can be decomposed in terms of a causal sequence $\mathcal{L}_+(d)$ and a strictly anticausal sequence $\tilde{\mathcal{L}}_-(d)$ as follows:

$$\mathcal{L}(d) = \mathcal{L}_+(d) + \tilde{\mathcal{L}}_-(d) \qquad (7.3\text{-}48)$$

$$\mathcal{L}_+(d) := Y(d)A_3^{-1}(d) - E(d)N_1(d)A^{-1}(d)D(d) \qquad (7.3\text{-}49)$$

Hence, by the same argument as in (4.1-24) to (4.1-28), we have

$$\begin{aligned}
\mathcal{C}_1 &= \mathrm{Tr}\left\langle \left[\mathcal{L}_+^*(d) + \tilde{\mathcal{L}}_-^*(d)\right]\left[\mathcal{L}_+(d) + \tilde{\mathcal{L}}_-(d)\right]\right\rangle \\
&= \mathrm{Tr}\langle\mathcal{L}_+^*(d)\mathcal{L}_+(d)\rangle + \mathcal{C}_3
\end{aligned} \qquad (7.3\text{-}50)$$

where

$$\mathcal{C}_3 := \langle\tilde{\mathcal{L}}_-^*(d)\tilde{\mathcal{L}}_-(d)\rangle \qquad (7.3\text{-}51)$$

With $\mathcal{C}_2$ as in (41d), assume that $\mathcal{C}_2 + \mathcal{C}_3$ is bounded. Then an optimal $N_1(d)$ is obtained by setting $\mathcal{L}_+(d) = O_{n \times p}$, that is,

$$\begin{aligned}
N_1(d) &= E^{-1}(d)Y(d)A_3^{-1}(d)D^{-1}(d)A(d) \\
&= E^{-1}(d)Y(d)D_2^{-1}(d) \qquad\qquad\qquad [(46)]
\end{aligned} \qquad (7.3\text{-}52)$$

Stability.   The remaining part of the transfer matrix of the optimal regulator, viz., the stable transfer matrix $M_1(d)$, can be found via (22b):

$$\begin{aligned}
M_1(d)A_2(d) &= I_m - N_1(d)B_2(d) \\
&= I_m - E^{-1}(d)Y(d)D_2^{-1}(d)B_2(d)
\end{aligned}$$

The problem here is that the Diophantine equation (45), even if solvable, need not have a unique solution $N_1(d)$. In such a case, some solutions may not yield stable transfer matrices $M_1(d)$ via the previous equation. The situation is similar to the one faced for the deterministic LQR problem as discussed in Sec. 4.3. By imposing stability of the closed-loop system, we obtain, under general conditions, uniqueness of the solution. To this end, we write

$$\begin{aligned}
M_1(d) &= E^{-1}(d)\bar{E}^{-1}(d)\left[\bar{E}(d)E(d) - \bar{E}(d)Y(d)D_2^{-1}(d)B_2(d)\right]A_2^{-1}(d) \\
&= E^{-1}(d)\bar{E}^{-1}(d)\left[\bar{E}(d)E(d) + Z(d)A_3(d)D_2^{-1}(d)B_2(d)\right. \\
&\qquad \left. - \bar{B}_2(d)\psi_y B_2(d)\right]A_2^{-1}(d) \qquad\qquad\qquad\qquad [(45a)] \\
&= E^{-1}(d)\bar{E}^{-1}(d)\left[\bar{A}_2(d)\psi_u + Z(d)B_3(d)D_1^{-1}(d)\right] \quad [(43),(46),(53)]
\end{aligned}$$

To get the last equality, we have set

$$B_3(d)D_1^{-1}(d) := D^{-1}(d)B(d) \tag{7.3-53}$$

with $B_3(d)$ and $D_1(d)$ right-coprime polynomial matrices. Hence,

$$M_1(d)D_1(d) = E^{-1}(d)\bar{E}^{-1}(d)\left[\bar{A}_2(d)\psi_u D_1(d) + Z(d)B_3(d)\right] \tag{7.3-54}$$

Recall that by (31) or (37), $D_1(d)$ is Hurwitz, and because $E(d)$ is Hurwitz, $\bar{E}(d)$, is anti-Hurwitz. Then a necessary condition for stability of $M_1(d)$ is that the polynomial matrix within brackets in (54) be divided on the left by $\bar{E}(d)$. That is, there must be a polynomial matrix $X(d)$ such as to satisfy the following equation:

$$\bar{E}(d)X(d) - Z(d)B_3(d) = \bar{A}_2(d)\psi_u D_1(d) \tag{7.3-55}$$

By the same argument used after (4.3-7), it follows that $X(d)$ is nonsingular. Recalling (1), (21), (52), (54), and (55), we conclude that in order to solve the steady-state LQSL-regulation problem, in addition to the spectral factorization problems (31) or (37) and (39), we have to find a solution $(X(d), Y(d), Z(d))$ with $\partial Z(d) < \partial \bar{E}(d)$ of the two bilateral Diophantine equations (45) and (55). Using (55) in (54), we find

$$M_1(d) = E^{-1}(d)X(d)D_1^{-1}(d) \tag{7.3-56}$$

This, together with (21) and (52), yields

$$\mathcal{K}(d) = D_1(d)X^{-1}(d)Y(d)D_2^{-1}(d) \tag{7.3-57}$$

We then see that $Z(d)$ in (45) and (55) plays the role of a "dummy" polynomial matrix. By eliminating $Z(d)$ in (45) and (55), we get

$$X(d)D_1^{-1}(d)A_2(d) + Y(d)D_2^{-1}(d)B_2(d) = E(d) \tag{7.3-58}$$

Then, in turn, setting

$$\begin{bmatrix} D_1^{-1}(d)A_2(d) \\ D_2^{-1}(d)B_2(d) \end{bmatrix} = \begin{bmatrix} A_4(d) \\ B_4(d) \end{bmatrix} D_3^{-1}(d) \tag{7.3-59}$$

with the expression on the R.H.S. an irreducible right MFD, it can be rewritten as

$$X(d)A_4(d) + Y(d)B_4(d) = E(d)D_3(d) \tag{7.3-60}$$

It is instructive to compare (60) with (2-22) and conclude that, in the absence of possible cancellations, the d-characteristic polynomial of the steady-state LQSL-regulated system is proportional to $\det E(d) \cdot \det C(d)$.

Problem 7.3-8   Show that a triplet $(X(d), Y(d), Z(d))$ is a solution of (45) and (55) if and only if it solves (45) and (58). (*Hint:* Prove sufficiency by using (43).)

Finally, the closed-loop feedback system (1) $\big((32)\big)$, (20) $\big((33)\big)$, (21), and (22) is internally stable if and only if (52) and (56) are stable transfer matrices. The following lemma sums up the previous results.

Lemma 7.3-1.   *Provided that:*

(i)   Equations (45a) and (55) [or (45a) and (60)] admit a solution $(X(d), Y(d), Z(d))$ with $\partial Z(d) < \partial \bar{E}(d)$; and

(ii)   transfer matrices (52) and (56) are both stable,

the steady-state LQSL regulator is given by the dynamic feedback compensator

$$u(t) = -D_1(d)X^{-1}(d)Y(d)D_2^{-1}(d)y(t) \tag{7.3-61}$$

The minimum cost achievable with the optimal regulation law (61) equals

$$\mathcal{C}_{\min} = \mathcal{C}_2 + \mathcal{C}_3 \tag{7.3-62}$$

Solvability.   It remains to establish conditions under which (45) and (55) are solvable.

Problem 7.3-9   Modify the proof of Lemma 4.4-1 to show that if, according to assumption (2c), the greatest common left divisors of $A(d)$ and $B(d)$ are strictly Hurwitz, there is a unique solution $(X(d), Y(d), Z(d))$ of (45) and (55) [or (45) and (60)].

In [HŠG87], the use of the single unilateral Diophantine equation (60), instead of both (45) and (55) or (60), was discussed and shown to be possible provided that $A(d)$ and $B(d)$ are left-coprime. This is the counterpart in the present stochastic setting of the result in Lemma 4.4-2.

The following theorem immediately follows from Lemma 1 and Problem 4.

Theorem 7.3-3.   (Steady–State LQSL Regulator) *Consider the CARMA plant (1) subject to the conditions (2). Let (40) be fulfilled. Then, (45) and (55) [or (45) and (60)] admit a unique solution $(X(d), Y(d), Z(d))$ of minimum degree w.r.t. $Z(d)$. Further, the linear time-invariant feedback compensator (61) yields an internally stable closed-loop system if and only if (52) and (56) are both stable transfer matrices. In such a case the steady-state LQSL regulator is given by (61) and yields the minimum cost*

$$\begin{aligned} \mathcal{C}_{\min} &= \min_{K(d)} \mathrm{Tr}[\psi_y \Psi_y + \psi_u \Psi_u] \tag{7.3-63}\\ &= \mathcal{C}_2 + \mathcal{C}_3 \end{aligned}$$

$$\Psi_y := \mathrm{Cov}\Big(y(t)\Big) \qquad \Psi_u := \mathrm{Cov}\Big(u(t)\Big)$$

Remark 7.3-3.   In Theorem 3, conditions for the existence of the steady-state LQSL regulator are implicit, and they can be only checked after computing transfer matrices (52) and (56). Explicit sufficient conditions for solvability and uniqueness of the steady-state LQSL regulator, in addition to (2) and (3), are the following:

(i) $D(d)$ is strictly Hurwitz; and

(ii) $E(d)$ is strictly Hurwitz

In the steady-state LQG-regulation problem, $\Psi_\zeta > 0$ is a stronger analogue of (i) (cf. also (37)), and (ii) is fulfilled in case $\psi_y > 0$ and $\psi_u > 0$ (cf. Problem 10, which follows). Taking into account the difference in the plant models that are adopted in the two cases, we can thus conclude that solvability and uniqueness conditions for both steady-state LQSL regulation and steady-state LQG regulators are basically the same.

Problem 7.3-10   Consider the right spectral factorization problem (39). Show that $E(d)$ is strictly Hurwitz if $\psi_u > 0$ and $\psi_y > 0$ (*Hint*: Prove that $E'\left(e^{-j\theta}\right) E\left(e^{j\theta}\right) > 0$, $\theta \in [0, 2\pi)$.)

Problem 7.3-11   Find the polynomial equations giving the steady-state LQSL regulator for the CAR plant $A(d)y(t) = B(d)u(t) + e(t)$. Compare these equations with the ones of Chapter 4 solving the problem in the deterministic case.

In applications it is often important to consider, instead of (19), a performance index involving filtered versions of y and u:

$$
\begin{aligned}
\mathcal{C} &= \mathcal{E}\left\{\|W_y(d)y(k)\|^2 + \|W_u(d)u(k)\|^2\right\} \qquad\qquad\qquad (7.3\text{-}64)\\
&= \mathrm{Tr}\langle W_y(d)\Psi_y(d)W_y^*(d)\rangle + \mathrm{Tr}\langle W_u(d)\Psi_u(d)W_u^*(d)\rangle
\end{aligned}
$$

In (64), $W_y(d)$ and $W_u(d)$ denote two stable transfer matrices that we represent by irreducible right MFDs:

$$
W_y(d) = B_y(d)A_y^{-1}(d) \qquad W_u(d) = B_u(d)A_u^{-1}(d) \qquad (7.3\text{-}65)
$$

Problem 7.3-12 (Cost with Polynomial Weights)   Consider cost (64) with $W_y(d) = B_y(d)$ and $W_u(d) = B_u(d)$. Show that the polynomial equations giving the related steady-state LQSL regulator are the same as the ones obtained for the simpler cost (19), once the following changes are adopted:

$$
\psi_y \mapsto B_y^*(d)B_y(d) \qquad \psi_u \mapsto B_u^*(d)B_u(d)
$$

Namely,

$$
\begin{aligned}
E^*(d)E(d) &= A_2^*(d)B_u^*(d)B_u(d)A_2(d) + B_2^*(d)B_y^*(d)B_y(d)B_2(d) & (7.3\text{-}66a)\\
\bar{E}(d)E(d) &= \overline{A_2(d)B_u(d)}B_u A_2(d) + \overline{B_2(d)B_y(d)}B_y(d)B_2(d) & (7.3\text{-}66b)
\end{aligned}
$$

where $\bar{E}(d) := d^q E^*(d)$, $\overline{A_2(d)B_u(d)} := d^q A_2^*(d)B_u^*(d)$, $\overline{B_2(d)B_y(d)} := d^q B_2^*(d)B_y^*(d)$,
$q := \max\{\partial B_u(d) + \partial A_2(d), \partial B_y(d) + \partial B_2(d), \partial E(d)\}$,

$$\bar{E}(d)Y(d) + Z(d)A_3(d) = \overline{B_2(d)B_y(d)}B_y(d)D_2(d) \tag{7.3-67}$$
$$\bar{E}(d)X(d) - Z(d)B_3(d) = \overline{A_2(d)B_u(d)}B_u(d)D_1(d) \tag{7.3-68}$$

with the optimal regulation law given again by (61).

Problem 7.3-13 Show that the steady-state LQSL-regulator problem for SISO CARMA plant (1) and cost (19) can be equivalently reformulated for the CAR plant

$$A(d)y_c(t) = B(d)u_c(t) + e(t)$$

$$C(d)y_c(t) = y(t) \qquad C(d)u_c(t) = u(t)$$

and the cost

$$\mathcal{E}\left\{\psi_y \left[C(d)y_c(t)\right]^2 + \psi_u \left[C(d)u_c(t)\right]^2\right\}$$

Then find the steady-state LQSL regulator by using the results of Problems 11 and 12.

Problem 7.3-14 (Cost with Dynamic Weights) Consider cost (64) with $W_y(d)$ and $W_u(d)$ as in (65). Exploit the solution of Problem 12 to show that the polynomial equations giving the related steady-state LQSL regulator are again as in (66) to (68) provided that the following definitions are adopted:

$$[A(d)A_y(d)]^{-1}B(d)A_u(d) = B_2(d)A_2^{-1}(d) \tag{7.3-69}$$

$$D^{-1}(d)A(d)A_y(d) = A_3(d)D_2^{-1}(d) \qquad D^{-1}(d)B(d)A_u(d) = B_3(d)D_1^{-1}(d) \tag{7.3-70}$$

with $A_2(d)$ and $B_2(d)$, $D_2(d)$ and $A_3(d)$, and $D_1(d)$ and $B_3(d)$ all right-coprime.
Finally, the optimal regulation law is as follows:

$$u(t) = -A_u(d)D_1(d)X^{-1}(d)Y(d)D_2^{-1}(d)A_y^{-1}(d)y(t) \tag{7.3-71}$$

(*Hint*: Define new variables $\gamma(t) := A_y^{-1}(d)y(t)$ and $\nu(t) := A_u^{-1}(d)u(t)$.)

Problem 7.3-15 (Stabilizing Minimum-Variance Regulation) Consider the solution of the steady-state LQSL-regulation problem when the control variable is not costed, viz., $\psi_u = O_{m\times m}$. The resulting regulation law will be referred to as *stabilizing minimum-variance* regulation because the polynomial solution ensures internal stability if $D(d)$ and $E(d)$ are strictly Hurwitz. Find the stabilizing minimum-variance regulation law for two SISO CARMA plants, of which one is minimum and the other nonminimum-phase. Compare these results with those pertaining to minimum-variance regulation. Contrast stabilizing minimum-variance regulation with minimum-variance regulation.

Problem 7.3-16 (LQSL-Regulator Information Pattern) Consider the LQSL regulator for a SISO CARMA plant with $A(d)$, $B(d)$, and $C(d)$ having unit gcd and ord $B(d) = 1 + \ell$. Show that the LQSL regulation law $X(d)u(t) = -Y(d)y(t)$ is such that

$$\partial X(d) = \begin{cases} \partial B(d) - 1 & \psi_u = 0 \\ \max\{\partial B(d) - 1, \partial C(d)\} & \psi_u > 0 \end{cases}$$

$$\partial Y(d) = \max\{\partial A(d) - 1, \partial C(d) - 1 - \ell\}$$

7.3.3 LQSL-Regulator Optimality among Nonlinear Compensators

Suppose that, instead of just assuming the innovations process e in the CARMA plant (1) white, we adopt for e the martingale difference properties (3). Then, as shown hereafter, the steady-state LQSL regulator of Theorem 3 turns out to be also optimal among all possibly nonlinear compensators.

Consider the CARMA *plant* (1) to (3), *cost* (19) and the *admissible regulation strategy* (4). We wish to consider the following problem.

> **Steady-State LQ Stochastic (LQS) Regulator.** Consider the CARMA plant (1) to (2), cost (19), and the admissible regulation strategy (4). Assume that innovations e have the martingale difference properties (3). Among all possibly nonlinear regulation strategies, find, whenever they exist, the ones that make the closed-loop system internally stable and minimize (19).

We point out that, if the steady-state LQSL regulator exists, the admissible regulation strategy (4) can be written as

$$
\begin{aligned}
u(t) &= -M_1^{-1}(d)N_1(d)y(t) + v(t) \\
&= -N_2(d)M_2^{-1}(d)y(t) + v(t)
\end{aligned}
\tag{7.3-72}
$$

where $N_1(d)$ and $M_1(d)$ are the stable transfer matrices in (52) and (56), whose ratio $\mathcal{K}(d) = M_1^{-1}(d)N_1(d)$ defines the steady-state LQSL-regulator transfer matrix; $N_2(d)$ and $M_2(d)$ are such that $\mathcal{K}(d) = N_2(d)M_2^{-1}(d)$ and satisfy (22a); and $v(t)$ is any wide-sense stationary process such that

$$
v(t) \in \sigma\left\{y^t\right\} = \sigma\left\{e^t\right\}
\tag{7.3-73}
$$

Thus, the previously stated steady-state LQS regulation problem amounts to finding a process v as in (73) such that the corresponding regulation law (72) stabilizes the plant and minimizes (19).

Under the feedback control law (72), y and u can be expressed in terms of e and v as follows:

$$
\begin{aligned}
y(t) &= y_e(t) + y_v(t) & \text{(7.3-74a)} \\
y_e(t) &:= M_2(d)A_1(d)A^{-1}(d)C(d)e(t) & \text{(7.3-74b)} \\
y_v(t) &:= B_2(d)M_1(d)v(t) & \text{(7.3-74c)}
\end{aligned}
$$

$$
\begin{aligned}
u(t) &= u_e(t) + u_v(t) & \text{(7.3-75a)} \\
u_e(t) &:= -A_2(d)N_1(d)A^{-1}(d)C(d)e(t) & \text{(7.3-75b)} \\
u_v(t) &:= A_2(d)M_1(d)v(t) & \text{(7.3-75c)}
\end{aligned}
$$

Problem 7.3-17 By using (22) and (3.2-32b), verify (74) and (75).

With reference to the previous decompositions, cost (19) can be split as follows:

$$\mathcal{C} = \mathcal{C}_{ee} + \mathcal{C}_{ev} + \mathcal{C}_{vv}$$

where

$$
\begin{aligned}
\mathcal{C}_{ee} &:= \mathcal{E}\left\{\|y_e(t)\|^2_{\psi_y} + \|u_e(t)\|^2_{\psi_u}\right\} \\
\mathcal{C}_{ev} &:= 2\mathcal{E}\left\{y'_e(t)\psi_y y_v(t) + u'_e(t)\psi_u u_v(t)\right\} \qquad (7.3\text{-}76) \\
\mathcal{C}_{vv} &:= \mathcal{E}\left\{\|y_v(t)\|^2_{\psi_y} + \|u_v(t)\|^2_{\psi_u}\right\} \\
&= \mathcal{E}\left\{\|E(d)M_1(d)v(t)\|^2\right\} \qquad (7.3\text{-}77)
\end{aligned}
$$

Problem 7.3-18 Using (25) and the results of Problem 5, show that

$$\mathcal{C}_{ev} = 2\operatorname{Tr}\langle\Psi_{ev}(d)\left[H^*_{yv}(d)\psi_y H_{ye}(d) + H^*_{uv}(d)\psi_u H_{ue}(d)\right]\rangle$$

$$
\begin{aligned}
H_{yv}(d) &:= B_2(d)M_1(d) \qquad & H_{ye}(d) &:= M_2(d)A_1(d)A^{-1}(d)C(d) \\
H_{uv}(d) &:= A_2(d)M_1(d) \qquad & H_{ue}(d) &:= -A_2(d)N_1(d)A^{-1}(d)C(d)
\end{aligned}
$$

where $\Psi_{ev}(d)$ is defined as follows. Let

$$
\begin{aligned}
K_{ev}(k) &:= \mathcal{E}\left\{e(t+k)v'(t)\right\} \qquad (7.3\text{-}78) \\
&= K'_{ve}(-k)
\end{aligned}
$$

Then

$$
\begin{aligned}
\Psi_{ev}(d) &:= \sum_{k=-\infty}^{\infty} K_{ev}(k)d^k \qquad (7.3\text{-}79) \\
&= \Psi'_{ve}(d^{-1}) =: \Psi^*_{ve}(d)
\end{aligned}
$$

Problem 7.3-19 Using the results of Problem 18, conclude that

$$\mathcal{C}_{ev} = 2\operatorname{Tr}\langle\bar{Z}(d)M_1(d)\Psi_{ve}(d)\rangle \qquad (7.3\text{-}80)$$

$$\bar{Z}(d) := d^q Z^*(d)$$

where q and $Z(d)$ are as in (42) and (45), respectively.

We next show that, from the martingale difference properties (3) of the innovations process e, it follows that $\mathcal{C}_{ev} = 0$. In fact, by the smoothing properties of conditional expectations (cf. Appendix D.3),

$$
\begin{aligned}
K_{ev}(k) &= \mathcal{E}\left\{\mathcal{E}\left\{e(t+k)\mid e^t\right\}v'(t)\right\} \\
&= O_{p\times m}, \quad \forall k \geq 1 \qquad [(3a)]
\end{aligned}
$$

Thus, $K_{ev}(\cdot)$ $[K_{ve}(\cdot)]$ is an anticausal [causal] matrix sequence. Because $M_1(d)$ is causal and, by (45b), $\bar{Z}(d)$ strictly causal, it follows that (cf. Sec. 4.2) (80) vanishes.

Lemma 7.3-2. *Consider cost $\mathcal{C}_{ev}$ (76) where the involved processes are as in (74) and (75). Let v satisfy (73) and e the martingale difference properties (3). Then $\mathcal{C}_{ev} = 0$.*

It then follows that the optimal process v to be used in (73) equals O_m a.s. We have thus established the desired result.

Theorem 7.3-4. (Steady–State LQS regulator) *Whenever it exists, the steady-state LQSL regulator of Theorem 3 is also optimal among all possibly non-linear regulation strategies (4), provided that the innovations process e enjoys the martingale difference properties (3).*

We finally point out that if the CARMA plant (1) to (3) is the innovations representation of the "physical" plant (2-14), in order that (3) hold, it is essentially required that the processes in the plant (2-14) be jointly Gaussian.

Main Points of the Section. The polynomial equation approach can be used to solve steady-state LQ stochastic regulation problems. The single-step stochastic regulator does not involve any spectral factorization problem and can be computed by solving a single Diophantine equation. As with its deterministic counterpart, applicability of single-step stochastic regulation is limited because it yields an internally stable feedback system only under restrictive assumptions on the plant and the cost weights. The polynomial solution for the steady-state LQ stochastic regulator of CARMA plants can be obtained along the same lines as the ones followed for the deterministic steady-state LQOR problem of Chapter 4. The case of dynamic weights in the cost can be accommodated in a straightforward way within the equations solving the standard case of constant weights.

7.4   MONOTONIC PERFORMANCE PROPERTIES OF LQ STOCHASTIC REGULATION

We report a discussion on some monotonicity properties of steady-state LQ stochastic regulation. As will be seen in due time, these properties are important for establishing local convergence results of adaptive LQ regulators with mean-square input constraints.

We consider a performance index parameterized by a positive real input weight ρ, $\rho > 0$,

$$\begin{aligned} \mathcal{C} &= \mathcal{E}\left\{\|y(k)\|^2 + \rho\|u(k)\|^2\right\} \\ &= \mathrm{Tr}\left[\Psi_y\right] + \rho\,\mathrm{Tr}\left[\Psi_u\right] \end{aligned} \tag{7.4-1}$$

where $\Psi_y = \langle\Psi_y(d)\rangle$ and $\Psi_u = \langle\Psi_u(d)\rangle$ are covariance matrices of wide-sense stationary processes y and u, respectively. More precisely, $\Psi_y = \Psi_y(\rho)$ and $\Psi_u = \Psi_u(\rho)$ are the covariance matrices of y and u in stochastic steady state when the plant is regulated by the steady-state LQ stochastic regulator optimal for the given

ρ. We then show that $\text{Tr}\left[\Psi_u(\rho)\right]$ $(\text{Tr}\left[\Psi_y(\rho)\right])$ is a strictly decreasing (increasing) function of ρ. To see this, consider two different values of ρ, $\rho_i, i = 1, 2$. Let Ψ_y^i and Ψ_u^i denote the stochastic steady-state covariance matrices pertaining to the steady-state LQ stochastic regulator minimizing (1) for $\rho = \rho_i$. Recall that under suitable assumptions (cf. Secs. 2 and 3), for every $\rho > 0$, there exists a unique optimal steady-state LQ stochastic regulator. Then, the following inequalities hold:

$$\text{Tr}\left[\Psi_y^1\right] + \rho_1 \, \text{Tr}\left[\Psi_u^1\right] \; < \; \text{Tr}\left[\Psi_y^2\right] + \rho_1 \, \text{Tr}\left[\Psi_u^2\right]$$
$$\text{Tr}\left[\Psi_y^2\right] + \rho_2 \, \text{Tr}\left[\Psi_u^2\right] \; < \; \text{Tr}\left[\Psi_y^1\right] + \rho_2 \, \text{Tr}\left[\Psi_u^1\right]$$

or, equivalently,

$$\rho_2 \, \text{Tr}\left[\Psi_u^2 - \Psi_u^1\right] \; < \; \text{Tr}\left[\Psi_y^1 - \Psi_y^2\right]$$
$$\text{Tr}\left[\Psi_y^1 - \Psi_y^2\right] \; < \; \rho_1 \, \text{Tr}\left[\Psi_u^2 - \Psi_u^1\right]$$

Hence,

$$\rho_2 > \rho_1 > 0 \Rightarrow \begin{cases} \text{Tr}\left[\Psi_u\left(\rho_2\right)\right] < \text{Tr}\left[\Psi_u\left(\rho_1\right)\right] \\ \text{Tr}\left[\Psi_y\left(\rho_2\right)\right] > \text{Tr}\left[\Psi_y\left(\rho_1\right)\right] \end{cases} \qquad (7.4\text{-}2)$$

Theorem 7.4-1. *Consider a steady-state LQ stochastic regulation problem with the quadratic performance index (1) parameterized by the positive input weight ρ. Assume the problem solvable. Let $\Psi_u(\rho)$ and $\Psi_y(\rho)$ be the stochastic steady-state covariance matrices of u and y when the plant is fed back by the steady-state LQ stochastic regulator optimal for the given ρ. Then, (2) holds, viz., $\text{Tr}\left[\Psi_u(\rho)\right]$ $(\text{Tr}\left[\Psi_y(\rho)\right])$ is a strictly decreasing (increasing) function of ρ.*

Problem 7.4-1 Extend the conclusion of Theorem 1 to the steady-state LQS regulator and cost

$$\mathcal{C} = \mathcal{E}\left\{\|y(k)\|_{\psi_y}^2 + \rho\|u(k)\|_{\psi_u}^2\right\}$$

Besides its use in the analysis of adaptive LQ regulators with mean-square input constraints, the previous monotonic performance properties are important for designing purposes, in that they allow us to trade between output and input covariance matrices.

We point out that the monotonic performance properties in Theorem 1 follow from the minimization of the *unconditional* expectation (1) achieved by steady-state LQ stochastic regulation. In contrast, single-step stochastic regulators, which in stochastic steady-state minimize the *conditional* expectation (3-5), in general do not possess similar monotonic performance properties.

Example 7.4-1 [MS82]
Consider the SISO CARMA plant (3-1) with

$$\begin{aligned} A(d) &= 1 - 2.75d + 2.61d^2 + 0.885d^3 \\ B(d) &= d - 0.5d^2 \\ C(d) &= 1 - 0.2d + 0.5d^2 - 0.1d^3 \end{aligned}$$

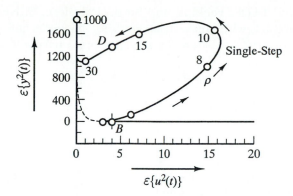

Figure 7.4-1: The relation between $\mathcal{E}\left\{u^2(k)\right\}$ and $\mathcal{E}\left\{y^2(k)\right\}$ parameterized by ρ for the plant of Example 1 under single step stochastic regulation (solid line) and steady-state LQS regulation (dotted line).

By using the single-step stochastic regulator optimal for cost

$$\mathcal{C} = \mathcal{E}\left\{y^2(k+1) + \rho u^2(k) \mid y^k\right\}$$

the plant under consideration gives rise to nonmonototic relationships between the stochastic steady-state input variance $\mathcal{E}\left\{u^2(k)\right\}$, the stochastic steady-state output variance $\mathcal{E}\left\{y^2(k)\right\}$, and the input weight ρ (Fig. 1). On the contrary, as guaranteed by Theorem 1, steady-state LQS regulation yields monotonic performance relationships (dashed line in Fig. 1).

Main Points of the Section. In contrast with single-step stochastic regulated systems, steady-state LQS regulated systems possess performance monotonicity properties that enable us, by varying an input weight knob, to trade off between output and input covariance matrices. These monotonicity properties turn out to be important to establish convergence results for self-tuning regulators with mean-square input constraints.

7.5 STEADY-STATE LQS TRACKING AND SERVO

7.5.1 Problem Formulation and Solution

We consider the CARMA plant (3-1) to (3-3) along with a wide-sense stationary *reference* process r, $\dim r(t) = \dim y(t) = p$. We assume that

$$r \text{ and } e \text{ are mutually independent processes} \qquad (7.5\text{-}1)$$

We wish to consider the following problem.

Steady-State LQS Tracking and Servo Problem. Given: the CARMA *plant* (3-1) to (3-2) with innovations satisfying the martingale difference properties (3-3); the *performance index* consisting of the unconditional expectation

$$\mathcal{E}\left\{ \|y(t) - r(t)\|^2_{\psi_y} + \|u(t)\|^2_{\psi_u} \right\} \tag{7.5-2}$$

with $\psi_y = \psi'_y \geq 0$, $\psi_u = \psi'_u \geq 0$, and r a wide-sense stationary reference; the *admissible control strategy*

$$u(t) \in \sigma\left\{ y^t, u^{t-1}, r^{t+\ell} \right\} = \sigma\left\{ y^t, r^{t+\ell} \right\} \tag{7.5-3}$$

find, among all the admissible strategies, the ones making the closed-loop system internally stable and minimizing (2).

It is clear from (3) that we are searching for an optimal 2-DOF controller (cf. Sec. 5.8). For now, we aim at solving the mathematical problem just formulated with no concern for the controller integral action or dynamic weights in (2). As will be shown, these issues can be accommodated in the basic theory by suitable simple modifications. Another point that we emphasize is that our admissible control strategy (3) allows us to select the present input $u(t)$ knowing the reference up to time $t+h$. If h is positive and very large, the situation appears as an extension of the deterministic 2-DOF LQ control problem of Theorem 5.8-1 to the present stochastic setting. We have already noticed the improvement in performance that can be achieved, especially with nonminimum-phase plants, by exploiting the knowledge of the reference future, provided that this is available to the controller. For the sake of generality, in this section, we assume that h can be any integer. According to the sign of h, we adopt two different names for the problem. We call it either a *tracking* problem, if the reference is known to the controller with a delay of $|h|$ steps, $h \leq 0$, or a *servo* problem, if $h > 0$. The solution that we are to find, allowing h to take any value, can be used for both the tracking and servo problems.

To begin with, let us first assume that the underlying steady-state LQS pure regulation problem, viz., the one with $r(t) \equiv O_p$, is solvable. Recall that its solution is given by Theorem 3-3 and Theorem 3-4 in the following form:

$$M_1(d)u(t) = -N_1(d)y(t)$$

with

$$M_1(d)A_2(d) + N_1(d)B_2(d) = I_m$$

We now follow a line similar to that adopted after (3-72). Thus, any admissible control law (3) can be written as

$$u(t) = -M_1^{-1}(d)N_1(d)y(t) + v(t) \tag{7.5-4}$$

with $v(t)$ any wide-sense stationary process such that

$$v(t) \in \sigma\left\{ y^t, r^{t+h} \right\} \tag{7.5-5}$$

Under the closed-loop control law (5), y and u can be expressed in terms of e and v, as in (3-74) and (3-75). Consequently, cost (2) can be split as follows:

$$C = C_{ee} + C_{ev} + C_{er} + C_{rr} + C_{vr} + C_{vv} \tag{7.5-6}$$

where

$$
\begin{aligned}
C_{ee} &:= \mathcal{E}\left\{\|y_e(t)\|^2_{\psi_y} + \|u_e(t)\|^2_{\psi_u}\right\} \\
C_{ev} &:= 2\mathcal{E}\left\{y'_e(t)\psi_y y_v(t) + u'_e(t)\psi_u u_v(t)\right\} \\
C_{er} &:= -2\mathcal{E}\left\{r'(t)\psi_y y_e(t)\right\} = O_{p\times p} \qquad [(1)]\\
C_{rr} &:= \mathcal{E}\left\{\|r(t)\|^2_{\psi_y}\right\} \\
C_{vr} &:= -2\mathcal{E}\left\{y'_v(t)\psi_y r(t)\right\} \\
C_{vv} &:= \mathcal{E}\left\{\|y_v(t)\|^2_{\psi_y} + \|u_v(t)\|^2_{\psi_u}\right\} \\
&= \mathcal{E}\left\{\|E(d)M_1(d)v(t)\|^2\right\} \qquad [(3\text{-}77)]
\end{aligned}
$$

We now show that the key property $C_{ev} = 0$, which was proved in Lemma 3-3 for the underlying LQ stochastic pure regulation problem, holds true if v satisfies (5).

Lemma 7.5-1. *Consider the cost C_{ev} where the involved processes are as in (3-74) and (3-75) with v satisfying (5). Let e have the martingale difference properties (3). Then $C_{ev} = 0$.*

Proof: Here (3-80) still holds true. We also have

$$
\begin{aligned}
K_{ev}(k) &= \mathcal{E}\left\{e(t+k)v'(t)\right\} \\
&= \mathcal{E}\left\{\mathcal{E}\left\{e(t+k) \mid y^t, r^{t+\ell}\right\}v'(t)\right\} \\
&= \mathcal{E}\left\{\mathcal{E}\left\{e(t+k) \mid e^t\right\}v'(t)\right\} \qquad [(1)]\\
&= O_{p\times m}, \quad \forall k \geq 1
\end{aligned}
$$

Hence, the proof follows by the same argument as in Lemma 3.3.

The main result can now be stated.

Theorem 7.5-1.   **(Steady–State LQS Tracking and Servo)** *Suppose that the underlying steady-state LQS pure regulation problem is solvable. Let the left spectral factor $E(d)$ in (3-39) be strictly Hurwitz. Then, the optimal control law for the steady-state LQS tracking and servo problems is given by*

$$M_1(d)u(t) = -N_1(d)y(t) + u^c(t) \tag{7.5-7}$$

where $M_1(d)$ and $N_1(d)$ are the stable transfer matrices solving the underlying steady-state LQS pure regulation problem, and $u^c(t)$ is the command or feedforward input defined by

$$u^c(t) = E^{-1}(d)\mathcal{E}\left\{E^{-*}(d)B_2^*(d)\psi_y r(t) \mid r^{t+h}\right\} \tag{7.5-8}$$

Proof: Because in (6) $\mathcal{C}_{ee}$ and $\mathcal{C}_{rr}$ are not affected by v, $\mathcal{C}_{er} = 0$, and $\mathcal{C}_{ev} = 0$ by Lemma 1, the optimal control law is given by (4) with $v(t) \in \sigma\left\{y^t, r^{t+h}\right\}$ minimizing, for $z(t) := M_1(d)v(t)$,

$$
\begin{aligned}
\mathcal{C}_{vv} + \mathcal{C}_{vr} &= \mathcal{E}\left\{\|E(d)z(t)\|^2 - 2\left[B_2(d)z(t)\right]' \psi_y r(t)\right\} \\
&= \mathcal{E}\left\{\|E(d)z(t)\|^2 - 2z'(t)\left[B_2^*(d)\psi_y r(t)\right]\right\} \\
&= \mathcal{E}\left\{\|E(d)z(t)\|^2 - 2\left[E(d)z(t)\right]'\left[E^{-*}(d)B_2^*(d)\psi_y r(t)\right]\right\} \\
&= \mathcal{E}\left\{\|E(d)z(t) - E^{-*}(d)B_2^*(d)\psi_y r(t)\|^2\right\} - \mathcal{E}\left\{\|E^{-*}(d)B_2^*(d)\psi_y r(t)\|^2\right\}
\end{aligned}
$$

Because the second term in the last line is not affected by $z(t)$, the minimum is attained at $z(t) = M_1(d)v(t) = u^c(t)$ with $u^c(t)$ as in (8).

Recalling (3-52) and (3-56), we have

$$
\mathcal{R}(d) \quad := \quad E(d)M_1(d) = X(d)D_1^{-1}(d) \tag{7.5-9}
$$

$$
\mathcal{S}(d) \quad := \quad E(d)N_1(d) = Y(d)D_2^{-1}(d) \tag{7.5-10}
$$

and (7) and (8) can be equivalently rewritten as follows:

$$
\mathcal{R}(d)u(t) = -\mathcal{S}(d)y(t) + v^c(t) \tag{7.5-11}
$$

$$
v^c(t) = \mathcal{E}\left\{E^{-*}(d)B_2^*(d)\psi_y r(t) \mid r^{t+h}\right\} \tag{7.5-12}
$$

Equations (11) and (12) should be compared with (5.8-26) and (5.8-27), which give the solution of the deterministic version of the present stochastic control problem. We refer the reader to the relevant part of Sec. 5.8 where the strictly "anticipative" nature of

$$
\begin{aligned}
v(t) &= E^{-*}(d)B_2^*(d)\psi_y r(t) \\
&= \bar{E}^{-1}(d)\bar{B}_2(d)\psi_y r(t)
\end{aligned}
$$

was thoroughly discussed.

Example 7.5-1   Consider again Example 5.8-2 where we found

$$
\begin{aligned}
w(t) \quad &:= \quad E^{-*}(d)B_2^*(d)\psi_y r(t) \\
&= \quad k^{-1}\left[r(t+1) + \left(1 - b^2\right)\sum_{j=1}^{\infty} b^{-j}r(t+j+1)\right]
\end{aligned}
$$

If $h > 0$, the feedforward input $v^c(t)$ can be decomposed as $\hat{v}^c(t) + \tilde{v}^c(t)$:

$$
\hat{v}^c(t) \quad := \quad k^{-1}\left[r(t+1) + \left(1 - b^2\right)\sum_{j=1}^{h-1} b^{-j}r(t+j+1)\right]
$$

$$
\tilde{v}^c(t) \quad := \quad k^{-1}\left(1 - b^2\right)b^{-h+1}\sum_{j=1}^{\infty} b^{-j}\mathcal{E}\left\{r(t+h+j) \mid r^{t+h}\right\}
$$

Note that of these two components, only $\tilde{v}^c(t)$ depends on the statistical properties of reference r.

Equation (12) can be further elaborated when a stochastic model for the reference is given. In this connection, let us assume that

$$r(t) = G_2(d)F_2^{-1}(d)n(t) \tag{7.5-13}$$

where n, $n(t) \in \mathbb{R}^p$, has the martingale difference properties:

$$\mathcal{E}\left\{n(k+1) \mid n^k\right\} = O_p \qquad \text{a.s.} \tag{7.5-14a}$$

$$\infty > \mathcal{E}\left\{n(k+1)n'(k+1) \mid n^k\right\} = \Psi_n > 0 \qquad \text{a.s.} \tag{7.5-14b}$$

$G_2(d)F_2^{-1}(d)$ is a right-coprime MFD with

$$G_2(d) \text{ and } F_2(d) \text{ both strictly Hurwitz} \tag{7.5-15}$$

Assumptions (15) entail no substantial limitation. In fact, because $r(t)$ is wide-sense stationary, $F_2(d)$ must be strictly Hurwitz. Further, $G_2(d)$ strictly Hurwitz, means that $G_2(d)F_2^{-1}(d)$ is stably invertible and, hence, (13) represents a standard innovations representation of r (cf. Theorem 6.2-4). Let

$$p := \max\left\{\partial E(d), \partial B_2(d) - (h \wedge 0)\right\} \tag{7.5-16}$$

where $\wedge$ denotes the minimum, and $\partial E(d)$ the degree of $E(d)$. Let

$$\underline{E}(d) := d^p E^*(d) \qquad ; \qquad \underline{B}_2(d) := d^{p+(h\wedge 0)} B_2^*(d) \tag{7.5-17}$$

Proposition 7.5-1. *Let reference r be modeled as in (13) to (15). Then, under the same assumptions as in Theorem 1, the optimal feedforward input (12) is given by*

$$v^c(t) = \Gamma(d)G_2^{-1}(d)r(t+h) \tag{7.5-18}$$

where $\Gamma(d)$ and $L(d)$ are $m \times p$ polynomial matrices given by the unique solution of minimum degree w.r.t. $L(d)$, that is, $\partial L(d) < \partial \underline{E}(d)$, of the following bilateral Diophantine equation

$$\underline{E}(d)\Gamma(d) + L(d)F_2(d) = d^{h\vee 0}\underline{B}_2(d)\psi_y G_2(d) \tag{7.5-19}$$

where $\vee$ denotes the maximum.

Proof: First, from the assumptions on $\underline{E}(d)$ and $F_2(d)$, and Results C.2-1 and C.2-2 in Appendix C, it follows that (19) has a unique solution $(\Gamma(d), L(d))$ with $\partial L(d) < \partial \underline{E}(d)$. Next, taking into account (13), (12) becomes

$$v^c(t) = \mathcal{E}\left\{E^{-*}(d)B_2^*(d)\psi_y G_2(d)F_2^{-1}(d)n(t) \mid n^{t+h}\right\}$$

because, $G_2(d)F_2^{-1}(d)$ being stably invertible, $\sigma\left\{r^t\right\} = \sigma\left\{n^t\right\}$. We consider separately the case $h > 0$ and $h \leq 0$.

Assume $h > 0$. Then, $p = \max\{\partial E(d), \partial B_2(d)\}$ and $\underline{B}_2(d) = d^p B_2^*(d)$. Hence,

$$
\begin{aligned}
v^c(t) &= \mathcal{E}\left\{\underline{E}^{-1}(d)\underline{B}_2(d)\psi_y G_2(d)F_2^{-1}(d)d^h n(t+h) \mid n^{t+h}\right\} \\
&= \mathcal{E}\left\{\underline{E}^{-1}(d)\left[\underline{E}(d)\Gamma(d) + L(d)F_2(d)\right]F_2^{-1}(d)n(t+h) \mid n^{t+h}\right\} \\
&= \mathcal{E}\left\{\left[\Gamma(d)F_2^{-1}(d) + \underline{E}^{-1}(d)L(d)\right]n(t+h) \mid n^{t+h}\right\} \\
&= \Gamma(d)F_2^{-1}(d)n(t+h)
\end{aligned}
$$

where the last equality follows from (14a) and the degree constraint on $L(d)$. In fact, $\underline{E}^{-1}(d)L(d)$ turns out to be a strictly anticausal matrix (cf. 4.2-10).

Assume $h \leq 0$. Then $p = \max\{\partial E(d), \partial B_2(d) + |h|\}$ and $\underline{B}_2(d) = d^{p-|h|}B_2^*(d)$. Hence,

$$
\begin{aligned}
v^c(t) &= \mathcal{E}\left\{\underline{E}^{-1}(d)\underline{B}_2(d)d^{|h|}\psi_y G_2(d)F_2^{-1}(d)n(t) \mid n^{t+h}\right\} \\
&= \mathcal{E}\left\{\underline{E}^{-1}(d)\underline{B}_2(d)\psi_y G_2(d)F_2^{-1}(d)n(t+h) \mid n^{t+h}\right\} \\
&= \mathcal{E}\left\{\left[\Gamma(d)F_2^{-1}(d) + \underline{E}^{-1}(d)L(d)\right]n(t+h) \mid n^{t+h}\right\} \\
&= \Gamma(d)F_2^{-1}(d)n(t+h)
\end{aligned}
$$

where the last equality follows by the same argument used in the $h > 0$ case.

Example 7.5-2   Consider again Example 1 and assume a first-order AR model for the reference

$$(1 - fd)r(t) = n(t) \qquad |f| < 1$$

Then, solving (19), we get for (18)

$$v^c(t) = \hat{v}^c(t) + \frac{k^{-1}(1 - b^2)b^{1-h}f}{b - f}r(t+h)$$

Examples 1 and 2 indicate one of the advantages of the more general expression (12) over (18). In fact, when a stochastic model for the reference is not available, but h is positive and large enough, from (12), a tight approximation of the optimal feedforward input can still be obtained simply by replacing $v^c(t)$ with $\hat{v}^c(t)$. In the earlier two examples, for $\tilde{v}^c(t) = v^c(t) - \hat{v}^c(t)$, we find

$$\mathcal{E}\left\{[\tilde{v}^c(t)]^2\right\} = \frac{f^2(1 - b^2)\Psi_n}{k^2|b|^{2(h-1)}(b - f)^2(1 - f^2)}$$

which, with $|b| > 1$, decays exponentially as h increases.

Problem 7.5-1 (Tracking a Predictable Reference)   Assume that the reference be a *predictable* process, viz., its realizations satisfy the equation

$$W(d)r(t) = O_p \qquad \text{a.s.} \tag{7.5-20}$$

where $W(d)$ is a $p \times p$ polynomial matrix such that all roots of $\det W(d)$ are on the unit circle and simple. Show that the optimal feedforward input $v^c(t)$ in (11) is given by the FIR filter:

$$v^c(t) = \Gamma(d)r(t) \tag{7.5-21}$$

where $\Gamma(d)$ and $L(d)$ are $m \times p$ polynomial matrices given by the unique solution of minimum degree w.r.t. $\Gamma(d)$, that is, $\partial\Gamma(d) < \partial W(d)$, of the following bilateral Diophantine equation

$$\underline{E}(d)\Gamma(d) + L(d)W(d) = \underline{B}_2(d)\psi_y \qquad (7.5\text{-}22)$$

where $\underline{E}(d)$ and $\underline{B}_2$ are as in (16) and (17) with $h = 0$.

Problem 7.5-2   Consider again Example 5.8-2. Assume that reference r satisfies (20) with $W(d) = 1 - d + d^2$. By using the results of Problem 1, prove that the optimal feedforward input $v^c(t)$ in (11) is given by

$$v^c(t) = \frac{kb}{b^2 - b + 1}[(2b - 1)r(t) + b(b - 2)r(t - 1)]$$

Problem 7.5-3 (1-DOF Steady-State LQS Tracking)   Consider again the steady-state LQS tracking with the admissible control strategy (3) replaced by

$$u(t) \in \sigma\left\{\varepsilon_y{}^t, u^{t-1}\right\}$$

where $\varepsilon_y(t) := y(t) - r(t)$. Assume that reference r is modeled as an ARMA process

$$F(d)r(t) = G(d)\nu(t)$$

with $F(d)$ and $G(d)$ both strictly Hurwitz, and ν satisfying (14). Show that the problem can be reformulated as a steady-state LQS pure regulation problem.

7.5.2   Use of Plant CARIMA Models

The direct use of the results so far obtained for the tracking and servo problems does not ensure asymptotic tracking of constant references and rejection of constant disturbances. In order to guarantee these properties, we can extend the approach of Sec. 5.8 to the present stochastic setting.

We assume that the CARMA plant (3-1) to (3-2) is affected by a constant disturbance n, $\Delta(d)n(t) = O_p$, and $\Delta(d) := \operatorname{diag}(1 - d)$, viz.,

$$A(d)y(t) = B(d)u(t) + C(d)e(t) + n(t) \qquad (7.5\text{-}23)$$

or, premultiplying by $\Delta(d)$,

$$\Delta(d)A(d)y(t) = B(d)\delta u(t) + \Delta(d)C(d)e(t) \qquad (7.5\text{-}24)$$

We see that, in the present stochastic case, the situation is complicated by the presence of factor $\Delta(d)$ in the innovations polynomial matrix. According to (3-60), such a presence indicates that in the generic case, some closed-loop eigenvalues of the steady-state LQS-regulated system equal one. More specifically, in such a case, the implicit Kalman filter embedded in the steady-state LQS regulator would exhibit undamped constant modes. To avoid such an undesirable situation, a heuristic approach consists of acting in designing the control law as if innovations e *were* a random walk $\Delta(d)e(t) = \nu(t)$ or

$$e(t) = e(t - 1) + \nu(t) \qquad (7.5\text{-}25)$$

with ν a zero-mean wide-sense stationary white process with a nonsingular covariance matrix. Note that (25) is unacceptable in that it yields a nonstationary process e with an ever-increasing covariance matrix. Nonetheless, plain substitution of (24) with the model

$$\Delta(d)A(d)y(t) = B(d)\delta u(t) + C(d)\nu(t) \qquad (7.5\text{-}26)$$

where ν has the properties stated before, leads us to recover acceptable closed-loop eigenvalues for the steady-state LQS-regulated system at the expense of a response degradation to the stochastic disturbances acting on the plant. The plant representation (26) is referred to as a *CARIMA* (controlled autoregressive integrated moving average) model. For control design of the plant (24), we use the model (26) along with the performance index

$$\mathcal{E}\left\{\|y(t) - r(t)\|^2_{\psi_y} + \|\delta u(t)\|^2_{\psi_u}\right\} \qquad (7.5\text{-}27)$$

and the admissible control strategy

$$\delta u(t) \in \sigma\left\{y^t, \delta u^{t-1}, r^{t+h}\right\} \qquad (7.5\text{-}28)$$

Hence, by Theorem 1, we find the corresponding steady-state LQS tracking and servo solutions.

Problem 7.5-4 Assume that the steady-state LQS regulator for the CARIMA model (26) yields an internally stable feedback system. Then, show that for any constant reference $r(t) \equiv r$, the 2-DOF steady-state LQS controller resulting from (26) to (28) applied to the plant (24) yields an offset-free closed-loop system and asymptotic rejection of constant disturbances, provided that $\dim y = \dim u$.

7.5.3   Dynamic Control Weight

We extend to the present stochastic case the considerations made at the end of Sec. 5.8 on the effects of filtered variables in the cost to be minimized. Specifically, instead of (27), we consider the cost

$$\mathcal{E}\left\{\|y_H(t) - r(t)\|^2_{\psi_y} + \|\delta u_H(t)\|^2_{\psi_u}\right\} \qquad (7.5\text{-}29)$$

where y_H and u_H are filtered versions of y and u, respectively:

$$y_H(t) = H(d)y(t) \qquad u_H(t) = H(d)u(t) \qquad (7.5\text{-}30)$$

We assume that $H(d)$ is a strictly Hurwitz polynomial and, for the sake of simplicity, the plant to be SISO. The model (26) can now be represented as

$$\Delta(d)A(d)y_H(t) = B(d)\delta u_H(t) + H(d)C(d)\nu(t) \qquad (7.5\text{-}31)$$

The 2-DOF steady-state LQS controller minimizing (29) for plant (31) is given by (7):

$$M_1(d)\delta u_H(t) = -N_1(d)y_H(t) + u^c(t)$$

Assuming $\Delta(d)A(d)$, $B(d)$, and $H(d)C(d)$ pairwise coprime, we find for the output of model (24) controlled according to the previous equation:

$$y(t) = \frac{B(d)}{H(d)} u^c(t) + \frac{X(d)}{\sigma_e E(d)} \frac{\Delta(d)}{H(d)} e(t) \qquad (7.5\text{-}32)$$

where $\sigma_e^2 := \mathcal{E}\left\{e^2(t)\right\}$, and $X(d)$ and $E(d)$ are the polynomials as in Sec. 3. Equation (32) shows that filtering $y(t)$ and $\delta u(t)$ as in (29) and (30) has the effect of filtering both the reference and innovations by $1/H(d)$. Notice, however, that the latter filtering action is only approximate in that the solution of (3-45) and (3-55), and hence polynomial $X(d)$, is implicitly affected by $H(d)$. According to such considerations, the use of a high-pass polynomial $H(d)$, such that $1/H(d)$ cuts off frequencies outside the desired closed-loop bandwidth, may turn out to be beneficial for both shaping the reference and attenuating high-frequency disturbances. Notice that in fact, if $e(t)$ is white, in (32), $\Delta(d)e(t)$ has most of its power at high frequencies.

Main Points of the Section. The optimal 2-DOF controller of Sec. 5.8 is extended to a steady-state LQ stochastic setting. The optimal feedforward action can be expressed in terms of the conditional expectation (8). This has the advantage of enabling us to approximately computing the optimal feedforward variable without using any reference stochastic model, provided that the reference is known a few steps in advance. CARIMA plant models whose inputs are plant input increments are often adopted in applications so as to asymptotically achieve tracking of constant references and rejection of constant disturbances. Dynamic cost weights in the performance index may be beneficial for both reference shaping and stochastic disturbance attenuation.

7.6 $\mathcal{H}_\infty$ AND LQ STOCHASTIC CONTROL

We have found that in steady-state LQ stochastic control, the characteristic polynomial of the optimally controlled system depends on the innovations polynomial $C(d)$. Obviously, if $C(d) = I_p$, the robust stability properties of LQ-regulated systems hold true because the results of Sec. 4.6 are still applicable. However, in general, stability robustness can deteriorate if an unfavorable $C(d)$ polynomial is used. Such a situation has been already met with plant (5-24). On that occasion, a reasonable heuristic approach was to design the "optimal" controller for a mismatched plant model where the $C(d)$ polynomial was suitably modified. A similar heuristic approach can be in general adopted so as to possibly recover the LQ robust stability properties: This is usually referred to as the *linear quadratic Gaussian/loop transfer recovery* technique ([AM90], [DS79], [DS81], [IT86b], [Kwa69], and [Mac85]).

A more systematic approach is to consider the following minimax regulation problem. Let the plant be represented by

$$y(t) = P(d)u(t) + Q(d)n(t) \tag{7.6-1}$$

where n is a p-dimensional zero-mean white disturbance with identity covariance matrix, and $P(d)$ and $Q(d)$ are rational transfer matrices such that $P(0) = O_{p \times m}$. We note that

$$\mathcal{E}\left\{\|Q(d)n(t)\|^2\right\} = \mathrm{Tr}\langle Q(d)\Psi_n(d)Q^*(d)\rangle \tag{7.6-2}$$
$$= \mathrm{Tr}\langle Q^*(d)Q(d)\rangle = \|Q(d)\|^2$$

where $\|Q(d)\|$ denotes the norm introduced in (3.1-12). Here $\|Q(d)\|^2$ equals the *power* of the disturbance $Q(d)n(t)$, that is, the sum of the variances of its components. The *minimax LQ stochastic linear regulation* problem consists of finding linear compensators:

$$u(t) = -K(d)y(t) \tag{7.6-3}$$

minimizing cost (3-19) for the worst possible disturbance of bounded power, viz.,

$$\inf_{K(d)} \sup_{\|Q(d)\| \le 1} \mathcal{E}\left\{\|y(t)\|^2_{\psi_y} + \|u(t)\|^2_{\psi_u}\right\} \tag{7.6-4}$$

In closed loop, we find

$$\left[\begin{array}{c} y(t) \\ u(t) \end{array}\right] = \left[\begin{array}{c} S(d) \\ -T(d) \end{array}\right] Q(d)n(t) \tag{7.6-5}$$

where

$$S(d) := [I_p + P(d)K(d)]^{-1} \qquad \text{and} \qquad T(d) := K(d)S(d) \tag{7.6-6}$$

are the sensitivity matrix and the power transfer matrix, respectively, of the feedback system. Then, if $\mathcal{C}$ denotes the expectation in (4), we have

$$\mathcal{C} = \mathrm{Tr}\langle Q^*(d)\mathcal{S}^*(d)\mathcal{S}(d)Q(d)\rangle \tag{7.6-7}$$
$$= \|\mathcal{S}(d)Q(d)\|^2$$

where $\mathcal{S}(d)$ is the mixed sensitivity matrix:

$$\mathcal{S}(d) := \left[\begin{array}{c} \varphi_y S(d) \\ \varphi_u T(d) \end{array}\right] \tag{7.6-8}$$

with

$$\varphi_y' \varphi_y = \psi_y \qquad \text{and} \qquad \varphi_u' \varphi_u = \psi_u$$

Thus, (4) becomes

$$\inf_{K(d)} \sup_{\|Q(d)\|\leq 1} \|\mathcal{S}(d)Q(d)\|^2 = \inf_{K(d)} \|\mathcal{S}(d)\|_\infty^2 \qquad (7.6\text{-}9)$$

where

$$\|\mathcal{S}(d)\|_\infty := \operatorname*{ess\,sup}_{0\leq\theta<2\pi} \bar{\sigma}\left[\mathcal{S}\left(e^{j\theta}\right)\right] \qquad (7.6\text{-}10)$$

is the so-called *H-infinity* ($\mathcal{H}_\infty$) *norm*[1] of $\mathcal{S}(d)$. The equality in (9) can be proved as in [DV75]. We see that the minimax LQ stochastic linear regulation problem amounts to finding compensator transfer matrices $K(d)$ minimizing the $\mathcal{H}_\infty$–norm of $\mathcal{S}(d)$, viz., the value of the frequency peak of the maximum singular value of $\mathcal{S}(e^{j\theta})$, $\theta \in [0, 2\pi)$.

A discussion on how to solve (1) to (10) would lead us too much afield, our main interest being in indicating that robust stability can be systematically obtained by the steady-state LQ stochastic linear regulator for the worst possible disturbance case, or, equivalently, for the worst possible dynamic weights in the cost to be minimized. To this end, we next show that (1) to (10) are equivalent to a deterministic minimax regulation problem that is used [Fra91] for systematically designing robust compensators. This regulation problem is called the $\mathcal{H}_\infty$ *mixed-sensitivity* regulation problem. Here the plant is given by

$$y(t) = P(d)u(t) + \gamma(t) \qquad (7.6\text{-}11)$$

where, in contrast with (1), the disturbance is represented by a vector-valued causal sequence γ of finite energy:

$$\|\gamma(\cdot)\|^2 = \sum_{k=0}^{\infty} \gamma(k)\gamma'(k) < \infty$$

The $\mathcal{H}_\infty$ mixed-sensitivity regulation problem ([Kwa85] and [VJ84]), is to find linear compensators (3) minimizing the cost

$$J = \sum_{k=0}^{\infty} \left[\|y(k)\|_{\psi_y}^2 + \|u(k)\|_{\psi_u}^2 \right] \qquad (7.6\text{-}12)$$

for the worst possible deterministic disturbance of bounded energy, viz.,

$$\inf_{K(d)} \sup_{\|\gamma(\cdot)\|\leq 1} J \qquad (7.6\text{-}13)$$

By the result in [DV75], this amounts again to finding $K(d)$ so as to minimize the $\mathcal{H}_\infty$–norm of $\mathcal{S}(d)$.

We state these results formally in the following theorem.

---

[1]$\mathcal{H}_\infty$ denotes the Hardy space consisting of all matrix-valued functions on $\mathbb{C}$ that are analytic and bounded in the open unit circle [Fra87].

Theorem 7.6-1.   *The minimax LQ stochastic linear regulation problem is equivalent to the $\mathcal{H}_\infty$ mixed-sensitivity regulation problem.*

$\mathcal{H}_\infty$ optimal sensitivity problems were ushered in control engineering by [Zam81] in the early eighties in order to cope systematically with the robust stability problem. The connection established in Theorem 1 is important in that it suggests that robust stability can be achieved in LQ stochastic control by suitably dynamically weighting the variables in the cost to be minimized. Indeed, if we consider the SISO CARMA plant

$$A(d)y(t) = B(d)u(t) + A(d)e(t) \qquad (7.6\text{-}14)$$

and the cost

$$\mathcal{C} = \mathcal{E}\left\{\psi_y y_f^2(t) + \psi_u u_f^2(t)\right\} \qquad (7.6\text{-}15)$$

where

$$y_f(t) := Q(d)y(t) \qquad \text{and} \qquad u_f(t) := Q(d)u(t)$$

with $Q(d)$ a stable and stably invertible transfer matrix, we see that the compensator

$$u(t) = -K(d)y(t)$$

solving, for the given CARMA plant, the minimax problem

$$\inf_{K(d)} \sup_{\|Q(d)\| \leq 1} \mathcal{C}$$

coincides with the $\mathcal{H}_\infty$ mixed-sensitivity compensator for the deterministic plant $A(d)y(t) = B(d)u(t)$.

Problem 7.6-1   Consider the minimax LQ stochastic linear regulation problem when $\mathcal{C}$ is as in (3-64). Discuss this dynamic weighted version of the problem by introducing in $\mathcal{C}$ filtered variables $y_f(t) = W_y(d)y(t)$ and $u_f(t) = W_u(d)u(t)$.

Main Points of the Section.   The $\mathcal{H}_\infty$ mixed-sensitivity compensators coincide with the ones solving the steady-state LQ stochastic linear regulation problem for the worst possible disturbance case, or, equivalently, for the worst possible dynamic weights in the cost to be minimized.

7.7   PREDICTIVE CONTROL OF CARMA PLANTS

7.7.1   Stochastic SIORHR

We wish to extend stabilizing I/O receding horizon regulation (SIORHR) to SISO CARMA plants. SIORHR was introduced and discussed in Chapter 5 within a deterministic setting. Here we assume that the plant to be regulated is represented by a SISO CARMA model:

$$A(d)y(t) = \mathcal{B}(d)u(t) + C(d)e(t) \qquad (7.7\text{-}1)$$

where $A(0) = C(0) = 1$ and

- $[1/A(d)] \begin{bmatrix} \mathcal{B}(d) & C(d) \end{bmatrix}$ is an irreducible transfer matrix $\qquad (7.7\text{-}2a)$

- $C(d)$ is strictly Hurwitz $\qquad (7.7\text{-}2b)$

- the gcd of $A(d)$ and $\mathcal{B}(d)$ is strictly Hurwitz $\qquad (7.7\text{-}2c)$

Except for the *strictly* Hurwitz condition of $C(d)$, these assumptions are the same as in (3-2). The strictly Hurwitz condition of $C(d)$ is here adopted in that it simplifies SIORHR synthesis. Similarly to (3-3), we also assume that the innovations process e satisfies the following martingale difference properties:

$$\mathcal{E}\left\{e(t+1) \mid e^t\right\} = 0 \qquad \text{a.s.} \qquad (7.7\text{-}3a)$$

$$\mathcal{E}\left\{e^2(t+1) \mid e^t\right\} = \sigma_e^2 > 0 \qquad \text{a.s.} \qquad (7.7\text{-}3b)$$

Finally, we assume that

$$\operatorname{ord}\mathcal{B}(d) = 1 + \ell \qquad (7.7\text{-}4a)$$

viz., the plant exhibits a dead time $\ell \in \mathbb{Z}_+$ in addition to the intrinsic one. Consequently (cf. (5.4-22)),

$$\mathcal{B}(d) = d^\ell B(d) \qquad (7.7\text{-}4b)$$

with $\mathcal{B}(d)$ as in (5.4-22).

It is convenient to introduce the filtered I/O variables:

$$\gamma(t) := \frac{1}{C(d)}y(t) \qquad \nu(t) := \frac{1}{C(d)}u(t) \qquad (7.7\text{-}5)$$

so as to represent (1) by the CAR model:

$$A(d)\gamma(t) = \mathcal{B}(d)\nu(t) + e(t) \qquad (7.7\text{-}6)$$

We are now ready to formally state the SIORHR problem for CARMA plants.

Stochastic SIORHR. Consider the CARMA plant (1) to (4) under the assumption that for all negative time steps, plant inputs $u(k)$ have been measurable w.r.t. the σ-field generated by $\{\gamma^k, \nu^{k-1}\}$:

$$u(k) \in \sigma\left\{\gamma^k, \nu^{k-1}\right\} \qquad (7.7\text{-}7)$$

with a.s. bounded γ^0, and ν^{-1}. Consider next the problem of finding, whenever it exists, an "open-loop" input sequence:

$$u(k) = f\left(k, \gamma^0, \nu^{-1}\right) \in \sigma\left\{\gamma^0, \nu^{-1}\right\}, \qquad k = 0, \cdots, T-1 \qquad (7.7\text{-}8)$$

minimizing the conditional expectation

$$\frac{1}{T} \sum_{k=0}^{T-1} \mathcal{E} \left\{ \psi_y y^2(k+\ell) + \psi_u u^2(k) \mid \gamma^0, \nu^{-1} \right\} \tag{7.7-9}$$

under the constraints

$$u_{T+n-2}^T = O_{n-1} \qquad \mathcal{E} \left\{ y_{T+\ell+n-1}^{T+\ell} \mid \gamma^0, \nu^{-1} \right\} = O_n \quad \text{a.s.} \tag{7.7-10}$$

Then, the feedback compensator

$$u(t) = f\left(0, \gamma^t, \nu^{t-1}\right) \tag{7.7-11}$$

is referred to as the stochastic SIORHR with prediction horizon T.

We next study how to solve the previous problem along similar lines as in Sec. 5.5. It is known (cf. Example 5.4-1) that (6) can be represented in state-space form by introducing state vector

$$s(t) := \left[\left(\gamma_t^{t-n_a+1} \right)' \quad \left(\nu_{t-1}^{t-\ell-n_b+1} \right)' \right]' \in \mathbb{R}^{n_a+\ell+n_b-1}$$

where $n_a = \partial A(d)$, and $n_b = \partial B(d)$, or $\ell + n_b = \partial B(d)$. In fact, we have

$$
\begin{aligned}
s(t+1) &= \Phi s(t) + G\nu(t) + Le(t+1) \\
\gamma(t) &= Hs(t)
\end{aligned}
$$

with (Φ, G, H) given similarly to (5.4-3) to (5.4-5), $L = \underline{e}_{n_a}$, and $\underline{e}_{n_a}$ is the n_ath vector of the natural basis of $\mathbb{R}^{n_a+\ell+n_b-1}$. The aim is now to construct a state-space representation for the initial CARMA plant (1). Note that, by (5),

$$
\begin{aligned}
\nu(t) &= u(t) - c_1\nu(t-1) - \cdots - c_{n_c}\nu(t-n_c) \\
&= u(t) - \left[c_{n_c} \quad \cdots \quad c_1 \right] \nu_{t-1}^{t-n_c} \tag{7.7-12}
\end{aligned}
$$

and

$$
\begin{aligned}
y(t) &= \gamma(t) + c_1\gamma(t-1) + \cdots + c_{n_c}\gamma(t-n_c) \\
&= \left[c_{n_c} \quad \cdots \quad c_1 \quad 1 \right] \gamma_t^{t-n_c} \tag{7.7-13}
\end{aligned}
$$

if

$$C(d) = 1 + c_1 d + \cdots + c_{n_c} d^{n_c}$$

Extend the previous state $s(t)$ as follows:

$$s_c(t) := \left[\left(\gamma_t^{t-n_\gamma} \right)' \quad \left(\nu_{t-1}^{t-n_\nu} \right)' \right]' \in \mathbb{R}^{n_\gamma+n_\nu+1} \tag{7.7-14a}$$

$$n_\gamma := \max\left(n_a - 1, n_c\right) \qquad n_\nu := \max\left(\ell + n_b - 1, n_c\right) \tag{7.7-14b}$$

Problem 7.7-1 Verify that if the variables γ and ν are related by (6), the following state-space representation holds for vector $s_c(t)$ in (14):

$$
\begin{aligned}
s_c(t+1) &= \Phi s_c(t) + G\nu(t) + Le(t+1) \\
\gamma(t) &= H s_c(t)
\end{aligned}
\tag{7.7-15}
$$

where

$$
\Phi = \begin{bmatrix}
O_{n_\gamma \times 1} & I_{n_\gamma} & O_{n_\gamma \times n_\nu} \\
-a_{n_\gamma+1} \quad \cdots \quad -a_1 & -\beta_{n_\nu+1} \quad \cdots \quad -\beta_2 \\
O_{(n_\nu-1)\times(n_\gamma+2)} & I_{n_\nu-1} \\
0 \quad \cdots & \cdots & \cdots \quad 0
\end{bmatrix}
$$

$$
G = b_1 \underline{e}_{n_\gamma+1} + \underline{e}_{n_\gamma+n_\nu+1} \qquad L = \underline{e}_{n_\gamma+1} \qquad H = \underline{e}'_{n_\gamma+1}
$$

and $a_{n_a+i} = \beta_{\ell+n_b+i} = 0$, $i = 1, 2, \cdots$, and $\underline{e}_i$ denotes the ith vector of the natural basis of $\mathbb{R}^{n_\gamma+n_\nu+1}$.

Write (12) as

$$
\nu(t) = u(t) - F_c s_c(t) \qquad F_c := \begin{bmatrix} O_{1\times(n_\gamma+1)} & c_{n_\nu} & \cdots & c_1 \end{bmatrix}
\tag{7.7-16a}
$$

and (13) as

$$
y(t) = H_c s_c(t) \qquad H_c := \begin{bmatrix} c_{n_\gamma} & \cdots & c_1 & 1 & O_{1\times n_\nu} \end{bmatrix}
\tag{7.7-16b}
$$

to get

$$
\left.
\begin{aligned}
s_c(t+1) &= \Phi_c s_c(t) + Gu(t) + Le(t+1) \\
y(t) &= H_c s_c(t)
\end{aligned}
\right\}
\tag{7.7-16c}
$$

$$
\Phi_c := \Phi - GF_c
\tag{7.7-16d}
$$

This is a state-space representation for the initial plant description (1). We have

$$
\begin{aligned}
y(k+\ell) &= w_{\ell+1}u(k-1) + \cdots + w_{\ell+k}u(0) + S_{\ell+k}s_c(0) \\
&\quad + e(k+\ell) + \cdots + g_{k+\ell-1}e(1)
\end{aligned}
$$

where

$$
w_k := H_c \Phi_c^{k-1} G
$$

is the kth sample of the impulse response associated with $\mathcal{B}(d)/A(d)$:

$$
S_k := H_c \Phi_c^k
$$

and

$$
g_k := H_c \Phi_c^k L
$$

Now

$$
\begin{aligned}
\hat{y}(k+\ell) &:= \mathcal{E}\left\{ y(k+\ell) \mid \gamma^0, \nu^{-1} \right\} \\
&= w_{\ell+1}u(k-1) + \cdots + w_{\ell+k}u(0) + S_{\ell+k}s_c(0)
\end{aligned}
$$

Further, because by (7), for $k \in \mathbb{Z}_+$, $\sigma\{\gamma^0, \nu^{-1}\} \subset \sigma\{e^k\}$, for $\tilde{y}(k) := y(k) - \hat{y}(k)$, the conditional expectations

$$\mathcal{E}\left\{\tilde{y}^2(k+\ell) \mid \gamma^0, \nu^{-1}\right\} = \mathcal{E}\left\{\mathcal{E}\left\{\tilde{y}^2(k+\ell) \mid e^{k+\ell-1}\right\} \mid \gamma^0, \nu^{-1}\right\}$$

are by (3) a.s. constant for $k \geq 1$. Hence, the conclusion is that the optimal sequence $u_{[0,T)}^0$ for the stochastic problem (7) to (11) is the same as if $e(t+1) \equiv 0$ in (16c). In particular, provided that $n = \hat{n}$, $\hat{n}$ being the McMillan degree of $B(d)/A(d)$, the SIORHR solution is given by (cf. (5.5-21))

$$u_{T-1}^0 = -M^{-1}\left[\psi_y\left(I_T - QLM^{-1}\right)W_1'\Gamma_1 + Q\Gamma_2\right]s_c(0) \qquad (7.7\text{-}17)$$

where M, Q, L, W_1, Γ_1, and Γ_2 are now related to the system (16). Consequently, the SIORHR law equals

$$u(t) = -\underline{e}_1' M^{-1}\left[\psi_y\left(I_T - QLM^{-1}\right)W_1'\Gamma_1 + Q\Gamma_2\right]s_c(t) \qquad (7.7\text{-}18)$$

The next theorem states the stabilizing properties of (18).

Theorem 7.7-1. *Let the CARMA plant (1) satisfy (2) to (4). Then, provided that $\psi_u > 0$, the SIORHR law (18) stabilizes the CARMA plant whenever*

$$T \geq n = \hat{n} \qquad (7.7\text{-}19)$$

where $\hat{n}$ is the McMillan degree of $B(d)/A(d)$. Further, for

$$T = n = \hat{n} \qquad (7.7\text{-}20)$$

(18) yields a closed-loop system whose observable-reachable part is state-deadbeat.

Proof: First, note that, because of (2), $\Sigma := (\Phi_c, G, H_c)$ is stabilizable and detectable. Next, by recalling the definitions of Γ_1 and Γ_2, it is seen that (18) is not affected by the unobservable states. As a consequence, the unobservable eigenvalues of Σ, which are stable, are left unchanged by the feedback action. Let Σ_0 be the observable subsystem resulting from any GK canonical observability decomposition of Σ. Next, let us consider any GK canonical reachability decomposition of Σ_0:

$$\hat{\Phi} = \begin{bmatrix} \Phi_r & \Phi_{r\bar{r}} \\ 0 & \Phi_{\bar{r}} \end{bmatrix} \qquad \hat{G} = \begin{bmatrix} G_r \\ 0 \end{bmatrix} \qquad \hat{H} = \begin{bmatrix} H_r & H_{\bar{r}} \end{bmatrix} \qquad (7.7\text{-}21)$$

with states $\hat{x} = \begin{bmatrix} x_r' & x_{\bar{r}}' \end{bmatrix}'$ and $\dim \Phi_r = \hat{n}$, the McMillan degree of $B(d)/A(d)$. The regulation law (18) can be written as $u(t) = Fs_c(t) = F_r x_r(t) + F_{\bar{r}} x_{\bar{r}}(t)$. With $\Phi_{\bar{r}}$ stable, the closed-loop system is stable if and only if $\Phi_r + G_r F_r$ is a stability matrix. To prove this, suppose temporarily that $x_{\bar{r}}(0) = 0$. In such a case, for all $k \geq 0$, $x_r(k+1) = \Phi_r x_r(k) + G_r u(k)$ and $y(k) = H_r x_r(k)$. Thus, by virtue of Theorem 5.3-2, $\Phi_r + G_r F_r$ is a stability matrix. That, under (20), the observable-reachable part of the closed-loop system exhibits the state-deadbeat property follows by the previous arguments and Theorem 5.3-2.

7.7.2 Stochastic SIORHC

We extend SIORHC to SISO CARIMA plants. SIORHC was introduced and treated in Sec. 5.8 within a deterministic setting. For a discussion on the motivations for considering CARIMA plant models, the reader is referred to Sec. 5. The plant to be controlled is therefore represented by the following CARIMA model (cf. (5-26)):

$$\Delta(d)A(d)y(t) = B(d)\delta u(t) + C(d)e(t) \tag{7.7-22}$$

where $\Delta(d) := 1 - d$, and (2) to (4) hold true. We also consider a reference sequence $r(\cdot)$ that is assumed to be known by the controller at time t up to time $t + \ell + T$. We wish to address the following 2-DOF servo problem.

Stochastic SIORHC. Consider the CARIMA plant (22). Let (2) to (4) and $\delta u(k) \in \sigma\left\{\gamma^k, \delta\nu^{k-1}\right\}$ hold true and the reference be known $\ell + T$ steps in advance. Find, whenever they exist, input increments $\widehat{\delta u}_{t+T}^{t} \in \sigma\left\{\gamma^t, \delta\nu^{t-1}\right\}$ minimizing the conditional expectation

$$\frac{1}{T} \sum_{k=t}^{t+T-1} \mathcal{E}\left\{\psi_y\varepsilon_y^2(k+\ell) + \psi_u\delta u^2(k) \mid \gamma^t, \delta\nu^{t-1}\right\} \tag{7.7-23a}$$

$$\varepsilon_y(k) := y(k) - r(k) \tag{7.7-23b}$$

under the constraints

$$\delta u_{t+T+n-2}^{t+T} = O_{n-1} \qquad \mathcal{E}\left\{y_{t+\ell+T+n-1}^{t+\ell+T} \mid \gamma^t, \delta\nu^{t-1}\right\} = \underline{r}(t+\ell+T) \qquad \text{a.s.} \tag{7.7-24}$$

with

$$\underline{r}(k) := \left[\begin{array}{ccc} r(k) & \cdots & r(k) \end{array}\right]' \in \mathbb{R}^n$$

Then, the plant increment at time t given by SIORHC equals

$$\delta u(t) = \widehat{\delta u}(t) \tag{7.7-25}$$

It is straightforward to find for the solution of (23) and (24)

$$\begin{aligned} \delta u_{t+T-1}^{t} &= -M^{-1}\Big\{\psi_y\left(I_T - QLM^{-1}\right)W_1'\left[\Gamma_1 s_c(t) - r_{t+\ell+T-1}^{t+\ell+1}\right] \\ &\quad + Q\Big[\Gamma_2 s_c(t) - \underline{r}(t+\ell+T)\Big]\Big\} \end{aligned} \tag{7.7-26}$$

provided that $n \leq \hat{n}$, $\hat{n}$ being here the McMillan degree of $B(d)/\Delta(d)A(d)$. In (26), $s_c(t)$ is the same as in (14) except for the replacement of n_a by $n_a + 1$, and $\nu_{t-1}^{t-n_\nu}$ by $\delta\nu_{t-1}^{t-n_\nu}$. Further, as in (14), all matrices are referred to the system (16).

Theorem 7.7-2.   *Under the same assumptions as in Theorem 1 with $A(d)$ replaced by $\Delta(d)A(d)$ and*

$$\mathcal{B}(1) = B(1) \neq 0 \tag{7.7-27}$$

the SIORHC law

$$
\begin{aligned}
\delta u(t) \;=\; & -\underline{e}_1' M^{-1} \Big\{ \psi_y \left(I_T - QLM^{-1} \right) W_1' \left[\Gamma_1 s_c(t) - r_{t+\ell+T-1}^{t+\ell+1} \right] \\
& + Q \Big[\Gamma_2 s_c(t) - \underline{r}(t+\ell+T) \Big] \Big\}
\end{aligned}
\tag{7.7-28}
$$

inherits all the stabilizing properties of stochastic SIORHR, whenever

$$T \geq n = \hat{n} \tag{7.7-29}$$

where $\hat{n}$ is the McMillan degree of $B(d)/[\Delta(d)A(d)]$. Further, whenever stabilizing, SIORHC yields, thanks to its integral action, asymptotic rejection of constant disturbances, and an offset–free closed–loop system.

Proof: The stabilizing properties of (28) follow directly from Theorem 1. Asymptotic rejection of constant disturbances is a consequence of the presence of the integral action in the loop. Finally offset-free behavior is proved as follows. First, rewrite (28) in polynomial form as $\delta u(t) = -R_1(d)\delta v(t-1) - S(d)\gamma(t) + Z(d)r(t+\ell+T)$ or, after straightforward manipulations, as $R(d)\delta u(t) = -S(d)y(t) + C(d)Z(d)r(t+\ell+T)$ with $R(d) := C(d) + R_1(d)$. Hence, if $r(t) \equiv r$, we have $\lim_{t\to\infty} \delta u(t) = 0$ and $y_\infty := \lim_{t\to\infty} y(t) = C(1)Z(1)S^{-1}(1)r$. That $S(1) = C(1)Z(1)$, and hence the closed-loop system has unit dc gain, can be shown along the same lines as in (5.8-40) to (5.8-44) by replacing 1 by $C(d)$ in the L.H.S. of (5.8-43).

We conclude this section by pointing out that the extension of SIORHC to the stochastic case can be carried out by using formally the same equations as in deterministic case, provided that state (5.8-32) be replaced with the C-filtered state:

$$s_c(t) := \left[\; \left(\gamma_t^{t-n_\gamma} \right)' \quad \left(\delta v_{t-1}^{t-n_\nu} \right)' \; \right]' \tag{7.7-30}$$

$$C(d)\gamma(t) = y(t) \qquad C(d)\delta v(t) = \delta u(t)$$

$$n_\gamma = \max\left(n_a, n_c\right) \qquad n_\nu = \max\left(\ell + n_b - 1, n_c\right)$$

and the matrices in (28) be referred to the system (16).

Problem 7.7-2 (GPC for CARIMA Plants)   Consider GPC as in Sec. 5-8. Formulate a 2-DOF GPC servo problem for a CARIMA plant. Find the related GPC law. Compare this result with (5.8-52).

Problem 7.7-3 (Stochastic SIORHC Information Pattern)   Consider the SIORHC law (28) for the CARIMA plant (22). Show that it can be written in polynomial form as follows:

$$R(d)\delta u(t) = -S(d)y(t) + C(d)v(t)$$

$$v(t) := Z(d)r(t+\ell+T)$$

Compute the maximum values of the degrees of polynomials $R(d)$, $S(d)$, and $Z(d)$ in terms of $\partial A(d)$, $\partial B(d)$, $\partial C(d)$, T, and ℓ.

It is interesting to point out the strict connection that exists between the LQ stochastic servo of Sec. 5 and SIORHC and GPC predictive controllers. In fact, whenever stabilizing, the latter approximate, as $T \to \infty$ for SIORHC and $N_1 = 0$ and $N_u \to \infty$ for GPC, the LQ stochastic servo behavior. This can be concluded by comparing the performance indices that the previous controllers minimize. The foregoing connection makes it possible to extend to predictive control the considerations made at the end of Sec. 5 on the benefits that can be acquired by costing suitable filtered I/O variables.

Problem 7.7-4 (Stochastic SIORHC and Dynamic Weights)   For the CARIMA plant (22), consider the problem of finding input increments that minimize the conditional expectation

$$\frac{1}{T} \sum_{k=t}^{t+T-1} \mathcal{E}\left\{ [W_y(d)y(k+\ell) - r(k)]^2 + [W_u(d)\delta u(k)]^2 \mid \gamma^t, \delta\nu^{t-1} \right\}$$

under the terminal constraints

$$W_u(d)\delta u(k) = 0 \qquad\qquad\qquad k = t+T, \cdots, t+T+\hat{n}-2$$
$$\mathcal{E}\left\{ W_y(d)y(k+\ell) \mid \gamma^t, \delta\nu^{t-1} \right\} = \underline{r}(t+\ell+T) \qquad k = t+T, \cdots, t+T+\hat{n}-1$$

with $\hat{n}$ a positive integer, $W_u(d) = B_u(d)/A_u(d)$, and $W_y(d) = B_y(d)/A_y(d)$, where $B_u(d)$, $A_u(d)$, $B_y(d)$, and $A_y(d)$ are strictly Hurwitz polynomials. Show that this problem reduces to the standard problem (23) to (25) once u and y are changed into

$$\delta u_f(t) := W_u(d)\delta u(t) \qquad\qquad y_f(t) := W_y(d)y(t)$$

and the plant (22) is replaced by

$$A(d)\Delta(d)A_y(d)B_u(d)y_f(t) = \mathcal{B}(d)A_u(d)B_y(d)\delta u_f(t) + C(d)B_y(d)B_u(d)e(t)$$

From a more practical point of view, however, there are significant differences between predictive controllers like SIORHC and GPC and steady-state LQ stochastic control. In fact, although predictive controllers of the receding-horizon type are amenable to be extended to nonlinear plants or to embody constraints on state or I/O variables, such requirements cannot be accommodated with acceptable computational load within steady-state LQ stochastic control.

Main Points of the Section.   Predictive controllers, like SIORHC and GPC, can be extended to CARIMA plants with no formal changes into the design equations, by simply modifying the plant representation as in (16) and filtering the I/O variables to be fed back by the inverse of the $C(d)$ innovations polynomial.

NOTES AND REFERENCES

LQ stochastic control is a topic widely and thoroughly discussed in standard textbooks. Besides the ones referenced in Chapter 2, see also [Åst70], [ÅW84], [Cai88], [DV85], [FR75], [GJ88], [GS84], [May79], [May82a], [May82b], and [MG90].

The certainty-equivalence principle first appeared in the economics literature [Sim56]. A rigorous proof of the separation principle for continuous-time LQG regulation was first given in [Won68].

The minimum-variance regulator was studied in [Åst70] and [Pet70], the latter in an adaptive setting. The generalized minimum-variance adaptive regulator was first presented in [CG75] and analyzed in [CG79]. Steady-state LQ stochastic regulation for $\psi_u = 0$, or stabilizing minimum-variance regulation, was first addressed and solved by [Pet72]. See also [SK86] and [PK92]. Steady-state LQ stochastic linear regulation was discussed in the monograph [Kuč79]. For an extension to more general system configurations, see [CM91], [HKŠ92], and [HŠK91]. The material showing optimality of the steady-state LQ stochastic linear regulator among possibly nonlinear regulators appears to be new. The monotonic properties of steady-state LQ stochastic regulation were discussed in [MLMN92].

The approach to LQ stochastic tracking and servo discussed in Sec. 5 first appeared in [MZ89b]. See also [Gri90] and [MG92]. For an extension of this approach to an $\mathcal{H}_\infty$ setting, see [MCG90]. Unlike other relevant contributions, in [MCG90] and [MZ89b], the future of the reference realizations is used in the controller. For SISO plants, a similar idea was adopted in [Sam82] though in a state-space representation setting.

$\mathcal{H}_\infty$ control theory was ushered by [Zam81]. For a general overview, see monographs [FFH+91] and [Fra87]. For an alternative approach, see [CD89] and [LPVD83].

Section 7 on predictive control of CARMA plants improves on [CM92a].

PART
3

ADAPTIVE CONTROL

CHAPTER 8

SINGLE-STEP-AHEAD SELF-TUNING CONTROL

In this chapter, we remove the assumption, which has been used so far, according to which a dynamical model of the plant is available for control design. We then combine recursive identification and optimal control methods to build adaptive control systems for unknown linear plants. Under some conditions, such systems behave asymptotically in an optimal way as if the control synthesis is made by using the true plant model.

In Sec. 1, we briefly discuss various control approaches for uncertain plants and describe the two basic groups of adaptive controllers, viz., model-reference adaptive controllers and self-tuning controllers. Section 2 points out the difficulties encountered in formulating adaptive control as on optimal stochastic control problem, and, in contrast, the possibility of adopting a simple suboptimal procedure by enforcing the certainty-equivalence principle. Section 3 presents some analytic tools for establishing global convergence of deterministic self-tuning control systems. In Section 4, we discuss the deterministic properties of the RLS identification algorithm that typically are not subject to persistency of excitation and, hence, applicable in the analysis of self-tuning systems. In Sec. 5, these RLS properties are used so as to construct a self-tuning control system based on the cheap control law for which global convergence can be established in a deterministic setting. Section 6 discusses a constant-trace RLS identification algorithm with data normalization and extends the global convergence result of Sec. 5 to a self-tuning cheap control system based on such an estimator. The finite memory length of the latter is important for time-varying plants. Self-tuning minimum-variance control is discussed in Sec. 7, where it is pointed out that implicit modeling of CARMA plants under minimum-variance control can be exploited so as to construct self-tuning minimum-variance control algorithms whose global convergence can be proved via the stochastic Lyapunov equation method. Section 8 shows that generalized minimum-variance control is equivalent to minimum-variance control of a modified plant, and, hence, globally convergent self-tuning algorithms based on the former control law can be developed by exploiting the previous equivalence and the results in Sec. 7. Section 9 ends the chapter by describing how to robustify self-tuning cheap control to counteract the presence of neglected dynamics.

We point out that all the results of this chapter pertain to single-step-ahead, or myopic, adaptive control. For this reason, applicability of these results is severely limited by the requirements that the plant be minimum-phase and its I/O delay exactly known. Nonetheless, the study of these adaptive controllers is important in that it introduces at a quite basic level ideas that, as will be shown in the next chapter, can be effectively used to develop adaptive multistep predictive controllers with wider application potential.

8.1 CONTROL OF UNCERTAIN PLANTS

In the remaining part of this book, we shall study how to use control and identification methods for controlling *uncertain plants*, viz., plants described by models whose structure and parameters are not all a priori known to the designer. This is a situation virtually always met in practice. It is then of paramount interest to approach this issue by using the tools introduced in the previous chapters where, apart from a few exceptions, we have permanently assumed that an exact plant representation—either deterministic or stochastic—is a priori available.

In practice, we meet many different situations that can be referred to under the "uncertain plant" heading. If it is known that the plant behaves approximately like a given nominal model, *robust control methods* can be used to design suitable feedback compensators (cf. Secs. 3.3, 4.6, and 7.6). In other cases, the plant may exhibit significant variations but auxiliary variables can be measured, yielding information on the plant dynamics. Then, the parameters of a feedback compensator can be changed according to the values taken on by the auxiliary variables. Whenever these variables provide no feedback from the actual performance of the closed-loop system that can compensate for an incorrect parameter setting, the approach is called *gain scheduling*. This name can be traced back to the early use of the method finalized to compensate for the changes in plant gain. Gain scheduling is used in flight-control systems where the Mach number and the dynamic pressure are measured and used as scheduling variables.

Adaptive control mainly pertains to uncertain plants that can be modelled as dynamic systems with some unknown constant, or slowly time-varying, parameters. Adaptive controllers are traditionally grouped into the two separate classes described hereafter.

Model-Reference Adaptive Controllers (MRACs).   In an MRAC system (Fig. 1), the specifications are given in terms of a reference model that indicates

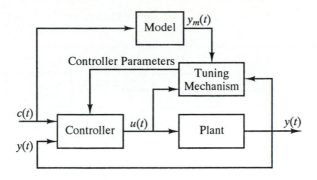

Figure 8.1-1: Block diagram of an MRAC system.

how the plant output should respond ideally to the command signal $c(t)$. The overall control system can be conceived as if it consists of two loops: an inner loop, the ordinary control system, composed of the plant and the controller; and an outer loop that comprises the parameter-adjustment, or tuning, mechanism. The controller parameters are adjusted by the outer loop so as to make plant output $y(t)$ close to model output $y_m(t)$.

Self-Tuning Controllers (STCs).   In an STC system (Fig. 2), the specifications are given in terms of a performance index, for example, an index involving a quadratic term in the tracking error $\varepsilon(t) = y(t) - r(t)$ between the plant output and the output reference plus an additional quadratic term in the control variable $u(t)$ or its increments $\delta u(t)$. As in an MRAC system, there are two loops: the inner loop consists of the ordinary control system and is composed by the plant and the controller; the outer loop consists of the parameter-adjustment mechanism. The latter, in turn, is made up by a recursive identifier, for example an RLS identifier (cf. Sec. 6.3) and a design block, for example a steady-state LQS tracking design block (cf. Sec. 7.5). The identifier updates an estimate of the unknown plant parameters according to which the controller parameters are tuned on-line by the design block. The control problem that is solved by the design block is the *underlying control problem*. If the identifier attempts to explicitly estimate an (open-loop) plant model, for example, a CARMA model, required for solving the underlying control problem, for example, a steady-state LQS tracking problem, the scheme is referred to as an *explicit* or *indirect* STC system. In contrast with the explicit scheme, some STCs do not attempt to explicitly identify the plant model required for solving *off-line* the underlying control problem. On the contrary, the tuning mechanism is designed in such a way that self-tuning occurs

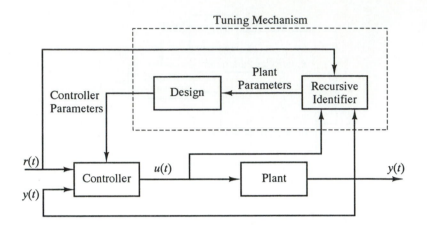

Figure 8.1-2: Block diagram of an STC system.

thanks to *identification in closed loop* of parameters that are relevant for *solving on-line* the underlying control problem. Typically, in such a case, combined spread-in-time iterations of both the identifier and the design block take place to yield at convergence the controller parameters solving the underlying control problem. A celebrated example of such a scheme is the original self-tuning minimum-variance controller of Åström and Wittenmark [ÅW73]. Here an RLS algorithm identifies a linear-regression model relating in closed loop the inputs and the outputs of a CARMA plant, and a minimum-variance control design (cf. Sec. 7.3) is carried out at each time step as if the plant coincides with the currently estimated CAR model. These STC schemes that do not explicitly identify the (open-loop) plant model are referred to as *implicit* STC systems. Whereas in indirect adaptive 2-DOF control systems, the feedforward law is computed from the estimated plant parameters in accordance with the underlying control law, in some implicit adaptive schemes, the feedforward law is estimated by the identifier itself in a direct or almost direct way (cf. 2-DOF MUSMAR in Sec. 9.4). This possibility is accounted in Fig. 2, where also the output reference enters the identifier. An extreme case within implicit STC systems are the *direct* STCs whereby the controller parameters are directly updated via the recursive identifier. In such a case, the block labeled "Design" in Fig. 2 disappears. Direct STCs can be obtained when the underlying control law is such that it allows one to reparameterize in a closed loop the model relating the plant inputs and outputs in terms of the controller parameters.

A closer comparison between Figs. 1 and 2 reveals the existence of a strong connection between MRACs and STCs. Their basic difference, in fact, resides in the block labeled "Model" present in Fig. 1 and absent in Fig. 2. Further, if in Fig. 2 we let $r(t) = M(d)c(t) =: y_m(t)$, with $M(d)$ a stable transfer matrix, we see that the STC system, provided that it makes the tracking error $\varepsilon(t) = y(t) - r(t) = y(t) - y_m(t)$ small, basically solves the same problem as in MRAC. In fact, under the previous choice, the STC system tends to make its output response $y(t)$ to command input $c(t)$ close to the desired model output $y_m(t)$. Therefore, it follows that MRAC and STC systems need not differ for either their ultimate goals or their implementative architectures. Moreover, the fact that originally MRACs have been developed mainly as direct adaptive controllers for continuous-time plants, whereas the majority of STCs were introduced as schemes for discrete-time plants, can be considered an accidental fact. Actually, there are indirect discrete-time MRACs as well as continuous-time STCs. Consequently, we conclude that the distinction between MRACs and STCs can be properly justified on the grounds of their different design methodologies.

In MRAC, there are basically three design approaches: the gradient approach, the Lyapunov function approach, and the passivity theory approach. The gradient approach is the original design methodology of MRAC. It consists of an adaptation mechanism that in the controller parameter space proceeds along the negative gradient of a scalar function of the error $y(t) - y_m(t)$. It was found that the gradient approach does not always yield stable closed-loop systems. This stimulated the application of stability theory, viz., Lyapunov stability theory and passivity theory, so as to obtain guaranteed stable MRAC systems.

The design methodology of the STCs basically consists of minimizing on-line a quadratic performance criterion for the currently identified plant model. This approach is therefore more akin to LQ and predictive control theory as presented in the previous chapters of this book. For this reason, in the subsequent part of this book, we shall concentrate merely on STCs.

Simple controllers are adequate in many applications. In fact, three-term compensators, for example, discrete-time PID controllers, generating the plant input increment $\delta u(t) := u(t) - u(t-1)$ in terms of a linear combination of the three most recent tracking errors $\varepsilon(t)$, $\varepsilon(t - 1)$, $\varepsilon(t - 2)$, $\varepsilon(t) := y(t) - r(t)$, are ubiquitous in industrial applications. Such controllers are traditionally tuned by simple empirical rules using the results of an experimental phase in which probing signals, such as steps or pulses, are injected into the plant. This way of setting the tuning knobs of a PID controller is called *auto-tuning*. Auto-tuners can be obtained [ÅW89] by using rules based on transient responses, relay feedback, or relay oscillations. Auto-tuning is generally well-accepted by industrial control engineers. In fact, typically, the auto-tuning phase is started and supervised by a human operator. The prior knowledge on the process dynamics is then allowed to be poorer and the "safety nets" simpler than when the controller parameters are adapted continuously, as in STC systems. Many adaptive control methods can be used so as to develop efficient auto-tuning techniques for a wide range of industrial applications. Auto-tuning is then a practically important application area of adaptive control techniques.

Other design methodologies for uncertain plant control that deserve to be mentioned are variable structure systems and universal controllers. In *variable-structure* systems ([Eme67], [Itk76], [Utk77], [Utk87], and [Utk92]), the controller forces the closed-loop system to evolve in a sliding mode along a sliding or switching surface, chosen in the state space. This can yield insensitivity to plant parameter variations. Drawbacks of variable-structure systems are the choice of the switching surface, the chattering associated to the sliding modes, and the required measurement of all plant state variables.

Universal controllers ([Mår85], [MM85], [Nus83], and [WB84]) have a structure that does not explicitly contain any parameter related to the plant. Hence, they can be used "universally" for any unknown linear plant for which it is known to exist a stabilizing fixed-gain controller of a given order. One drawback of universal controllers is that they are liable to exhibit very violent transients after the operation is started.

We have so far intentionally avoided to define what is meant by adaptive control. This is a quite difficult task. In fact, a meaningful and widely accepted definition, which would make it possible to look at a given controller and decide whether it is adaptive or not, is still lacking. As it emerges from the earlier description of MRACs and STCs, we adhere to the pragmatic viewpoint that adaptive control consists of a special type of nonlinear feedback control system in which the states can be separated into two sets corresponding to two different time scales. Fast time-varying states are the ones pertaining to ordinary feedback (inner loop in Figs. 1 and 2); slow time-varying states are regarded as parameters and consist of the estimated plant model parameters or controller parameters (outer loop in Figs. 1 and 2). This implies that linear time-invariant feedback compensators, for example, constant-gain robust controllers, are not adaptive controllers. We also assume that in an adaptive control system, some feedback action exists from the performance of the closed-loop system. Hence, gain scheduling is not an adaptive control technique, because its controller parameters are determined by a schedule without any feedback from the actual performance of the closed-loop system.

Main Points of the Section. There are several alternative approaches to the control problem of uncertain plants. They include robust control, gain scheduling, adaptive control, and variable-structure systems. In practice, the appropriate choice of a specific approach is dictated by the application at hand. Adaptive control, which is traditionally subdivided into MRACs and STCs, becomes appropriate whenever plant variations are large to such an extent as to jeopardize the stability or reduce to an unacceptable level the performance of the system compensated by nonadaptive methods.

8.2   BAYESIAN AND SELF-TUNING CONTROL

It would be conceptually appealing to formulate the adaptive control problem as an optimal stochastic control problem. We illustrate the point by discussing a simple example.

Example 8.2-1 (Bayesian Formulation)

Consider the SISO plant

$$y(k) = \theta u(k-1) + \zeta(k) \tag{8.2-1}$$

$k \in \mathbb{Z}_1$, where θ is an unknown parameter, and ζ is a white Gaussian disturbance with mean zero and variance σ^2, $\zeta \sim N(0, \sigma^2)$. The goal is to choose $u_{[t,t+T)}$, $u(k) \in \sigma\{y^k\}$, so as to minimize the performance index:

$$C = \mathcal{E}\left\{ \frac{1}{T} \sum_{k=t+1}^{t+T} [y(k) - r(k)]^2 \right\} \tag{8.2-2}$$

where r is a given reference. One way to proceed is to embed (1) and (2) in a stochastic control problem. This can be done by modeling the unknown parameter θ as a Gaussian random variable independent on ζ with prior distribution $\theta \sim N\left(\theta_0, \sigma_\theta^2\right)$, where θ_0 is the nominal value of θ. Under such an assumption, we can rewrite (1) as follows:

$$\left\{ \begin{array}{rcl} \theta(k+1) & = & \theta(k) \qquad\qquad\qquad \theta(1) \sim N\left(\theta_0, \sigma_\theta^2\right) \\ y(k) & = & u(k-1)\theta(k) + \zeta(k) \end{array} \right. \tag{8.2-3}$$

This is a nonlinear dynamic stochastic system with state $\theta(k)$ and observations $y(k)$. Then, (2) and (3) make up a stochastic control problem with partial state information. For such a problem, the conditional or posterior probability density function $p\left(\theta(k) \mid y^{k-1}\right)$ of $\theta(k)$ given the observed past $\left(y^{k-1}, u^{k-2}\right)$, or, equivalently, y^{k-1}, is Gaussian with conditional mean $\hat{\theta}(k) := \theta(k \mid k-1)$ and conditional variance $\sigma_\theta^2(k) = \mathcal{E}\left\{\tilde{\theta}^2(k) \mid y^{k-1}\right\}$, $\tilde{\theta}(k) := \theta(k) - \hat{\theta}(k)$,

$$p\left(\theta(k) \mid y^{k-1}\right) = N\left(\hat{\theta}(k), \sigma_\theta^2(k)\right)$$

The last two quantities can be computed in accordance to the conditionally Gaussian Kalman filter (cf. Fact 6.2-1). We get

$$\hat{\theta}(k+1) = \hat{\theta}(k) + \frac{\sigma_\theta^2(k)u(k-1)}{u^2(k-1)\sigma_\theta^2(k) + \sigma^2}\left[y(k) - u(k-1)\hat{\theta}(k)\right] \tag{8.2-4a}$$

$$\sigma_\theta^2(k+1) = \sigma_\theta^2(k) - \frac{u^2(k-1)\sigma_\theta^4(k)}{u^2(k-1)\sigma_\theta^2(k) + \sigma^2} \tag{8.2-4b}$$

with $\hat{\theta}(1) = \theta_0$, and $\sigma_\theta^2(1) = \sigma_\theta^2$. Further, we can write

$$y(k) = u(k-1)\hat{\theta}(k) + \nu(k) \tag{8.2-4c}$$

where $\nu(k) := u(k-1)\tilde{\theta}(k) + \zeta(k)$ conditionally on y^{k-1} is Gaussian:

$$\nu(k) \sim N\left(0, u^2(k-1)\sigma_\theta^2(k) + \sigma^2\right)$$

In (4), vector $\chi(k) := \left[\begin{array}{cc} \hat{\theta}(k) & \sigma_\theta^2(k) \end{array}\right]' \in \mathbb{R}^2$ can be regarded as a state, which makes it possible to update $p\left(\theta(k) \mid y^{k-1}\right)$ and express the observations as in (4c). Vector $\chi(k)$ is called the *hyperstate*. The optimal stochastic control problem (2) and (4) has been reduced to a complete state information problem. It can be solved via stochastic dynamic programming using Theorem 7.1-1. However, the system (4) is nonlinear, and no explicit closed form for the optimal control law can be obtained, except for the $T = 1$ case. The

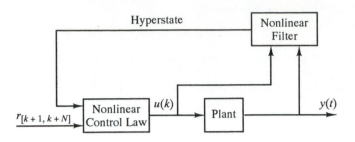

Figure 8.2-1: Block diagram of an adaptive controller as the solution of an optimal stochastic control problem.

latter is called a *myopic controller*, because it is short-sighted and looks only one step ahead. For $T = 1$, we have

$$\mathcal{E}\left\{[y(t+1) - r(t+1)]^2\right\} = \mathcal{E}\left\{\mathcal{E}\left\{[y(t+1) - r(t+1)]^2 \mid y^t\right\}\right\}$$

$$= \mathcal{E}\left\{\left[u(t)\hat{\theta}(t+1) - r(t+1)\right]^2 + u^2(t)\sigma_\theta^2(t+1) + \sigma^2\right\}$$

Minimization w.r.t. $u(t)$ yields the one-step-ahead optimal control law:

$$u(t) = \frac{\hat{\theta}(t+1)}{\hat{\theta}^2(t+1) + \sigma_\theta^2(t+1)} r(t+1) \qquad (8.2\text{-}5)$$

Note that if $\sigma_\theta^2 = 0$, that is, we know a priori that θ equals its nominal value θ_0, we have $\sigma_\theta^2(t+1) = 0$, $\hat{\theta}(t+1) = \theta_0$, and

$$u(t) = \frac{r(t+1)}{\hat{\theta}(t+1)} = \frac{r(t+1)}{\theta_0} \qquad (8.2\text{-}6)$$

The one-step-ahead controller (5) is sometimes called *cautious* [ÅW89] because, as its comparison with (6) shows, it takes into account the parameter uncertainty.

Although in the multistep case $T > 1$, the optimal stochastic control problem (2) and (4) admits no explicit solution, Example 1 leads us to make the following general remarks. First, because hyperstate $\chi(k)$ is an accessible state of the stochastic nonlinear dynamic system to be controlled, $u(k)$ is expected to be a nonlinear function of $\chi(k)$. Figure 1 illustrates the situation. Second, by (4b), the choice of $u(k)$ influences $\sigma_\theta^2(k+2)$, the posterior uncertainty on θ based on y^{k+1}. Thus, it might be advantageous to select large values of $u^2(k)$ so as to reduce $\sigma_\theta^2(k+2)$. On the other hand, this has to be balanced against the disadvantage of increasing $[y(k+1) - r(k+1)]^2$. Thus, the control here has a dual effect (cf. Sec. 7.2). As a consequence, the control problem of Example 1 is an excerpt of *dual control* [ÅW89]. Because its solution is based on assigning prior distributions to the unknown parameters and on-line computation of posterior distributions from the available observations, the optimal stochastic control problem formulated in Example 1 is also referred to as a *Bayesian adaptive control* problem [KV86].

Apart from the previous general important hints, Example 1 points out the difficulties of resorting in adaptive control to optimal stochastic control theory, except for the myopic control case. Because the latter has ultimately all the limitations inherent to cheap control and single-step regulation (cf. Secs. 2.6 and 2.7), optimal stochastic control theory is of little practical use in adaptive control. This is one of the reasons why most of the time suboptimal approaches are adopted. A very popular approach to adaptive control is the one described in the following example.

Example 8.2-2 (Enforced Certainty Equivalence)
Consider again the plant (1) where ζ is a zero-mean white disturbance, and θ is a nonzero constant but unknown parameter for which no prior probability density is assigned or assumed. The aim is to choose $u(k) \in \sigma\left\{y^k\right\}$ so as to make $y(k) \approx r(k) \equiv r$. If we know θ, according to (6), we could simply choose

$$u(t) = \frac{r}{\theta} \tag{8.2-7}$$

This, the minimum-variance (MV) control law (cf. Theorem 7.3-2), minimizes

$$\mathcal{C} = \mathcal{E}\left\{[y(t+1) - r]^2\right\} \tag{8.2-8}$$

for every t. When θ is unknown, we can proceed by enforced certainty equivalence: We estimate on-line θ via LS estimation:

$$\theta(t) = \left[\sum_{k=0}^{t-1} u^2(k)\right]^{-1} \sum_{k=0}^{t-1} u(k)y(k+1) \tag{8.2-9}$$

and we set at each $t \in \mathbb{Z}_1$,

$$u(t) = \frac{r}{\theta(t)} \tag{8.2-10}$$

with $u(0) \neq 0$. In other terms, to compute the control variable, we use the current estimate $\theta(t)$ as it were the true parameter θ.

The controller (9) to (10) is an adaptive controller of self-tuning type for the plant (1) and the performance index (8). *Enforced certainty equivalence (ECE)* is a simple procedure for designing adaptive controllers. ECE-based adaptive controllers compute the control variable by solving the underlying control problem using the current estimate $\theta(t)$ of the unknown parameter vector θ as if $\theta(t)$ were the true θ. If the adaptive controller achieves the same cost as the minimum that could be achieved if θ was a priori known, we say that the adaptive system is *self-optimizing*. Whenever the control law asymptotically approaches the one solving the underlying control problem, we say that *self-tuning* occurs. Further, we say that the adaptive controller is *weakly self-optimizing*, and/or that w.s. (weak-sense) *self–tuning* occurs, if the previous properties hold under the assumption that the adaptive control law converges.

Example 8.2-3

Consider again the self-tuning controller (9) and (10) applied to the plant (1). Let ζ be a possibly non-Gaussian zero-mean white disturbance satisfying the following martingale difference properties:

$$\mathcal{E}\left\{\zeta(t+1)\mid\zeta^t\right\}=0\quad,\qquad\text{a.s.}\qquad\qquad(8.2\text{-}11a)$$

$$\mathcal{E}\left\{\zeta^2(t+1)\mid\zeta^t\right\}=\sigma^2\quad,\qquad\text{a.s.}\qquad\qquad(8.2\text{-}11b)$$

The following result is useful to study the properties of the adaptive system.

Result 8.2-1. [LW82] **(Martingale Local Convergence)** *Let $\{\zeta(k),\mathcal{F}_k\}$ be a martingale difference sequence such that*

$$\sup_k\mathcal{E}\left\{\zeta^2(k+1)\mid\mathcal{F}_k\right\}<\infty\qquad\text{a.s.}\qquad\qquad(8.2\text{-}12)$$

Let u be a process adapted to $\mathcal{F}_k$, that is, $u(k)\in\mathcal{F}_k$. Then

$$\sum_{k=0}^{\infty}u^2(k)<\infty\Longrightarrow\sum_{k=0}^{\infty}u(k)\zeta(k+1)\qquad\text{converges a.s.}\qquad\qquad(8.2\text{-}13a)$$

$$\sum_{k=0}^{\infty}u^2(k)=\infty\Longrightarrow\sum_{k=0}^{t-1}u(k)\zeta(k+1)=o\left(\sum_{k=0}^{t-1}u^2(k)\right)\qquad\text{a.s.}\quad(8.2\text{-}13b)$$

To apply this result, set $\mathcal{F}_k:=\sigma\left\{\zeta^k\right\}$. By induction, we can check that $u(k)\in\mathcal{F}_k$. Further, (11b) implies (12). Substituting (1) into (9), we get

$$\theta(t)=\theta+\left[\sum_{k=0}^{t-1}u^2(k)\right]^{-1}\sum_{k=0}^{t-1}u(k)\zeta(k+1)\qquad\qquad(8.2\text{-}14)$$

We next show that $\sum_{k=0}^{\infty}u^2(k)=\infty$ a.s. In fact, we have a.s.

$$\sum_{k=0}^{\infty}u^2(k)<\infty\implies\sum_{k=0}^{t-1}u(k)\zeta(k+1)\qquad\text{converges}$$

$$\implies\left[\sum_{k=0}^{t-1}u^2(k)\right]^{-1}\sum_{k=0}^{t-1}u(k)\zeta(k+1)\qquad\text{converges}$$

$$\implies\theta(t)\quad\text{converges}\implies\lim\frac{|r|}{|\theta(t)|}>0$$

$$\implies\lim|u(t)|>0\implies\sum_{k=0}^{\infty}u^2(k)=\infty$$

Hence, $\sum_{k=0}^{\infty}u^2(k)=\infty$ a.s. It then follows from (13b) and (14) that

$$\lim_{t\to\infty}\theta(t)=\theta\qquad\text{a.s.}\qquad\qquad(8.2\text{-}15)$$

Therefore, in the adaptive controller (9) and (10), applied to the plant (1), self-tuning occurs.

In order to see if the adaptive control system is self-optimizing, we write

$$\frac{1}{t}\sum_{k=1}^{t}[y(k)-r]^2 \;=\; \frac{1}{t}\sum_{k=1}^{t}[\theta u(k-1)+\zeta(k)-r]^2$$

$$=\; \frac{1}{t}\sum_{k=1}^{t}\left\{[\theta u(k-1)-r]^2 + \zeta^2(k) - 2r\zeta(k) + 2\theta u(k-1)\zeta(k)\right\}$$

Now, because self-tuning occurs:

- $[\theta u(k-1)-r]\to 0$

- $\displaystyle\sum_{k=1}^{t} r\zeta(k) = o(t)$ by (13b)

- $\displaystyle\sum_{k=1}^{t} u(k-1)\zeta(k) = o\left(\sum_{k=1}^{t} u^2(k)\right)$ and $\displaystyle\sum_{k=1}^{t} u^2(k) = O(t)$ because of (15).

We can then conclude that

$$\frac{1}{t}\sum_{k=1}^{t}[y(k)-r]^2 \;\longrightarrow\; \frac{1}{t}\sum_{k=1}^{t}\zeta^2(k) \qquad\qquad (8.2\text{-}16)$$

as $t\to\infty$. If we adopt the additional assumption

$$\mathcal{E}\left\{\zeta^4(t+1)\mid \zeta^t\right\} = M < \infty \qquad\qquad (8.2\text{-}17)$$

by Lemma D.6-1 in Appendix D, we find from (16)

$$\bar{C} := \lim_{t\to\infty}\frac{1}{t}\sum_{k=1}^{t}[y(k)-r]^2 = \sigma^2 \qquad \text{a.s.} \qquad\qquad (8.2\text{-}18)$$

On the other hand, by the discussion leading to (16), we also see that if θ is known, minimum cost $\bar{C}$ equals the R.H.S. of (18). Then, we conclude that, under (17), the adaptive system is self-optimizing for cost $\bar{C}$. Note that self-optimization cannot be claimed for cost (8) for any finite t, because only the asymptotic behavior of the adaptive system can be analyzed.

Problem 8.2-1   Prove that under all the assumptions used in Example 3, the adaptive control system (1), (9), and (10) is self-optimizing for the asymptotic MV cost:

$$\lim_{t\to\infty}\mathcal{E}\left\{[y(t)-r]^2\right\}$$

Main Points of the Section.   A systematic optimal stochastic control approach based on a Bayesian reformulation of nonmyopic adaptive control problems leads to dual control. This is, however, so awkward to compute that the systematic approach turns out to be of little practical use. Enforced certainty equivalence (ECE) is a nonoptimal but simple procedure for designing adaptive controllers. ECE-based adaptive controllers exist for which, under given conditions, self-tuning and self-optimization occur.

8.3   GLOBAL CONVERGENCE TOOLS FOR DETERMINISTIC STCS

In this section, we present the main analytic tools that have been originally used [GRC80] to establish some desirable convergence properties of adaptive cheap controllers, viz., deterministic STCs whose underlying control problem is cheap control (cf. Sect. 2.6). One reason for an in-depth study of these tools is that, as will be seen in the next chapter, they can be extended to analyze asymptotic properties of multistep predictive STCs as well. By "convergence," we mean that some of the control objectives are asymptotically achieved and all the system variables remain bounded for the given set of initial conditions. Some of the reasons for which convergence theory is important are as follows:

- A convergence proof, though based on ideal assumptions, makes us more confident on the practical applicability of the algorithm.

- Convergence analysis helps in distinguishing between good and bad algorithms.

- Convergence analysis may suggest ways in which an algorithm might be improved.

For these reasons, there has been considerable research effort on the question of convergence of adaptive control algorithms. However, the nonlinearity of the adaptive control algorithms has turned out to be a major stumbling block in establishing convergence properties. In fact, taking into account that the algorithms are nonlinear and time-varying, it is quite surprising that convergence proofs can be obtained at all. Faced by the complexity of the convergence question, researchers initially concentrated on algorithms for which the control synthesis task is simple, viz., single-step-ahead STC systems. Even for this simple class of algorithms, convergence analysis turned out to be very difficult. It took the combined efforts of many researchers over about two decades to solve the convergence problem for single-step-ahead STC systems at the end of the seventies.

We assume that the plant to be controlled with inputs $u(k) \in \mathbb{R}$ is exactly represented as in (6.3-1). Hence, similarly to (6.3-2), for $k \in \mathbb{Z}_1$,

$$y(k) = \varphi'(k-1)\theta \tag{8.3-1a}$$

$$\varphi(k-1) := \left[\ \left(-y_{k-\hat{n}_a}^{k-1} \right)' \quad \left(u_{k-\hat{n}_b}^{k-1} \right)' \ \right]' \in \mathbb{R}^{\hat{n}_\theta} \tag{8.3-1b}$$

$$\theta := \left[\ a_1 \cdots a_{\hat{n}_a} \quad b_1 \cdots b_{\hat{n}_b} \ \right] \tag{8.3-1c}$$

with $\hat{n}_\theta := \hat{n}_a + \hat{n}_b$. Here $\hat{n}_a$ and $\hat{n}_b$ denote two known upper bounds for $n_a = \partial A^o(d)$ and $n_b = \partial B^o(d)$, respectively, $\partial A^o(d)$ and $\partial B^o(d)$ being the degrees of the two coprime polynomials in the irreducible plant transfer function $B^o(d)/A^o(d)$, $A^o(0) = 1$. In general, vector θ in (1) is not unique. In fact, $B(d)/A(d) :=$

$p(d)B^o(d)/[p(d)A^o(d)] = B^o(d)/A^o(d)$, where $p(d)$ is intended here to be any monic polynomial with $\partial p(d) =: \nu \leq \min(\hat{n}_a - n_a, \hat{n}_b - n_b)$. To any such pair $(A(d), B(d))$,

$$
\begin{aligned}
A(d) &= 1 + a_1 d + \cdots + a_{n_a+\nu} d^{n_a+\nu} = p(d) A^o(d) \\
B(d) &= b_1 d + \cdots + b_{n_b+\nu} d^{n_b+\nu} = p(d) B^o(d)
\end{aligned}
$$

we can associate parameter vector $\theta \in \mathbb{R}^{\hat{n}_\theta}$, $\theta \sim (A(d), B(d))$,

$$
\theta := \begin{cases}
\begin{bmatrix} a_1 \cdots a_{\hat{n}_a} & b_1 \cdots b_{n_b+\nu} & \underbrace{0 \cdots 0}_{\hat{n}_b - n_b - \nu} \end{bmatrix}' & (\nu = \hat{n}_a - n_a) \\[3ex]
\begin{bmatrix} a_1 \cdots a_{n_a+\nu} & \underbrace{0 \cdots 0}_{\hat{n}_a - n_a - \nu} & b_1 \cdots b_{\hat{n}_b} \end{bmatrix}' & (\nu = \hat{n}_b - n_b)
\end{cases}
$$

In particular, $\theta^o \sim (A^o(d), B^o(d))$ if

$$
\theta^o = \begin{bmatrix} a_1^o \cdots a_{n_a}^o & \underbrace{0 \cdots 0}_{\hat{n}_a - n_a} & b_1^o \cdots b_{n_b}^o & \underbrace{0 \cdots 0}_{\hat{n}_b - n_b} \end{bmatrix}'
$$

The set Θ of all vectors θ satisfying (1) consists of the linear variety [Lue69], or affine subspace, in $\mathbb{R}^{\hat{n}_\theta}$,

$$
\Theta = \theta^o + V \tag{8.3-2}
$$

where V is the ν-dimensional linear subspace parameterized, according to the previous, by the ν-free coefficients of polynomial $p(d)$. Θ will be referred to as the *parameter variety*. Despite the nonuniqueness of θ in (1), standard recursive identifiers, like the modified projection and the RLS algorithm, enjoy the following set of properties.

Properties P1

- *Uniform boundedness of the estimates*

$$
\|\theta(t)\| < M_\theta < \infty \quad, \qquad \forall t \in \mathbb{Z}_1 \tag{8.3-3}
$$

- *Vanishing normalized prediction error*

$$
\lim_{t \to \infty} \frac{\varepsilon^2(t)}{1 + c\|\varphi(t-1)\|^2} = 0 \tag{8.3-4}
$$

for some $c \geq 0$.

Here $\theta(t)$ denotes the estimate based on the observations $y^t := \{y(k)\}_{k=1}^t$ and regressors $\varphi^{t-1} := \{\varphi(k-1)\}_{k=1}^t$, and $\varepsilon(t) := y(t) - \varphi'(t-1)\theta(t-1)$ is the prediction error.

Property (3) is essential for STC implementation. In fact, boundedness of $\{\theta(t)\}$ is necessary to possibly compute the controller parameters at every t. On the other hand, (3) is not sufficient to design a controller with bounded parameters. As the next example shows, difficulties may arise from possible common divisors of $A(t, d)$ and $B(t, d)$.

Example 8.3-1 (Adaptive Pole Assignment)

Consider an STC system wherein the underlying control problem is pole assignment. Let $\theta(t) \sim (A(t, d), B(t, d))$ be the plant parameter estimate at t. The corresponding pole-assignment controller should generate the next input $u(t)$ in accordance with the difference equation

$$R(t, d)u(t) = -S(t, d)\varepsilon_y(t) \tag{8.3-5}$$

where $\varepsilon_y(t) := y(t) - r(t)$ is the tracking error, $\{r(t)\}$ being an assigned reference sequence, and $R(t, d)$ and $S(t, d)$ are polynomials solving the following Diophantine equation for a given strictly Hurwitz polynomial $\chi_{cl}(d)$:

$$A(t, d)R(t, d) + B(t, d)S(t, d) = \chi_{cl}(d) \tag{8.3-6}$$

where $\chi_{cl}(d)$ equals the desired closed-loop characteristic polynomial. Let $p(t, d)$ be the gcd of $A(t, d)$ and $B(t, d)$. Then, from Result C.1-1, we know that (6) is solvable if and only if $p(t, d) \mid \chi_{cl}(d)$. In practice, (6) becomes ill–conditioned whenever any root of $A(t, d)$ approaches a root of $B(t, d)$ that is far from the roots of $\chi_{cl}(d)$. In such a case, the magnitude of some coefficients of $R(t, d)$ and $S(t, d)$ becomes increasingly large.

The next problem points out that in adaptive cheap control, the underlying control design is always solvable irrespective of the gcd of $A(t, d)$ and $B(t, d)$, provided that $b_1(t) \neq 0$.

Problem 8.3-1 (Adaptive Cheap Control)

Recall that given the plant parameter vector $\theta(t) \sim (A(t, d), B(t, d)$ with $b_1(t) \neq 0$, in cheap control, the closed-loop d-characteristic polynomial equals $\chi_{cl}(t, d) = B(t, d)/[b_1(t)d]$. Check then that in such a case, (6) is always solvable, and find its minimum-degree solution $(R(t, d), S(t, d))$.

The previous discussion shows that some provisions have to be taken in STCs different from adaptive cheap control in order to make the underlying control problem solvable at each finite t. Further, in order to ensure successful operation as $t \to \infty$, in some adaptive schemes, the following additional identifier properties become important.

Properties P2

- *Slow asymptotic variations*

$$\lim_{t \to \infty} \|\theta(t) - \theta(t - k)\| = 0 \quad , \qquad \forall k \in \mathbb{Z}_1 \tag{8.3-7}$$

- *Convergence*

$$\lim_{t \to \infty} \theta(t) = \theta_\infty \in \mathbb{R}^{\hat{n}_\theta} \tag{8.3-8}$$

- *Convergence to the parameter variety*

$$\lambda_{\min}\left[P^{-1}(t)\right] = \infty \quad \Longrightarrow \quad \theta_\infty \in \Theta \tag{8.3-9}$$

- *Linear boundedness condition*

$$\|\varphi(t-1)\| \leq c_1 + c_2 \max_{k \in [1,t]} |\varepsilon(k)| \tag{8.3-10}$$

$0 \leq c_1 < \infty,\ 0 \leq c_2 < \infty.$

Note that (7) does not imply that $\{\theta(t)\}$ is a Cauchy sequence and, hence, that it converges. For example, $\theta(t) = \sin\left[\frac{2\pi t}{10\ln(t+1)}\right]$ satisfies (7) but does not converge. Whereas property (7) holds true for both the modified projection algorithm and RLS, the stronger properties (8) and (9) hold only for RLS. In (9), $P^{-1}(t)$ denotes the positive definite matrix given by (6.3-13g) for RLS. Another crucial property that, in addition to P1 and possibly P2, is required to establish global convergence of STC systems is (10). As will be seen, in adaptive cheap control, (10) is satisfied if the plant is nonminimum-phase, whereas in other schemes, like adaptive SIORHC, it follows from the underlying control and the other properties in P2.

The use of (4) and (10) in the analysis of STC systems is based on the following lemma.

Lemma 8.3-1. [GRC80] **(Key Technical Lemma)** *If*

$$\lim_{t \to \infty} \frac{\varepsilon^2(t)}{\alpha(t) + c(t)\|\varphi(t-1)\|^2} = 0 \tag{8.3-11}$$

where $\{\alpha(t)\}$, $\{c(t)\}$, and $\{\varepsilon(t)\}$ are real-valued sequences and $\{\varphi(t-1)\}$ a vector-valued sequence; then subject to: the uniform boundedness condition

$$0 \leq \alpha(t) < K < \infty \quad \text{and} \quad 0 \leq c(t) < K < \infty \tag{8.3-12}$$

for all $t \in \mathbb{Z}_1$, and the linear boundedness condition (10), it follows that

$$\|\varphi(t-1)\| \quad \text{is uniformly bounded} \tag{8.3-13}$$

for $t \in \mathbb{Z}_1$ and

$$\lim_{t \to \infty} \varepsilon(t) = 0 \tag{8.3-14}$$

Proof: If $|\varepsilon(t)|$ is uniformly bounded, from (10) it follows that $\|\varphi(t-1)\|$ is uniformly bounded as well. Then (14) follows from (11). By contradiction, assume that $|\varepsilon(t)|$ is not uniformly bounded. Then there exists a subsequence $\{t_n\}$ such that $\lim_{t_n \to \infty} |\varepsilon(t_n)| = \infty$ and $|\varepsilon(t)| \leq |\varepsilon(t_n)|$, for $t \leq t_n$. Now

$$\frac{|\varepsilon(t_n)|}{[\alpha(t) + c(t)\|\varphi(t_n-1)\|^2]^{1/2}} \geq \frac{|\varepsilon(t_n)|}{[K + K\|\varphi(t_n-1)\|^2]^{1/2}}$$

$$\geq \frac{|\varepsilon(t_n)|}{K^{1/2} + K^{1/2}\|\varphi(t_n-1)\|}$$

$$\geq \frac{|\varepsilon(t_n)|}{K^{1/2} + K^{1/2}[c_1 + c_2|\varepsilon(t_n)|]} \qquad [(10)]$$

Hence,

$$\lim_{t_n \to \infty} \frac{|\varepsilon(t_n)|}{[\alpha(t) + c(t)\|\varphi(t_n - 1)\|^2]^{1/2}} \geq \frac{1}{K^{1/2}c_2} > 0$$

This contradicts (11). Hence, $|\varepsilon(t)|$ must be uniformly bounded.

Lemma 1 presupposes that $|\varepsilon(t)|$ and $\|\varphi(t-1)\|$ are bounded for every finite $t \in \mathbb{Z}_1$. As long as we consider an STC system as in Fig. 1-2, where the plant and the controller are linear, and the tuning mechanism guarantees boundedness of the controller parameters at every $t \in \mathbb{Z}_1$, there is *no* chance of having a *finite escape time*. Hence, Lemma 1 is applicable to STC systems, even if they are highly nonlinear, to show that their variables remain bounded as $t \to \infty$.

Main Points of the Section. In an STC system it is required that the tuning mechanism provides the controller with bounded parameters. This must be insured jointly by the identifier and the underlying control law. Once this is accomplished, no finite escape time is possible, and uniform boundedness of all the involved variables can be established by using the Key Technical Lemma along with the asymptotic properties of the identifier.

8.4   RLS DETERMINISTIC PROPERTIES

We next derive some properties of the RLS algorithm, like P1 and P2 of Sec. 3, which are important in the analysis of STC systems. In contrast with Sec. 6.4, here we are not so much concerned about convergence to the true value of θ or to the parameter variety Θ, our interest being instead mainly directed to complementary properties of RLS that hold true even when no persistency of excitation is ensured. We rewrite the RLS algorithm (6.3-13):

$$\begin{aligned}
\theta(t) &= \theta(t-1) + \frac{P(t-1)\varphi(t-1)}{1 + \varphi'(t-1)P(t-1)\varphi(t-1)} \\
&\quad \times [y(t) - \varphi'(t-1)\theta(t-1)] \qquad\qquad\qquad\text{(8.4-1a)} \\
&= \theta(t-1) + P(t)\varphi(t-1)[y(t) - \varphi'(t-1)\theta(t-1)] \qquad\text{(8.4-1b)}
\end{aligned}$$

$$P(t) = P(t-1) - \frac{P(t-1)\varphi(t-1)\varphi'(t-1)P(t-1)}{1 + \varphi'(t-1)P(t-1)\varphi(t-1)} \qquad\qquad\text{(8.4-1c)}$$

$$P^{-1}(t) = P^{-1}(t-1) + \varphi(t-1)\varphi'(t-1) \qquad\qquad\qquad\text{(8.4-1d)}$$

with $t \in \mathbb{Z}_1$, $\theta(0) \in \mathbb{R}^{\hat{n}_\theta}$, and $P(0) = P'(0) > 0$.

Theorem 8.4-1.   (Deterministic Properties of RLS) *Consider the RLS algorithm (1) with*

$$y(t) = \varphi'(t-1)\theta \quad , \qquad \theta \in \Theta \subset \mathbb{R}^{\hat{n}_\theta} \qquad\qquad\text{(8.4-2)}$$

where Θ is the parameter variety (3-2). For any $\theta \in \Theta$, let $\tilde{\theta}(t) := \theta(t) - \theta$ and

$$
\begin{aligned}
\varepsilon(t) &:= y(t) - \varphi'(t-1)\theta(t-1) \\
&= -\varphi'(t-1)\tilde{\theta}(t-1)
\end{aligned}
\tag{8.4-3}
$$

Then, it follows that:

(i) There exists

$$
\lim_{t \to \infty} P(t) = P(\infty) = P'(\infty) \geq 0
\tag{8.4-4}
$$

(ii) As $t \to \infty$, $\theta(t)$ converges to $\theta(\infty)$:

$$
\theta(\infty) = \theta + P(\infty)P^{-1}(0)\tilde{\theta}(0)
\tag{8.4-5}
$$

from which for any $k \in \mathbb{Z}$

$$
\lim_{t \to \infty} \|\theta(t) - \theta(t-k)\| = 0
\tag{8.4-6}
$$

(iii)

$$
\|\tilde{\theta}(t)\|^2 \leq k_1 \|\tilde{\theta}(0)\|^2
\tag{8.4-7}
$$

$$
k_1 = \frac{\lambda_{\max}\left[P^{-1}(0)\right]}{\lambda_{\min}\left[P^{-1}(0)\right]} = \frac{\lambda_{\max}\left[P(0)\right]}{\lambda_{\min}\left[P(0)\right]}
$$

$\qquad$: $\quad$ the condition number of $P(0)$

(iv)

$$
\lim_{t \to \infty} \sum_{k=1}^{t} \frac{\varepsilon^2(k)}{1 + \varphi'(k-1)P(k-1)\varphi(k-1)} < \infty
\tag{8.4-8}
$$

and this implies

(a)

$$
\lim_{t \to \infty} \frac{\varepsilon^2(t)}{1 + k_2\|\varphi(t-1)\|^2} = 0
\tag{8.4-9}
$$

$$
k_2 = \lambda_{\max}[P(0)]
$$

(b)

$$
\lim_{t \to \infty} \sum_{k=1}^{t} \|\theta(k) - \theta(k-1)\|^2 < \infty
\tag{8.4-10}
$$

or more generally

(c)

$$
\lim_{t \to \infty} \sum_{k=i}^{t} \|\theta(k) - \theta(k-i)\|^2 < \infty
\tag{8.4-11}
$$

for every positive integer i.

Proof:

(i) Because $\{P(t)\}_{t=0}^{\infty}$ is a symmetric nonnegative-definite monotonically nonincreasing matrix sequence, (4) follows from Lemma 2.4-1.

(ii) We have

$$
\begin{aligned}
P^{-1}(t)\tilde{\theta}(t) &= P^{-1}(t-1)\tilde{\theta}(t-1) \qquad [(6.4\text{-}2f)] \\
&= P^{-1}(0)\tilde{\theta}(0)
\end{aligned}
$$

Hence,

$$
\theta(t) = \theta + P(t)P^{-1}(0)\tilde{\theta}(0)
$$

Taking the limit for $t \to \infty$ and using (4), we get (5).

(iii) We recall that in (6.4-5) for $V(t) := \tilde{\theta}'(t)P^{-1}(t)\tilde{\theta}(t)$, we found

$$
V(t) = V(t-1) - \frac{\varepsilon^2(t)}{1 + \varphi'(t-1)P(t-1)\varphi(t-1)} \tag{8.4-12}
$$

Thus, $V(t)$ is monotonically nonincreasing and, hence,

$$
\tilde{\theta}'(t)P^{-1}(t)\tilde{\theta}(t) \le \tilde{\theta}'(0)P^{-1}(0)\tilde{\theta}(0) \tag{8.4-13}
$$

Now from (1d),

$$
\lambda_{\min}\left[P^{-1}(t)\right] \ge \lambda_{\min}\left[P^{-1}(t-1)\right] \ge \lambda_{\min}\left[P^{-1}(0)\right]
$$

Then

$$
\begin{aligned}
\lambda_{\min}\left[P^{-1}(0)\right]\|\tilde{\theta}(t)\|^2 &\le \lambda_{\min}\left[P^{-1}(t)\right]\|\tilde{\theta}(t)\|^2 \\
&\le \tilde{\theta}'(t)P^{-1}(t)\tilde{\theta}(t) \\
&\le \tilde{\theta}'(0)P^{-1}(0)\tilde{\theta}(0) \qquad [(13)] \\
&\le \lambda_{\max}\left[P^{-1}(0)\right]\|\tilde{\theta}(0)\|^2
\end{aligned}
$$

This establishes (7).

(iv) Summing (12) from 1 to N, we get

$$
V(N) = V(0) - \sum_{k=1}^{N} \frac{\varepsilon^2(k)}{1 + \varphi'(k-1)P(k-1)\varphi(k-1)}
$$

Because $V(N)$ converges as $N \to \infty$ (cf. Theorem 6.4-1), (8) immediately follows.

(a) Equation (9) is a consequence of (8) because

$$
\lambda_{\max}\left[P(t-1)\right] \le \lambda_{\max}\left[P(t-2)\right] \le \lambda_{\max}[P(0)]
$$

(b) Equation (10) is a consequence of (8) because

$$
\begin{aligned}
\|\theta(k) &- \theta(k-1)\|^2 \\
&= \frac{\varphi'(k-1)P^2(k-1)\varphi(k-1)}{[1 + \varphi'(k-1)P(k-1)\varphi(k-1)]^2}\varepsilon^2(k) \qquad [(1)] \\
&\le \lambda_{\max}\left[P(k-1)\right]\frac{\varphi'(k-1)P(k-1)\varphi(k-1)}{[1 + \varphi'(k-1)P(k-1)\varphi(k-1)]^2}\varepsilon^2(k) \\
&\le \lambda_{\max}\left[P(0)\right]\frac{\varepsilon^2(k)}{1 + \varphi'(k-1)P(k-1)\varphi(k-1)}
\end{aligned}
$$

(c) We have

$$\|\theta(k) - \theta(k-i)\|^2 \quad = \quad \left\| \sum_{r=k-i+1}^{k} [\theta(r) - \theta(r-1)] \right\|^2$$

$$\leq \quad i \sum_{r=k-i+1}^{k} \|\theta(r) - \theta(r-1)\|^2$$

In fact, setting $v(r) := \theta(r) - \theta(r-1)$, we find

$$\left\| \sum_r v(r) \right\|^2 \quad = \quad \sum_r \|v(r)\|^2 + \sum_{r \neq s} v'(r)v(s)$$

$$\leq \quad \sum_r \|v(r)\|^2 + \sum_{r \neq s} \|v(r)\|\|v(s)\|$$

$$\leq \quad \sum_r \|v(r)\|^2 + \frac{1}{2} \sum_{r \neq s} \left[\|v(r)\|^2 + \|v(s)\|^2 \right]$$

$$= \quad i \sum_r \|v(r)\|^2$$

where the first inequality follows by the Schwarz inequality.

For a similar analysis of the modified projection algorithm (6.3-11), the reader is referred to [GS84].

Problem 8.4-1 (Uniqueness of the Asymptotic Estimate) Prove that $\theta(\infty)$ given by (5) is not affected by v, where $v = \theta - \theta^0 \in V$, V being the subspace of $\mathbb{R}^{n_\theta}$ in (3-2). (*Hint*: Show that $v \in V$ implies $\varphi'(t-1)v = 0$ and, hence, $P(0)P^{-1}(t)v = v$.)

Problem 8.4-2 Show that in the modified projection algorithm (6.3-11), we have

$$\|\tilde{\theta}(t)\| \leq \|\tilde{\theta}(t-1)\| \leq \|\tilde{\theta}(0)\|$$

and

$$\lim_{t \to \infty} \sum_{k=1}^{t} \frac{\varepsilon^2(k)}{1 + \|\varphi(k-1)\|^2} < \infty$$

Main Points of the Section. In a deterministic setting, the RLS algorithm fulfills Properties 1 and 2 of Sec. 3 required for establishing global convergence of STC systems. These results hold true in the absence of any persistency of excitation condition.

8.5   SELF-TUNING CHEAP CONTROL

We shall consider a SISO plant of the form

$$A^o(d)y(k) = B^o(d)u(k) + c \qquad (8.5\text{-}1)$$

with I/O delay τ:

$$\tau := \operatorname{ord} B^o(d) \geq 1$$

where $A^o(d)$ and $B^o(d)$ coprime, $\partial A^o(d) = n_a$, $\partial B^o(d) = \tau + n_b - 1$, and c a constant disturbance. By setting

$$A(d) := \Delta(d)A^o(d) \qquad \text{and} \qquad B(d) := \Delta(d)B^o(d)$$

with $\Delta(d) := 1 - d$, (1) can be rewritten as follows:

$$
\begin{aligned}
A(d)y(k) &= B(d)u(k) \qquad &(8.5\text{-}2)\\
&= d^\tau B_o(d)u(k)
\end{aligned}
$$

where, similarly to (7.3-6), $d^\tau B_o(d) = B(d)$, with $\partial B_o(d) = n_b$. Let $(Q_\tau(d), G_\tau(d))$, the minimum-degree solution w.r.t. $Q_\tau(d)$ of the Diophantine equation:

$$
\left.
\begin{aligned}
1 &= A(d)Q_\tau(d) + d^\tau G_\tau(d)\\
\partial Q_\tau(d) &\leq \tau - 1
\end{aligned}
\right\} \qquad (8.5\text{-}3)
$$

Then, we have

$$
\begin{aligned}
y(t+\tau) &= G_\tau(d)y(t) + Q_\tau(d)B_o(d)u(t) \qquad &(8.5\text{-}4a)\\
&= \alpha(d)y(t) + \beta(d)u(t)
\end{aligned}
$$

where, provided that ℓ, $\ell := \tau - 1$, denotes the I/O transport delay:

$$
\begin{aligned}
\alpha(d) &:= G_\tau(d)\\
&= \alpha_0 + \alpha_1 d + \cdots + \alpha_{n_a}d^{n_a} \qquad &(8.5\text{-}4b)\\
\beta(d) &:= Q_\tau(d)B_o(d)\\
&= \beta_0 + \beta_1 d + \cdots + \beta_{n_b+\ell}d^{n_b+\ell} \qquad \beta_0 \neq 0 \qquad &(8.5\text{-}4c)
\end{aligned}
$$

Let $\hat{n}_a \geq n_a$ and $\hat{n}_b \geq n_b$. Then we can write

$$y(t+\tau) = \varphi'(t)\theta \qquad (8.5\text{-}5a)$$

$$\varphi(t) := \left[\; (y_{t-\hat{n}_a}^t)' \quad (u_{t-\hat{n}_b-\ell}^t)' \;\right]' \quad \in \mathbb{R}^{\hat{n}_\theta} \qquad (8.5\text{-}5b)$$

with $\theta \in \Theta$, the parameter variety as in (3-2), $\Theta = \theta^o + V$,

$$\theta^o = \left[\; \alpha_0 \cdots \alpha_{n_a} \quad \underbrace{0 \cdots 0}_{\hat{n}_a - n_a} \quad \beta_0 \cdots \beta_{n_b+\ell} \quad \underbrace{0 \cdots 0}_{\hat{n}_b - n_b} \;\right]'$$

Define the output tracking error:

$$\varepsilon_y(t+\tau) \ := \ y(t+\tau) - r(t+\tau) \tag{8.5-6}$$
$$= \ \varphi'(t)\theta - r(t+\tau)$$

where r is the output reference. If we know θ, cheap control chooses $u(t)$, $t \in \mathbb{Z}$, so as to satisfy

$$\varphi'(t)\theta = r(t+\tau) \tag{8.5-7}$$

Note that for every $\theta \in \Theta$, the $\hat{n}_a^o + 1$ component equals $\beta_0 \neq 0$. This makes it possible to solve (7) w.r.t. $u(t)$. The control law given implicitly by (7) makes the tracking error (6) identically zero, and the closed-loop system internally stable (cf. Sec. 3.2) if the plant (1) is minimum-phase.

Hereafter, we assume that θ is unknown. We then proceed to design a *self-tuning cheap controller (STCC)*, of *implicit* type, viz., an enforced certainty equivalence adaptive controller whereby θ is replaced by its RLS estimate $\theta(t)$ and the underlying control law is cheap control. In this way, if $\tau > 1$, we do not estimate the explicit plant model (1) or (2) but instead the τ-step-ahead output-prediction model (4) that is directly related to the underlying control law (7). For this reason, the adjective "implicit" is associated to the STCC defined in what follows.

8.5.1   Implicit STCC

Parameter vector θ is estimated via the following RLS algorithm, $t \in \mathbb{Z}_1$,

$$\theta(t) = \theta(t-1) + \frac{a(t)P(t-\tau)\varphi(t-\tau)}{1 + a(t)\varphi'(t-\tau)P(t-\tau)\varphi(t-\tau)}\varepsilon(t) \tag{8.5-8a}$$

$$\varepsilon(t) = y(t) - \varphi'(t-\tau)\theta(t-1) \tag{8.5-8b}$$

$$P(t-\tau+1) = P(t-\tau) - \frac{a(t)P(t-\tau)\varphi(t-\tau)\varphi'(t-\tau)P(t-\tau)}{1 + a(t)\varphi'(t-\tau)P(t-\tau)\varphi(t-\tau)} \tag{8.5-8c}$$

or

$$P^{-1}(t-\tau+1) = P^{-1}(t-\tau) + a(t)\varphi(t-\tau)\varphi'(t-\tau) \tag{8.5-8d}$$

In (8), $a(t)$ is a positive real number chosen according to the following rule:

$$a(t) = \begin{cases} 1, & \text{if the } [(\hat{n}_a+1)\text{th component of the R.H.S. of (8a)} \\ & \text{evaluated using } a(t) = 1] \neq 0 \\ a, & \text{otherwise, } a \neq 1 \text{ a fixed positive real} \end{cases} \tag{8.5-8e}$$

Such a rule guarantees that the $(\hat{n}_a+1)$th component $\theta_{\hat{n}_a+1}(t)$ of $\theta(t)$ is nonzero, as required by the adopted control law:

$$\varphi'(t)\theta(t) = r(t+\tau) \tag{8.5-9}$$

The STCC algorithm is initialized from any $\theta(0)$ with $\theta_{\hat{n}_a+1}(0) \neq 0$, and any $P(1-\tau) = P'(1-\tau) > 0$.

Problem 8.5-1   Consider the RLS algorithm (8) embodying the constraint that $\theta_{\hat{n}_a+1}(t) \neq 0$. Let $\tilde{\theta}(t) := \theta(t) - \theta$ and $V(t) := \tilde{\theta}'(t)P^{-1}(t - \tau + 1)\tilde{\theta}(t)$. Show that

$$V(t) = V(t - 1) - \frac{a(t)\varepsilon^2(t)}{1 + a(t)\varphi'(t - \tau)P(t - \tau)\varphi(t - \tau)}$$

(cf. (4-12)). Use this result to show that the algorithm still enjoys the same properties as in Theorem 4-1.

We now analyze the implicit STCC (8) and (9) by the tools discussed in Sec. 3 under the following assumptions.

ASSUMPTION 8.5-1

- The plant I/O delay τ is known.

- Upper bounds $\hat{n}_a$ and $\hat{n}_b$ for the degrees of the polynomials in (4) are known.

- The plant (1) is minimum-phase, that is, $\mathcal{B}^o(d)/d^\tau$ is strictly Hurwitz.

- The reference sequence $\{r(t)\}_{t=1}^\infty$ is bounded:

$$|r(t)| \leq R < \infty$$

We first point out that by the first two items of Assumption 1, $\theta(t)$ and the controller parameters are bounded. Therefore, $u(t)$ and $y(t)$ are bounded for every finite t (no finite escape time). Further, (4-9) holds true for the estimation algorithm (8) (cf. Problem 1):

$$\lim_{t\to\infty} \frac{\varepsilon^2(t)}{1 + k_2\|\varphi(t - \tau)\|^2} = 0 \qquad (8.5\text{-}10)$$

Moreover,

$$\lim_{t\to\infty} \|\theta(t) - \theta(t - k)\| = 0 \qquad (8.5\text{-}11)$$

Look now at the tracking error:

$$\begin{aligned} \varepsilon_y(t) &:= y(t) - r(t) \\ &= \varphi'(t - \tau)\theta - \varphi'(t - \tau)\theta(t - \tau) \qquad [(5)\&(9)] \\ &= -\varphi'(t - \tau)\tilde{\theta}(t - \tau) \qquad\qquad (8.5\text{-}12) \end{aligned}$$

Thus,

$$\frac{-\varepsilon_y(t)}{[1 + k_2\|\varphi(t - \tau)\|^2]^{1/2}} = \frac{\varphi'(t - \tau)\tilde{\theta}(t - 1) - \varphi'(t - \tau)\left[\tilde{\theta}(t - 1) - \tilde{\theta}(t - \tau)\right]}{[1 + k_2\|\varphi(t - \tau)\|^2]^{1/2}}$$

Because $\varphi'(t-\tau)\tilde{\theta}(t-1) = -\varepsilon(t)$, the limit of the R.H.S. is clearly zero from (10) and (11). Hence,

$$\lim_{t\to\infty} \frac{\varepsilon_y^2(t)}{1 + k_2\|\varphi(t-\tau)\|^2} = 0 \qquad (8.5\text{-}13)$$

The aim is now to apply the key technical lemma, Lemma 3-1, with $\varepsilon(t)$ changed into $\varepsilon_y(t)$. To this end, we need to establish the linear boundedness condition:

$$\|\varphi(t-\tau)\| \leq c_1 + c_2 \max_{k\in[1,t]} |\varepsilon_y(k)| \qquad (8.5\text{-}14)$$

for some nonnegative reals c_1 and c_2. To prove (14), we rewrite (1) as follows:

$$\frac{\mathcal{B}^o(d)}{d^\tau} u(k-\tau) = A^o(d)y(k) - c$$

We see that $u(k-\tau)$ can be seen as the output and $y(k)$ and c as inputs of a time-invariant linear dynamic system that by the third item of Assumption 1 is asymptotically stable. Then, there exist [GS84] nonnegative reals m_1 and m_2 such that, for all $k \in [1,t]$,

$$|u(k-\tau)| \leq m_1 + m_2 \max_{i\in[1,t]} |y(i)| \qquad (8.5\text{-}15)$$

with m_1 depending upon the initial condition $\{y(0), \cdots, y(-n_a+1), u(-\tau), \cdots, u(-n_b-\tau+1)\}$. Therefore, by (5b),

$$\|\varphi(t-\tau)\| \leq \hat{n}_\theta \left\{ m_3 + \left[\max(1, m_2) \right] \max_{i\in[1,t]} |y(i)| \right\}$$

On the other hand, by boundedness of the reference,

$$|\varepsilon_y(t)| \geq |y(t)| - |r(t)| \geq |y(t)| - R$$

Hence,

$$\begin{aligned}
\|\varphi(t-\tau)\| &\leq \hat{n}_\theta \left\{ m_3 + \left[\max(1, m_2) \right] \max_{k\in[1,t]} \left[|\varepsilon_y(k)| + R \right] \right\} \\
&= c_1 + c_2 \max_{k\in[1,t]} |\varepsilon_y(k)|
\end{aligned}$$

for nonnegative reals c_1 and c_2. The previous discussion and Problem 2, which follows, establish global convergence of the implicit STCC.

Theorem 8.5-1. **(Global Convergence of the Implicit STCC)** *Provided that Assumption 1 is satisfied, the implicit STCC algorithm (8) and (9), when applied to the plant (1), for any possible initial condition yields:*

(i)   *$\{y(t)\}$ and $\{u(t)\}$ are bounded sequences*   (8.5-16)

(ii)
$$\lim_{t \to \infty} [y(t) - r(t)] = 0 \qquad (8.5\text{-}17)$$

(iii)
$$\lim_{t \to \infty} \sum_{k=\tau}^{t} [y(k) - r(k)]^2 < \infty \qquad (8.5\text{-}18)$$

Problem 8.5-2   Prove (18) by using Theorem 4.1, part *(iv)*:

$$\lim_{t \to \infty} \sum_{k=1}^{t} \frac{\varepsilon^2(k)}{1 + \varphi'(k-\tau)P(k-\tau)\varphi(k-\tau)} < \infty \qquad (8.5\text{-}19)$$

$$\lim_{t \to \infty} \sum_{k=i}^{t} \|\theta(t) - \theta(t-i)\|^2 < \infty \qquad (8.5\text{-}20)$$

the relationship between $\varepsilon(k)$ and $\varepsilon_y(k)$:

$$\varepsilon_y(k) = \varepsilon(k) - \varphi'(k-\tau)\left[\tilde{\theta}(k-1) - \tilde{\theta}(k-\tau)\right] \qquad (8.5\text{-}21)$$

the Schwarz inequality, and the boundedness of $\{\varphi(k-\tau)\}$ as implied by (16).

Theorem 1 is important in that it establishes that irrespective of the initial conditions, for the implicit STCC system:

- closed-loop stability is achieved

- the output tracking error asymptotically vanishes

- the convergence rate for the square of the output tracking error is faster than $1/t$

Note that such conclusions are obtained without assuming convergence of the estimated parameter vector $\theta(t)$ to the parameter variety Θ. In fact, no claim that self-tuning occurs can be advanced. On the other hand, the minimum-phase assumption is quite restrictive and, as we know, an unavoidable consequence of the adopted underlying control. We can try to remove such a restriction by using a long-range predictive control law. In the next chapter, we shall see that this is surprisingly complicated if global convergence of the resulting adaptive system has to be guaranteed.

8.5.2   Direct STCC

We assume hereafter that the reference to be tracked is a constant set point. Then, $r(k+1) = r(k) = r$, $k \in \mathbb{Z}_1$. Hence, denoting by $\varepsilon_y(k) := y(k) - r$ the output tracking error, similarly to (2), we can write

$$\begin{aligned} A(d)\varepsilon_y(k) &= B(d)u(k) \\ &= d^\tau B_o(d)u(k) \end{aligned} \qquad (8.5\text{-}22)$$

Hence, similarly to (3) and (4), we have the predictive model:

$$\varepsilon_y(k + \tau) = \alpha(d)\varepsilon_y(k) + \beta(d)u(k) \tag{8.5-23}$$

Here cheap control consists of choosing $u(k)$ so as to make the L.H.S. of the preceding equation equal to zero:

$$\begin{aligned}
u(t) &= -\frac{\alpha(d)}{\beta_0}\varepsilon_y(t) - \frac{\beta(d) - \beta_0}{\beta_0}u(t) \\
&= -f's(t) \tag{8.5-24}
\end{aligned}$$

$$f' := \left[\begin{array}{ccccc} \dfrac{\alpha_0}{\beta_0} & \cdots & \dfrac{\alpha_{n_a}}{\beta_0} & \dfrac{\beta_1 - \beta_0}{\beta_0} & \cdots & \dfrac{\beta_{n_b+\ell} - \beta_0}{\beta_0} \end{array}\right] \tag{8.5-25}$$

$$s'(t) := \left[\begin{array}{cc} \left(\varepsilon_{y\,t-n_a}^{t}\right)' & \left(u_{t-n_b-\ell}^{t-1}\right)' \end{array}\right] \tag{8.5-26}$$

We then see that (23) can be reparameterized in terms of the cheap control feedback vector f:

$$\frac{\varepsilon_y(k + \tau)}{\beta_0} - u(k) = f's(k) \tag{8.5-27}$$

Then, when β_0 is known, a direct STC algorithm can be obtained as follows. From the observations,

$$\begin{aligned}
z(t) &:= \frac{\varepsilon_y(t)}{\beta_0} - u(t - \tau) \tag{8.5-28} \\
&= s'(t - \tau)f, \qquad t \in \mathbb{Z}_1
\end{aligned}$$

recursively estimate the cheap control feedback vector f. Let $f(t)$ be the estimate of f based on z^t. For example, such an estimate can be obtained by the modified projection or the RLS algorithm. Specifically, in the latter case,

$$\begin{aligned}
f(t) &= f(t - 1) \tag{8.5-29} \\
&\quad + \frac{P(t - \tau)s(t - \tau)}{1 + s'(t - \tau)P(t - \tau)s(t - \tau)}\left[z(t) - s'(t - \tau)f(t - 1)\right]
\end{aligned}$$

$$P(t - \tau + 1) = P(t - \tau) - \frac{P(t - \tau)\varphi(t - \tau)\varphi'(t - \tau)P(t - \tau)}{1 + s'(t - \tau)P(t - \tau)s(t - \tau)} \tag{8.5-30}$$

with $f(0)$ and $P(1 - \tau) = P'(1 - \tau) > 0$ arbitrary. For the next input $u(t)$, set

$$u(t) = -f'(t)s(t) \tag{8.5-31}$$

We can still try to use the preceding direct STCC if, though β_0 is not exactly known, a nominal value $\hat{\beta}_0$ of β_0, $\hat{\beta}_0 \approx \beta_0$, is available. In such a case, to estimate f in (29), we use, instead of $z(t)$, the observations:

$$\hat{z}(t) := \frac{\varepsilon_y(t)}{\hat{\beta}_0} - u(t - \tau), \qquad t \in \mathbb{Z}_1 \tag{8.5-32}$$

How close to β_0 should $\hat{\beta}_0$ be in order to possibly make the direct STCC work? To find an answer, we resort to the following simple argument. Multiply each term of (29) by $p(t - \tau) := s'(t - \tau)P(t - \tau)s(t - \tau)$ to obtain

$$p(t - \tau)s'(t - \tau)f(t) + s'(t - \tau)\left[f(t) - f(t-1)\right] = p(t-\tau)\hat{z}(t)$$

Assuming that $f(t) \cong f(t-1)$, we get

$$s'(t-\tau)f(t) \quad \cong \quad \frac{1}{\hat{\beta}_0}\left[\varepsilon_y(t) - \hat{\beta}_0 u(t-\tau)\right] \qquad\qquad [(32)]$$

$$= \quad \frac{1}{\hat{\beta}_0}\left[\left(\beta_0 - \hat{\beta}_0\right)u(t-\tau) + \beta_0 s'(t-\tau)f\right] \qquad\qquad [(27)]$$

$$= \quad s'(t-\tau)\left[\frac{\hat{\beta}_0 - \beta_0}{\hat{\beta}_0}f(t-\tau) + \frac{\beta_0}{\hat{\beta}_0}f\right] \qquad\qquad [(31)]$$

This equation is valid in a closed loop and yields an updating equation for $f(t)$ with each term premultiplied by $s'(t-\tau)$. We can conjecture from such an equation that the evolution of $f(t)$ is stable provided that $|(\hat{\beta}_0 - \beta_0)/\hat{\beta}_0| < 1$, or, equivalently,

$$0 < \frac{\beta_0}{\hat{\beta}_0} < 2 \tag{8.5-33}$$

Therefore, (33) looks like a stability condition that must be guaranteed to the direct STCC. This condition was in fact pointed out in [ÅBLW77], and later in [GS84], to be required for global convergence of the direct STCC algorithm under the additional assumption that the plant has minimum phase and its I/O delay τ is known. Note that (33) is equivalent to

$$\left.\begin{array}{ll} \dfrac{\beta_0}{2} < \hat{\beta}_0 < \infty & (\beta_0 > 0) \\[4mm] -\infty < \hat{\beta}_0 < \dfrac{\beta_0}{2} & (\beta_0 < 0) \end{array}\right\} \tag{8.5-34}$$

In particular, it requires that $\hat{\beta}_0$ and β_0 have the same sign. Simulation experience [GS84] suggests, however, that a practical range for $\beta_0/\hat{\beta}_0$ be (0.5, 1.5).

Main Points of the Section. An implicit STCC system with a global convergence property can be constructed, even if no claim that self-tuning occurs for its feedback gain can be advanced. It is restricted to minimum-phase plants with known I/O delay. The direct STCC system (29) to (31) has the additional restriction that the sign of the first nonzero coefficient of the plant $B(d)$ polynomial must be known together with its approximate size.

8.6 CONSTANT-TRACE NORMALIZED RLS AND STCC

An important reason for using adaptive control in practice is to achieve good performance with time-varying plants. In this case, the use of an identifier with a finite data memory is required; cf. (6.3-40) to (6.3-43). To this end, we first discuss in detail the properties of a constant-trace RLS algorithm. We next show that an STCC system equipped with such a finite data memory identifier still enjoys the compatibility property of being globally convergent when applied to time-invariant plants.

8.6.1 Constant-Trace Normalized RLS (CT-NRLS)

With reference to the data-generating mechanism (4-2), we define the normalization factor

$$m(t-1) := \max\left\{m, \|\varphi(t-1)\|\right\}, \qquad m > 0 \qquad (8.6\text{-}1a)$$

and the normalized data

$$\gamma(t) := \frac{y(t)}{m(t-1)} \qquad x(t-1) := \frac{\varphi(t-1)}{m(t-1)} \qquad (8.6\text{-}1b)$$

Estimate $\theta(t)$ based on γ^t and x^{t-1}, $t \in \mathbb{Z}_1$, is given by (cf. (6.3-22) and (6.3-43))

$$
\begin{aligned}
\theta(t) &= \theta(t-1) + \frac{P(t-1)x(t-1)}{1 + x'(t-1)P(t-1)x(t-1)}\bar{\varepsilon}(t) & (8.6\text{-}1c)\\
&= \theta(t-1) + \lambda(t)P(t)x(t-1)\bar{\varepsilon}(t)
\end{aligned}
$$

$$\bar{\varepsilon}(t) = \gamma(t) - x'(t-1)\theta(t-1) \qquad (8.6\text{-}1d)$$

$$P(t) = \frac{1}{\lambda(t)}\left[P(t-1) - \frac{P(t-1)x(t-1)x'(t-1)P(t-1)}{1 + x'(t-1)P(t-1)x(t-1)}\right] \qquad (8.6\text{-}1e)$$

$$P^{-1}(t) = \lambda(t)\left[P^{-1}(t-1) + x(t-1)x'(t-1)\right] \qquad (8.6\text{-}1f)$$

$$\lambda(t) = 1 - \frac{1}{\text{Tr}[P(0)]}\frac{x'(t-1)P^2(t-1)x(t-1)}{1 + x'(t-1)P(t-1)x(t-1)} \qquad (8.6\text{-}1g)$$

with $\theta(0) \in \mathbb{R}^{n_\theta}$ arbitrary, and $P(0) = P'(0) > 0$.

As discussed in connection with (6.3-43), algorithm (1) has the constant-trace property $\text{Tr}[P(t)] = \text{Tr}[P(0)]$. Further, data normalization is adopted so as to ensure that $\lambda(t)$ be lower-bounded away from zero.

Lemma 8.6-1. *In the CT-NRLS algorithm (1), we have*

$$\frac{1}{1 + \text{Tr}[P(0)]} \leq \lambda(t) \leq 1 \qquad (8.6\text{-}2)$$

Proof: For the sake of brevity, we omit the argument $t-1$ throughout the proof. First, note that, because $P = P'$, P can be written as $L'L$. Further, $\text{sp}(L'L) = \text{sp}(LL')$ and by normalization, $\|x\| \leq 1$, $\text{sp}(M)$ denoting the set of the eigenvalues of square matrix M. Hence,

$$\frac{x'P^2x}{\text{Tr}[P(0)](1+x'Px)} \leq \frac{\lambda_{\max}(P)x'Px}{\text{Tr}[P(0)](1+x'Px)} \leq \frac{x'Px}{1+x'Px} \leq \frac{\text{Tr}[P(0)]}{1+\text{Tr}[P(0)]}$$

Then,

$$1 - \frac{x'P^2x}{\text{Tr}[P(0)](1+x'Px)} \geq 1 - \frac{\text{Tr}[P(0)]}{1+\text{Tr}[P(0)]} = \frac{1}{1+\text{Tr}[P(0)]}$$

Problem 8.6-1 Consider the CT-NRLS algorithm (1) with $y(t)$, $t \in \mathbb{Z}_1$, satisfying (4-2). Let $\tilde{\theta}(t) := \theta(t) - \theta$ and $V(t) := \tilde{\theta}'(t)P^{-1}(t)\tilde{\theta}(t)$. Show that

$$V(t) = \lambda(t)\left[V(t-1) - \frac{\bar{\varepsilon}^2(t)}{1+x'(t-1)P(t-1)x(t-1)}\right] \tag{8.6-3}$$

Conclude that

$$\|\tilde{\theta}(t)\|^2 \leq \left[\prod_{j=1}^{t}\lambda(j)\right]\text{Tr}[P(0)]V(0) \tag{8.6-4}$$

and $\{V(t)\}$ converges.

Problem 8.6-2 Under the same assumptions as in Problem 1, show that

$$P^{-1}(t)\tilde{\theta}(t) = \lambda(t)P^{-1}(t-1)\tilde{\theta}(t-1)$$

and, hence,

$$\tilde{\theta}(t) = \left[\prod_{j=1}^{t}\lambda(j)\right]P(t)P^{-1}(0)\tilde{\theta}(0) \tag{8.6-5}$$

Let

$$\delta(t) := \prod_{j=1}^{t}\lambda(j) \tag{8.6-6}$$

and

$$\delta(\infty) := \lim_{t\to\infty}\delta(t) \tag{8.6-7}$$

Then, from (4), it follows that

$$\delta(\infty) = 0 \quad\Longrightarrow\quad \theta(\infty) := \lim_{t\to\infty}\theta(t) = \theta \tag{8.6-8}$$

If, on the contrary, $\delta(\infty) > 0$, Lemma 2, which follows, establishes the existence of $P(\infty) := \lim_{t\to\infty}P(t)$. Hence, from (5),

$$\delta(\infty) > 0 \quad\Longrightarrow\quad \theta(\infty) = \theta + \delta(\infty)P(\infty)P^{-1}(0)\tilde{\theta}(0) \tag{8.6-9}$$

Lemma 8.6-2. *Consider the CT-NRLS algorithm (1). Then, there exists*

$$\lim_{t \to \infty} P(t) =: P(\infty) = P'(\infty) \geq 0$$

whenever $\delta(\infty) > 0$.

Proof: Let $P_{ij}(t)$ be the (i,j)th entry of $P(t)$. Then, because $P(t) = P'(t)$,

$$P_{ij}(t) = e_i' P(t) e_j = \frac{1}{2} \left\{ (e_i + e_j)' \, P(t) \, (e_i + e_j) - e_i' P(t) e_i - e_j' P(t) e_j \right\}$$

where $\{e_i\}_{i=1}^{\hat{n}_\theta}$ denotes the natural basis of $\mathbb{R}^{\hat{n}_\theta}$. Consequently, $P_{ij}(t)$ converges if the quadratic form $z' P(t) z$ converges for any $z \in \mathbb{R}^{\hat{n}_\theta}$. Now

$$z' P(t) z = \frac{1}{\lambda(t)} \left[z' P(t-1) z - r(t-1) \right] \qquad\qquad [(1e)]$$

where

$$r(t-1) := \frac{[z' P(t-1) x(t-1)]^2}{1 + x'(t-1) P(t-1) x(t-1)} \geq 0$$

By iterating the previous expression,

$$z' P(t) z \;=\; \frac{1}{\lambda(t)} \left\{ \frac{1}{\lambda(t-1)} \left[z' P(t-2) z - r(t-2) \right] - r(t-1) \right\}$$

$$=\; \frac{z' P(0) z}{\displaystyle\prod_{j=i}^{t} \lambda(j)} - \sum_{j=1}^{t} \left[\frac{r(j-1)}{\displaystyle\prod_{i=j}^{t} \lambda(i)} \right] \geq 0$$

Because $\delta(\infty) > 0$, it remains to show that the last term converges. This can be proved as follows:

$$\sum_{j=1}^{t} \left[\frac{r(j-1)}{\displaystyle\prod_{i=j}^{t} \lambda(i)} \right] = \frac{1}{\displaystyle\prod_{i=1}^{t} \lambda(i)} \sum_{j=1}^{t} \left\{ \left[\prod_{i=1}^{j-1} \lambda(i) \right] r(j-1) \right\}$$

This together with the previous equation shows that

$$\sum_{j=1}^{t} \left\{ \left[\prod_{i=1}^{j-1} \lambda(i) \right] r(j-1) \right\} \leq z' P(0) z$$

Now the L.H.S. of the previous inequality is nondecreasing with t and upper bounded by $z' P(0) z$. Hence, it converges.

Problem 8.6-3 Consider the CT-NRLS algorithm. Use the convergence of $\{V(t)\}$ established in Problem 1 to show that

$$\lim_{t \to \infty} \sum_{k=1}^{t} \lambda(k) \frac{\bar{\varepsilon}^2(k)}{1 + x'(k-1) P(k-1) x(k-1)} < \infty$$

and

$$\lim_{t\to\infty} \sum_{k=1}^{t} \bar{\varepsilon}^2(k) < \infty$$

Problem 8.6-4 For the CT-NRLS algorithm, establish that

$$\lim_{t\to\infty} \sum_{k=1}^{t} \|\theta(k) - \theta(k-1)\|^2 < \infty$$

and, more generally,

$$\lim_{t\to\infty} \sum_{k=i}^{t} \|\theta(k) - \theta(k-i)\|^2 < \infty$$

for every positive integer i.

The foregoing results are summed up in the following theorem.

Theorem 8.6-1. **(Deterministic Properties of CT-NRLS)** *Consider the CT-NRLS algorithm (1) with*

$$y(t) = \varphi'(t-1)\theta \quad, \qquad \theta \in \Theta \subset \mathbb{R}^{\hat{n}_\theta} \tag{8.6-10}$$

where Θ is the parameter variety (3-2). Then, for any $\theta \in \Theta$, $\tilde{\theta}(t) := \theta(t) - \theta$, and $\delta(\infty)$ as in (8), we have

(i) $\delta(\infty) > 0 \implies \displaystyle\lim_{t\to\infty} P(t) = P(\infty) = P'(\infty) \geq 0$ $\tag{8.6-11}$

(ii) as $t \to \infty$, $\theta(t)$ converges to $\theta(\infty)$, where

$$\theta(\infty) = \begin{cases} \theta & \delta(\infty) = 0 \\ \theta + \delta(\infty)P(\infty)P^{-1}(0)\tilde{\theta}(0) & , \quad \delta(\infty) > 0 \end{cases} \tag{8.6-12}$$

from which for any $k \in \mathbb{Z}$,

$$\lim_{t\to\infty} \|\theta(t) - \theta(t-k)\| = 0 \tag{8.6-13}$$

(iii) $$\|\tilde{\theta}(t)\|^2 \leq \left[\prod_{j=i}^{t} \lambda(j)\right] \mathrm{Tr}[P(0)]V(0) \tag{8.6-14}$$

(iv) $$\lim_{t\to\infty} \sum_{k=1}^{t} \bar{\varepsilon}^2(k) < \infty \tag{8.6-15}$$

and this implies

(a) $$\lim_{t\to\infty} \bar{\varepsilon}(t) = 0 \tag{8.6-16}$$

$$(b) \qquad \lim_{t \to \infty} \sum_{k=1}^{t} \|\theta(k) - \theta(k-1)\|^2 < \infty \qquad (8.6\text{-}17)$$

or, more generally,

$$(c) \qquad \lim_{t \to \infty} \sum_{k=i}^{t} \|\theta(k) - \theta(k-i)\|^2 < \infty \qquad (8.6\text{-}18)$$

for every positive integer i.

Problem 8.6-5   Consider the CT-NRLS algorithm (1). Let

$$\tilde{P}^{-1}(t) := P^{-1}(0) + \sum_{k=0}^{t-1} x(k)x'(k)$$

Then show that $\lambda_{\min}\left[\tilde{P}^{-1}(t)\right] \to \infty$ as $t \to \infty$ implies that $\delta(\infty) = 0$ and, hence, provided that (8) holds, $\theta \in \Theta$. (*Hint:* First show that, for $t \in \mathbb{Z}_1$, $\delta^{-1}(t)P^{-1}(t) = P^{-1}(0) + \sum_{k=0}^{t-1} \delta^{-1}(k)x(k)x'(k)$.)

Note that some of the properties of Theorem 1 would not hold true for a constant-trace RLS algorithm with no data normalization. In fact, data normalization is essential to guarantee that $\lambda(t) > 0$ in (2), and, consequently, the result of Problem 5 and $\{\bar{\varepsilon}(k)\} \in \ell_2$ in (15), and, hence, (16) to (18).

8.6.2   Implicit STCC with CT–NRLS

Hereafter, we shall consider a variant of the implicit STCC of Sec. 8.5 whereby the RLS estimates are replaced by estimates $\theta(t)$ supplied by a CT-NRLS identifier modified so as to keep $\theta_{\hat{n}_a+1}(t) \neq 0$. Specifically, parameter vector θ is estimated via the following CT-NRLS algorithm, $t \in \mathbb{Z}_1$:

$$\theta(t) = \theta(t-1) + \frac{a(t)P(t-\tau)x(t-\tau)}{1 + a(t)x'(t-\tau)P(t-\tau)x(t-\tau)}\bar{\varepsilon}(t) \qquad (8.6\text{-}19a)$$

$$P(t-\tau+1) = \frac{1}{\lambda(t)}\left[P(t-\tau) - \frac{a(t)P(t-\tau)x(t-\tau)x'(t-\tau)P(t-\tau)}{1 + a(t)x'(t-\tau)P(t-\tau)x(t-\tau)}\right] \qquad (8.6\text{-}19b)$$

$$\lambda(t) = 1 - \frac{1}{\mathrm{Tr}[P(0)]}\frac{a(t)x'(t-\tau)P^2(t-\tau)x(t-\tau)}{1 + a(t)x'(t-\tau)P(t-\tau)x(t-\tau)} \qquad (8.6\text{-}19c)$$

where $a(t)$ is as in (5-8e),

$$\varepsilon(t) \quad = \quad y(t) - \varphi'(t-\tau)\theta(t-1) \qquad (8.6\text{-}19d)$$

$$\bar{\varepsilon}(t) \quad = \quad \frac{\varepsilon(t)}{m(t-\tau)} \qquad (8.6\text{-}19e)$$

$$= \quad \gamma(t) - x'(t-\tau)\theta(t-1)$$

$$m(t-\tau) \quad := \quad \max\left\{m, \|\varphi(t-\tau)\|\right\} , \qquad m > 0 \qquad (8.6\text{-}19f)$$

As in (5-9), the control law is given by

$$\varphi'(t)\theta(t) = r(t+\tau) \qquad (8.6\text{-}20)$$

The algorithm is initialized from any $\theta(0)$ with $\theta_{\hat{n}_a+1} \neq 0$, and any $P(1-\tau) = P'(1-\tau) > 0$.

Problem 8.6-6   Consider the CT-NRLS algorithm (19) embodying the constraint that $\theta_{\hat{n}_a+1}(t) \neq 0$. Let $V(t) := \tilde{\theta}'(t)P^{-1}(t-\tau+1)\tilde{\theta}(t)$, $\tilde{\theta}(t) := \theta(t) - \theta$. Show that

$$V(t) = \lambda(t)\left[V(t-1) - \frac{a(t)\bar{\varepsilon}^2(t)}{1 + a(t)x'(t-\tau)P(t-\tau)x(t-\tau)}\right]$$

(cf. (3)). Use this result to show that the algorithm still enjoys the same properties as in Theorem 1.

The conclusion of Problem 5 allows us to follow similar lines as the ones leading to Theorem 5-1 to prove the next global convergence result.

Theorem 8.6-2.   **(Global Convergence of the Implicit STCC with CT-NRLS)**   *Provided that Assumption 5-1 is satisfied, the implicit STCC algorithm based on the CT-NRLS (19) and the control law (20), when applied to the plant (5-1), for any possible initial condition, yields*

(*i*) $\{y(t)\}$ *and* $\{u(t)\}$ *are bounded sequences* $\qquad (8.6\text{-}21)$

(*ii*) $\lim_{t\to\infty}[y(t) - r(t)] = 0$ $\qquad (8.6\text{-}22)$

(*iii*) $\lim_{t\to\infty}\sum_{k=\tau}^{t}[y(k) - r(k)]^2 < \infty$ $\qquad (8.6\text{-}23)$

Problem 8.6-7   Prove Theorem 2 (*Hint*: Apply first the key technical lemma, Lemma 3-1, to prove (21) and (22). Next, exploit (21) to prove (23).)

Main Points of the Section.   In order to achieve good performance with time-varying plants, it is advisable that adaptive controllers be equipped with identifiers with finite data memory, for example, constant-trace RLS. The implicit STCC controller (19) and (20), based on a constant-trace RLS identifier with data normalization, when applied to the plant (3-1) enjoys the same global convergence properties as the implicit STCC system (8) and (9), based on the standard RLS identifier.

8.7 MINIMUM-VARIANCE SELF-TUNING CONTROL

8.7.1 Implicit Linear-Regression Models and ST Regulation

We now turn to consider self-tuning control of a SISO CARMA plant

$$A(d)y(t) = B(d)u(t) + C(d)e(t) \tag{8.7-1}$$

satisfying all the assumptions of Theorem 7.3-2. In particular, we focus our attention on the *minimum-variance* (MV) regulator given by (7.3-17) for $\rho = 0$:

$$Q_\tau(d)B_0(d)u(t) = -G_\tau(d)y(t) \tag{8.7-2}$$

where

$$C(d) = A(d)Q_\tau(d) + d^\tau G_\tau(d) \quad , \qquad \partial Q_\tau(d) \le \tau - 1$$

and, as usual, τ denotes the plant I/O delay. We remind you of the prediction form (7.3-9) of (1):

$$C(d)y(t+\tau) = Q_\tau(d)B_0(d)u(t) + G_\tau(d)y(t) + C(d)Q_\tau(d)e(t+\tau) \tag{8.7-3}$$

Now, take into account that by the regulation law (2) in a closed loop

$$y(t) = Q_\tau(d)e(t) \tag{8.7-4}$$

to write

$$
\begin{aligned}
y(t+\tau) &= Q_\tau(d)B_0(d)u(t) + G_\tau(d)y(t) \\
&\quad + [1 - C(d)]\,y(t+\tau) + C(d)Q_\tau(d)e(t+\tau) \\
&= Q_\tau(d)B_0(d)u(t) + G_\tau(d)y(t) \\
&\quad + \{[1 - C(d)]\,Q_\tau(d) + C(d)Q_\tau(d)\}\,e(t+\tau) \\
&= Q_\tau(d)B_0(d)u(t) + G_\tau(d)y(t) + Q_\tau(d)e(t+\tau) \tag{8.7-5}
\end{aligned}
$$

Therefore, we conclude that the MV-regulated system admits the output-prediction model (5). Note that the term $v(t+\tau) := Q_\tau(d)e(t+\tau)$ is a linear combination of $e(t+\tau), \cdots, e(t+1)$. Then, $\mathcal{E}\{\varphi(t)v(t+\tau)\} = 0$ if $\varphi(t)$ is any vector with components from y^t and u^t. Hence, by the same argument as in (6.4-13), we can conclude that (5) is a *linear-regression model* in that the coefficients of the polynomials $Q_\tau(d)B_0(d)$ and $G_\tau(d)$ can be consistently estimated via linear-regression algorithms, such as the RLS algorithm. Note also that the model (5) includes the same polynomials that are relevant for the MV-regulation law (2). Thus, if the plant (1) is under MV regulation, it is reasonable to attempt to estimate parameter vector θ of the coefficients of polynomials $Q_\tau(d)B_0(d)$ and $G_\tau(d)$ in the linear-regression model (5) via a recursive linear-regression identifier, for example, standard RLS. This is a significant simplification over the explicit or indirect approach consisting of identifying the CARMA model (1) via pseudo-linear-regression algorithms.

This route is the transposition to the present stochastic setting of the one followed in the deterministic case that led us to consider the implicit STCCs of the last two sections. We insist on pointing out that (5) is not a representation of the plant. In fact, it only yields a correct description of the evolution of the closed-loop system in stochastic steady state provided that MV regulation is used and the regulated system is internally stable. Such a closed-loop representation will be referred to as an *implicit model* and the related adaptive MV regulator briefly alluded to after (5) as an *implicit MV self-tuning (MV ST)* regulator. *We see that MV regulation acts in such a way that the original CARMA plant (1) behaves in closed-loop as the implicit linear-regression model (5).* MV regulation is not the only regulation law under which a CARMA plant admits an implicit linear-regression model. In the next chapter, we shall see that this holds true for LQS regulation as well. This implicit modeling property is important in that it can be used to construct implicit self-tuning LQS regulators.

It is not obvious that the implicit MV ST regulator based on the earlier optimistic reasoning will actually work. It is instructive to answer this issue by analyzing in detail an MV ST regulator of implicit type. For the sake of simplicity, we assume throughout that the plant I/O delay τ equals 1. In such a case, we have

$$\left.\begin{aligned}
A(d) &= 1 + a_1 d + \cdots + a_{n_a} d^{n_a} \\
B(d) &= \quad\;\; b_1 d + \cdots + b_{n_b} d^{n_b}, \quad (b_1 \neq 0) \\
C(d) &= 1 + c_1 d + \cdots + c_{n_c} d^{n_c}
\end{aligned}\right\} \tag{8.7-6}$$

$$Q_1(d) = 1 \qquad \text{and} \qquad dG_1(d) = C(d) - A(d) \tag{8.7-7}$$

the MV regulation law:

$$u(t) = -\frac{1}{b_1}\left[\sum_{i=1}^{\hat{n}}(c_i - a_i)y(t+1-i) + \sum_{i=2}^{\hat{n}}b_i u(t+1-i)\right] \tag{8.7-8}$$

and the implicit CAR model:

$$\begin{aligned}
y(t) &= (c_1 - a_1)y(t-1) + \cdots + (c_{\hat{n}} - a_{\hat{n}})y(t-\hat{n}) \\
&\quad + b_1 u(t-1) + \cdots + b_{\hat{n}}u(t-\hat{n}) + e(t) \\
&= \varphi'(t-1)\theta + e(t) \tag{8.7-9a}
\end{aligned}$$

where

$$\hat{n} \geq \max\{n_a, n_b, n_c\} \tag{8.7-9b}$$

$$\varphi(t-1) := \left[\; (y_{t-\hat{n}}^{t-1})' \quad (u_{t-\hat{n}}^{t-1})' \;\right]' \tag{8.7-9c}$$

$$\theta := \left[\; c_1 - a_1 \quad \cdots \quad c_{\hat{n}} - a_{\hat{n}} \quad b_1 \quad \cdots \quad b_{\hat{n}} \;\right]' \tag{8.7-9d}$$

8.7.2 Implicit RLS+MV ST Regulation

To show that the previous approach may lead to the desired result, we consider an example.

Example 8.7-1 (An Implicit RLS+MV ST Regulator)
Consider the CARMA plant

$$y(t+1) = -ay(t) + bu(t) + ce(t) + e(t+1) \tag{8.7-10}$$

where e is a zero-mean white noise with $\mathcal{E}\{e^2(t)\} = \sigma^2$ and $|c| < 1$. We assume that RLS with regressor $\varphi(t) := \begin{bmatrix} y(t) & u(t) \end{bmatrix}'$ are used to estimate $\theta := \begin{bmatrix} \alpha & b \end{bmatrix}'$ in the implicit CAR model:

$$y(t) = \begin{bmatrix} y(t-1) & u(t-1) \end{bmatrix} \begin{bmatrix} \alpha \\ b \end{bmatrix} + e(t) \tag{8.7-11}$$

$$\alpha := c - a$$

which results from (10) under MV regulation:

$$u(t) = -\frac{c-a}{b}y(t) \tag{8.7-12}$$

Let $\theta(t) = \begin{bmatrix} \alpha(t) & b(t) \end{bmatrix}'$ be the RLS estimate of θ based on $\left(y^t, u^{t-1}\right)$. Then the next input $u(t)$ is chosen, according to the enforced certainty equivalence, so as to satisfy $\varphi'(t)\theta(t) = 0$ or, explicitly,

$$u(t) = -\frac{\alpha(t)}{b(t)}y(t) \tag{8.7-13}$$

Such an adaptive regulator will be referred to as the *RLS+MV ST regulator*.

Assume that all variables stay bounded and $\theta(t)$ converges to $\theta(\infty) := \begin{bmatrix} \hat{a} & \hat{b} \end{bmatrix}'$. Then, by (6.3-14), $\theta(\infty)$ satisfies the normal equations

$$\lim_{t\to\infty} \frac{1}{t}\sum_{k=0}^{t-1} \begin{bmatrix} y^2(k) & y(k)u(k) \\ y(k)u(k) & u^2(k) \end{bmatrix} \begin{bmatrix} \hat{a} \\ \hat{b} \end{bmatrix} = \lim_{t\to\infty} \frac{1}{t}\sum_{k=0}^{t-1} \begin{bmatrix} y(k)y(k+1) \\ u(k)y(k+1) \end{bmatrix} \tag{8.7-14}$$

Now the L.H.S. of this equation under the asymptotic regulation law

$$u(t) = -\frac{\hat{a}}{\hat{b}}y(t) \tag{8.7-15}$$

is found to be zero. Hence, under (15), the R.H.S. of (14) is zero and this reduces to the condition

$$\lim_{t\to\infty} \frac{1}{t}\sum_{k=0}^{t-1} y(k)y(k+1) = 0 \tag{8.7-16}$$

To find the implications of (16), we use (15) into (10) to get

$$\left.\begin{array}{rcl} y(k+1) & = & -\hat{a}y(k) + ce(k) + e(k+1) \\ \hat{a} & := & a + \frac{b}{\hat{b}}\hat{a} \end{array}\right\} \tag{8.7-17}$$

Multiplying each term by $y(k)$ and taking the time average and next the limit, we get

$$
\lim_{t \to \infty} \frac{1}{t} \sum_{k=0}^{t-1} y(k)y(k+1) \;=\; \lim_{t \to \infty} \frac{1}{t} \sum_{k=0}^{t-1} \left[-\hat{a}y^2(k) + ce(k)y(k) + e(k+1)y(k) \right]
$$

$$
=\; -\hat{a}\sigma_y^2 + c\sigma^2 \tag{8.7-18}
$$

Here we have used the fact that

$$
\lim_{t \to \infty} \frac{1}{t} \sum_{k=0}^{t-1} e(k+1)y(k) = 0
$$

because of whiteness of e,

$$
\sigma_y^2 := \lim_{t \to \infty} \frac{1}{t} \sum_{k=0}^{t-1} y^2(k) \tag{8.7-19}
$$

and by (17),

$$
e(k)y(k) = \hat{a}e(k)y(k-1) + ce(k)e(k-1) + e^2(k)
$$

and, hence,

$$
\lim_{t \to \infty} \frac{1}{t} \sum_{k=0}^{t-1} e(k)y(k) = \sigma^2
$$

From (16) and (18), it follows that

$$
-\hat{a}\sigma_y^2 + c\sigma^2 = 0 \tag{8.7-20}
$$

To evaluate σ_y^2, we use (17) to get

$$
\sigma_y^2 = \hat{a}^2\sigma_y^2 - 2\hat{a}c\sigma^2 + (1+c^2)\sigma^2
$$

or

$$
\sigma_y^2 = \frac{1 - 2\hat{a}c + c^2}{1 - \hat{a}^2}\sigma^2 \tag{8.7-21}
$$

Substituting (21) in (20), we get the following quadratic equation in $\hat{a}$:

$$
-\hat{a}\frac{1 - 2\hat{a}c + c^2}{1 - \hat{a}^2} + c = 0
$$

whose roots are $\hat{a} = c$ and $\hat{a} = 1/c$. The only solution corresponding to a stable closed-loop system is

$$
\hat{a} = c \tag{8.7-22}
$$

Then, it follows from (17),

$$
\frac{\hat{\alpha}}{\hat{b}} = \frac{c - a}{b} \tag{8.7-23}
$$

which, according to (12) and (15), is the MV regulation gain. Thus, the conclusion is that if the RLS+MV adaptive regulation law (13) converges to a stabilizing regulation law, then self-tuning occurs. Note, however, that $\theta(t) = \begin{bmatrix} \alpha(t) & b(t) \end{bmatrix}'$ need not converge to $\begin{bmatrix} c - a & b \end{bmatrix}'$. In fact, it is enough if the ratio converges to $(c - a)/b$. This is ensured if $v(t)$ converges to a random multiple of θ, viz., $\lim_{t \to \infty} \theta(t) = v\theta$, where v is a scalar random variable.

8.7.3 Implicit SG+MV ST Regulation

The results of Example 1 are encouraging in that they indicate that in the RLS+MV ST regulator, self-tuning occurs whenever the adaptive regulator converges to a stabilizing compensator. This is the celebrated *w.s. self-tuning property of the adaptive RLS+MV regulator*, which, for the first time, was pointed out in the seminal paper [ÅW73]. However, no insurance of convergence is given. We next turn to discuss global convergence of an adaptive MV regulator based on a stochastic gradient (SG) identifier. The reason for this choice is that global convergence analysis for the RLS+MV ST regulator is a difficult task. For some results on this subject, see [Joh92].

Consider then (1) to (9). Let parameter vector

$$\theta = \begin{bmatrix} \alpha_1 & \cdots & \alpha_{\hat{n}} & b_1 & \cdots & b_{\hat{n}} \end{bmatrix}'$$

be estimated via the SG algorithm:

$$\theta(t) = \theta(t-1) + \frac{a\varphi(t-1)}{q(t)}\varepsilon(t) \quad , \quad a > 0 \tag{8.7-24a}$$

$$\varepsilon(t) = y(t) - \varphi'(t-1)\theta(t-1) \tag{8.7-24b}$$

$$q(t) = q(t-1) + \|\varphi(t-1)\|^2 \tag{8.7-24c}$$

with $t \in \mathbb{Z}_1$, $\theta(0) \in \mathbb{R}^{n_\theta}$, $n_\theta := 2\hat{n}$, and $q(0) > 0$. Then $u(t)$ is chosen according to the enforced certainty equivalence as follows:

$$\varphi'(t)\theta(t) = 0 \tag{8.7-25a}$$

or

$$\left.\begin{array}{l} u(t) = -\dfrac{1}{b_1(t)} \begin{bmatrix} \alpha_1(t) & \cdots & \alpha_{\hat{n}}(t) & b_2(t) & \cdots & b_{\hat{n}}(t) \end{bmatrix} s(t) \\[2mm] s(t) := \begin{bmatrix} \left(y_{t-\hat{n}+1}^t\right)' & \left(u_{t-\hat{n}+1}^{t-1}\right)' \end{bmatrix}' \in \mathbb{R}^{n_\theta - 1} \end{array}\right\} \tag{8.7-25b}$$

The algorithm (24) and (25) will be referred to as the *SG+MV ST regulator*. To analyze the algorithm, we make the following stochastic assumptions. The process $\{\varphi(0), e(1), e(2), \cdots\}$ is defined on an underlying probability space $(\Omega, \mathcal{F}, \mathbb{P})$, and we define $\mathcal{F}_0$ to be the σ-field generated by $\varphi(0)$. Further, for all $t \in \mathbb{Z}_1$, $\mathcal{F}_t$ denotes the σ-field generated by $\{\varphi(0), e(1), \cdots, e(t)\}$. The following martingale difference assumptions are adopted in process e:

$$\mathcal{E}\left\{e(t) \mid \mathcal{F}_{t-1}\right\} = 0 \quad \text{a.s.} \tag{8.7-26a}$$

$$\mathcal{E}\left\{e^2(t) \mid \mathcal{F}_{t-1}\right\} = \sigma^2 \quad \text{a.s.} \tag{8.7-26b}$$

$$\mathcal{E}\left\{e^4(t) \mid \mathcal{F}_{t-1}\right\} \leq M < \infty \quad \text{a.s.} \tag{8.7-26c}$$

$$e(t) \text{ has a strictly positive probability density} \tag{8.7-26d}$$

The last condition implies ([KV86] and [MC85]) that the event $\{b_1(t) = 0\}$ has zero probability, and hence the control law (25b) is well-defined a.s.. In the global convergence proof of the SG+MV ST regulator of next theorem, Theorem 1, which is based on the stochastic Lyapunov function method (cf. Sec. 6.4), we shall avail of the following lemma.

Lemma 8.7-1. [Cai88] **(Passivity and Positive Reality)** *Consider a time-invariant linear system with $p \times p$ transfer matrix $H(d)$. Let $H(d)$ be PR (cf. (6.4-34)). Then, the system is passive, that is, for all input sequences $\{u(k)\}_{k=0}^{\infty}$ and corresponding outputs $\{y(k)\}_{k=0}^{\infty}$:*

$$\sum_{k=0}^{N-1} u'(k)y(k) \geq K$$

for some constant K.

Theorem 8.7-1. **(Global Convergence of the SG+MV ST Regulator)** *Consider the CARMA plant (1) and (6), where the innovations e satisfy (26), regulated by the SG+MV algorithm (24) and (25), with $\hat{n}$ as in (9b). Suppose that*

$$\text{the plant is minimum-phase} \tag{8.7-27}$$

and

$$C(d) - \frac{a}{2} \text{ is SPR} \tag{8.7-28}$$

Then

$$\|\theta(t) - \theta\| \text{ converges a.s.} \tag{8.7-29}$$

the input sample paths satisfy

$$\limsup_{t \to \infty} \frac{1}{t} \sum_{k=0}^{t-1} u^2(k) < \infty \qquad a.s. \tag{8.7-30}$$

and the adaptive system is self-optimizing, that is,

$$\lim_{t \to \infty} \frac{1}{t} \sum_{k=1}^{t} y^2(k) = \sigma^2 \qquad a.s. \tag{8.7-31}$$

Proof: Let $V(k) := \|\tilde{\theta}(k)\|^2$, $\tilde{\theta}(k) := \theta(k) - \theta$, with θ as in (9d). Considering (25a), we find $\tilde{\theta}(k+1) = \tilde{\theta}(k) + aq^{-1}(k+1)\varphi(k)y(k+1)$. Hence,

$$V(k+1) = V(k) + \frac{2a}{q(k+1)}\varphi'(k)\tilde{\theta}(k)y(k+1) + \frac{a^2}{q^2(k+1)}\|\varphi(k)\|^2 y^2(k+1)$$

By taking conditional expectations,

$$\mathcal{E}\left\{V(k+1) \mid \mathcal{F}_k\right\} = V(k) + \frac{2a}{q(k+1)} \varphi'(k)\tilde{\theta}(k)\mathcal{E}\left\{y(k+1) \mid \mathcal{F}_k\right\}$$

$$+ \frac{a^2}{q^2(k+1)}\|\varphi(k)\|^2\mathcal{E}\left\{y^2(k+1) \mid \mathcal{F}_k\right\}$$

Further, from (1),

$$y(k+1) = \mathcal{E}\left\{y(k+1) \mid \mathcal{F}_k\right\} + e(k+1) \tag{8.7-32}$$

and, in turn,

$$\mathcal{E}\left\{y^2(k+1) \mid \mathcal{F}_k\right\} = \mathcal{E}^2\left\{y(k+1) \mid \mathcal{F}_k\right\} + \sigma^2 \tag{8.7-33}$$

Moreover,

$$\begin{aligned} C(d)\mathcal{E}\left\{y(k+1) \mid \mathcal{F}_k\right\} &= C(d)\left[y(k+1) - e(k+1)\right] & [(32)] \\ &= \left[C(d) - A(d)\right]y(k+1) + B(d)u(k+1) & [(1)] \\ &= \varphi'(k)\theta = -\varphi'(k)\tilde{\theta}(k) & [(25a)] \end{aligned}$$

Therefore, setting

$$z(k) := \mathcal{E}\left\{y(k+1) \mid \mathcal{F}_k\right\} \tag{8.7-34}$$

we get

$$\begin{aligned} \mathcal{E}\left\{V(k+1) \mid \mathcal{F}_k\right\} &= V(k) - \frac{2a}{q(k+1)}\left\{\left[C(d) - \frac{a}{2}\frac{\|\varphi(k)\|^2}{q^2(k+1)}\right]z(k)\right\}z(k) \\ &\quad + \frac{a^2}{q^2(k+1)}\|\varphi(k)\|^2\sigma^2 \\ &\leq V(k) - \frac{2a}{q(k+1)}\left\{\left[C(d) - \frac{a}{2}\right]z(k)\right\}z(k) \\ &\quad + \frac{a^2\sigma^2}{q^2(k+1)}\|\varphi(k)\|^2 \qquad \left[\frac{\|\varphi(k)\|}{q(k+1)} \leq 1\right] \\ &= V(k) - \frac{2a}{q(k+1)}\left\{\left[C(d) - \frac{a+\gamma}{2}\right]z(k)\right\}z(k) \\ &\quad - \frac{a\gamma}{q(k+1)}z^2(k) + \frac{a^2\sigma^2}{q^2(k+1)}\|\varphi(k)\|^2 \end{aligned}$$

where $\gamma > 0$ is such that $C(d) - (a+\gamma)/2$ is PR. Let, for an appropriate K,

$$S_k := 2a\sum_{i=0}^{k}\left\{\left[C(d) - \frac{a+\gamma}{2}\right]z(i)\right\}z(i) + K > 0$$

Such a K exists because $C(d) - (a+\gamma)/2$ is PR and by virtue of Lemma 1. Using S_k, we can write

$$\mathcal{E}\left\{V(k+1) \mid \mathcal{F}_k\right\} \leq \left[V(k) - \frac{S_k - S_{k-1}}{q(k+1)}\right] - \frac{a\gamma z^2(k)}{q(k+1)} + \frac{a\sigma^2}{q^2(k+1)}\|\varphi(k)\|^2$$

or setting

$$M(k) := V(k) + \frac{S_{k-1}}{q(k)} > 0 \tag{8.7-35}$$

because $q(k) \leq q(k+1)$, we obtain the following *near-supermartingale inequality*:

$$\mathcal{E}\left\{M(k+1) \mid \mathcal{F}_k\right\} \leq M(k) + \frac{a\sigma^2}{q^2(k+1)}\|\varphi(k)\|^2 - \frac{a\gamma}{q(k+1)}z^2(k) \qquad (8.7\text{-}36)$$

To exploit the martingale convergence theorem, Theorem D.6-1, we must check that

$$\sum_{k=0}^{\infty} \frac{\|\varphi(k)\|^2}{q^2(k+1)} < \infty \qquad \text{a.s.}$$

This follows because

$$\sum_{k=0}^{N} \frac{\|\varphi(k)\|^2}{q^2(k+1)} = \sum_{k=0}^{N} \frac{q(k+1) - q(k)}{q^2(k+1)} \leq \sum_{k=0}^{N} \frac{q(k+1) - q(k)}{q(k+1)q(k)}$$

$$= \sum_{k=0}^{N}\left[\frac{1}{q(k)} - \frac{1}{q(k+1)}\right] = \frac{1}{q(0)} - \frac{1}{q(N+1)}$$

$$\leq \frac{1}{q(0)} < \infty$$

Applying the martingale convergence theorem to (36), we obtain

$$M(k) \to M(\infty) \qquad \text{a.s.} \qquad (8.7\text{-}37)$$

and also

$$\sum_{k=0}^{\infty} \frac{z^2(k)}{q(k+1)} < \infty \qquad \text{a.s.} \qquad (8.7\text{-}38)$$

By (37), (29) follows. Because $q(k+1) > 0$, if

$$\lim_{k \to \infty} q(k) = \infty \qquad \text{a.s.} \qquad (8.7\text{-}39)$$

by Kronecker's lemma (Result D.6-1), we conclude that

$$\lim_{N \to \infty} \frac{1}{q(N+1)} \sum_{k=0}^{N} z^2(k) = 0 \qquad \text{a.s.} \qquad (8.7\text{-}40)$$

We show that (39) holds by contradiction. Suppose that $\lim_{k \to \infty} q(k) < \infty$. This implies that $\lim_{k \to \infty} y^2(k) = \lim_{k \to \infty} u^2(k) = 0$. From this, because of (1), it follows that $\lim_{k \to \infty} e(k) = 0$. But this happens only on a set in Ω of zero probability measure because by (26)

$$\lim_{k \to \infty} \frac{1}{N} \sum_{k=1}^{N} e^2(k) = \sigma^2 \qquad \text{a.s.} \qquad (8.7\text{-}41)$$

Hence, (39) follows.

We next want to show that

$$\left\{\frac{q(t)}{t+1}\right\}_{t=0}^{\infty} \qquad \text{is bounded a.s.} \qquad (8.7\text{-}42)$$

from which (30) follows at once. The plant is minimum-phase, so we can apply the same argument used after (5-14) to show that

$$\frac{1}{t}\sum_{k=0}^{t-1}u^2(k) \;\leq\; \frac{1}{t}\sum_{k=0}^{t-1}\left[c_1y^2(k+1)+c_2e^2(k+1)\right]$$

$$\leq\; \frac{c_1}{t}\sum_{k=0}^{t-1}y^2(k+1)+c_3 \qquad [(41)]$$

Therefore,

$$\frac{q(t+1)}{t+1} \;\leq\; \frac{c_4}{t+1}\sum_{k=0}^{t}y^2(k+1)+c_5$$

$$\leq\; \frac{c_6}{t+1}\sum_{k=0}^{t}z^2(k)+c_7$$

where the last inequality with $c_6 > 0$ follows from (32), (34), and (41). Then we have

$$\frac{1}{q(t+1)}\sum_{k=0}^{t}z^2(k) \geq \frac{q(t+1)-(t+1)c_7}{c_6q(t+1)}$$

Suppose now that (42) is not true. Then along some subsequence $\{t_k\}$, we get

$$\lim_{k\to\infty}\frac{1}{q(t_k+1)}\sum_{i=0}^{t_k}z^2(i) \geq \frac{1}{c_6} > 0$$

which contradicts (40). Then (42) holds.

We finally prove (31).

$$\sum_{k=1}^{t}y^2(k) = \sum_{k=1}^{t}\left[z^2(k-1)+2z(k-1)e(k)+e^2(k)\right] \qquad [(32)\text{ and }(34)]$$

By the Schwarz inequality,

$$\sum_{k=1}^{t}z(k-1)e(k) \leq \left[\sum_{k=1}^{t}z^2(k-1)\right]^{1/2}\left[\sum_{k=1}^{t}e^2(k)\right]^{1/2}$$

By (40) and (42),

$$\lim_{t\to\infty}\frac{1}{t}\sum_{k=1}^{t}z^2(k-1) = 0 \qquad \text{a.s.} \qquad (8.7\text{-}43)$$

(31) follows by virtue of (41).

Theorem 8.7-2. [KV86] *Under the same assumptions as in Theorem 1 with the only exception of (28) replaced by*

$$C(d) \text{ is SPR} \qquad (8.7\text{-}44)$$

the SG+MV ST regulator for every a ≠ 0 in (24) has all the properties of Theorem 1 except possibly for (29).

Problem 8.7-1 Prove Theorem 2. (*Hint*: Use Theorem 1, the fact that the control law (25) is invariant if $\theta(t)$ is changed into $\alpha\theta(t)$, $\alpha \in \mathbb{R}$, and that if $\theta(0)$ and a are changed into $\alpha\theta(0)$ and αa, respectively, $\theta(t)$ is changed into $\alpha\theta(t)$ in the SG+MV ST regulated system, as can be proved by induction on t.)

Conditions under which self-tuning occurs follow.

Fact 8.7-1. [KV86] *Assume the conditions of Theorem 2 and that the plant has no reduced order MV regulator than (8). Then, for the SG+MV ST regulator, we have*

$$\lim_{t\to\infty} \theta(t) = v\theta \tag{8.7-45}$$

for some random variable v.

Note that the order of the MV regulator (8) can be always reduced under (9b) unless $\hat{n} = \max(n_a, n_b, n_c)$.

Minimum-Variance Self-Tuning Trackers   We consider again the SISO CARMA plant (1), (6) and (26). Our interest is to find the control law that minimizes in stochastic steady state the performance index:

$$\mathcal{C} = \mathcal{E}\left\{ [y(t+\tau) - r(t+\tau)]^2 \mid y^t, r^{t+\tau} \right\} \tag{8.7-46}$$

where r is a preassigned output reference. Along the lines of Sec. 7.3, we find for the optimal control law

$$Q_\tau(d)B_0(d)u(t) = -G_\tau(d)y(t) + C(d)r(t+\tau) \tag{8.7-47}$$

which reduces to (2) in the pure regulation problem $r(t) \equiv 0$. The control law (47), which will be referred to as *minimum-variance (MV) control*, yields (cf. Problem 7.3-1) a stable closed-loop system if and only if $B_0(d)$ is strictly Hurwitz, that is, the plant is minimum-phase. This is, therefore, an assumption that we shall adopt throughout the section.

As with MV regulation, the next step is to derive an implicit model for the controlled system. To this end, recall the prediction form (3) of (1). Take into account that by (47) in a closed loop,

$$y(t) = r(t) + Q_\tau(d)e(t) \tag{8.7-48}$$

to find

$$
\begin{aligned}
y(t+\tau) &= Q_\tau(d)B_0(d)u(t) \\
&\quad + G_\tau(d)y(t) + [1 - C(d)]\,r(t+\tau) + Q_\tau(d)e(t+\tau)
\end{aligned}
\tag{8.7-49}
$$

or denoting the tracking error by

$$\varepsilon_y(t) := y(t) - r(t) \tag{8.7-50}$$

$$\begin{aligned}
\varepsilon_y(t+\tau) &= Q_\tau(d)B_0(d)u(t) \\
&\quad + G_\tau(d)y(t) - C(d)r(t+\tau) + Q_\tau(d)e(t+\tau)
\end{aligned} \tag{8.7-51}$$

Therefore, we conclude that the MV–controlled system evolves in accordance with the linear-regression model (51), where reference r appears as an exogenous input. The coefficient of the polynomials of this model can be estimated by a linear-regression algorithm, for example, SG or RLS. In fact, (51) can be rewritten as

$$\varepsilon_y(t) = \varphi'(t-\tau)\theta + Q_\tau(d)e(t) \tag{8.7-52a}$$

$$\varphi(t) := \left[\ \left(y_{t-n_y}^t\right)'\ \ \left(u_{t-n_u}^t\right)'\ \ \left(-r_{t+\tau-n_c}^{t+\tau}\right)\ \right]' \tag{8.7-52b}$$

$$\theta := \left[\ \alpha_0\ \ \cdots\ \ \alpha_{n_y}\ \ \beta_0\ \ \cdots\ \ \beta_{n_u}\ \ 1\ \ c_1\ \ \cdots\ \ c_{n_c}\ \right]' \tag{8.7-52c}$$

and (47) is equivalent to

$$\varphi'(t)\theta = 0 \tag{8.7-53}$$

In (52), n_y, n_u, and n_c denote the degrees of polynomials $G_\tau(d)$, $Q_\tau(d)B_0(d)$, and $C(d)$, respectively, and α_i and β_i the coefficients of $G_\tau(d)$ and $Q_\tau(d)B_0(d)$, respectively. The previous discussion, along with the convergence results on SG+MV ST regulation, motivates the following ST controller:

$$\theta(t) = \theta(t-1) + \frac{a\varphi(t-\tau)}{q(t-\tau+1)}\varepsilon(t)\ \ ,\ \ a > 0 \tag{8.7-54a}$$

$$\varepsilon(t) = \varepsilon_y(t) - \varphi'(t-\tau)\theta(t-1) \tag{8.7-54b}$$

$$q(t) = q(t-1) + \|\varphi(t-1)\|^2 \tag{8.7-54c}$$

with $t \in \mathbb{Z}_1$, $\theta(0) \in \mathbb{R}^{n_\theta}$, and $q(1-\tau) > 0$. Further, $u(t)$ is chosen according to the enforced certainty equivalence as follows:

$$\varphi'(t)\theta(t) = 0 \tag{8.7-55a}$$

or

$$u(t) = -\frac{1}{\beta_0(t)} \tag{8.7-55b}$$

$$\times \left[\ \alpha_0(t)\ \ \cdots\ \ \alpha_{n_y}(t)\ \ \beta_1(t)\ \ \cdots\ \ \beta_{n_u}(t)\ \ c_0(t)\ \ c_1(t)\ \ \cdots\ \ c_{n_c}(t)\ \right]' s(t)$$

where

$$s(t) := \left[\ \left(y_{t-n_y}^t\right)'\ \ \left(u_{t-n_u}^{t-1}\right)'\ \ \left(-r_{t+\tau-n_c}^{t+\tau}\right)'\ \right]' \tag{8.7-55c}$$

The algorithm (54) and (55) will be referred to as the *SG+MV ST controller*. By using the stochastic Lyapunov function method as in Theorem 1, it can be shown that this adaptive controller applied to the CARMA plant is self-optimizing.

Theorem 8.7-3. **(Global Convergence of the SG+MV ST Controller)** *Provided that $\{r(k)\}_{k=1}^{\infty}$ is a bounded sequence, under the same conditions of Theorem 2 with $a > 0$, we have for the SG+MV ST algorithm applied to the CARMA plant:*

- $\|\theta(t) - \theta\|$ *converges a.s.*   (8.7-56)

- *The input paths satisfy*

$$\limsup_{t \to \infty} \frac{1}{t} \sum_{k=0}^{t-1} u^2(k) < \infty \qquad a.s. \qquad (8.7\text{-}57)$$

- *Self-optimization occurs:*

$$\lim_{t \to \infty} \frac{1}{t} \sum_{k=1}^{t} [y(k) - r(k)]^2 = \sigma^2 \qquad a.s. \qquad (8.7\text{-}58)$$

Note that no convergence result for $\theta(t)$ is included in Theorem 3. Nonetheless, a result similar to the one in Fact 1 can be proved. However, because the regressor of the SG+MV control algorithm includes reference samples, here conditions under which self-tuning occurs, viz., $\lim_{t \to \infty} \theta(t) = v\theta$ for some random variable v, involve that the reference trajectory be sufficiently rich in an appropriate sense [KV86].

Main Points of the Section. CARMA plants under minimum-variance control admit implicit models of a linear-regression type. This fact can be exploited so as to properly construct implicit minimum-variance ST control algorithms whose global convergence can be proved via the stochastic Lyapunov equation method.

8.8 GENERALIZED MINIMUM-VARIANCE SELF-TUNING CONTROL

We next focus our attention on self-tuning schemes whose underlying control problem is the generalized minimum-variance (GMV) control. For a SISO CARMA plant,

$$A(d)y(t) = d^\tau B_0(d)u(t) + C(d)e(t) \qquad (8.8\text{-}1)$$

we found in Sec. 7.3 that the GMV control law equals

$$\left[\frac{\rho}{b_\tau} C(d) + Q_\tau(d) B_0(d) \right] u(t) = -G_\tau(d)y(t) + C(d)r(t + \tau) \qquad (8.8\text{-}2)$$

and the d-characteristic polynomial of the GMV-controlled system is given by

$$\chi_{cl}(d) = \gamma \left[\frac{\rho}{b_\tau} A(d) + B_0(d) \right] C(d) \qquad (8.8\text{-}3)$$

Provided that $\chi_{cl}(d)$ is strictly Hurwitz, (2) minimizes in stochastic steady state the conditional expectation

$$\mathcal{C} = \mathcal{E}\left\{ [y(t+\tau) - r(t+\tau)]^2 + \rho u^2(t) \mid y^t, r^{t+\tau} \right\} \tag{8.8-4}$$

In order to exploit the result obtained on SG+MV ST control, we show next that *GMV control is equivalent to MV control for a suitably modified plant*. To see this, by using (7-3), rewrite the polynomial on the L.H.S. of (2) as follows:

$$\left[\frac{\rho}{b_\tau}C(d) + Q_\tau(d)B_0(d)\right] = Q_\tau(d)\left[B_0(d) + \frac{\rho}{b_\tau}A(d)\right] + \frac{\rho}{b_\tau}d^\tau G_\tau(d) \tag{8.8-5}$$

Hence, the GMV control (2) is equivalently given by

$$\left.\begin{aligned} Q_\tau(d)\left[B_0(d) + \frac{\rho}{b_\tau}A(d)\right]u(t) &= -G_\tau(d)\bar{y}(d) + C(d)r(t+\tau) \\[2mm] \bar{y}(t) &:= y(t) + \frac{\rho}{b_\tau}u(t-\tau) \end{aligned}\right\} \tag{8.8-6}$$

Further,

$$\begin{aligned} A(d)\bar{y}(t) &= A(d)\left[y(t) + \frac{\rho}{b_\tau}u(t-\tau)\right] \\[2mm] &= d^\tau\left[B_0(d) + \frac{\rho}{b_\tau}A(d)\right]u(t) + C(d)e(t) \end{aligned} \tag{8.8-7}$$

By comparison with (7-47), it is immediate to recognize that (6) is the same as the MV control law for the modified plant (7) with output $\bar{y}(t)$. This conclusion is depicted in Fig. 1.

The equivalence between GMV and MV can be used to design globally convergent adaptive GMV controllers. If b_τ is a priori known, this can be done as follows. Modify the plant as shown in Fig. 1. Then use the SG+MV ST algorithm (7-55) and (7-56), for the modified plant, viz., replace in all equations $y(t)$ with the new output variable $\bar{y}(t)$. Hence, for the adaptive pure regulation problem, the conclusions of Theorems 7-1 and 7-2 are directly applicable to this new situation. Notice, however, that the minimum-phase plant condition means here that the polynomial $[B_0(d) + (\rho/b_\tau)A(d)]$ is strictly Hurwitz. For adaptive GMV tracking, see Problem 2 which follows.

Main Points of the Section. GMV control is equivalent to MV control of a suitably modified CARMA plant. If the leading coefficient b_τ of $B(d)$ is known, this fact can be exploited to construct a globally convergent SG+GMV control algorithm by the direct use of the results of Sec. 7.

Problem 8.8-1 (Global Convergence of the SG-GMV Controller)   Express the conclusions of the analog of Theorem 7-3 for the SG-GMV controller, based on the equivalence between GMV and MV, applied to the CARMA plant (1).

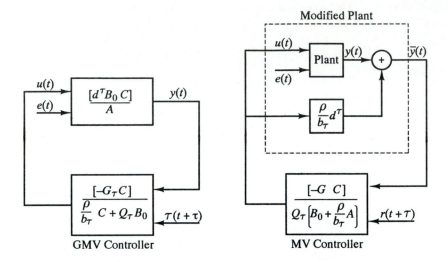

Figure 8.8-1: The original CARMA plant controlled by the GMV controller on the L.H.S. is equivalent to the modified CARMA plant controlled by the MV controller on the R.H.S.

Problem 8.8-2 (SG-GMV Controller with Integral Action) Following the discussion throughout (7.5-23) to (7.5-28), construct a globally convergent SG-GMV controller for the CARMA plant (7.5-26) yielding at convergence an offset-free closed-loop system and rejection of a constant disturbance.

Problem 8.8-3 Suppose that b_τ is unknown and, hence, an assumed $\hat{b}_\tau$ is used in (6) instead of b_τ. Find the closed-loop d-characteristic polynomial and the cost minimized in stochastic steady state by this new control law applied to the plant (1).

Problem 8.8-4 (MV Regulation with Polynomial Weights) Find the MV regulation problem equivalent to the following GMV regulation problem: Minimize in stochastic steady state the conditional cost $\mathcal{C} = \mathcal{E}\left\{[W_y(d)y(t+\tau)]^2 + [W_u(d)u(t)]^2 \mid y^t\right\}$, where $W_y(d)$ and $W_u(d)$ are polynomials such that $W_y(0) = 1$ and $W_u(0) \neq 0$ for the CARMA plant (1).

8.9   ROBUST SELF-TUNING CHEAP CONTROL

As pointed out earlier in this chapter, the motivation behind the development of adaptive control was the need to account for uncertainty in the structure and parameters of the physical plants. So far we have found that for self-tuners with underlying myopic control several reassuring results are available, provided that the physical plant is exactly described by the adopted linear system model once its unknown parameters are properly adjusted. These, together with similar results for MRAC systems, were mainly obtained in the late 1970s–early 1980s. At the beginning of the 1980s, it became clear that an adaptive control system designed for the case of an exact plant model can become unstable in the presence of unmodeled external disturbances or small modeling errors. As a result, the issue of robustness of adaptive systems has become of crucial interest.

In order to obtain improvements in stability, various modifications of the algorithms originally designed for the ideal case have been proposed. In this connection, projection of the parameter estimates onto a given fixed convex set can be adopted to prove stability properties. This, however, requires that appropriate prior knowledge on the plant is available. Consequently, the use of projection is hampered whenever such a prior knowledge is unavailable. Another way to deal with modeling errors is to combine in the identifier data normalization and the dead zone. Data normalization is used to transform a possibly unbounded dynamics component into a bounded disturbance. The dead zone facility is used to switch off the estimate update whenever the prediction error is small comparatively to the expected size of a disturbance upper bound.

In this section, we discuss a possible approach to the robustification of STCC based on data normalization and the dead zone. Further, data prefiltering for identification and dynamic weights for control synthesis, which are very important in practice, are also described in some detail. STCC has been discussed under model-matching conditions in Secs. 5 and 6. Similar robustification tools will be used in Sec. 9.1, where we shall consider indirect adaptive predictive control in the presence of bounded disturbances and neglected dynamics.

8.9.1   Reduced-Order Models

We consider a SISO plant of the form

$$A^o(d)y(t) = B^o(d)u(t) + A^o(d)\left[\pi(t) + \omega(t)\right] \qquad (8.9\text{-}1a)$$

where $\pi(t)$ and $\omega(t)$ denote a predictable and a stochastic disturbance, respectively. In particular, we assume that

$$\Pi(d)\pi(t) = 0 \qquad (8.9\text{-}1b)$$

for a monic polynomial $\Pi(d)$ with all its roots of unit multiplicity and on the unit circle. In order to take into account constant disturbances, we also assume that $\Pi(1) = 0$, that is,

$$\Delta(d) \mid \Pi(d) \qquad (8.9\text{-}1c)$$

with $\Delta(d) = 1 - d$. Equation (1a) can be rewritten as follows:

$$A_o(d)y(t) = B_o(d)u(t) + A_o(d)\omega(t) \qquad (8.9\text{-}2a)$$

where

$$A_o(d) := A^o(d)\Pi(d) \qquad\qquad B_o(d) := B^o(d)\Pi(d) \qquad (8.9\text{-}2b)$$

We consider the situation wherein (2) is represented by a lower-order model as follows:

$$A(d)y(t) = B(d)u(t) + n(t) \qquad (8.9\text{-}3a)$$

where

$$A_o(d) = A(d)A^u(d) \qquad B_o(d) = B(d)B^u(d) \qquad (8.9\text{-}3b)$$

and

$$n(t) = \frac{B(d)\left[B^u(d) - A^u(d)\right]}{A^u(d)}u(t) + A(d)\omega(t) \qquad (8.9\text{-}3c)$$

In (3b), $A^u(d)$ and $B^u(d)$ are monic polynomials and superscript u stands for *unmodeled*. Because $B^u(d)$ is monic, $\operatorname{ord}B(d) = \operatorname{ord}\mathcal{B}_o(d)$. Hence, plant I/O delay is retained by the modeled part of (3a). We point out that the model (3a) to (3c) is another way of writing (2). However, we shall use (3a) in adaptive control without taking into account (3c). Specifically, the idea is to identify the parameters of the *reduced-order model* (3a), viz., polynomials $A(d)$ and $B(d)$, via a standard recursive identification algorithm, and for control design, use only the estimated $A(d)$ and $B(d)$ in place of $A_o(d)$ and $\mathcal{B}_o(d)$ and possibly the properties of process ω. In this way, we design a reduced-complexity controller by neglecting the *unmodeled dynamics* of the plant. In (3a), the latter are accounted for by the term $n(t)$ as given in (3c).

8.9.2 Prefiltering the Data

It is crucial to realize that the factorizations (3b) are not unique. Then, it follows that there are many candidate polynomial pairs $A(d)$ and $B(d)$ for the reduced-order model (3a). To each candidate pair, there corresponds an unmodeled dynamics disturbance $n(t)$ via (3c). Because our ultimate goal is to use (3a) for control design, our preference must go to those polynomial pairs making $n(t)$ small in the useful frequency band. In fact, it is to be expected that the closer the approximation of $A(d)$ and $B(d)$ to $A_o(d)$ and $\mathcal{B}_o(d)$ inside the useful frequency band, the better the reduced-complexity controller will behave. Because $A(d)$ and $B(d)$ are obtained via a recursive identification algorithm, for example, RLS, whereby the output prediction error is minimized in a mean-square sense (cf. (6.3-32)), an effective reduction of the mean-square value of $n(t)$ within the useful frequency band can be accomplished via the following filtering procedure.

The data to be sent to the identifier $y_L(t)$ and $u_L(t)$ are obtained by prefiltering the plant I/O variables

$$y_L(t) = L(d)y(t) \qquad u_L(t) = L(d)u(t) \qquad (8.9\text{-}4a)$$

where $L(d)$ is a stable low-pass transfer function that rolls off at frequencies beyond the useful band. In such a way, the identifier fits a model to the I/O process $\{y_L(t), u_L(t-\tau)\}$, τ being the plant I/O delay, described by the difference equation

$$A(d)y_L(t) = B(d)u_L(t) + n_L(t) \qquad (8.9\text{-}4b)$$

where $n_L(t) := L(d)n(t)$. Note that $n_L(t)$ has most of its power within the useful frequency band. Hence, the identifier, choosing $A(d)$ and $B(d)$ so as to minimize the overall power of $n_L(t)$, is forced by the prefiltering action of $L(d)$ to select, amongst the candidate polynomial pairs, one that can satisfactorily fit $A_o(d)$ and $B_o(d)$ within the useful frequency band.

8.9.3 Dynamic Weights

Having the identified polynomials $A(d)$ and $B(d)$ at a given time , we could proceed to compute the control law according to the enforced certainty equivalence procedure by referring to the model (3a) under the assumption that $n(t) = n_L(t)/L(d)$ is negligible or white, viz., its power being equally distributed over all frequencies. However, because our strategy has been to select $A(d)$ and $B(d)$ in order to well approximate $A_o(d)$ and $B_o(d)$ within the useful frequency band, we must expect that $n(t)$ is large at high frequencies. Then, in order to reduce the effect of the neglected dynamics on the controlled system, we take advantage of the considerations made after (7.5-29). To this end, we consider the filtered variables

$$y_H(t) := H(d)y(t) \qquad u_H(t) := H(d)u(t) \qquad (8.9\text{-}5a)$$

with $H(d)$ a monic strictly Hurwitz high-pass polynomial, and the related model

$$A(d)y_H(t) = B(d)u_H(t) + H(d)n(t) \qquad (8.9\text{-}5b)$$

We know that $n(t)$ is large at high frequencies. Nevertheless, for control design purposes, we act as if n were a zero-mean white noise. We then compute the MV control law minimizing in stochastic steady state

$$\mathcal{E}\left\{ \left[y_H(t) - H(1)r(t) \right]^2 \,\middle|\, y_H^t \right\} \qquad (8.9\text{-}6)$$

for the plant (5) as if it were a CARMA model. Notice that this is the same as computing the MV control law for the *dynamically weighted output* $y_H(t)$ and the model (3a) with n assumed to be white. Assuming also, for the sake of simplicity, the plant I/O delay τ equal to 1, by Theorem 7.3-2, we find the MV control law:

$$B(d)u_H(t) + [H(d) - A(d)]\,y_H(t) = H(d)H(1)r(t) \qquad (8.9\text{-}7)$$

It then follows that, for the output of the controlled system (5) and (7), in stochastic steady state,

$$y_H(t) = H(1)r(t) + n(t)$$

or

$$y(t) = \frac{H(1)}{H(d)}r(t) + \frac{n(t)}{H(d)} \qquad (8.9\text{-}8)$$

From (8), we see that the use of a high-pass Hurwitz polynomial $H(d)$, such that $1/H(d)$ rolls off at frequencies outside the desired closed-loop bandwidth, is beneficial for both shaping the reference and attenuating the effect of the neglected dynamics.

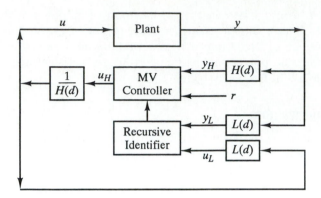

Figure 8.9-1: Block scheme of a robust adaptive MV control system involving a low-pass filter $L(d)$ for identification, and a high-pass filter $H(d)$ for the control synthesis.

Figure 1 depicts a robust adaptive MV control system designed in accordance with the previous criteria and involving a low-pass filter $L(d)$ for identification and a high-pass filter $H(d)$ for control synthesis.

8.9.4   CT-NRLS with Dead Zone and STCC

The adoption of data prefiltering for identification and dynamic control weights as discussed so far can be very effective for counteracting a plant order underestimation. This situation is almost the rule in practice, being usually the physical system to be controlled more complex than the model adopted for control synthesis purposes. Under such circumstances, however, the earlier two filtering actions are insufficient to construct globally convergent adaptive control systems. We now see that a self-tuning cheap control system can be made globally stable in the presence of neglected dynamics by using an RLS identifier equipped with both data normalization and a dead zone. The latter facility is used to freeze the estimates whenever the absolute value of the prediction error becomes smaller than a disturbance upper bound. For the sake of simplicity, we do not explicitly use data prefiltering or dynamic weights in the scheme that is adopted for analysis, because it is always possible to add suitable filtering actions so as to make robust the adaptive control system, as indicated earlier in this section. As an additional simplification, we assume that the plant I/O delay τ equals 1. Then, we consider the plant

$$A(d)y(t) = B(d)u(t) + n(t) \qquad (8.9\text{-}9a)$$

where

$$A(d) \quad := \quad 1 + a_1 d + \cdots + a_{n_a} d^{n_a} \qquad (8.9\text{-}9b)$$

$$B(d) \quad := \quad b_1 d + \cdots + b_{n_b} d^{n_b} \qquad (8.9\text{-}9\text{c})$$

and, similarly to (3a), $n(t)$ includes the effects of unmodeled dynamics. To assume from the outset that $n(t)$ is uniformly bounded would be very limiting in that (cf. (3c)) $n(t)$ depends on unmodeled dynamics and the control law. We assume instead that

$$|n(t)| \leq \mu m_o(t-1), \quad t \in \mathbb{Z}_1 \qquad (8.9\text{-}9\text{d})$$

where the nonnegative real $m_o(t)$ is given by

$$m_o(t) = \alpha m_o(t-1) + \|\varphi(t)\| \qquad (8.9\text{-}9\text{e})$$

for $\alpha \in [0,1)$, $0 < \mu < \infty$, and

$$\varphi(t-1) := \left[\; \left(-y_{t-n_a}^{t-1}\right)' \quad \left(u_{t-n_b}^{t-1}\right)' \; \right]' \qquad (8.9\text{-}9\text{f})$$

The next problem proves that if $n(t)$ is related to $u(t)$ (cf. (3c)) by a stable transfer function and $\omega(t)$ is uniformly bounded, then (9) is satisfied.

Problem 8.9-1 Consider a disturbance $n(t)$ as in (3c) with $\omega(t)$ uniformly bounded and $A^u(d)$ strictly Hurwitz. Show that

$$|n(t)| \leq \mu_\omega + \mu m_u(t-1)$$

$$m_u(t) = \alpha m_u(t-1) + |u(t-1)|$$

for $\alpha \in [0,1)$, and μ_ω and μ positive bounded reals.

Setting

$$\theta := \left[\; a_1 \quad \cdots \quad a_{n_a} \quad b_1 \quad \cdots \quad b_{n_b} \; \right]' \qquad (8.9\text{-}10\text{a})$$

(9a) to (9c) become

$$y(t) = \varphi'(t-1)\theta + n(t) \qquad (8.9\text{-}10\text{b})$$

If as in (8.6-1) we introduce the normalization factor

$$m(t-1) := \max\{m, m_o(t-1)\}, \quad m > 0 \qquad (8.9\text{-}11\text{a})$$

and the normalized data

$$\gamma(t) := \frac{y(t)}{m(t-1)} \qquad x(t-1) := \frac{\varphi(t-1)}{m(t-1)} \qquad \bar{n}(t) := \frac{n(t)}{m(t-1)} \qquad (8.9\text{-}11\text{b})$$

we get

$$\gamma(t) = x'(t-1)\theta + \bar{n}(t) \qquad (8.9\text{-}11\text{c})$$

with

$$|\bar{n}(t)| = \left| \frac{n(t)}{m(t-1)} \right| \leq \left| \frac{n(t)}{m_o(t-1)} \right| \leq \mu \qquad (8.9\text{-}12)$$

Thus, data normalization and property (9d) allow us to transform (10b) with the possibly unbounded sequence $\{n(t)\}$ into (11c) with the uniformly bounded disturbance $\bar{n}(t)$.

We next consider the following identification algorithm.

CT-NRLS with Relative Dead Zone (RDZ-CT-NRLS).   (Cf. (6-19).)

$$\theta(t) = \theta(t-1) + \kappa(t)\frac{a(t)P(t-1)x(t-1)}{1 + a(t)x'(t-1)P(t-1)x(t-1)}\bar{\varepsilon}(t) \qquad (8.9\text{-}13a)$$

$$P(t) = \frac{1}{\lambda(t)}\left[P(t-1) - \kappa(t)\frac{a(t)P(t-1)x(t-1)x'(t-1)P(t-1)}{1 + a(t)x'(t-1)P(t-1)x(t-1)}\right] \qquad (8.9\text{-}13b)$$

$$\lambda(t) = 1 - \frac{\kappa(t)}{\text{Tr}[P(0)]}\frac{a(t)x'(t-1)P^2(t-1)x(t-1)}{1 + a(t)x'(t-1)P(t-1)x(t-1)} \qquad (8.9\text{-}13c)$$

where $a(t)$ is as in (5-8c),

$$\varepsilon(t) = y(t) - \varphi'(t-1)\theta(t-1) \qquad (8.9\text{-}13d)$$

$$\bar{\varepsilon}(t) = \frac{\varepsilon(t)}{m(t-1)} = \gamma(t) - x'(t-1)\theta(t-1) \qquad (8.9\text{-}13e)$$

and

$$\kappa(t) = \begin{cases} \kappa \in \left(0, \frac{\epsilon}{1+\epsilon}\right), & \text{if } |\bar{\varepsilon}(t)| \ge (1+\epsilon)^{1/2}\mu \\ 0, & \text{otherwise} \end{cases} \qquad (8.9\text{-}13f)$$

with $\epsilon > 0$. The algorithm is initialized from any $\theta(0)$ with $\theta_{n_a+1}(0) \ne 0$ and any $P(0) = P'(0) > 0$. The dead-zone facility (13f) is also called a *relative dead zone (RDZ)* in that it freezes the estimate whenever the absolute value of the prediction error becomes smaller than a quantity depending on the norm of the regression vector.

Problem 8.9-2   Consider the estimation algorithm (13). Let

$$V(t) := \tilde{\theta}'(t)P^{-1}(t)\tilde{\theta}(t) \qquad (8.9\text{-}14)$$

and $\tilde{\theta}(t) := \theta(t) - \theta$. Show that

$$V(t) = \lambda(t)\left\{V(t-1) - \kappa(t)a(t)\left[\frac{\bar{\varepsilon}^2(t)}{1 + Q(t)} - \frac{\bar{n}^2(t)}{1 + [1 - \kappa(t)]Q(t)}\right]\right\} \qquad (8.9\text{-}15a)$$

with

$$Q(t) := a(t)x'(t-1)P(t-1)x(t-1) \qquad (8.9\text{-}15b)$$

(*Hint*: Use (5.3-16) to show that

$$P^{-1}(t) = \lambda(t)\left[P^{-1}(t-1) + \kappa(t)a(t)\frac{x(t-1)x'(t-1)}{1 + [1 - \kappa(t)]Q(t)}\right] \qquad (8.9\text{-}16)]$$

It is easy to check that (15a) can be rewritten as follows:

$$
\begin{aligned}
V(t) \;=\; & \lambda(t)\Big\{ V(t-1) \\[4pt]
& - \kappa(t)a(t)\left[\left(\frac{\epsilon}{1+\epsilon}\right)\frac{1+\left[1-\left(\frac{1+\epsilon}{\epsilon}\right)\kappa(t)\right]Q(t)}{\left[1+Q(t)\right]\left\{1+\left[1-\kappa(t)\right]Q(t)\right\}}\,\bar{\varepsilon}^2(t)\right. \\[4pt]
& \left. + \frac{\bar{\varepsilon}^2(t)-(1+\epsilon)\bar{n}^2(t)}{(1+\epsilon)\left\{1+\left[1-\kappa(t)\right]Q(t)\right\}}\right]\Big\}
\end{aligned}
\tag{8.9-17}
$$

This equation shows that for every $\epsilon > 0$, the dead-zone facility (13f) ensures that $\{V(t)\}_{t=0}^{\infty}$ is monotonically nonincreasing:

$$
V(t) \leq \lambda(t)V(t-1) \leq V(t-1)
\tag{8.9-18}
$$

where the latter inequality follows from (13c). Hence, $V(t)$ converges as $t \to \infty$. We can then establish the following result.

Proposition 8.9-1. (RDZ-CT-NRLS) *Consider the RDZ-CT-NRLS algorithm (13) with the data-generating mechanism (10) to (12). Then, the following properties hold:*

(i) *Uniform boundedness of the estimates:*

$$
\|\tilde{\theta}(t)\|^2 \leq \frac{\mathrm{Tr}[P(0)]}{\lambda_{\min}[P(0)]}\|\tilde{\theta}(0)\|^2
\tag{8.9-19}
$$

(ii) *Finite-time convergence inside the dead zone, viz., there is a finite integer T_1 such that for all $t > T_1$,*

$$
\kappa(t) = 0
\tag{8.9-20}
$$

or

$$
|\bar{\varepsilon}(t)| < (1+\epsilon)^{1/2}\mu
\tag{8.9-21}
$$

(iii) *Estimate convergence in finite time:*

$$
\theta(t) = \theta_{\infty} := \lim_{k\to\infty}\theta(k), \quad \forall t > T_1
\tag{8.9-22}
$$

Proof:

(i) Equation (19) follows by the same argument used to get (4-7).

(ii) Let

$$
\mathcal{T} := \left\{ t \in \mathbb{Z}_{+} \mid \kappa(t) \underset{\neq}{>} 0 \right\}
$$

It will be proved by contradiction that $\mathcal{T}$ has a finite number of elements. Suppose that this is untrue. Then there is a sequence $\{t_k\}_{k=1}^{\infty}$ in $\mathcal{T}$ with $\lim_{k\to\infty} t_k = \infty$. Then from (18), it follows that $\lim_{k\to\infty} \bar{\varepsilon}^2(t_k) = 0$. This in turn implies that there is a $t_1 \in \mathcal{T}$ such that for all $t_k \in \mathcal{T}$, $t_k > t_1$, $|\bar{\varepsilon}(t_k)| < (1+\epsilon)^{1/2}\mu$. Consequently, $\kappa(t_k) = 0$ or $t_k \bar{\in} \mathcal{T}$: a contradiction.

(iii) Equation (22) follows trivially from (ii).

STCC.   Given the estimate $\theta(t)$ of θ obtained by the dead-zone CT-NRLS algorithm, control variable $u(t)$ is selected by solving w.r.t. $u(t)$ the equation

$$\varphi'(t)\theta(t) = r(t+1) \qquad\qquad (8.9\text{-}23)$$

We shall refer to the algorithm (13) and (23), when applied to (9), as the *reduced-order STCC* or the *RDZ-CT-NRLS+CC* algorithm. By virtue of (23), (21) yields

$$\frac{|y(t) - r(t)|}{m(t-1)} < (1+\epsilon)^{1/2}\mu \quad , \qquad \forall t > T_1$$

If $m(t-1)$, $t \in \mathbb{Z}_1$, is uniformly bounded, the last inequality indicates that the tracking error is upper bounded by $(1+\epsilon)^{1/2}$ times the disturbance upper bound $\mu m(t-1)$, for $n(t)$ as given by (12).

In order to prove the crucial issue of boundedness, we resort to a variant of the key technical lemma based on (21). We first show that the linear boundedness condition (3-10) holds under the following assumption. For the plant (1), among the reduced-order models (9) with all the stated properties, there is one such that

$$\frac{B(d)}{dA(d)} \quad \text{is stably invertible} \qquad\qquad (8.9\text{-}24)$$

As indicated after (5-15), this condition, and because by (23), $\varepsilon(t) = y(t) - r(t)$, implies that

$$\|\varphi(t-1)\| \le c_1 + c_2 \max_{k \in [1,t]} |\varepsilon(k)| \qquad\qquad (8.9\text{-}25)$$

for nonnegative reals c_1 and c_2. Then

$$m_o(t-1) \;\le\; m_o(-1) + (1-\alpha)^{-1} \max_{k \in [0,t]} \|\varphi(k-1)\| \qquad [(9c)]$$

$$\le\; c_3 + c_4 \max_{k \in [1,t]} |\varepsilon(k)| \qquad\qquad (8.9\text{-}26)$$

We are now ready to establish global convergence of the adaptive system in the presence of neglected dynamics.

Theorem 8.9-1.   **(Global Convergence of the RDZ-CT-NRLS+CC algorithm)** *Let the minimum-phase condition (24) be satisfied and output reference $r(t)$ be uniformly bounded. Suppose that for some nonnegative real c_4 as in (26),*

$$(1+\epsilon)^{1/2}\mu < \frac{1}{c_4} \qquad\qquad (8.9\text{-}27)$$

Then, the reduced-order self-tuning control algorithm (13) and (23), when applied to the plant (1), for any possible initial condition, yields that:

(i) *y(t) and u(t) are uniformly bounded*   (8.9-28)

(ii) *There exists a finite T_1 after which the controller parameters self-tune in such a way that*

$$|y(t) - r(t)| < (1 + \epsilon)^{1/2} \mu m(t-1), \quad \forall t > T_1 \qquad (8.9\text{-}29)$$

Proof: We use (21) and (25) to (27) to prove (i). If $\varepsilon(t)$ is uniformly bounded, the uniform boundedness of $\|\varphi(t-1)\|$ follows at once from (25). Assume that $\{\varepsilon(t)\}_{i=1}^{\infty}$ is unbounded. Then, arguing as in the second part of the proof of Lemma 3-1, along a subsequence $\{t_n\}$ such that $\lim_{t_n \to \infty} |\varepsilon(t_n)| = \infty$ and $|\varepsilon(t)| \le |\varepsilon(t_n)|$, for $t \le t_n$, we have

$$\kappa(t_n) = 0 \qquad \forall t_n > T_1$$

or

$$\frac{\varepsilon^2(t_n)}{m^2(t_n - 1)} < (1 + \epsilon)\mu^2 \quad \forall t_n > T_1 \qquad (8.9\text{-}30)$$

On the other hand,

$$\frac{|\varepsilon(t_n)|}{m(t_n - 1)} \ge \frac{|\varepsilon(t_n)|}{m + m_o(t_n - 1)} \qquad [(11a)]$$

$$\ge \frac{|\varepsilon(t_n)|}{m + c_3 + c_4|\varepsilon(t_n)|} \qquad [(26)]$$

which implies that

$$\lim_{t_n \to \infty} \frac{|\varepsilon(t_n)|}{m(t_n - 1)} \ge \frac{1}{c_4} \qquad (8.9\text{-}31)$$

This inequality contradicts (30) whenever (27) is satisfied.

It must be pointed out that (27), in order to be satisfied for a large μ, entails that (26) holds for a small c_4. Tracing back the meaning of c_4 (cf. (5-15)), we see that, in this respect, the more stably invertible the transfer function in (24), the smaller c_4 turns out to be. On the other hand, given the transfer function in (24), the adaptive system stays stable provided that disturbance $n(t)$ is linearly bounded as in (9d) and (9e) and μ is small enough to satisfy the complementary condition (27). Being proportional to μ, the relative dead zone must also be made small enough in agreement to (27). To sum up, the practical relevance of Theorem 1 is that it suggests that the use of a relative dead zone, small enough to satisfy (27) for given μ and c_4, can make the RDZ-CT-NRLS+CC self-tuning controller robust against the class of neglected dynamics for which the upper bound (9) holds.

Main Points of the Section.   Low-pass prefiltering of data is instrumental for forcing the identifier to choose, among all possible reduced-order models of the plant, the ones for which the unmodeled dynamics have reduced effects inside the useful frequency band. Further, high-pass dynamic weighting in control design turns out to be beneficial for both reference shaping and attenuating the response to the neglected dynamics high-frequency disturbances.

Under some conditions for the unmodeled dynamics, self-tuning cheap control systems can be made robustly globally convergent by equipping the RLS identifier with both data normalization and a relative dead-zone facility, the latter to freeze the estimates whenever the absolute value of the prediction error becomes smaller than a disturbance upper bound. The latter is required, however, to be small enough so as not to destabilize the controlled system.

Problem 8.9-3 (STCC with Bounded Disturbance)   Consider the plant (9) with (9d) replaced by

$$|n(t)| \leq N < \infty$$

Next, redefine the normalization factor as in (6-1a), and modify the dead-zone mechanism in the CT-NRLS identifier (13) as follows:

$$\kappa(t) = \begin{cases} \kappa \in \left(0, \frac{\epsilon}{1+\epsilon}\right), & \text{if } |\varepsilon(t)| \geq (1+\epsilon)^{1/2} N \\ 0, & \text{otherwise} \end{cases}$$

Show that, with the previous modifications for the STCC algorithm (13) and (25) and the plant (1), (28) and

$$|y(t) - r(t)| < (1+\epsilon)^{1/2} N$$

both hold under all the validity conditions of Theorem 1 with no need of (27).

NOTES AND REFERENCES

Thorough presentations and studies of MRAC systems are available in [ÅW89], [NA89], [But92], and [SB89]. See also [Cha87] and [Lan79]. The early MRAC approach based on the gradient method dates back to about 1958. The difficulties with stability of the gradient method were analyzed in [Par66]. The innovative approach of [Mon74], whereby the feedback gains are directly estimated and the use of pure differentiators avoided, was thereafter adopted to produce technically sound MRAC systems for continuous-time minimum-phase plants. However, global stability of the previously mentioned MRAC systems, subject to additional assumptions on the available prior information, was established not earlier than 1978 by [Ega79], [FM78], [Mor80], and [NV78]. [Ega79] also gives a unification of MRAC and STC systems. For Bayesian stochastic adaptive control, see [DUL73] and [LDU72]; for the continuous-time case, [Hij86].

At different levels of mathematical detail, self–tuning control systems are presented in [ÅW89], [Che85], [CG91], [DV85], [GS84], [ILM92], [KV86], and [WZ91]. For a monograph on continuous-time self-tuning control, see [Gaw87]. One of the earlier works on the self-tuning idea is [Kal58] whereby least-squares estimation with deadbeat control was proposed. Two similar schemes, [Pet70] and [WW71], combining RLS estimation and MV regulation were presented at an IFAC symposium in 1970. The first thorough analysis of the RLS+MV self-tuning regulator was reported at the 5th IFAC World Congress in 1972 [ÅW73], showing that in this scheme, w.s. self-tuning and weak self-optimization occur. GMV self-tuning control was presented in [CG75]. However, it was not until 1980 that global convergence of the STCC, as in Theorem 5-1, was reported by [GRC80]. The global convergence proof of Theorem 6-2 of the implicit STCC algorithm (19) and (20), based on the constant-trace RLS identifier with data normalization of [LG85] is reported here for the first time. For global convergence of other STCC algorithms based on finite-memory identifiers, see also [BBC90b]. In a stochastic setting, the global convergence proof of the SG+MV algorithm, as in Theorem 7-1, was first given in [GRC81], where MIMO plants with an I/O delay $\tau \geq 1$ are also considered. [GRC81] can be considered the first rigorous stochastic adaptive control result in a self-tuning framework. The extension of Theorem 7-1 in Theorem 7-2 is reported in [KV86]. In [KV86], the geometric properties of the SG estimation algorithm are used to establish the self-tuning property of the SG+MV regulator in Fact 7-1. For self-tuning trackers, see also [KP87]. A global convergence analysis of an indirect stochastic self-tuning MV control based on an RELS(PO) estimation algorithm is given in [GC91].

At the late 1970s–early 1980s, it became clear that violation of the exact modeling conditions can cause adaptive control algorithms to become unstable. This phenomenon was pointed out, among others, by [Ega79], [IK84], [RVAS81], [RVAS82], and [RVAS85]. To counteract instability and improve robustness w.r.t. bounded disturbances and unmodeled dynamics, various modifications to the basic algorithms have been proposed. Some overview of these techniques can be found in [ABJ$^+$86], [Åst87], [ID91], [IS88], [MGHM88], and [OY87]. The major modifications consist of data normalizations with parameter projection ([Pra83] and [Pra84]), σ-modifications ([IK83] and [IT86a]), relative dead zones with parameter projection [KA86]. Another technique to enhance robustness is to inject a perturbation signal into the plant so as make the regression vector persistently exciting ([LN88], [NA89], and [SB89]). For robust STCC, our analysis in Sec. 9 shows that the choice can be limited to data normalization and a relative dead zone. Most of the previous robustification techniques and studies do not use stochastic models for disturbances. These are in fact merely assumed to be bounded or originate from plant undermodeling. For an exception to this, see [CG88], [PLK89], and [Yds91], and the stochastic analysis of an STCC made robust via parameter projection onto a compact convex set reported in [RM92].

For data prefiltering in identification for robust control design, see [RPG92] and [SMS92].

CHAPTER 9

ADAPTIVE PREDICTIVE CONTROL

In this chapter, we shall study how to combine identification methods and multistep predictive control to develop adaptive predictive controllers with nice properties. The main motivation for using underlying multistep predictive control laws in self-tuning control is to extend the field of possible applications beyond the restrictions pertaining to single-step-ahead controllers. In Sec. 1, we first study how to construct a globally convergent adaptive predictive control system under ideal model-matching conditions. To this end, the use of a self-excitation mechanism, though of a vanishingly small intensity, turns out to be essential to guarantee that the controller self-tunes on a stabilizing control law. We next study how to robustify the controller for the bounded disturbances and neglected dynamics case. In this case, along with a self-excitation of high-intensity low-pass filtering, normalization of the data entering the identifier and the use of a dead zone become of fundamental importance.

From Secs. 2 to 6, we deal with *implicit* adaptive predictive control. In Sec. 2, we show how the implicit description of CARMA plants in terms of linear-regression models, which is known to hold under minimum-variance control, can be extended to more complex control laws, such as those of predictive type. In Secs. 3 and 4, this property is exploited so as to construct implicit adaptive predictive controllers. In Sec. 5, one of such controllers, MUSMAR, which possesses attractive local self-optimizing properties, is studied via the ODE (ordinary differential equation) approach to analyzing recursive stochastic algorithms. Finally, Sec. 6 deals with two extensions of the MUSMAR algorithm: the first imposes a mean-square input constraint to the controlled system; the second is finalized to exactly recover the steady-state LQ stochastic regulation law as an equilibrium point of the algorithm.

9.1 INDIRECT ADAPTIVE PREDICTIVE CONTROL

9.1.1 The Ideal Case

Consider the SISO plant:

$$A(d)y(t) = B(d)u(t) + A(d)c(t) \qquad (9.1\text{-}1a)$$

where $y(t)$ and $u(t)$ are the output and the manipulable input, respectively, $c(t) \equiv c$ is an unknown constant disturbance, and

$$\left. \begin{aligned} A(d) &= 1 + a_1 d + \cdots + a_{n_a} d^{n_a} & a_{n_a} \neq 0 \\ B(d) &= b_1 d + \cdots + b_{n_b} d^{n_b} & b_{n_b} \neq 0 \end{aligned} \right\} \qquad (9.1\text{-}1b)$$

Note that here the leading coefficients of $B(d)$ are allowed to be zero, and hence an unknown plant I/O delay τ, $1 \leq \tau < n_b$, can be present. The plant can be also represented in terms of input increments $\delta u(t) := u(t) - u(t-1)$ by the model

$$A(d)\,\Delta(d)y(t) = B(d)\,\delta u(t) \qquad (9.1\text{-}1c)$$

with $\Delta(d) = 1 - d$. The main goal is to develop a globally convergent adaptive controller based on SIORHC, the predictive control law of Sec. 5.8, which, for convenience, is restated.

Given the state

$$s(t) := \left[\ (y_t^{t-n_a})' \quad (\delta u_{t-1}^{t-n_b+1})' \ \right]' \qquad (9.1\text{-}2)$$

find, whenever it exists, the input increment sequence $\widehat{\delta u}_{[t,t+T)}$ to the plant (1c) minimizing

$$J\left(s(t), \delta u_{[t,t+T)}\right) = \sum_{k=t}^{t+T-1} \left[\varepsilon_y^2(k+1) + \rho \delta u^2(k) \right] \quad , \quad \rho > 0 \qquad (9.1\text{-}3)$$

under the constraints

$$\delta u_{t+T+n-2}^{t+T} = O_{n-1} \qquad y_{t+T+n-1}^{t+T} = \underline{r}(t+T) \qquad (9.1\text{-}4)$$

In the previous equations, $\varepsilon_y(k) := y(k) - r(k)$, $r(k)$ is the output reference to be tracked, and $\underline{r}(k) := \left[\ r(k) \quad \cdots \quad r(k) \ \right]' \in \mathbb{R}^n$. Then the plant input increment $\delta u(t)$ at time t is set equal to $\widehat{\delta u}(t)$:

$$\delta u(t) = \widehat{\delta u}(t) \qquad (9.1\text{-}5)$$

The SIORHC law is given by (5.8-45)

$$\begin{aligned} \delta u(t) &= -e_1' M^{-1} \Big\{ \left(I_T - QLM^{-1} \right) W_1' \left[\Gamma_1 s(t) - r_{t+T}^{t+1} \right] \\ &\quad + Q\left[\Gamma_2 s(t) - \underline{r}(t+T) \right] \Big\} \end{aligned} \qquad (9.1\text{-}6)$$

with the properties stated in Theorem 5.8-2. In particular, in order to ensure that (6) is well-defined and that it stabilizes the plant (1), we shall adopt the following assumptions.

ASSUMPTION 9.1-1

- $A(d)\Delta(d)$ and $B(d)$ are coprime polynomials $\hspace{3cm}$ (9.1-7)

- $n := \max\{n_a + 1, n_b\}$ is a priori known $\hspace{3cm}$ (9.1-8)

- $T \geq n$ $\hspace{8cm}$ (9.1-9)

It is instructive to compare these assumptions with Assumption 8.5-1 adopted for the implicit STCC (8.5-8) and (8.5-9). First, here there is no need of assuming that the plant is minimum-phase and that its I/O delay is known. As repeatedly remarked, these are limitative assumptions that strongly reduce the range of possible applications. On the other hand, here plant order n, as opposed to an upper-bound $\hat{n}$, is supposed to be a priori known. This is a quite limitative assumption as well. At the end of this section, however, we shall study how to modify the adaptive algorithm so as to deal with the common practical case of a plant whose true order n exceeds the presumed plant order. It is, nonetheless, convenient to begin with studying adaptive predictive control in the present ideal setting.

Estimation Algorithm. In order to possibly deal with slowly time-varying plants, an estimation algorithm with finite data-memory length is considered. In particular, the CT-NRLS of Sec. 8.6 is chosen because of the nice properties established in Theorem 8.6-1. The aim is to recursively estimate plant polynomials $A(d)$ and $B(d)$ and use these estimates to compute the input increment $\delta u(t)$ in (6) at each time step. In order to suppress the effect of the constant disturbance c on the estimates, the plant *incremental model* is considered for parameter estimation:

$$A(d)\delta y(t) = B(d)\delta u(t) \quad , \quad t \in \mathbb{Z}_1 \qquad (9.1\text{-}10a)$$

with $\delta y(t) := y(t) - y(t-1)$. Defining

$$\varphi(t-1) := \left[\ \left(-\delta y_{t-n+1}^{t-1}\right)' \ \ \left(\delta u_{t-n}^{t-1}\right)' \ \right]' \qquad (9.1\text{-}10b)$$

$$\theta^* := \left[\ a_1 \ \cdots \ a_{n-1} \ b_1 \ \cdots \ b_n \ \right]' \in \mathbb{R}^{n_\theta} \qquad (9.1\text{-}10c)$$

$$m(t-1) := \max\left\{ m, \|\varphi(t-1)\| \right\} \quad , \quad m > 0 \qquad (9.1\text{-}11a)$$

and the normalized data

$$\delta\gamma(t) := \frac{\delta y(t)}{m(t-1)} \qquad x(t-1) := \frac{\varphi(t-1)}{m(t-1)} \qquad (9.1\text{-}11b)$$

we have

$$\delta y(t) = \varphi'(t-1)\theta^* \qquad \delta\gamma(t) = x'(t-1)\theta^* \qquad (9.1\text{-}12)$$

Note that, in contrast with θ, which was used in Chapter 8 to denote any vector in the parameter variety $\Theta \subset \mathbb{R}^{n_\theta}$, here the symbol θ^* denotes the unique vector in $\mathbb{R}^{n_\theta}$ satisfying (12), with $n_\theta = 2n - 1$. The following CT-NRLS algorithm (cf. Sec. 8.6) is used to estimate θ^*:

$$\begin{aligned} \theta(t) &= \theta(t-1) + \frac{P(t-1)x(t-1)}{1 + x'(t-1)P(t-1)x(t-1)}\bar{\varepsilon}(t) \\ &= \theta(t-1) + \lambda(t)P(t)x(t-1)\bar{\varepsilon}(t) \end{aligned} \tag{9.1-13a}$$

$$\bar{\varepsilon}(t) = \delta\gamma(t) - x'(t-1)\theta(t-1) \tag{9.1-13b}$$

$$P(t) = \frac{1}{\lambda(t)}\left[P(t-1) - \frac{P(t-1)x(t-1)x'(t-1)P(t-1)}{1 + x'(t-1)P(t-1)x(t-1)} \right] \tag{9.1-13c}$$

$$\lambda(t) = 1 - \frac{1}{\text{Tr}[P(0)]}\frac{x'(t-1)P^2(t-1)x(t-1)}{1 + x'(t-1)P(t-1)x(t-1)} \tag{9.1-13d}$$

The algorithm is initialized from any $P(0) = P'(0) > 0$ and any $\theta(0) \in \mathbb{R}^{n_\theta}$ such that $A_0(d)\Delta(d)$ and $B_0(d)$ are coprime, and $A_0(d)$ and $B_0(d)$ are the polynomials in the plant model (10a) corresponding to parameter vector $\theta(0)$. Algorithm (13) still enjoys the same properties as the standard CT-NRLS in Theorem 8.6-1 (cf. Problem 8.6-5). For convenience, we list the ones that will be used in what follows.

- $\|\theta(t)\| < M_\theta < \infty, \quad \forall t \in \mathbb{Z}_+$ \hfill (9.1-14)

- $\lim_{t\to\infty} \theta(t) = \theta(\infty)$ \hfill (9.1-15)

- $\lim_{t\to\infty} \|\theta(t) - \theta(t-k)\| = 0, \quad \forall k \in \mathbb{Z}_+$ \hfill (9.1-16)

- $\lim_{t\to\infty} \delta(t) = 0 \quad \Longrightarrow \quad \theta(\infty) = \theta^*$ \hfill (9.1-17)

$$\delta(t) := \prod_{j=1}^{t} \lambda(j)$$

- $\lim_{t\to\infty} \bar{\varepsilon}(t) = 0$ \hfill (9.1-18)

According to the standard indirect self-tuning procedure based on the enforced certainty equivalence, the desired adaptive control algorithm should be completed as follows. After estimating

$$\theta(t) = \begin{bmatrix} a_1(t) & \cdots & a_{n-1}(t) & b_1(t) & \cdots & b_n(t) \end{bmatrix}' \tag{9.1-19a}$$

construct polynomials $A_t(d)\Delta(d)$ and $B_t(d)$, where

$$\left.\begin{aligned} A_t(d) &:= 1 + a_1(t)d + \cdots + a_{n-1}(t)d^{n-1} \\ B_t(d) &:= b_1(t)d + \cdots + b_n(t)d^n \end{aligned}\right\} \tag{9.1-19b}$$

Next, from $A_t(d)\Delta(d)$ and $B_t(d)$, compute matrices $W(t)$ and $\Gamma(t)$ as in (5.5-9) and (5.5-10), respectively. Partition $W(t)$ and $\Gamma(t)$ as indicated in (5.5-11) and (5.5-12) to obtain $W_1(t)$, $W_2(t)$, $\Gamma_1(t)$, $\Gamma_2(t)$, $M(t)$, and $Q(t)$. Finally, determine the next input increment $\delta u(t)$ by using (6) once W_1, W_2, L, Γ_1, Γ_2, M, and Q are replaced by $W_1(t)$, $W_2(t)$, $L(t)$, $\Gamma_1(t)$, $\Gamma_2(t)$, $M(t)$, and $Q(t)$, respectively.

This route, however, cannot be safely adopted without modifications. One reason is that the estimation algorithm (13) does not ensure that $A_t(d)\Delta(d)$ and $B_t(d)$ be coprime at every $t \in \mathbb{Z}$, and, hence, boundedness of matrix $Q(t)$. In fact, if $A_t(d)\Delta(d)$ and $B_t(d)$ are not coprime, $L(t)$ does not have full row rank, and, in turn, $Q(t) = L'(t)\left[L(t)M^{-1}(t)L'(t)\right]^{-1}$ does not exist. Even if the control law is modified so as to ensure boundedness of $Q(t)$ when $A_t(d)\Delta(d)$ and $B_t(d)$ become noncoprime, boundedness of controller parameters must be also guaranteed asymptotically. One approach that has been often suggested to this end is to constrain $\theta(t)$ to belong to a convex admissible subset of $\mathbb{R}^{n_\theta}$ containing θ^* and whose elements give rise to coprime polynomials $A_t(d)\Delta(d)$ and $B_t(d)$. This can be achieved by suitably projecting $\theta(t)$ in (13) onto the previous admissible subset. In most practical cases, the choice of such a subset appears artificial, so we shall follow a different approach. It consists of injecting into the plant, along with the control variable, an additive *dither* input whenever estimate $\theta(t)$ turns out to yield a control law close to singularity. In the ideal case, such an action, if well-designed, turns out to be very effective. In fact, the dither input, despite that it turns off forever after a finite time, drives estimate $\theta(t)$ to converge far from any singular vector. Such a mode of operation will be referred to as a *self-excitation* mechanism because the dither is switched on only when the controller judges its current state close to singularity. The resulting control philosophy is then of the dual control type (cf. Sec. 8.2).

We next proceed to construct a self-excitation mechanism suitable for the adopted underlying control law. To this end, we have to define a syndrome according to which the controller detects its pathological states. Consider assumption (7). It implies that

$$\Xi(L) := (n)^n \frac{\det(LL')}{[\operatorname{Tr}(LL')]^n} = \varsigma^* \in (0,1] \qquad (9.1\text{-}20a)$$

if L is the matrix in the SIORHC law (6) corresponding to the true plant parameter vector θ^*. Let now ς be any positive real such that

$$0 < \varsigma < \varsigma^* \qquad (9.1\text{-}20b)$$

In practice, if no prior information is given on the size of ς^*, ς can be chosen to equal any positive number arbitrarily close but greater than the zero of the digital processor implementing the adaptive controller. Given an estimate $\theta(t)$ of θ^*, the related *syndrome* can be defined to be $\Xi(\theta(t)) := \Xi(L(t))$, if $L(t)$ denotes the matrix in (6) when the SIORHC law is computed from $\theta(t)$. The estimate $\theta(t)$ is judged to be pathologic whenever $\Xi(\theta(t)) \leq \varsigma$. In this way, the set of admissible $\theta(t)$, viz., the ones for which $\Xi(\theta(t)) > \varsigma$, includes θ^* and all estimates yielding a bounded SIORHC law. We are now ready to proceed to construct the remaining part of the adaptive controller.

Controller with Self-Excitation. The control law is retuned every N sampling steps with

$$N \geq 4n - 1 \tag{9.1-21}$$

even if plant parameter estimate $\theta(t)$ is updated at every sampling time. The main reason for doing this is to keep the analysis of the adaptive system as simple as possible. For the sake of simplicity, we shall indicate that a matrix M depends on estimate $\theta(t)$ by using the notation $M(t)$ in place of $M(\theta(t))$.

If $t = (k - 1)N + 1$, $k \in \mathbb{Z}_1$:

1. Form $A_t(d)\Delta(d)$ and $B_t(d)$ by using the current estimate $\theta(t)$.

2. Compute matrices $W(t)$ and $\Gamma(t)$ via (5.5-9) and (5.5-10), respectively.

3. Partition $W(t)$ and $\Gamma(t)$ so as to obtain $W_1(t)$, $L(t)$, $W_2(t)$, $\Gamma_1(t)$, $\Gamma_2(t)$, $M(t)$, and $Q(t)$.

4. If

$$\Xi(\theta(t)) > \varsigma \tag{9.1-22}$$

compute the plant input increments $\delta u(\tau)$, $t \leq \tau < t + N$, by using (6) with W_1, W_2, L, Γ_1, Γ_2, M, and Q replaced by $W_1(t)$, $W_2(t)$, $L(t)$, $\Gamma_1(t)$, $\Gamma_2(t)$, $M(t)$, and $Q(t)$, respectively. If

$$\Xi(\theta(t)) \leq \varsigma \tag{9.1-23}$$

self-excitation is turned on, viz., set

$$\delta u(\tau) = \delta u^o(\tau) + \eta(\tau) \quad , \quad t \leq \tau < t + N \tag{9.1-24}$$

$$\eta(\tau) = \nu(k)\delta_{\tau, kN - 2n + 1} \tag{9.1-25}$$

where $\delta u^o(\tau)$ is either given by (6), computing $Q(t)$ via any pseudo-inverse (cf. (5.5-2)), or, if $L(t) = 0$, by the same control law over the previous time interval $((k - 2)N, (k - 1)N]$; $\eta(\tau)$ is the dither component due to self-excitation with $\delta_{\tau, i}$ the Kronecker symbol and $\nu(k)$ a scalar specified by the next Lemma 1.

The algorithm (13) and (21) to (25), whose mode of operation is depicted in Fig. 1, will be referred to as *adaptive SIORHC*. It generates plant input increments as follows:

$$\left. \begin{aligned} R_t(d)\delta u(t) &= -S_t(d)y(t) + v(t) + \eta(t) \\ v(t) &:= Z_t(d)r(t + T) \end{aligned} \right\} \tag{9.1-26}$$

where $R_t(d)$, $S_t(d)$, $Z_t(d)$, with $R_t(0) = 1$ and $\partial Z_t(d) = T - 1$, are polynomials corresponding to the SIORHC law (6) at time t.

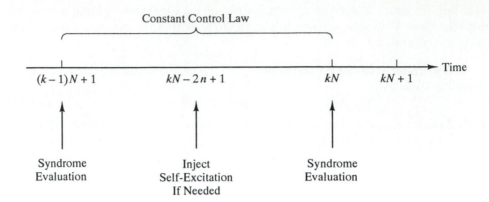

Figure 9.1-1: Illustration of the mode of operation of adaptive SIORHC.

Remark 9.1-1. In the algorithm (13) and (21) to (25), the plant parameter-vector estimate is updated every time step, whereas the controller parameters are kept constant for N time steps. This mode of operation is only adopted for keeping the analysis of the algorithm as simple as possible. However, we point out that the adaptive SIORHC can operate in a more efficient way—with no change in the conclusions of the subsequent convergence analysis—if the controller parameters are updated at each time step, except at the times where (23) holds. At such times, the controller parameters are to be computed according to (24) and (25), and frozen for the subsequent $N \geq 4n - 1$ time steps.

In order to establish which value to assign to $\nu(k)$, it is convenient to introduce the following square Toeplitz matrix of dimension $2n$ whose entries are the feedforward input increments defined in (26):

$$V(k) := \begin{bmatrix} v(kN - 2n + 1) & v(kN - 2n) & \cdots & v(kN - 4n + 2) \\ v(kN - 2n + 2) & v(kN - 2n + 1) & \cdots & v(kN - 4n + 1) \\ \vdots & \vdots & \ddots & \vdots \\ v(kN) & v(kN - 1) & \cdots & v(kN - 2n + 1) \end{bmatrix} \quad (9.1\text{-}27)$$

Lemma 9.1-1. *Assume that $\nu(k)$ in (25) is not an eigenvalue of $V(k)$:*

$$\nu(k) \,\overline{\in}\, \mathrm{sp}[V(k)] \quad (9.1\text{-}28)$$

Then if (23) holds for infinitely many $t = kN + 1$, it follows that the estimate $\theta(t)$ converges to the true plant parameter vector θ^*:

$$\lim_{t \to \infty} \theta(t) = \theta^* \quad (9.1\text{-}29)$$

Remark 9.1-2. Equation (28) indicates that the self-excitation signal must be chosen by taking into account the feedforward signal. The reason is that we have to consider the interaction between self-excitation and feedforward so as to avoid the latter annihilating the effects of the former.

Proof of Lemma 1: Equation (13c) yields

$$P^{-1}(t) = \lambda(t)\left[P^{-1}(t-1) + x(t-1)x'(t-1)\right]$$

Thus, setting $\delta(t) := \prod_{i=1}^{t}\lambda(i)$, because $0 < \lambda(t) \leq 1$ by Lemma 8.6-1, it follows that

$$\delta^{-1}(t)P^{-1}(t) \geq I(t) := P^{-1}(0) + \sum_{i=1}^{t}x(i-1)x'(i-1)$$

With $I(t)$ monotonically nondecreasing,

$$\lim_{t\to\infty} I(t) = \lim_{k\to\infty} I(kN)$$

Let

$$\Phi(t) := \left[\begin{array}{ccc} x(t) & \cdots & x(t-N+1) \end{array}\right] \in \mathbb{R}^{n_\theta \times N}$$

Then

$$I(kN) = P^{-1}(0) + x(0)x'(0) + \sum_{i=1}^{k}\Phi(iN)\Phi'(iN)$$

Consequently,

$$\Phi(iN)\Phi'(iN) > 0 \text{ for infinitely many } i \implies \lim_{t\to\infty}\lambda_{\min}[I(t)] = \infty \tag{9.1-30}$$

Next, we constructively show that if $\nu(t)$ is chosen so as to fulfill (28), the L.H.S. of (30) holds whenever (23) is satisfied for infinitely many $t = kN + 1$.

Consider the following "partial-state" representation of the plant:

$$A(d)\Delta(d)\xi(t) = \delta u(t)$$
$$y(t) = B(d)\xi(t)$$

Let

$$\mathcal{Z}(t) := \xi_{t-2n+1}^{t} \in \mathbb{R}^{2n}$$

and

$$\varphi_1(t) := \left[\begin{array}{cc} \left(y_{t-n+1}^{t}\right)' & \left(\delta u_{t-n+1}^{t}\right)' \end{array}\right]' \in \mathbb{R}^{2n} \tag{9.1-31}$$

Then $\varphi(t) = \Gamma\varphi_1(t)$ with Γ a full row-rank matrix, and $\varphi_1(t) = S\mathcal{Z}(t)$, where S is the Sylvester resultant matrix [Kai80] associated with polynomials $A(d)\Delta(d)$ and $B(d)$, viz.,

$$S = \left[\begin{array}{ccccccc} 0 & b_1 & \cdots & b_n & & & \\ & 0 & b_1 & \cdots & b_n & & \\ & & & \vdots & & & \\ & & 0 & b_1 & \cdots & b_n & \\ \hline 1 & \alpha_1 & \cdots & \alpha_n & & & \\ & 1 & \alpha_1 & \cdots & \alpha_n & & \\ & & & \vdots & & & \\ & & 1 & \alpha_1 & \cdots & \alpha_n & \end{array}\right]$$

if $A(d)\Delta(d) = 1 + \alpha_1 d + \cdots + \alpha_n d^n$. Because $x(t) = \varphi(t)/m(t)$, Γ is full row rank, and by (7a) S is nonsingular,

$$\sum_{t=iN-4n+2}^{iN} \mathcal{Z}(t)\mathcal{Z}'(t) > 0$$

implies

$$\Phi(iN)\Phi'(iN) = \sum_{t=(i-1)N+1}^{iN} x(t)x'(t) > 0$$

because $N \geq 4n - 1$. Rewrite the control law (26) as follows:

$$\Psi'(t)S\mathcal{Z}(t) = v(t) + \eta(t)$$

where

$$\Psi'(t) := \begin{bmatrix} s_0(t) \cdots & s_{n-1}(t) & 1 & r_1(t) & \cdots & r_{n-1}(t) \end{bmatrix}$$

is the row vector whose components are the coefficients of $R_t(d)$ and $S_t(d)$. For $t \in ((i-1)N, iN]$, $\Psi(t) = \Psi(iN)$. Hence, for $t = iN, \cdots, iN - 2n + 1$,

$$\Psi'(iN)S\Lambda(t) = \mathbf{v}'(t) + \nu(i)\mathbf{f}'(t)$$

$$
\begin{aligned}
\Lambda(t) &:= \begin{bmatrix} \mathcal{Z}(t) & \cdots & \mathcal{Z}(t-2n+1) \end{bmatrix} \\
\mathbf{v}'(t) &:= \begin{bmatrix} v(t) & \cdots & v(t-2n+1) \end{bmatrix} \\
\mathbf{f}'(t) &:= \begin{bmatrix} \delta_{t,iN-2n+1} & \cdots & \delta_{t-2n+1,iN-2n+1} \end{bmatrix}
\end{aligned}
\qquad (9.1\text{-}32)
$$

Let w be a vector in $\mathbb{R}^{2n}$ of unit norm. Then

$$\left[\Psi'(iN)S\Lambda(t)w\right]^2 = \left[\mathbf{v}'(t)w + \nu(i)\mathbf{f}'(t)w\right]^2$$

By denoting the inner product by $\langle \cdot, \cdot \rangle$, the L.H.S. of the previous equation equals

$$\left[\langle \Lambda(t)w, S'\Psi(iN)\rangle\right]^2 \leq \|S'\Psi(iN)\|^2 \|\Lambda(t)w\|^2$$

where the upper bound follows by the Schwarz inequality. Summing over $t = iN - 2n + 1, \cdots, iN$, we get

$$\|S'\Psi(iN)\|^2 \sum_{t=iN-2n+1}^{iN} \|\Lambda(t)w\|^2 \geq \sum_{t=iN-2n+1}^{iN} \left[\mathbf{v}'(t)w + \nu(i)\mathbf{f}'(t)w\right]^2 \qquad (9.1\text{-}33)$$

Let

$$\mathbf{F} := \begin{bmatrix} \mathbf{f}(iN) & \cdots & \mathbf{f}(iN-2n+1) \end{bmatrix}'$$

By (32), $\mathbf{F} = \begin{bmatrix} e_{2n+1} & \cdots & e_1 \end{bmatrix}$, where e_i denotes the ith vector of the natural basis of $\mathbb{R}^{2n}$. Hence, $\mathbf{F} = \mathbf{F}'$ and $\mathbf{F}^2 = I$. Then the R.H.S. of (33) equals $\|(\nu\mathbf{F} + \mathbf{V})w\|^2 = \|(\nu I + \mathbf{F}\mathbf{V})w\|^2$, where $\mathbf{V} := \begin{bmatrix} \mathbf{v}(iN) & \cdots & \mathbf{v}(iN-2n+1) \end{bmatrix}'$ and $\mathbf{F}\mathbf{V} = V(k)$ with

$V(k)$ as in (27). It follows that the choice (28) makes the R.H.S. of (33) positive. On the other hand, by using the symmetry of $\Lambda(t)$,

$$\sum_{t=iN-2n+1}^{iN} \|\Lambda(t)w\|^2 = \sum_{t=iN-2n+1}^{iN} \sum_{j=0}^{2n-1} \left[w' \mathcal{Z}(t-j)\right]^2$$

$$\leq 2n \sum_{t=iN-4n+2}^{iN} \left[w' \mathcal{Z}(t)\right]^2$$

Because $0 < \|\mathcal{S}'\Psi(iN)\|^2 < \infty$, it follows that

$$\sum_{t=iN-4n+2}^{iN} \mathcal{Z}(t)\mathcal{Z}'(t) > 0$$

Hence, $\lambda_{\min}[I(t)] \to \infty$. On the other hand,

$$\lambda_{\min}[I(t)] \leq \delta^{-1}(t)\lambda_{\min}\left[P^{-1}(t)\right] = [\delta(t)\lambda_{\max}[P(t)]]^{-1} \leq \left[\delta(t)\frac{\text{Tr}[P(0)]}{n_\theta}\right]^{-1}$$

Hence, $\lim_{t \to \infty} \delta(t) = 0$ and, by (17), $\lim_{t \to \infty} \theta(t) = \theta^*$.

The next lemma points out that, if the self-excitation signal equals (28), after a finite time, the self-excitation mechanism turns off forever and, henceforth, the estimate is secured to be nonpathologic.

Lemma 9.1-2.   *For the adaptive SIORHC algorithm applied to the plant (1) the following self-excitation stopping time property holds. Let the self-excitation take place as in (28). Then, for the adaptive SIORHC algorithm (13), (21) to (25), applied to the plant (1), (7) and (8), there is a finite integer T_1 such that $\Xi(\theta(t)) > \varsigma$, and hence $\eta(t) = 0$, for all $t > T_1$.*

Proof (by contradiction): Assume that no T_1 exists with the stated property. Thus, there is an infinite subsequence $\{t_i\}$ such that $\Xi(\theta(t_i)) \leq \varsigma$. From Lemma 1, it follows that $\theta(t)$ converges to θ^*. Because $\varsigma < \varsigma^*$, there is a finite T_1 such that $\Xi(\theta(t)) > \varsigma$, for every $t > T_1$. This contradicts the assumption.

We are now ready to prove global convergence of the adaptive control system. To this end, we recall that the adaptive controller generates $\delta u(t)$ as in (26) with

$$\left.\begin{array}{l} R_t(d) = R_{t-1}(d) \\ S_t(d) = S_{t-1}(d) \\ Z_t(d) = Z_{t-1}(d) \end{array}\right\} \quad t \in \left[kN+2, (k+1)N\right]$$

From (13b), it follows that

$$A_{t-1}(d)\Delta(d)y(t) = B_{t-1}(d)\delta u(t) + \varepsilon(t) \tag{9.1-34}$$

with $\varepsilon(t) = m(t-1)\bar{\varepsilon}(t)$. By using (26) and (34), the following closed-loop system is obtained (d is omitted):

$$
\begin{bmatrix} A_{t-1}\Delta & -B_{t-1} \\ S_{kN+1} & R_{kN+1} \end{bmatrix} \begin{bmatrix} y(t) \\ \delta u(t) \end{bmatrix} = \begin{bmatrix} 1 \\ 0 \end{bmatrix} \varepsilon(t) + \begin{bmatrix} 0 \\ Z_{kN+1} \end{bmatrix} r(t+T) + \begin{bmatrix} 0 \\ 1 \end{bmatrix} \eta(t)
$$

$$(9.1\text{-}35)$$

where $t \in (kN, (k+1)N]$. By (14), the coefficients of polynomials $A_t\Delta$ and B_t are bounded and the same is true for R_t, S_t, and Z_t. Further, from (16), $(A_{t-1} - A_{kN+1}) \to 0$ and $(B_{t-1} - B_{kN+1}) \to 0$, $t \in (kN, (k+1)N]$, as $t \to \infty$. Hence, as $t \to \infty$, the d-characteristic polynomial of the system (35)

$$
\begin{aligned}
\chi_t(d) \quad &:= \quad A_{t-1}\Delta R_{kN+1} + B_{t-1}S_{kN+1} \\
&= \quad A_{kN+1}\Delta R_{kN+1} + B_{kN+1}S_{kN+1} \\
&\quad + (A_{t-1} - A_{kN+1})\Delta R_{kN+1} + (B_{t-1} - B_{kN+1})S_{kN+1}
\end{aligned}
$$

and

$$
\chi_k(d) := A_{kN+1}\Delta R_{kN+1} + B_{kN+1}S_{kN+1}
$$

as $k \to \infty$, behave in the same way. Now, by virtue of Lemma 2, the latter is strictly Hurwitz for all k such that $kN + 1 > T_1$, T_1 being a finite integer. Consequently, there exists a finite time such that for all subsequent times, the d-characteristic polynomial $\chi_t(d)$ of the slowly time-varying system (35) is strictly Hurwitz. Then it follows from Theorem A.2-3 that (35) is exponentially stable. Consequently, by Lemma 2 being $\eta(t) \equiv 0$, for all $t > T_1$, and assuming that $|r(t)| < M_r < \infty$, the linear boundedness condition (cf. (8.3-10))

$$
\|\varphi_1(t-1)\| \le c_1 + c_2 \max_{i \in [1,t)} |\varepsilon(i)|
$$

holds for bounded nonnegative reals c_1 and c_2. Here $\varphi_1(t)$ denotes the vector in (31).

By (18), we also have

$$
\begin{aligned}
0 \quad = \quad \lim_{t \to \infty} \bar{\varepsilon}^2(t) \quad &= \quad \lim_{t \to \infty} \frac{\varepsilon^2(t)}{[\max\{m, \|\Gamma\varphi_1(t-1)\|\}]^2} \\
&\ge \quad \lim_{t \to \infty} \frac{\varepsilon^2(t)}{m^2 + c_3\|\varphi_1(t-1)\|^2}
\end{aligned}
$$

where Γ, as noted after (31), is such that $\varphi(t) = \Gamma\varphi_1(t)$. Hence, the last limit is zero. We can then apply the key technical lemma (8.3-1) to conclude that $\{\|\varphi_1(t)\|\}$ is a bounded sequence and $\lim_{t \to \infty} \varepsilon(t) = 0$. In particular, boundedness of $\{\|\varphi_1(t)\|\}$ is equivalent to boundedness of $\{y(t)\}$ and $\{\delta u(t)\}$. To show that $\{u(t)\}$ is also bounded, we use the following argument. By (7a) and (B.1-10), there exist polynomials $X(d)$ and $Y(d)$ satisfying the Bezout identity:

$$A(d)\Delta(d)X(d) + B(d)Y(d) = 1$$

Therefore,

$$
\begin{aligned}
u(t) &= A(d)X(d)\Delta(d)u(t) + B(d)Y(d)u(t) \\
&= A(d)X(d)\delta u(t) + Y(d)A(d)\left[y(t) - c\right] \qquad [(1)]
\end{aligned}
$$

Hence, $u(t)$ is expressed in terms of a linear combination of a finite number of terms from δu^t and y^t, plus the constant term $Y(d)A(d)c$. Boundedness of $\{u(t)\}$ thus follows for that of $\{\delta u(t)\}$, $\{y(t)\}$, and c.

The previous results are summed up in the next theorem in which additional convergence properties of the adaptive system are also stated.

Theorem 9.1-1. **(Global Convergence of Adaptive SIORHC)** *Consider the adaptive SIORHC algorithm (13), (21) to (25), applied to the plant (1), (7) and (8). Let the output reference sequence $\{r(t)\}$ be bounded and the self-excitation signal be chosen so as to fulfill (28). Then, the resulting adaptive system is globally convergent. Specifically:*

(i) *$u(t)$ and $y(t)$ are uniformly bounded.*

(ii) *The controller parameters self–tune to a stabilizing control law in such a way that after a finite number of steps $\Xi(\theta(t)) > \varsigma$ and henceforth self-excitation turns off.*

(iii) *The multistep-ahead output prediction errors asymptotically vanish, viz.,*

$$\lim_{t\to\infty}\left[\hat{y}^{t+1}_{t+T+n-1} - y^{t+1}_{t+T+n-1}\right] = 0 \qquad (9.1\text{-}36a)$$

where

$$\hat{y}^{t+1}_{t+T+n-1} := W(t)\delta u^t_{t+T+n-2} + \Gamma(t)s(t) \qquad (9.1\text{-}36b)$$

(iv) *The adaptive system is asymptotically offset-free, that is,*

$$r(t) \equiv r \implies \lim_{t\to\infty} y(t) = r \quad and \quad \lim_{t\to\infty} \delta u(t) = 0 \qquad (9.1\text{-}37)$$

and yields asymptotic rejection of constant disturbances.

Proof: It remains to prove (iii) and (iv). As shown in [MZ91], (36) follows because $\lim_{t\to\infty} \varepsilon(t) = 0$ and $\lim_{t\to\infty} \|\theta(t) - \theta(t-1)\| = 0$. As for (37), if $r(t) \equiv r$ from (35), (15), and taking into account that $\varepsilon(t) \to 0$, $\eta(t) \to 0$, $\chi_t(d) \to \chi_\infty(d) = A_\infty(d)\Delta(d)R_\infty(d) + B_\infty(d)S_\infty(d)$ with $\chi_\infty(d)$ strictly Hurwitz, we conclude that $\delta u(t) \to 0$ and

$$
\begin{aligned}
\lim_{t\to\infty} y(t) &= \frac{B_\infty(1)Z_\infty(1)}{A_\infty(1)\Delta(1)R_\infty(1) + B_\infty(1)S_\infty(1)}r \\
&= \frac{Z_\infty(1)}{S_\infty(1)}r \quad = \quad r
\end{aligned}
$$

where the last equality follows because $Z_t(1) = S_t(1)$ by the same arguments preceding Theorem 5.8-2.

Remark 9.1-3.   The self-excitation condition (28) can be easily fulfilled whenever matrix $\mathbf{V}(k)$ is known at time $kN - 2n + 1$. In such a case, in fact, one has to check that $\nu(k)$ is not one of the $2n$ roots of the characteristic polynomial of $\mathbf{V}(k)$. Consequently, at most $2n + 1$ attempts suffice to satisfy (28) with an arbitrarily small $|\nu(k)|$. In particular, $\nu(k)$ can be chosen to be zero whenever $\det \mathbf{V}(k) \neq 0$. In fact, $\det \mathbf{V}(k) \neq 0$ can be interpreted as a condition of excitation over the interval $[kN - 4n + 2, kN]$ caused by the command input $v_{kN}^{kN-4n+2}$. However, knowledge of $V(k)$ at time $kN - 2n + 1$ implies knowledge of the reference up to time $kN + T$, viz., $T + 2n$ steps in advance. This is, in fact, the case in some applications where the desired future output profile is known a few steps in advance.

If $\mathbf{V}(k)$ is unknown at time $kN - 2n + 1$ and $\{r(t)\}$ is upper-bounded by M_r:

$$|r(t)| \leq M_r$$

(28) can be guaranteed (cf. Problem 1) by taking

$$|\nu(k)| > 2nT^{1/2}M_r\|Z_{kN}(d)\| \geq \bar{\sigma}(V(k)) \qquad (9.1\text{-}38)$$

with $\|Z_{kN}(d)\|^2 = \sum_{i=0}^{T-1}(z_{kN,i})^2$ if $Z_{kN}(d) = \sum_{i=0}^{T-1}z_{kN,i}d^i$ and $\bar{\sigma}(V(k))$ denotes the maximum singular value for $V(k)$. Note that (38) is quite conservative w.r.t. (28).

Problem 9.1-1   Prove that (28) is guaranteed if the self-excitation signal $\nu(k)$ satisfies (38).   (*Hint*:   Show first that $|\nu(k)| > \bar{\sigma}(V(k))$ suffices. Next prove that $\bar{\sigma}(V(k))$ is upper-bounded as in (38).)

Problem 9.1-2   Modify the adaptive SIORHC algorithm so as to construct for the plant (1) a globally convergent adaptive pole-positioning regulator with self-excitation, whose underlying control problem consists of selecting the regulation law $R(d)\delta u(t) = -S(d)y(t)$, polynomials $R(d)$ and $S(d)$ solving the Diophantine equation

$$A(d)\Delta(d)R(d) + B(d)S(d) = Q(d)$$

with $Q(d)$ strictly Hurwitz and such that $\partial Q(d) = 2n - 1$.

9.1.2   The Bounded Disturbance Case

Here, unlike (1), the plant is given by

$$A(d)y(t) = B(d)\left[u(t) + \omega_u(t)\right] + A(d)\left[\omega_y(t) + c\right] \qquad (9.1\text{-}39\text{a})$$

where polynomials $A(d)$ and $B(d)$ as well as $y(t)$, $u(t)$, and c are as in (1). Further, $\omega_u(t)$ and $\omega_y(t)$ denote respectively input and output bounded disturbances such that

$$|\omega_u(t)| \leq \Omega_u < \infty \qquad |\omega_y(t)| \leq \Omega_y < \infty \qquad (9.1\text{-}39\text{b})$$

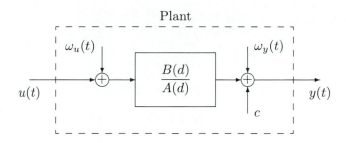

Figure 9.1-2: Plant with input and output bounded disturbances.

where Ω_u and Ω_y are two nonnegative real numbers. Figure 2 depicts the situation. As with (1c), here we find

$$A(d)\Delta(d)y(t) = B(d)\left[\delta u(t) + \delta\omega_u(t)\right] + A(d)\Delta(d)\omega_y(t) \tag{9.1-39c}$$

We again adopt Assumption 1 so as to guarantee that the SIORHC law (6), constructed from (39c) with $\delta\omega_u(t) \equiv \omega_y(t) \equiv 0$, stabilizes the plant.

We now go into the details of constructing and analyzing an adaptive predictive controller for the plant (39). The controller is obtained by combining a CT-NRLS with a dead zone and the SIORHC law in such a way to make the resulting adaptive control system globally convergent.

Identification Algorithm. The identification algorithm is finalized to identify polynomials $A(d)$ and $B(d)$ in the plant incremental model:

$$A(d)\delta y(t) = B(d)\delta u(t) + \delta\omega(t) \tag{9.1-40a}$$

$$\delta\omega(t) := B(d)\delta\omega_u(t) + A(d)\delta\omega_y(t) \tag{9.1-40b}$$

Note that for a nonnegative real number Ω, we have

$$|\delta\omega(t)| \leq \Omega < \infty \tag{9.1-41a}$$

In fact,

$$|\delta\omega(t)| \leq \Omega := 2n\left[\Omega_u \max_{1\leq i\leq n_b} |b_i| + \Omega_y \max_{1\leq i\leq n_b} |a_i|\right] \tag{9.1-41b}$$

Using the same notations as in (10), we have

$$\delta y(t) = \varphi'(t-1)\theta^* + \delta\omega(t) \tag{9.1-42}$$

or, in terms of normalized data,

$$\delta\gamma(t) = x'(t-1)\theta^* + \delta\bar{\omega}(t) \tag{9.1-43a}$$

$$\delta\bar{\omega}(t) := \frac{\delta\omega(t)}{m(t-1)} \tag{9.1-43b}$$

We next consider the following identification algorithm.

CT-NRLS with a Dead Zone (DZ-CT-NRLS).   This is the same as the RDZ-CT-NRLS algorithm (8.9-13) with one change. It consists of replacing the relative dead zone (8.9-13f) with an "absolute" dead zone. Specifically, the DZ-CT-NRLS algorithm is defined by (8.9-13a) to (8.9-13d) with

$$\kappa(t) = \begin{cases} \kappa \in \left(0, \dfrac{\epsilon}{1+\epsilon}\right), & \text{if } |\varepsilon(t)| \geq (1+\epsilon)^{1/2}\,\Omega \\ 0, & \text{otherwise} \end{cases} \tag{9.1-44}$$

The algorithm is initialized for any $P(0) = P'(0) > 0$ and any $\theta(0) \in \mathbb{R}^{n_\theta}$ such that $A_0(d)\Delta(d)$ and $B_0(d)$ are coprime.

The rationale for the dead-zone facility (44) is suggested by an equation similar to (8.9-17) that can be obtained by adapting the solution of Problem 8.9-2 to the present context. The dead-zone mechanism (44) freezes the estimate whenever the absolute value of the prediction error becomes smaller than an upper bound for the disturbance $\delta\omega(t)$ in (42). The properties of this algorithm, which will be used in the sequel, follow.

Result 9.1-1.   (DZ-CT-NRLS)   *Consider the DZ-CT-NRLS algorithm (8.9-13a) to (8.9-13e) and (44) along with the data-generating mechanism (40) and (41). Then, the following properties hold:*

(*i*) Uniform boundedness of the estimates

$$\|\theta(t)\| < M_\theta < \infty \qquad \forall t \in \mathbb{Z}_+ \tag{9.1-45}$$

(*ii*) Vanishing normalized prediction error

$$\lim_{t\to\infty} \kappa(t)\bar{\varepsilon}^2(t) = 0 \tag{9.1-46}$$

(*iii*) Slow asymptotic variations

$$\lim_{t\to\infty} \|\theta(t) - \theta(t-k)\| = 0 \qquad \forall k \in \mathbb{Z}_1 \tag{9.1-47}$$

Problem 9.1-3   Prove Result 1.   (*Hint:*   Adapt Problem 8.9-2 to the present case.)

For the same reasons discussed before (20a), we next proceed to adopt a self-excitation mechanism. To this end, we define a syndrome according to which the controller detects its pathological condition as in (20) and (22). We now construct the remaining part of the adaptive controller.

Controller with Self-Excitation.   The tuning mechanism of the controller parameters is the same as in the ideal case, viz., (21) to (26). See also Fig. 1. Remark 1 applies to the present context as well. The algorithm (8.9-13a) to (8.9-13d), (44), and (21) to (26) will be referred to by the acronym DZ-CT-NRLS+SIORHC.

Convergence Analysis.   We analyze the DZ-CT-NRLS+SIORHC algorithm applied to the plant (39). The following lemma replaces Lemma 1 in the present case.

Lemma 9.1-3.   *For the DZ-CT-NRLS+SIORHC algorithm applied to the plant (39), the following property holds. Given any nonnegative β, there is a nonnegative bounded real number $\underline{\nu}(k, \beta)$ such that*

$$|\nu(k)| > \underline{\nu}(k, \beta) \implies \sum_{t=kN-4n+2}^{kN} \varphi(t)\varphi'(t) > \beta^2 I_{n_\theta} \qquad (9.1\text{-}48)$$

where $|\nu(k)|$ denotes the intensity of the dither injected into the plant according to the self-excitation mechanism (21) to (26). Further, the implication (48) is fulfilled if $\underline{\nu}(k, \beta)$ is chosen as follows:

$$\underline{\nu}(k, \beta) = \frac{\bar{\sigma}(\Gamma)\bar{\sigma}(\mathcal{S})}{\underline{\sigma}(\Gamma')\underline{\sigma}(\mathcal{S})} \left[\sqrt{2n} \frac{\|\Psi(kN)\|}{\bar{\sigma}(\Gamma)} \beta_1 + \bar{\sigma}(V(k)) + \bar{\sigma}(V_\omega(k)) \right] \qquad (9.1\text{-}49a)$$

where

$$\beta_1 := \beta + \bar{\sigma}(\Gamma) \left[n\, (4n-1)\, (\Omega_y^2 + 4\Omega_u^2) \right]^{1/2} \qquad (9.1\text{-}49b)$$

$\mathcal{S}$ denotes the Sylvester resultant matrix defined after (31); $\Psi(kN)$ is the vector

$$\Psi(kN) := \begin{bmatrix} s_0(kN) & \cdots & s_{n-1}(kN) & 1 & r_1(kN) & \cdots & r_{n-1}(kN) \end{bmatrix}' \qquad (9.1\text{-}49c)$$

whose components are the coefficients of $R_{kN}(d)$ and $S_{kN}(d)$ of the SIORHC law at the time kN; Γ is the matrix defined after (31); $V(k)$ is as in (27);

$$V_\omega(k) := \begin{bmatrix} \zeta(kN - 2n + 1) & \zeta(kN - 2n) & \cdots & \zeta(kN - 4n + 2) \\ \vdots & \vdots & \ddots & \vdots \\ \zeta(kN) & \zeta(kN - 1) & \cdots & \zeta(kN - 2n + 1) \end{bmatrix} \qquad (9.1\text{-}49d)$$

$$\zeta(t) := \Psi'(kN)\varphi_\omega(t)$$

with $\varphi_\omega(t)$ as in (53b) to (53d); and $\bar{\sigma}(M)$ and $\underline{\sigma}(M)$ denote respectively the maximum and the minimum singular value of matrix M.

Proof: Consider any unit norm vector $w \in \mathbb{R}^{2n-1}$. Then the R.H.S. of (48) is equivalent to the inequality

$$\beta^2 < w' \left[\sum_{t=kN-4n+2}^{kN} \varphi(t)\varphi'(t) \right] w = w_1' \left[\sum_{t=kN-4n+2}^{kN} \varphi_1(t)\varphi_1'(t) \right] w_1 \qquad (9.1\text{-}50)$$

$$w_1 := \Gamma' w \in \mathbb{R}^{2n}$$

In fact, as discussed in the proof of Lemma 1,

$$\varphi(t) = \Gamma \varphi_1(t) \qquad (9.1\text{-}51)$$

with

$$\varphi_1(t) := \left[\ \left(y_{t-n+1}^t\right)' \quad \left(\delta u_{t-n+1}^t\right)' \ \right]'$$

and Γ a full row-rank matrix.

Consider the following "partial-state" representation of the plant (39c):

$$A(d)\Delta(d)\xi(t) = \delta u(t) + \delta \omega_u(t)$$
$$y(t) = B(d)\xi(t) + \omega_y(t)$$

Define next the partial-state vector

$$\mathcal{Z}(t) := \xi_{t-2n+1}^t \in \mathbb{R}^{2n} \qquad (9.1\text{-}52)$$

Then we find the following relationship between $\varphi_1(t)$ and $\mathcal{Z}(t)$:

$$\varphi_1(t) = \mathcal{S}\mathcal{Z}(t) + \varphi_\omega(t) \qquad (9.1\text{-}53a)$$

where

$$\varphi_\omega(t) = \varphi_y(t) - \varphi_u(t) \qquad (9.1\text{-}53b)$$

and

$$\varphi_y(t) := \left[\ \omega_y(t) \quad \cdots \quad \omega_y(t-n+1) \quad \underbrace{0 \quad \cdots \quad 0}_{n} \ \right]' \qquad (9.1\text{-}53c)$$

$$\varphi_u(t) := \left[\ \underbrace{0 \quad \cdots \quad 0}_{n} \quad \delta\omega_u(t) \quad \cdots \quad \delta\omega_u(t-n+1) \ \right]' \qquad (9.1\text{-}53d)$$

and, as in the proof of Lemma 1, $\mathcal{S}$ denotes the Sylvester resultant matrix associated with polynomials $A(d)\Delta(d)$ and $B(d)$. Substituting (53a) into (50), we get

$$\sum_{t=kN-4n+2}^{kN} \left[w_1'\mathcal{S}\mathcal{Z}(t) + w_1'\varphi_\omega(t) \right]^2 > \beta^2 \qquad (9.1\text{-}54)$$

Now

$$\sum \left[w_1'\mathcal{S}\mathcal{Z}(t) + w_1'\varphi_\omega(t) \right]^2 \geq \left[\left(\sum \left[w_1'\mathcal{S}\mathcal{Z}(t) \right]^2 \right)^{1/2} - \left(\sum \left[w_1'\varphi_\omega(t) \right]^2 \right)^{1/2} \right]^2$$

Further

$$\left[w_1'\tilde{\varphi}_\omega(t) \right]^2 \ \leq \ \|w_1\|^2 \|\varphi_\omega(t)\|^2 \ = \ \|\Gamma'w\|^2 \|\varphi_\omega(t)\|^2$$
$$\leq \ \bar{\sigma}^2(\Gamma) n \Omega_1^2$$

where

$$\Omega_1^2 := \Omega_y^2 + 4\Omega_u^2 \tag{9.1-55}$$

and $\bar{\sigma}(\Gamma')$ denotes the maximum singular value of Γ'. We then conclude that (54), and hence (50), is satisfied provided that

$$\sum \left[w_2' \mathcal{Z}(t)\right]^2 > \beta_1^2 \tag{9.1-56a}$$

$$\beta_1 := \beta + [n(4n-1)]^{1/2} \bar{\sigma}(\Gamma)\Omega_1 \tag{9.1-56b}$$

$$w_2 := \mathcal{S}' w_1 = \mathcal{S}' \Gamma' w \tag{9.1-56c}$$

Rewrite the control law (26) as follows:

$$\begin{aligned} v(t) + \eta(t) &= \Psi'(t)\varphi_1(t) \\ &= \Psi'(t) \left[\mathcal{S}\mathcal{Z}(t) + \varphi_\omega(t)\right] \end{aligned} \tag{[53a]}$$

where

$$\Psi'(t) := \begin{bmatrix} s_0(t) & \cdots & s_{n-1}(t) & 1 & r_1(t) & \cdots & r_{n-1}(t) \end{bmatrix}$$

is the row vector whose components are the coefficients of $R_t(d)$ and $S_t(d)$. Then

$$\Psi'(t)\mathcal{S}\mathcal{Z}(t) = v(t) - \Psi'(t)\varphi_\omega(t) + \eta(t)$$

Defining

$$\mathbf{v}'(t) := \begin{bmatrix} v(t) & \cdots & v(t-2n+1) \end{bmatrix}$$

$$\mathbf{z}'(t) := \Psi'(kN) \begin{bmatrix} \varphi_\omega(t) & \cdots & \varphi_\omega(t-2n+1) \end{bmatrix}$$

$$\tilde{\mathbf{v}}'(t) := \mathbf{v}'(t) - \mathbf{z}(t)$$

and proceeding exactly as in the proof of Lemma 1 after (32), similarly to (33), we find

$$\left\|\mathcal{S}'\Psi(kN)\right\|^2 \sum_{t=kN-2n+1}^{kN} \|\Lambda(t)w_2\|^2 \geq \sum_{t=kN-2n+1}^{kN} \left[\tilde{\mathbf{v}}'(t)w_2 + \nu(k)\mathbf{f}'(t)w_2\right]^2$$

$$= \left\|\left[\nu(k)I + \tilde{V}(k)\right]w_2\right\|^2 \tag{9.1-57a}$$

where

$$\tilde{V}(k) := V(k) - V_\omega(k) \tag{9.1-57b}$$

with $V(k)$ as in (27), and $V_\omega(k)$ as in (49d). Further, as after (33),

$$\sum_{t=kN-2n+1}^{kN} \|\Lambda(t)w_2\|^2 \leq 2n \sum_{t=kN-4n+2}^{kN} \left[w_2'\mathcal{Z}(t)\right]^2$$

Combining this inequality with (57), we get

$$w_2' \left[\sum \mathcal{Z}(t)\mathcal{Z}'(t)\right] w_2 \geq \frac{\left\|\left[\nu(k)I + \tilde{V}(k)\right]w_2\right\|^2}{2n \left\|\mathcal{S}'\Psi(kN)\right\|^2}$$

Therefore, comparing the latter inequality with (56a), we see that (56a), and hence (50), is satisfied provided that

$$\left\|\left[\nu(k)I + \tilde{V}(k)\right]w_2\right\|^2 > 2n \left\|\mathcal{S}'\Psi(kN)\right\|^2 \beta_1^2$$

Now, provided that all the differences in the next inequalities are nonnegative, we have

$$\left\| \nu(k)w_2 + \tilde{V}(k)w_2 \right\| \geq |\nu(k)| \, \|w_2\| - \bar{\sigma}\left(\tilde{V}(k)\right) \|w_2\|$$

$$\geq |\nu(k)| \underline{\sigma}\left(\mathcal{S}\right) \underline{\sigma}\left(\Gamma'\right) - \bar{\sigma}\left(\tilde{V}(k)\right) \bar{\sigma}\left(\mathcal{S}\right) \bar{\sigma}\left(\Gamma\right)$$

$$\geq |\nu(k)| \underline{\sigma}\left(\mathcal{S}\right) \underline{\sigma}\left(\Gamma'\right) - \left[\bar{\sigma}\left(V(k)\right) + \bar{\sigma}\left(V_\omega(k)\right)\right] \bar{\sigma}\left(\mathcal{S}\right) \bar{\sigma}\left(\Gamma\right)$$

Then, (50) is satisfied if

$$|\nu(k)| > \frac{(2n)^{1/2} \left\| \mathcal{S}' \Psi(kN) \right\| \beta_1 + \bar{\sigma}\left(\mathcal{S}\right) \bar{\sigma}\left(\Gamma\right) \left[\bar{\sigma}\left(V(k)\right) + \bar{\sigma}\left(V_\omega(k)\right)\right]}{\underline{\sigma}\left(\mathcal{S}\right) \underline{\sigma}\left(\Gamma'\right)} \qquad (9.1\text{-}58)$$

Problem 9.1-4   Consider (58). Show that

$$\bar{\sigma}\left(V_\omega(k)\right) \leq 2n^{3/2} \|\Psi(kN)\| \Omega_1$$

with Ω_1 as in (55). Recalling (38), check that the R.H.S. of (58) can be upper-bounded by

$$\sqrt{2n} \|\Psi(kN)\| \frac{\bar{\sigma}\left(\mathcal{S}\right) \bar{\sigma}\left(\Gamma\right)}{\underline{\sigma}\left(\mathcal{S}\right) \underline{\sigma}\left(\Gamma'\right)} \left[\frac{\beta}{\bar{\sigma}\left(\Gamma\right)} + \sqrt{n_1} \Omega_1 + \sqrt{2nT} \frac{\|Z_{kN}(d)\|}{\|\Psi(kN)\|} M_r \right]$$

$$\sqrt{n_1} := \sqrt{n}\left(\sqrt{4n-1} + \sqrt{2n}\right)$$

The next lemma points out that if the intensity of the self-excitation dither is high enough, after a finite time, the self-excitation mechanism turns off forever and, henceforth, the estimate is secured to be nonpathologic.

Lemma 9.1-4.   *For the DZ-CT-NRLS+SIORHC algorithm applied to the plant (39), the following self-excitation stopping time property holds. For large enough β and $\underline{\nu}(k, \beta)$ as in Lemma 3, there exists a finite integer T_1 such that*

$$|\nu(k)| > \underline{\nu}(k, \beta) \quad \Longrightarrow \quad \Xi(\theta(t)) > \varsigma, \quad \forall t > T_1 \qquad (9.1\text{-}59)$$

and hence $\eta(t) = 0$, for every $t > T_1$.

Proof: Suppose by contradiction that $\Xi(\theta(t)) \leq \varsigma$ infinitely often irrespective of how large β is chosen. Assume first that $\kappa(t) = \kappa > 0$ infinitely often and $\{m(t)\}$ unbounded. Then, by (46) and (41a), there is a subsequence $\{t_i\}_{i=1}^\infty$ along which $\lim_{i\to\infty} \|x(t_i)\| = 1$ and $\lim_{i\to\infty} \tilde{\theta}(t_i) = O_{n_\theta}$. The latter, together with (47), contradicts that $\Xi(\theta(t)) \leq \varsigma$ infinitely often because $\Xi(\theta^*) = \varsigma^* > \varsigma$ by (20). Therefore, if $\{m(t)\}$ is unbounded, a finite self-excitation stopping time must exist. To exhaust the possible alternatives, assume that either $\kappa(t) \equiv 0$ or, infinitely often, $\kappa(t) = \kappa > 0$ with $\{m(t)\}$ bounded. In both cases, by (46), we find that after a finite time,

$$\Omega_2 := \left[(1+\epsilon)^{1/2} + 1\right] \Omega > \left| \varphi'(t)\tilde{\theta}(t) \right| = \left| \varphi'(t)\tilde{\theta}(kN) + \varphi'(t)\left[\tilde{\theta}(t) - \tilde{\theta}(kN)\right] \right|$$

By (47), this implies that
$$\left| \varphi'(t)\tilde{\theta}(kN) \right| < \Omega_3(t)$$

with
$$\lim_{t\to\infty} \Omega_3(t) = \Omega_2$$

Squaring and summing both sides of the last inequality for $t = kN - 4n + 2, \cdots, kN$, $(k-1)N + 1$ being a time at which the syndrome turns on, we find

$$\Omega_4^2(k) \;:=\; \sum_{t=kN-4n+2}^{kN} \Omega_3^2(t) \;>\; \tilde{\theta}'(kN) \left[\sum_{t=kN-4n+2}^{kN} \varphi(t)\varphi'(t) \right] \tilde{\theta}(kN)$$

$$>\; \beta^2 \left\| \tilde{\theta}(kN) \right\|^2 \qquad\qquad [(48)]$$

Hence,

$$\left\| \tilde{\theta}(kN) \right\|^2 < \frac{\Omega_4^2(k)}{\beta^2} \qquad\qquad (9.1\text{-}60)$$

where for k large enough, $\Omega_4^2(k) < (4n-1)\Omega_2^2 + \delta^2$ with $\delta^2 > 0$. Thus, for k large enough, $\left\| \tilde{\theta}(kN) \right\|$ can be made as small as we wish by choosing β sufficiently large. As noted before, this contradicts the assumption that $\Xi(\theta(t)) \geq \varsigma$ infinitely often.

Remark 9.1-4. Equation (49) is an interesting expression in that it unveils how the different factors affect a lower bound for the required dither intensity $\underline{\nu}(k, \beta)$. First, the bound depends on the plant via the ratio $\bar{\sigma}(\mathcal{S})/\underline{\sigma}(\mathcal{S})$, which can be regarded as a quantitative measure of the reachability of the plant state-space representation for the state $\varphi_1(t)$ in (31), and via the disturbance bounds and $\bar{\sigma}(V_\omega(k))$. Second, the dependence on the controller action is explicit in $\Psi(kN)$ and implicit in $V_\omega(k)$ and $V(k)$, the latter accounting for the feedforward action. Note that if $\beta = \Omega_u = \Omega_y = 0$, (49) reduces to

$$\left. \underline{\nu}(k, \beta) \right|_{\substack{\beta=0 \\ \Omega_u=\Omega_y=0}} = \frac{\bar{\sigma}(\mathcal{S})\,\bar{\sigma}(\Gamma)}{\underline{\sigma}(\mathcal{S})\,\underline{\sigma}(\Gamma')}\bar{\sigma}(V(k))$$

which is a conservative version of the condition (28) valid for the ideal case.

We are now ready to prove the main result for the adaptive system in the presence of bounded disturbances.

Theorem 9.1-2. (**Global Convergence of DZ-CT-NRLS+SIORHC**) *Consider the DZ-CT-NRLS+SIORHC applied to the plant (39). Let the output reference $\{r(t)\}$ be bounded and the self-excitation intensity be chosen, according to Lemma 4, large enough to guarantee a finite self-excitation stopping time. Then, the resulting adaptive system is globally convergent. Specifically:*

(i) $u(t)$ and $y(t)$ are uniformly bounded.

(ii) After a finite time T_2, the parameter estimate equals

$$\theta(\infty) \sim (A_\infty(d), B_\infty(d)) \qquad\qquad (9.1\text{-}61)$$

and the controller self-tunes on a control law:

$$R_\infty(d)\delta u(t) = -S_\infty(d)y(t) + Z_\infty(d)r(t+T), \quad \forall t > T_2 \qquad (9.1\text{-}62)$$

which stabilizes the system and such that

$$\chi_\infty(d) := A_\infty(d)\Delta(d)R_\infty(d) + S_\infty(d)B_\infty(d) \qquad (9.1\text{-}63)$$

is strictly Hurwitz.

(iii) After the time T_2, the prediction error remains inside the dead zone:

$$|\varepsilon(t)| < (1+\epsilon)^{1/2}\Omega, \quad \forall t > T_2 \qquad (9.1\text{-}64)$$

(iv) If $r(t) \equiv r$, we have

$$y(t) - r \quad \xrightarrow[(t\to\infty)]{} \quad \frac{R_\infty(d)}{\chi_\infty(d)}\varepsilon(t) \qquad (9.1\text{-}65a)$$

$$\delta u(t) \quad \xrightarrow[(t\to\infty)]{} \quad \frac{-S_\infty(d)}{\chi_\infty(d)}\varepsilon(t) \qquad (9.1\text{-}65b)$$

Proof:

(i) We proceed along the same lines as after Lemma 2 to find that for all $t > T_3$, T_3 being a finite time greater than T_1 in Lemma 4, the following linear boundedness condition

$$\|\varphi_1(t-1)\| \le c_1 + c_2 \max_{i\in[1,t)} |\varepsilon(i)| \qquad (9.1\text{-}66)$$

holds for bounded nonnegative reals c_1 and c_2. Now if $\{\varepsilon(t)\}$ is bounded, from the preceding inequality, it follows that $\{\varphi_1(t)\}$ is also bounded. Suppose on the contrary that $\{\varepsilon(t)\}$ is unbounded. Then there is a subsequence $\{t_i\}$ along which $\lim_{t_i\to\infty} |\varepsilon(t_i)| = \infty$ and $|\varepsilon(t)| \le |\varepsilon(t_i)|$, for $t \le t_i$. Further, $\kappa(t_i) = \kappa > 0$. Thus,

$$
\begin{aligned}
\sqrt{\kappa(t_i)}\,|\bar\varepsilon(t_i)| &= \sqrt{\kappa}\,\frac{|\varepsilon(t_i)|}{m(t_i-1)} \\
&\ge \sqrt{\kappa}\,\frac{|\varepsilon(t_i)|}{m + \|\Gamma\varphi_1(t_i-1)\|} & [(51)] \\
&\ge \sqrt{\kappa}\,\frac{|\varepsilon(t_i)|}{c_3 + c_4|\varepsilon(t_i)|} & [(66)]
\end{aligned}
$$

This implies that

$$\lim_{t_i\to\infty} \sqrt{\kappa(t_i)}\,|\bar\varepsilon(t_i)| \ge \frac{\kappa^{1/2}}{c_4} > 0$$

which contradicts (46). Then $\{\varphi_1(t)\}$ is bounded. This is equivalent to boundedness of $\{y(t)\}$ and $\{\delta u(t)\}$. To prove boundedness of $\{u(t)\}$, we can use the same Bezout identity argument introduced before Theorem 1.

(ii) and (iii) Because $\{\varphi(t)\}$ is bounded by (i), (46) yields

$$\lim_{t\to\infty} \kappa(t)\varepsilon^2(t) = 0$$

Hence,

$$\limsup_{t\to\infty} |\varepsilon(t)| < (1+\epsilon)^{1/2}\,\Omega$$

which implies (64) and, hence, (ii) because of the finite self-excitation stopping time.

(iv) By using (62), that $Z_t(1) = S_t(1)$ and the fact that for $t > T_2$,

$$A_\infty(d)\Delta(d)y(t) = B_\infty(d)\delta u(t) + \varepsilon(t)$$

(65) follows.

It is difficult to find a sharp estimate of self-excitation intensity $|\nu(k)|$ that can guarantee the condition (59). On the other hand, even a conservative estimate of this intensity, such as that in (49) and (60), would depend in practice on a priori unknown parameters (the ratio between the maximum and minimum singular value of the transpose of the Sylvester resultant matrix associated to the true plant, how small $\|\tilde{\theta}(t)\|$ must be in order to guarantee that $\Xi(\theta(t)) > \varsigma$, and so on). Therefore, the practical relevance of Theorem 2 is to indicate that the combination of a high-intensity self-excitation dither with a CT-NRLS with a relative dead zone can make the adaptive system capable of self-tuning on a stable behavior in the presence of bounded disturbances.

9.1.3 The Neglected Dynamics Case

In this section, we discuss qualitatively some frequency-domain and related filtering ideas that turn out to be important in the neglected dynamics case. The discussion parallels the one on the same subject presented in the first part of Sec. 8.9 for STCC. Here we extend the ideas to adaptive SIORHC. However, we shall refrain from embarking on elaborating any globally convergent adaptive multistep predictive controller for the neglected dynamics case. This is in fact an issue in the realm of current research. For some results on this point, see [CMS91].

We assume that the plant to be controlled is again given by (8.9-1). Here, however, (8.9-1) is modified as follows:

$$A^o(d)\mathrm{II}(d)y(t) = B^o(d)\delta u(t) + A^o(d)\Pi(d)\omega(t) \qquad (9.1\text{-}67\text{a})$$

where

$$\delta u(t) := \Pi(d)u(t) \qquad (9.1\text{-}67\text{b})$$

In this way, the presence of the common divisor $\Pi(d)$ of $A^o(d)$ and $B^o(d)$ as in (8.9-2) is ruled out. This is important because, unlike cheap control, SIORHC design equations cannot be easily managed in the presence of the previous common divisor and, more important, stability of the controlled system is not guaranteed. Notice that (67b) generalizes our usual notational convention of denoting simply by $\delta u(t)$ an input increment. The reader should realize before proceeding any further that a SIORHC with no formal changes is fully compatible with (67), in that its terminal input constraints are still meaningful for the notion (67b) of *generalized input increments*. From (67), we obtain the reduced-order model (cf. (8.9-3))

$$A(d)\Pi(d)y(t) = B(d)\delta u(t) + n(t) \qquad (9.1\text{-}68\text{a})$$

$$A^o(d) = A(d)A^u(d) \qquad B^o(d) = B(d)B^u(d) \qquad (9.1\text{-}68\text{b})$$

and

$$n(t) = \frac{B(d)\left[B^u(d) - A^u(d)\right]}{A^u(d)} \delta u(t) + A(d)\Pi(d)\omega(t) \qquad (9.1\text{-}68c)$$

In order to possibly identify a reduced-order model that adequately fits the plant within the useful frequency band, we proceed in accordance with the guidelines given in Sec. 8.9. Here we select a low-pass stable and stably invertible transfer function $L(d)$ that rolls off beyond the useful frequency band.

By prefiltering the plant I/O variables with $L(d)$, we arrive at the following model:

$$A(d)\Pi(d)y_L(t) = B(d)\delta u_L(t) + n_L(t) \qquad (9.1\text{-}69a)$$

$$y_L(t) := L(d)y(t) \qquad \delta u_L(t) := L(d)\delta u(t) \qquad (9.1\text{-}69b)$$

and $n_L(t) := L(t)n(t)$. This is formally the same as (8.9-4). There is, however, a difference between the two models in that the route that we have now followed to arrive at (69) has been deliberately finalized to avoid the introduction of $\Pi(d)$ as a common divisor of $A(d)\Pi(d)$ and $B(d)$. The polynomials to be identified are $A(d)$ and $B(d)$ via the use of the filtered variables $\delta y_L(t) := \Pi(d)y_L(t)$ and $\delta u_L(t)$. These are related by the system

$$A(d)\delta y_L(t) = B(d)\delta u_L(t) + n_L(t) \qquad (9.1\text{-}70)$$

We next focus on how to make robust the control system by introducing suitable dynamic weights in the underlying control problem. This is done by adopting a procedure similar to the one in (8.9-5) to (8.9-8). Specifically, we consider the plant I/O filtered variables

$$y_H(t) := H(d)y(t) \qquad \delta u_H(t) := H(d)\delta u(t) \qquad (9.1\text{-}71a)$$

with $H(d)$ a monic high-pass strictly Hurwitz polynomial, and the model

$$A(d)\Pi(d)y_H(t) = B(d)\delta u_H(t) + H(d)n(t) \qquad (9.1\text{-}71b)$$

where $n(t)$ is assumed to be a zero-mean white noise. Then we compute the SIORHC law related to the cost

$$\frac{1}{N} \sum_{k=t}^{t+T-1} \mathcal{E}\left\{\varepsilon_y^2(k+1) + \Psi_u \delta u_H^2(k) \mid y^t, \delta u^{t-1}\right\} \qquad (9.1\text{-}72a)$$

$$\varepsilon_y(k) := y_H(k) - H(1)r(k) \qquad (9.1\text{-}72b)$$

and the constraints

$$\left. \begin{array}{ll} \delta u_H(k) = 0 & t+T \le k \le t+T+\hat{n}-1 \\ \mathcal{E}\left\{y_H(k) \mid y^t, \delta u^{t-1}\right\} = H(1)r(t+T+1) & t+T+1 \le k \le t+T+\hat{n} \end{array} \right\}$$

$$(9.1\text{-}72c)$$

where $T \geq \hat{n}$, and $\hat{n}$ is the McMillan degree of $B(d)/A(d)$. For the necessary details on the previous SIORHC law, we refer the reader to Sec. 7.7. The rationale for introducing the filtered variables is similar to the one discussed in (8.9-5) to (8.9-8). See also (7.5-29) to (7.5-32).

In the previous considerations, we pointed out that, in contrast with the ideal case, in the neglected dynamics case, it is essential to identify a reduced-order model using low-pass prefiltered I/O variables. In particular, identification of an incremental model as (1c) with no prefiltering is by all means unadvisable, in that the $\Delta(d)$ polynomial enhances the high-frequency components of the equation error. The second important point is that, in order to robustify the controlled system, it is advisable that the control synthesis be carried out relatively to high-pass filtered I/O variables as in (71) and (72).

Main Points of the Section. By using a self-excitation mechanism finalized to avoid possible singularities, a globally convergent adaptive SIORHC algorithm based on the CT-NRLS can be constructed. The constant-trace feature of the estimator makes adaptive SIORHC suitable for slowly time-varying plants. In order to choose the self-excitation signal, the interaction between self-excitation and feed-forward must be considered. If this is properly done, for a time-invariant plant, the self-excitation turns off forever after a finite time.

In the ideal case, the intensity of the dither due to self-excitation can be chosen vanishingly small. In the case of bounded disturbances, where the CT-NRLS identifier is equipped with a dead-zone facility, the dither intensity is required to be high enough so as to force the final plant estimated parameters to be nonpathologic. Low-pass filtering for identification and high-pass filtering for control synthesis are often procedures of paramount importance to favor successful operation in the presence of neglected dynamics.

9.2 IMPLICIT MULTISTEP PREDICTION MODELS OF LINEAR-REGRESSION TYPE

In Sec. 8.7, we showed that the output prediction of a MV-controlled CARMA plant can be described in terms of a linear-regression model. This fact was exploited to construct an implicit stochastic ST controller based on the MV control law. The aim of this section is to show that a similar property holds also for more general control laws. This is of interest in that it allows us to construct implicit adaptive controllers for CARMA plants with underlying control laws of wider applicability than MV control. In this respect, particular attention will be devoted to underlying long-range predictive control laws. As will be seen, the resulting implicit adaptive predictive controllers exhibit advantages and disadvantages over the explicit ones. One disadvantage is that there is no available proof of a globally convergent implicit adaptive predictive control scheme. The only possible exception to this is [Loz89], which is, however, solely focused on the adaptive stabilization problem with no performance-related goal. On the positive side, implicit adaptive predictive controllers can exhibit excellent local self-optimizing properties in the presence of neglected dynamics. This makes them attractive for auto-tuning simple controllers of highly complex plants.

The starting point of our study is the SISO CARMA plant:

$$A(d)y(t) = B(d)u(t) + C(d)e(t) \qquad (9.2\text{-}1a)$$

with $A(0) = C(0) = 1$. And

- $n := \max\{\partial A(d), \partial B(d), \partial C(d)\}$ \hfill (9.2-1b)

- $A(d)$, $B(d)$, and $C(d)$ have unit gcd \hfill (9.2-1c)

- $C(d)$ is strictly Hurwitz \hfill (9.2-1d)

- $A(d)$ and $B(d)$ have strictly Hurwitz gcd \hfill (9.2-1e)

Further, the innovations process e is zero-mean wide-sense stationary white with variance

$$\sigma_e^2 := \mathcal{E}\left\{e^2(t)\right\} > 0 \qquad (9.2\text{-}1f)$$

and such that

$$\mathcal{E}\left\{u(t)e(t+i)\right\} = 0 \quad , \quad t \in \mathbb{Z}, i \in \mathbb{Z}_1 \qquad (9.2\text{-}1g)$$

Irrespective of the actual plant I/O delay $\tau = \text{ord}\, B(d)$, we can follow similar lines to the ones that led us to (7.3-9) so as to find, for $k \in \mathbb{Z}_1$,

$$y(t+k) = \frac{Q_k(d)B(d)}{C(d)}u(t+k) + \frac{G_k(d)}{C(d)}y(t) + Q_k(d)e(t+k) \qquad (9.2\text{-}2)$$

where $(Q_k(d), G_k(d))$ is the minimum-degree solution w.r.t. $Q_k(d)$ of the following Diophantine equation:

$$\left.\begin{array}{r}C(d) = A(d)Q_k(d) + d^k G_k(d) \\ \partial Q_k(d) \le k - 1\end{array}\right\} \qquad (9.2\text{-}3)$$

The problem that we wish to study next is to possibly find conditions on the "past" input sequence u^{t-1} under which (2) simplifies to the form

$$y(t+k) = w_1 u(t+k-1) + \cdots + w_k u(t) + S_k s(t) + Q_k(d)e(t+k) \qquad (9.2\text{-}4)$$

for every possible "future" input sequence $u_{[t,t+k)}$. In (4), $s(t)$ denotes a vector with a *finite* number of components from y^t and u^{t-1}, and S_k a row vector of compatible dimension. Further, because of the degree constraint in (3), $Q_k(d)e(t+k)$ is a linear combination of future innovations samples in $e_{[t+1,t+k]}$. The difference between (4) and (2) is that the latter, due to the presence of $C(d)$ at the denominator, involves an infinite number of terms from y^t and u^{t-1}. Using a terminology similar to that adopted in Sec. 8.7, we shall call (4) an *implicit prediction model of linear-regression type*.

To solve the previous problem, we first find the minimum-degree solution $(W_k(d), L_k(d))$ w.r.t. $W_k(d)$ of the following Diophantine equation:

$$\left.\begin{array}{c} Q_k(d)B(d) = C(d)W_k(d) + d^{k+1}L_k(d) \\ \partial W_k(d) \leq k \end{array}\right\} \qquad (9.2\text{-}5)$$

Using (5) into (2), we get

$$y(t+k) = W_k(d)u(t+k) + \frac{L_k(d)}{C(d)}u(t-1) + \frac{G_k(d)}{C(d)}y(t) + Q_k(d)e(t+k) \quad (9.2\text{-}6)$$

Note that from (3),

$$\begin{aligned} C(d)B(d) &= A(d)Q_k(d)B(d) + d^k G_k(d)B(d) \\ &= C(d)A(d)W_k(d) + d^{k+1}\left[A(d)L_k(d) + \frac{G_k(d)B(d)}{d}\right] \end{aligned}$$

Then it follows that the coefficients of $W_k(d)$ coincide with the first k terms of the long division of $B(d)/A(d)$, viz.,

$$W_k(d) = w_1 d + \cdots + w_k d^k \qquad (9.2\text{-}7)$$

where w_i, $i = 1, \cdots, k$, are the first k samples of the impulse response associated with transfer function $B(d)/A(d)$.

We see that, in order to rewrite (6) in the form (4), two polynomials $U_k(d)$ and $\Gamma_k(d)$ must exist such as to satisfy

$$\frac{L_k(d)}{C(d)}u(t-1) + \frac{G_k(d)}{C(d)}y(t) = U_k(d)u(t-1) + \Gamma_k(d)y(t) \qquad (9.2\text{-}8)$$

where the equality has to be intended in a mean-square sense. For an arbitrary stochastic input process u, the previous equation is not solvable w.r.t. $U_k(d)$ and $\Gamma_k(d)$. On the other hand, in a regulation problem, we are only interested in input sequences generated up to time $t-1$ by a time-invariant nonanticipative linear feedback compensator of the form

$$R(d)u(i) = -S(d)y(i) \quad , \quad t-n \leq i \leq t-1 \qquad (9.2\text{-}9a)$$

where $R(d)$ and $S(d)$ are polynomials with $R(0) \neq 0$ and such that

$$\left.\begin{array}{c} \partial R(d) = n \quad , \quad \partial S(d) = n-1 \\ R(d) \text{ and } S(d) \text{ coprime} \end{array}\right\} \qquad (9.2\text{-}9b)$$

The lower-bound $t-n$ on time index i in (9a) indicates that the stated regulation law need not be used before time $t-n$ (see Fig. 1). We point out that the degree

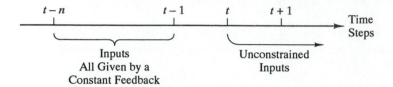

Figure 9.2-1: Visualization of the constraint (9a).

assumptions on $R(d)$ and $S(d)$ are consistent with both steady-state LQ stochastic regulation (cf. Problem 7.3-16) and stochastic predictive regulation (cf. Problem 7.7-3). Let us multiply each term of (6) by $C(d)R(d)$ to get

$$
\begin{aligned}
C(d)R(d)y(t+k) \;=\; & C(d)R(d)\left[W_k(d)u(t+k)+Q_k(d)e(t+k)\right] \\
& + L_k(d)R(d)u(t-1)+G_k(d)R(d)y(t)
\end{aligned}
$$

Because $\partial L_k(d) \le n-1$, the third additive term on the R.H.S. of the last equation only involves input variables comprised in (9a). Thus,

$$
L_k(d)R(d)u(t-1)+G_k(d)R(d)y(t) = [R(d)G_k(d)-dS(d)L_k(d)]\,y(t) \qquad (9.2\text{-}10)
$$

In order to fulfill (8), this quantity must coincide in a mean-square sense with

$$
\begin{aligned}
C(d)\left[U_k(d)R(d)u(t-1)+\Gamma_k(d)R(d)y(t)\right] & \qquad\qquad (9.2\text{-}11) \\
= C(d)\left[R(d)\Gamma_k(d)-dS(d)U_k(d)\right]y(t) &
\end{aligned}
$$

where the equality follows from (9a) provided that $\partial U_k(d) \le n-1$. Because by (1a), (1f), and (1g), process y contains an additive white component with nonzero variance, (10) equals (11) in a mean-square sense if and only if the following Diophantine equation

$$
C(d)\left[R(d)\Gamma_k(d)-dS(d)U_k(d)\right] = R(d)G_k(d)-dS(d)L_k(d) \qquad (9.2\text{-}12)
$$

admits a polynomial solution $U_k(d)$ and $\Gamma_k(d)$. With $R(d)$ and $S(d)$ coprime, the solvability of (12) is equivalent to require that the R.H.S. of (12) be divided by $C(d)$:

$$
C(d) \mid [R(d)G_k(d)-dS(d)L_k(d)] \qquad (9.2\text{-}13)
$$

Now, using (3) and (5), we find

$$
dS(d)L_k(d)-R(d)G_k(d) = \frac{Q_k(d)\chi_{cl}(d)-C(d)\left[R(d)+S(d)W_k(d)\right]}{d^k} \qquad (9.2\text{-}14)
$$

where

$$
\chi_{cl}(d) := A(d)R(d)+B(d)S(d)
$$

Therefore, (13) is satisfied provided that

$$C(d) \mid \chi_{cl}(d) \tag{9.2-15}$$

Further, by degree considerations, we see that, under (15), the minimum-degree solution of (12) is such that

$$\partial U_k(d) \leq n - 1 \qquad \partial \Gamma_k(d) \leq n - 1$$

We sum up the previous results in the following theorem.

Theorem 9.2-1. *Assume that the CARMA plant (1) be fed over the time interval $[t - n, t - 1]$ by the linear feedback compensator (9). Then, irrespective of $u_{[t,t+k)}$, the following implicit prediction models of linear-regression type*

$$\begin{aligned}
y(t + k) \;=\; & w_1 u(t + k - 1) + \cdots + w_k u(t) \\
& + S_k s(t) + Q_k(d) e(t + k) \qquad \forall k \in \mathbb{Z}_1
\end{aligned} \tag{9.2-16a}$$

$$s(t) := \left[\; \left(y_{t-n+1}^{t}\right)' \;\; \left(u_{t-n}^{t-1}\right)' \;\right]' \in \mathbb{R}^{n_s} \tag{9.2-16b}$$

hold provided that

$$C(d) \mid [A(d)R(d) + B(d)S(d)] \tag{9.2-17}$$

The extension of Theorem 1 to 2-DOF controllers is given by the next problem.

Problem 9.2-1 Assume that for $i \in [t - n, t - 1]$, the inputs to the plant of Proposition 1 are in accordance with the following difference equation:

$$R(d)u(i) = -S(d)y(i) + C(d)v(i) \tag{9.2-18}$$

where $R(d)$ and $S(d)$ are as in (9) and satisfy (17), and v denotes an exogenous random sequence possibly related to a reference to be tracked by the plant output (cf. (7.5-11)). Then, show that (16a) still holds provided that $s(t)$ be redefined as follows:

$$s(t) := \left[\; \left(y_t^{t-n+1}\right)' \;\; \left(u_{t-1}^{t-n}\right)' \;\; \left(v_{t-1}^{t-n}\right)' \;\right]' \tag{9.2-19}$$

Remark 9.2-1. Vector $s(t)$ in (16b) or (19) will be referred to as the *pseudo-state* because, under the stated past input conditions, $s(t)$ is a sufficient statistic to predict $y(t + k)$ in a MMSE sense on the basis of y^t, u^{t+k-1} (cf. Theorem 7.3-1).

Remark 9.2-2. Conditions (17) to (19) have the following state-space interpretation (cf. (7.5-4) and successive considerations). Equations (17) to (19) are equivalent to assuming that the control action on the plant over the time interval $i \in [t - n, t - 1]$ is given by

$$u(i) = \mathcal{F}x(i \mid i) + v(i) \tag{9.2-20}$$

the R.H.S. of (20) being the sum of $v(i)$ with a constant feedback from the steady-state Kalman filtered estimate $x(i \mid i)$ of a plant state $x(i)$.

Main Points of the Section. CARMA plants admit implicit multistep prediction models of linear-regression type, provided that their inputs over a finite past are given by feedback compensation from the steady-state Kalman filtered estimate of a plant state.

Problem 9.2-2 [MZ89c] Consider (16a), where $s(t)$ is possibly given by (19). Let

$$\frac{A(d)}{C(d)} = \sum_{i=0}^{\infty} \alpha_i d^i \quad \text{and} \quad \frac{B(d)}{C(d)} = \sum_{i=1}^{\infty} \beta_i d^i$$

Show that

$$w_i \quad = \quad \beta_i - \alpha_1 w_{i-1} - \cdots - \alpha_{i-1} w_1$$
$$\epsilon_i(t+i) \quad := \quad Q_i(d)e(t+i)$$
$$= \quad e(t+i) - \alpha_1 \epsilon_{i-1}(t+i-1) - \cdots - \alpha_{i-1}\epsilon_1(t+1)$$

with

$$w_1 = \beta_1 \quad \text{and} \quad \epsilon_1(t+1) = e(t+1)$$

9.3   USE OF IMPLICIT PREDICTION MODELS IN ADAPTIVE PREDICTIVE CONTROL

The interest in the multistep implicit prediction models (2-16) to (2-19) is that they can be exploited in adaptive predictive control schemes as if the pseudo-state $s(t)$ were a true plant state, irrespective of the innovations polynomial $C(d)$. However, unlike the implicit linear-regression model of Sec. 8.7, where prediction step k equals the plant I/O delay τ, the parameters of the multistep implicit prediction models (2-16) to (2-19) cannot be directly identified via recursive linear-regression algorithms because, for $k > 1$, $Q_k(d)e(t + k)$ can be correlated with u_{t+k-1}^{t+1}. Nevertheless, defining new "observations" by

$$z_t(t+k) \quad := \quad y(t+k) - w_1 u(t+k-1) - \cdots - w_{k-1}u(t+1)$$
$$= \quad w_k u(t) + S_k s(t) + \nu_k(t+k) \qquad [(2\text{-}16a)] \qquad (9.3\text{-}1)$$

we find that the equation error

$$\nu_k(t+k) := Q_k(d)e(t+k)$$

by (2-1g) is uncorrelated with the regressor $\varphi(t) := \begin{bmatrix} u(t) & s'(t) \end{bmatrix}'$. Thus, we can introduce the following identification scheme where any linear-regression algorithm could replace the RLS algorithm to which we shall refer from here on.

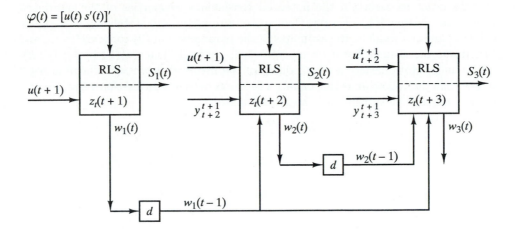

Figure 9.3-1: Signal flow in the interlaced identification scheme.

9.3.1 Interlaced Identification Scheme

At each time step t, and for each $k = 1, 2, \cdots, T + n - 1$:

1. Compute

$$\hat{z}_t(t + k) \quad := \quad y(t + k)$$
$$- w_1(t - 1)u(t + k - 1) - \cdots - w_{k-1}(t - 1)u(t + 1) \quad (9.3\text{-}2)$$

where $w_i(t - 1)$, $i \le k - 1$, is the RLS estimate of w_i based on the regressors $\varphi^{t-1} = \{u^{t-1}, s^{t-1}\}$.

2. Update the RLS estimates $w_k(t - 1)$ and $S_k(l - 1)$ so as to get $w_k(t)$ and $S_k(t)$ based on the new observation $\hat{z}_t(t + k)$ and the model

$$z_t(t + k) = w_k u(t) + S_k s(t) + \nu_k(t + k) \quad (9.3\text{-}3)$$

3. Cycle through steps 1 and 2.

Figure 1 depicts the signal flows in the interlaced identification scheme for $T + n - 1 = 3$. The upper arrow refers to the regressor $\begin{bmatrix} u(t) & s'(t) \end{bmatrix}'$, and the regressand $\hat{z}_t(t + k)$ is indicated at the bottom of each individual RLS identifier.

    Remark 9.3-1.   As shown in Fig. 1, the previous identification scheme comprises $T + n - 1$ *separate* RLS estimators all using a common regressor and, hence, a common updating gain. This feature moderates the overall computational burden of the interlaced identification scheme.

In order to ascertain the potential consistency properties of the interlaced estimation scheme, let us assume that, under a constant and stabilizing control law allowing implicit models, the estimates of the parameters in (3) converge to $\hat{w}_k$ and $\hat{S}_k$. Moreover, assume that $\hat{w}_j = w_j, \forall j = 1, 2, \cdots, k-1$; hence, $\hat{z}_t(t+k) = z_t(t+k)$. Thus, under stationariety and ergodicity of the involved processes, Theorem 6.4-2 shows that the following orthogonality condition between estimation residuals and regressors must be satisfied for the kth estimator:

$$
\begin{aligned}
0 &= \mathcal{E}\left\{\left[\begin{array}{cc} u(t) & s'(t) \end{array}\right]' \left(z_t(t+k) - \hat{w}_k u(t) - \hat{S}_k s(t)\right)\right\} \\
&= \mathcal{E}\left\{\left[\begin{array}{cc} u(t) & s'(t) \end{array}\right]' \left(\tilde{w}_k u(t) + \tilde{S}_k s(t) + \nu_k(t+k)\right)\right\} \\
&= \mathcal{E}\left\{\left[\begin{array}{cc} u(t) & s'(t) \end{array}\right]' \left(\tilde{w}_k u(t) + \tilde{S}_k s(t)\right)\right\} \quad (9.3\text{-}4)
\end{aligned}
$$

where $\tilde{w}_k := w_k - \hat{w}_k$, $\tilde{S}_k := S_k - \hat{S}_k$, and the last equality follows because

$$
\mathcal{E}\left\{\left[\begin{array}{cc} u(t) & s'(t) \end{array}\right]' \nu_k(t+k)\right\} = 0
$$

Note that, according to (2-18), $u(t) = Fs(t) + v(t)$. Assume that $v(t)$ has a component with nonzero variance and uncorrelated with $s(t)$. Then from (4), it follows that $\hat{w}_k = w_k$. Hence, $\hat{w}_j = w_j, \forall j = 1, 2, \cdots, k-1$, implies $\hat{w}_k = w_k$. Because $\hat{z}_t(t+1) = z_t(t-1)$, $\hat{w}_1 = w_1$. Therefore, by induction, the estimates of the w_k's are potentially consistent. Moreover, from (4), we also get $\mathcal{E}\left\{s(t)s'(t)\right\}\tilde{S}'_k = 0$ that yields $\hat{S}_k = S_k$ if $\Psi_s := \mathcal{E}\left\{s(t)s'(t)\right\} > 0$.

9.3.2 Implicit Adaptive SIORHC

We can exploit the earlier interlaced identification scheme to construct an implicit adaptive SIORHC algorithm for the CARMA plant

$$
A(d)\Delta(d)y(t) = B(d)\delta u(t) + C(d)e(t) \quad (9.3\text{-}5)
$$

having all the properties as in (2-1) once $A(d)$ is changed into $A(d)\Delta(d)$. In this case, $n = \max\{\partial A(d) + 1, \partial B(d), \partial C(d)\}$, and the pseudo-state $s(t)$ is as in (2-19) with u_{t-n}^{t-1} replaced by δu_{t-n}^{t-1}. If we consider SIORHC as the underlying control problem, by the enforced certainty equivalence, we can set the control variable to the plant input at time $\tau := t + T + n - 1$ to be given by

$$
\left.\begin{array}{rcl}
R_t(d)\delta u(\tau) &=& -S_t(d)y(\tau) + v(\tau) + \eta(\tau) \\
v(\tau) &=& \mathcal{Z}_t(d)r(\tau + T)
\end{array}\right\} \quad (9.3\text{-}6)
$$

where $R_t(d)$, $S_t(d)$, $\mathcal{Z}_t(d)$, with $R_t(0) = 1$, and $\partial \mathcal{Z}_t(d) = T - 1$ are the polynomials corresponding to the SIORHC law (1-6) computed by using the RLS estimates $w_k(t)$ and $S_k(t)$ from the interlaced identification scheme.

The last term $\eta(\tau)$ in the first equation of (6) is an additive *dither*, viz., an exogenous variable introduced so as to second parameter identifiability. For example, η can be a zero mean w.s. stationary white random sequence uncorrelated with both e and r, and with variance $\sigma_\eta^2 > 0$ smaller than σ_e^2. To sum up, once the pseudo-state $s(t)$ is formed and the interlaced estimation scheme used, we can use the RLS estimates $w_k(t)$ and $S_k(t)$ to compute $\delta u(t)$ via (1-6) as if the $C(d)$ innovations polynomial were equal to 1.

9.3.3 Implicit Adaptive TCI: MUSMAR

A noticeable simplification in the use of implicit models in adaptive predictive control can be achieved by adopting underlying control problems compatible with constraints on the future inputs $u_{(t,t+T)}$ in addition to the constraints (9) on the past inputs $u_{[t-n,t)}$. We shall describe this by focusing on the pure regulation problem. Consider again the CARMA plant (2.1). Let

$$n := \max\{\partial A(d), \partial B(d), \partial C(d)\}$$

and

$$s(t) := \left[\ (y_{t-n+1}^t)'\ \ (u_{t-n}^{t-1})'\ \right]'$$

Assume also that

$$R_p(d)u(i) = -S_p(d)y(i) \qquad t - n \le i \le t - 1 \tag{9.3-7a}$$

or

$$u(i) = F_p's(i) \qquad t - n \le i \le t - 1 \tag{9.3-7b}$$

where

$$\left.\begin{array}{c} \partial R_p(d) = n, \qquad \partial S_p(d) = n - 1 \\ R_p(d) \text{ and } S_p(d) \text{ coprime} \end{array}\right\} \tag{9.3-7c}$$

subscript p is appended to quantities related to the "past" regulation law, and F_p' is a row vector whose components are the coefficients of polynomials $R_p(d)$ and $S_p(d)$. Assuming that

$$C(d) \mid A(d)R_p(d) + B(d)S_p(d) \tag{9.3-8}$$

from Theorem 2-1 it follows that

$$\begin{aligned} y(t + k) &= w_1 u(t + k - 1) + \cdots + w_k u(t) \\ &\quad + S_k s(t) + Q_k(d)e(t + k) \qquad \forall k \in \mathbb{Z}_1 \end{aligned} \tag{9.3-9}$$

Suppose now that in addition to the "past" constraints, the following constraints are adopted for the "future" regulation law:

$$R_f(d)u(i) = -S_f(d)y(i) \qquad t + 1 \le i \le t + T - 1 \tag{9.3-10a}$$

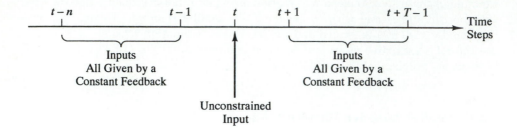

Figure 9.3-2: Visualization of the constraints (7) and (10).

or

$$u(i) = F_f's(i) \qquad t+1 \leq i \leq t+T-1 \qquad (9.3\text{-}10b)$$

The situation related to (7) and (9) is depicted in Fig. 2. There it is shown that although both the past and the future inputs are generated via constant feedback laws, the input at "current" time t is unconstrained. Figure 2 should be compared with Fig. 2-1 in order to visualize the additional constraints that we are now adopting.

We now go back to (9) to write successively

$$y(t+1) = w_1 u(t) + S_1 s(t) + e(t+1)$$

$$\begin{aligned} u(t+1) &= F_f' s(t+1) \\ &= \mu_2 u(t) + \Lambda_2' s(t) + \nu_2(t+1) \end{aligned}$$

$$\begin{aligned} y(t+2) &= w_1 u(t+1) + w_2 u(t) + S_2 s(t) + Q_2(d)e(t+2) \\ &= \theta_2 u(t) + \Gamma_2' s(t) + \epsilon_2(t+2) \end{aligned}$$

Note that $\nu_2(t+1)$ is proportional to $e(t+1)$, and $\epsilon_2(t+2)$ is a linear combination of $e(t+1)$ and $e(t+2)$. By induction, we can thus prove the following result.

Theorem 9.3-1. *Consider the CARMA plant (2-1). Let its past inputs $u_{[t-n,t)}$ satisfy (7), and its future inputs $u_{(t,t+T)}$ be given as in (10). Then, irrespective of $u(t)$, the following implicit prediction models of linear-regression type hold for $1 \leq i \leq T$*

$$y(t+i) = \theta_i u(t) + \Gamma_i' s(t) + \epsilon_i(t+i) \qquad (9.3\text{-}11a)$$

$$u(t+i-1) = \mu_i u(t) + \Lambda_i' s(t) + \nu_i(t+i-1) \qquad (9.3\text{-}11b)$$

where

$$\mu_1 = 1 \qquad and \qquad \Lambda_1 = 0 \qquad (9.3\text{-}11c)$$

$$\mathcal{E}\left\{\epsilon_i(t+i)\left[\begin{array}{cc} u(t) & s'(t) \end{array}\right]\right\} = \mathcal{E}\left\{\nu_{i+1}(t+i)\left[\begin{array}{cc} u(t) & s'(t) \end{array}\right]\right\} = 0 \qquad (9.3\text{-}11d)$$

θ_i and μ_i depend on the future regulation law, and Γ_i' and Λ_i' depend on both the past and future regulation laws.

The implicit prediction models (11) make it easy to solve the following problem. Under the validity conditions of Theorem 1, find the input variable

$$u(t) \in [s(t)] := \mathrm{Span}\{s(t)\} \qquad (9.3\text{-}12)$$

minimizing the performance index given by the conditional expectation

$$\mathcal{C}_T = \mathcal{E}\left\{J_T \mid s(t)\right\} \qquad (9.3\text{-}13\text{a})$$

$$J_T := \frac{1}{T}\sum_{i=1}^{T}\left[y^2(t+i) + \rho u^2(t+i-1)\right] \qquad (9.3\text{-}13\text{b})$$

In (12), $[s(t)]$ denotes the subspace of all random variables given by linear combinations of the components of $s(t)$ (cf. (6.1-20)). Recalling (D.2-1) and using the notation introduced in (6.4-10a), we get

$$\mathcal{E}\left\{y^2(t+i) \mid s(t)\right\} = \hat{y}_t^2(t+i) + \mathcal{E}\left\{\tilde{y}^2(t+i) \mid s(t)\right\}$$

$$
\begin{aligned}
\hat{y}_t(t+i) \quad &:= \quad \hat{\mathcal{E}}\left\{y(t+i) \mid s(t)\right\} \\
&= \quad \theta_i u(t) + \Gamma_i' s(t) & [(11\text{a})] \\
\tilde{y}_t(t+i) \quad &:= \quad y(t+i) - \hat{y}_t(t+i) \\
&= \quad \epsilon_i(t+i) & [(11\text{a})]
\end{aligned}
$$

$$\mathcal{E}\left\{u^2(t+i-1) \mid s(t)\right\} = \hat{u}_t^2(t+i-1) + \mathcal{E}\left\{\tilde{u}^2(t+i-1) \mid s(t)\right\}$$

$$
\begin{aligned}
\hat{u}_t(t+i-1) \quad &:= \quad \hat{\mathcal{E}}\left\{u(t+i-1) \mid s(t)\right\} \\
&= \quad \mu_i u(t) + \Lambda_i' s(t) & [(11\text{b})] \\
\tilde{u}_t(t+i-1) \quad &:= \quad u(t+i-1) - \hat{u}_t(t+i-1) \\
&= \quad \nu_i(t+i-1) & [(11\text{b})]
\end{aligned}
$$

Because $\mathcal{E}\left\{\tilde{y}^2(t+i) \mid s(t)\right\}$ and $\mathcal{E}\left\{\tilde{u}^2(t+i-1) \mid s(t)\right\}$ are unaffected by $s(t)$, we find for the optimal input at time t,

$$u(t) = F' s(t) \qquad (9.3\text{-}14\text{a})$$

$$F = -\Xi^{-1}\sum_{i=1}^{T}\left(\theta_i \Gamma_i + \rho \mu_i \Lambda_i\right) \qquad (9.3\text{-}14\text{b})$$

$$\Xi := \sum_{i=1}^{T}\left(\theta_i^2 + \rho \mu_i^2\right) \qquad (9.3\text{-}14\text{c})$$

Note that, by virtue of (11c), (14) are well-defined whenever $\rho > 0$.

Remark 9.3-2.

- The reader should compare (14) with (5.7-28) and (5.7-29), and more generally, the problem we have tackled before with TCI in Chapter 5, so as to be convinced that the equations of (14) represent the stochastic counterpart of TCI.

- An important point not to be overlooked is that the feedback-gain vector F in (14) depends on the past and future regulation laws as specified by Theorem 1.

- The equations of (14) are the SISO version of (5.7-28) and (5.7-29). We can conjecture that (14) can be extended to cover the MIMO case. This is true, as shown in the next section. There we shall see that the MIMO extension of (14) is *mutatis mutandis* formally the same as (5.7-28) and (5.7-29).

The previous discussion suggests a way to construct an implicit adaptive regulator whose underlying regulation law consists of TCI. We call such an adaptive regulator *MUSMAR* (**mu**ltistep **m**ultivariable **a**daptive **r**egulator) by the acronym under which it was first referred to in the literature [MM80].

9.3.4 MUSMAR (SISO Version)

Assume that the plant has been fed by inputs $u(k) = F'(k)s(k)$, $k \in \mathbb{Z}_+$, up to time $t - 1$. Let $u(t - 1) = F'(t - 1)s(t - 1)$ with $F(t - 1)$ based on estimates

$$\Theta_i(t - 1) := \left[\begin{array}{cc} \Gamma_i'(t - 1) & \theta_i(t - 1) \end{array} \right]' \qquad (9.3\text{-}15a)$$

$$M_i(t - 1) := \left[\begin{array}{cc} \Lambda_i'(t - 1) & \mu_i(t - 1) \end{array} \right]' \qquad (9.3\text{-}15b)$$

for $i = 1, 2, \cdots, T$, with $M_1(t - 1) \equiv \left[\begin{array}{cc} O_{n_s}' & 1 \end{array} \right]'$.

1. Update the estimates via the RLS algorithm:

$$\Theta_i(t) = \Theta_i(t - 1) + P(t - T + 1)\varphi(t - T) \qquad (9.3\text{-}16a)$$
$$\times \left[y(t - T + i) - \varphi'(t - T)\Theta_i(t - 1) \right]$$

$$M_i(t) = M_i(t - 1) + P(t - T + 1)\varphi(t - T) \qquad (9.3\text{-}16b)$$
$$\times \left[u(t - T + i - 1) - \varphi'(t - T)M_i(t - 1) \right]$$

$$\varphi(t - T) := \left[\begin{array}{cc} s'(t - T) & u(t - T) \end{array} \right]' \qquad (9.3\text{-}16c)$$

$$P^{-1}(t - T + 1) = P^{-1}(t - T) + \varphi(t - T)\varphi'(t - T) \qquad (9.3\text{-}16d)$$

with $P^{-1}(1 - T) = P^{-T}(1 - T) > 0$.

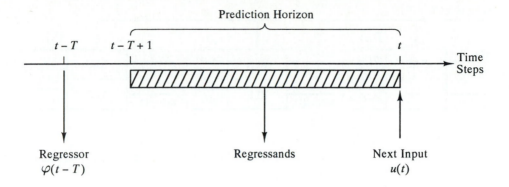

Figure 9.3-3: Time steps for the regressor and regressands when the next input to be computed is $u(t)$.

2. Compute the next input

$$u(t) = F'(t)s(t) \qquad (9.3\text{-}17a)$$

with

$$F(t) = -\Xi^{-1}(t) \sum_{i=1}^{T} [\theta_i(t)\Gamma_i(t) + \rho\mu_i(t)\Lambda_i(t)] \qquad (9.3\text{-}17b)$$

$$\Xi(t) = \sum_{i=1}^{T} [\theta_i^2(t) + \rho\mu_i^2(t)] \qquad (9.3\text{-}17c)$$

3. Cycle through steps 1 and 2.

Figure 3 shows the time steps corresponding to regressor $\varphi(t-T)$ and regressands $y(t-T+i)$ and $u(t-T+i-1)$, when the input to be computed is $u(t)$. Figure 4 depicts the signal flows in the RLS identifiers of MUSMAR when $T=3$.

Remark 9.3-3.

- As shown in Fig. 4, MUSMAR identifiers are made up by T *separate* RLS estimators all using a common regressor and hence a common updating gain. This feature moderates the overall computational burden. Notice that, in contrast with the scheme of Fig. 1, MUSMAR identifiers have no interlacing.

- Notice that we estimate the $M_i(t)$. These, however, could be alternatively computed from $\Theta_j(t)$ and $F(t-T+j)$, $j=1,\cdots,i-1$. Though it looks hazardous, the former alternative is suggested for the related positive working experience and its lower computational burden.

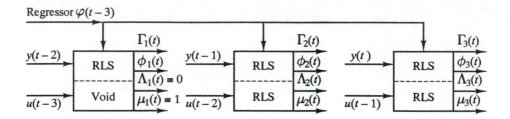

Figure 9.3-4: Signal flows in the bank of parallel MUSMAR RLS identifiers when $T = 3$ and the next input is $u(t)$.

- Besides RLS estimation of the parameters of suitable implicit prediction models of linear-regression type, MUSMAR performs on-line *spread-in-time* truncated cost iterations.

Main Points of the Section. The implicit multistep prediction models of linear-regression type can be exploited to single out implicit adaptive multistep predictive controllers based on SIORHC, GPC, or, by considering additional constraints to simplify the models, MUSMAR, the latter performing on-line spread-in-time truncated cost iterations. In contrast with the explicit schemes wherein the open-loop plant CARMA model is identified, the implicit adaptive multistep predictive controllers perform the separate identification of the parameters of closed-loop multistep-ahead prediction models of linear-regression type.

9.4   MUSMAR AS AN ADAPTIVE REDUCED-COMPLEXITY CONTROLLER

In this section, we derive the MUSMAR algorithm for MIMO CARMA plants following a quite different viewpoint from the one of the previous section, which is based on implicit multistep prediction models. The reason for doing this is to show that MUSMAR can be looked at as an adaptive reduced-complexity controller requiring little prior information on the plant.

9.4.1   A Delayed RHR Problem

We now consider the following regulation problem. A plant with inputs $u(t) \in \mathbb{R}^m$ and outputs $y(t) \in \mathbb{R}^p$ is to be regulated. It is known that $y(t)$ and $u(t)$ are at least locally related by the CARMA system

$$A(d)y(t) = B(d)u(t) + C(d)e(t) \qquad (9.4\text{-}1)$$

In (1), e is a p-vector-valued wide-sense stationary zero-mean white innovations sequence with positive-definite covariance; $A(d)$, $B(d)$, and $C(d)$ are polynomial matrices; $A(d)$ has dimension $p \times p$ and all other matrices have compatible dimensions. Further, $\mathcal{B}(0) = O_{p \times m}$, viz., the plant exhibits I/O delays at least equal to 1.

We assume the following:

- $A^{-1}(d) \begin{bmatrix} \mathcal{B}(d) & C(d) \end{bmatrix}$ is an irreducible left MFD (9.4-2a)

- $C(d)$ is strictly Hurwitz (9.4-2b)

- the gcld's of $A(d)$ and $\mathcal{B}(d)$ are strictly Hurwitz (9.4-2c)

We point out that, in view of Theorem 6.2-4, (2b) entails no limitation, and (2c) is a necessary condition for the existence of a linear compensator, acting on the manipulated input u only, capable of making the resulting feedback system internally stable.

Though it is known that the plant is representable as in (1), no, or only incomplete, information is available on the entries of the previous polynomial matrices. Then the structure, degrees, and coefficients of their polynomial entries, and, hence, the associated I/O delays are either unknown or only partially a priori given. Despite our ignorance on the plant, we assume that it is a priori known that there exist feedback-gain matrices F such that the plant can be stabilized by the regulation law

$$u(t) = F's(t) \tag{9.4-3}$$

where

$$s(t) := \left[\begin{array}{cc} \left(y_t^{t-n_y} \right)' & \left(u_{t-1}^{t-n_u} \right)' \end{array} \right]' \in \mathbb{R}^{n_s} , \quad n_s := n_u + n_y + 1 \tag{9.4-4}$$

Hereafter, $s(t)$ will be referred to as the pseudo-state or *regulator–regressor of complexity* (n_y, n_u). A priori knowledge of a suitable regulator–regressor complexity can be inferred from the physical characteristics of the plant. This happens to be frequently true in applications.

In the SISO case, we may know that, in the useful frequency band, (1) is an accurate enough description of the plant provided that

$$A(d) = 1 + a_1 d + \cdots + a_{\partial A} d^{\partial A} \qquad (a_{\partial A} \neq 0) \tag{9.4-5}$$

$$\mathcal{B}(d) = d^{\ell} B(d) = d^{\ell} \left[b_1 d + \cdots + b_{\partial B} d^{\partial B} \right] \qquad (b_1 \neq 0, b_{\partial B} \neq 0) \tag{9.4-6}$$

with ∂A and ∂B given and the I/O transport delay ℓ, $\ell = 0, 1, \cdots$, unknown, possibly time-varying, and such that

$$0 \leq \ell \leq \ell_M \tag{9.4-7}$$

where ℓ_M is the largest possible I/O delay. In such a case, if ∂C denotes the degree of $C(d)$, the regulator–regressor (4) corresponding to steady-state LQS regulation of (1) fulfills the following prescriptions (cf. Problem 7.3-16):

$$n_y = \max\{\partial A - 1, \partial C - \ell - 1\} \tag{9.4-8}$$

$$n_u = \max\{\partial B + \ell - 1, \partial C\} \tag{9.4-9}$$

which, in turn, should ℓ be in the uncertainty range (7), safely become

$$n_y \quad = \quad \max\{\partial A, \partial C\} - 1 \tag{9.4-10}$$

$$n_u \quad = \quad \max\{\partial B + \ell_M - 1, \partial C\} \tag{9.4-11}$$

It is worth saying that in practice, n_y and n_u seldom follow the previous prescriptions, but, more often, reflect a compromise between the complexity of the adaptive regulator and the ideally achievable performance of the regulated system.

The problem is how to develop a self-tuning algorithm capable of selecting a satisfactory feedback-gain matrix F. To do this, we only stipulate that, whatever plant structure and dimensionality might be, an (n_y, n_u) pseudo-state complexity is adequate. Let the performance index be

$$\mathcal{C}_T(t) \quad = \quad \mathcal{E}\{J_T(t) \mid s\} \tag{9.4-12a}$$

$$J_T(t) \quad = \quad \frac{1}{T} \sum_{k=t-T}^{t-1} \left[\|y(k+1)\|_{\psi_y}^2 + \|u(k)\|_{\psi_u}^2 \right] \tag{9.4-12b}$$

$$s \quad := \quad s(t-T) \tag{9.4-12c}$$

Assume that, except for the first, all inputs in (12b) are given by

$$u(k) = F'(k)s(k) \qquad t - T < k \leq t - 1 \tag{9.4-13}$$

with known feedback-gain matrices $F(k)$. Then the next feedback-gain matrix $F(t)$ is found as follows. Consider the performance index (12) subject to the constraints (13). As usual, let $[s]$ be the subspace of all random vectors whose components are linear combinations of the components of s. Find

$$u^o := u^o(t-T) \in [s] \qquad \text{or} \qquad u^o = F^{o\prime}s \tag{9.4-14}$$

such that

$$u^o = \arg \min_{u \in [s]} \mathcal{C}_T(t) \tag{9.4-15}$$

Then, set

$$u(t) = F^{o\prime}s(t) \tag{9.4-16}$$

and move to $\mathcal{C}_T(t+1)$ so as to compute $u(t+1)$.

The rule (12) to (16) is reminiscent of a receding-horizon regulation (RHR) scheme in a stochastic setting. A RHR rule would select input $u(t)$ at time t that, together with the subsequent input sequence $u^o_{[t+1,t+T)}$, minimizes some index and fulfills possible constraints. Then the input $u(t)$ would be applied and $u^o_{[t+1,t+T)}$ discarded. A similar operation is repeated with t replaced by $t+1$ in order to find $u(t+1)$. The adoption of a strict RHR procedure requires to explicitly use a plant description such as (1) in the related optimization stage. Being prevented from it by ignorance, we are forced to adopt the delayed scheme (12) to (16). There we find

$u(t)$ at time t, by considering the plant behavior over the "past" interval $[t - T, t)$. In order to distinguish the crucial differences between strict RHR and the adopted scheme (12) to (16), the latter will be referred to as a *delayed RHR scheme with prediction horizon T*.

9.4.2 The Algorithm

In order to solve (15), set

$$\mathcal{C}_T(t) \quad = \quad \hat{\mathcal{C}}_T(t) + \tilde{\mathcal{C}}_T(t) \tag{9.4-17a}$$

$$\hat{\mathcal{C}}_T(t) \quad := \quad \frac{1}{T} \sum_{k=t-T}^{t-1} \mathcal{E}\left\{ \|\hat{y}(k+1)\|_{\psi_y}^2 + \|\hat{u}(k)\|_{\psi_u}^2 \mid s \right\} \tag{9.4-17b}$$

$$\tilde{\mathcal{C}}(t) \quad := \quad \frac{1}{T} \sum_{k=t-T}^{t-1} \left[\operatorname{Tr} \psi_y \mathcal{E}\left\{ \tilde{y}(k+1)\tilde{y}'(k+1) \mid s \right\} \right. \tag{9.4-17c}$$

$$\left. + \operatorname{Tr} \psi_u \mathcal{E}\left\{ \tilde{u}(k)\tilde{u}'(k) \mid s \right\} \right]$$

where Tr denotes trace, and $\hat{y}(k+1)$ and $\hat{u}(k)$ are orthogonal projections of $y(k+1)$ and $u(k)$ onto $[u, s]$, the linear subspace of random vectors spanned by $\{u, s\}$:

$$\hat{y}(k+1) \quad := \quad \operatorname{Projec}\left[y(k+1) \mid [u, s] \right] \tag{9.4-17d}$$

$$\tilde{y}(k+1) \quad := \quad y(k+1) - \hat{y}(k+1)$$

and

$$\hat{u}(k) \quad := \quad \operatorname{Projec}\left[u(k) \mid [u, s] \right] \tag{9.4-17e}$$

$$\tilde{u}(k) \quad := \quad u(k) - \hat{u}(k)$$

Note that ideally, $[u, s] = [s]$ because $u \in [s]$. In practice, because we are not going to perform the optimization analytically but on-line using real data, we cannot rule out the possibility that the *actual* $u(t - T)$ does not belong to $[s]$. This indeed happens whenever the inputs are of the form

$$u(k) = F'(k)s(k) + \eta(k) \tag{9.4-18}$$

where $\eta(k)$ is either an undesirable disturbance or an intentional dither. In this respect, we shall assume that η is a wide-sense zero-mean white noise, uncorrelated with e. It is to be pointed out that even if the constraints (13) become in practice (18) for $k \in [t - T + 1, t)$, y_{t-T+1}^t and u_{t-T+1}^{t-1} depend *linearly* on u. Hence, $\tilde{y}_{t-T+1}^t$ and $\tilde{u}_{t-T+1}^{t-1}$ *are unaffected* by u. Further, note that

$$\left\{ \begin{array}{rcl} \hat{y}(k) & = & \mathcal{E}\left\{ y(k)\varphi' \right\} \mathcal{E}^{-1}\left\{ \varphi\varphi' \right\} \varphi \\ \hat{u}(k) & = & \mathcal{E}\left\{ u(k)\varphi' \right\} \mathcal{E}^{-1}\left\{ \varphi\varphi' \right\} \varphi \\ \varphi & := & \begin{bmatrix} s' & u' \end{bmatrix}' \end{array} \right. \tag{9.4-19}$$

In conclusion, our problem (15) simplifies as follows:

$$u^o := u^o(t - T) = \arg \min_{u \in [s]} \hat{J} \tag{9.4-20a}$$

$$\hat{J} := \frac{1}{T} \sum_{k=t-T}^{t-1} \left[\|\hat{y}(k+1)\|_{\psi_y}^2 + \|\hat{u}(k)\|_{\psi_u}^2 \right] \tag{9.4-20b}$$

Let, for $i = 1, 2, \cdots, T$,

$$\hat{y}(t - T + i) = \Theta_i' \varphi = \theta_i u + \Gamma_i' s \tag{9.4-21a}$$

where

$$\begin{aligned} \Theta_i' &:= \mathcal{E} \left\{ y(t - T + i)\varphi \right\} \mathcal{E}^{-1} \left\{ \varphi\varphi' \right\} \\ &= \begin{bmatrix} \Gamma_i' & \theta_i \end{bmatrix} \end{aligned} \tag{9.4-21b}$$

Similarly,

$$\hat{u}(t - T + i - 1) = M_i' \varphi = \mu_i u + \Lambda_i' s \tag{9.4-22a}$$

where

$$\begin{aligned} M_i' &:= \mathcal{E} \left\{ u(t - T + i)\varphi \right\} \mathcal{E}^{-1} \left\{ \varphi\varphi' \right\} \\ &= \begin{bmatrix} \Lambda_i' & \mu_i \end{bmatrix} \end{aligned} \tag{9.4-22b}$$

Obviously, in (22a),

$$\mu_1 = I_m \quad \text{and} \quad \Lambda_1 = O_{n_s \times m} \tag{9.4-22c}$$

Using (21) and (22) in (20), we find

$$u^o = F^{o\prime} s \tag{9.4-23a}$$

$$F^{o\prime} = -\Xi^{-1} \sum_{i=1}^{T} [\theta_i' \psi_y \Gamma_i' + \mu_i' \psi_u \Lambda_i'] \tag{9.4-23b}$$

$$\Xi := \sum_{i=1}^{T} [\theta_i' \psi_y \theta_i + \mu_i' \psi_u \mu_i] \tag{9.4-23c}$$

Note that by (22c), whenever $\psi_u > 0$, $\Xi = \Xi' > 0$, and, hence, (20) is uniquely solved by (23).

In order to compute Θ_i and M_i, we use Proposition 6.3-1 on the relationship between RLS updates and normal equations as follows. Let $\Theta_i(t)$, $i = 1, 2, \cdots, T$, $t = T, T+1, \cdots$, and $\dim \Theta_i(t) = \dim \varphi$ be given by the RLS updates:

$$\left. \begin{aligned} \Theta_i(t) &= \Theta_i(t-1) + P(t - T + 1)\varphi(t - T) \\ &\quad \times \left[y'(t - T + i) - \varphi'(t - T)\Theta_i(t-1) \right] \\ P^{-1}(t - T + 1) &= P^{-1}(t - T) + \varphi(t - T)\varphi'(t - T) \\ P(0) &= P'(0) > 0 \end{aligned} \right\} \tag{9.4-24}$$

Then $\Theta_i(t)$ satisfies the normal equations

$$\left[\sum_{k=0}^{t-T} \varphi(k)\varphi'(k)\right]\Theta_i(t) = \sum_{k=0}^{t-T} \varphi(k)y'(k+i) + P^{-1}(0)\left[\Theta_i(T-1) - \Theta_i(t)\right] \quad (9.4\text{-}25)$$

The following result then follows directly from Theorem 6.4-2.

Proposition 9.4-1. *Let the I/O joint process $\{u(k-1), y(k)\}$ be strictly stationary and ergodic with bounded $\Psi_\varphi := \mathcal{E}\{\varphi\varphi'\} > 0$. Let $\Theta_i(t)$ be given by the RLS updates (24). Then*

$$\lim_{t\to\infty} \Theta_i(t) = \mathcal{E}^{-1}\{\varphi\varphi'\}\,\mathcal{E}\{\varphi y'(t-T+i)\} \qquad a.s. \qquad (9.4\text{-}26)$$

Similarly, let $M_i(t)$ be given by the following RLS updates for $i = 2, 3, \cdots, T$:

$$\left.\begin{aligned}
M_i(t) &= M_i(t-1) + P(t-T+1)\varphi(t-T) \\
&\quad \times \left[u'(t-T+i-1) - \varphi'(t-T)M_i(t-1)\right] \\
M_1'(t) &= M_1'(t-1) = \begin{bmatrix} I_m & O_{m\times n_s} \end{bmatrix}
\end{aligned}\right\} \qquad (9.4\text{-}27)$$

Then

$$\lim_{t\to\infty} M_i(t) = \mathcal{E}^{-1}\{\varphi\varphi'\}\,\mathcal{E}\{\varphi u'(t-T+i-1)\} \qquad a.s. \qquad (9.4\text{-}28)$$

Putting together the foregoing results, we arrive at a recursive regulation algorithm, a candidate for solving on-line the delayed RHR problem (12) to (16).

Theorem 9.4-1. **(MUSMAR)** *Consider the delayed RHR problem (12) to (16) for the multivariable CARMA plant (1) having unknown structure (state dimension, I/O deadtimes, etc.) and parameters. Then the following recursive algorithm, for $t = T, T+1, \cdots$:*

$$P^{-1}(t-T+1) = P^{-1}(t-T) + \varphi(t-T)\varphi'(t-T) \qquad (9.4\text{-}29\text{a})$$

$$\begin{aligned}
\Theta_i(t) &= \Theta_i(t-1) \\
&\quad + P(t-T+1)\varphi(t-T)\left[y'(t-T+i) - \varphi'(t-T)\Theta_i(t-1)\right]
\end{aligned} \qquad (9.4\text{-}29\text{b})$$

$$\begin{aligned}
M_i(t) &= M_i(t-1) + \\
&\quad P(t-T+1)\varphi(t-T)\left[u'(t-T+i+1) - \varphi'(t-T)M_i(t-1)\right]
\end{aligned} \qquad (9.4\text{-}29\text{c})$$

$$\Theta_i = \begin{bmatrix} \Gamma_i'(t) & \theta_i(t) \end{bmatrix}' \qquad M_i = \begin{bmatrix} \Lambda_i'(t) & \mu_i(t) \end{bmatrix}' \qquad (9.4\text{-}29\text{d})$$

$$F'(t) = -\Xi^{-1}(t)\sum_{i=1}^{T}\left[\theta_i'(t)\psi_y\Gamma_i'(t) + \mu_i'(t)\psi_u\Lambda_i'(t)\right] \qquad (9.4\text{-}29\text{e})$$

$$\Xi(t) = \sum_{i=1}^{T}\left[\theta_i'(t)\psi_y\theta_i(t) + \mu_i'(t)\psi_u\mu_i(t)\right] \qquad (9.4\text{-}29\text{f})$$

$$u(t) = F'(t)s(t) \qquad (9.4\text{-}29\text{g})$$

with $P(0) > 0$ and $\Lambda_1(t) \equiv O_{n_s \times m}$, $\mu_1 = I_m$, as $t \to \infty$, solves the stated delayed RHR problem, whenever the joint process φ becomes strictly stationary and ergodic, with bounded $\mathcal{E}\{\varphi\varphi'\} > 0$.

Theorem 1 justifies the use of the recursive algorithm (29) to solve on-line the delayed RHR problem (12) to (16) for any unknown plant. However, it does not tell whether the algorithm—assuming that it converges—yields a satisfactory closed-loop system. This issue will be studied in depth in Sect. 5.

9.4.3   2-DOF MUSMAR

Our interest is to modify the pure regulation algorithm (29) so as to make the plant output capable of tracking a reference $r(t) \in \mathbb{R}^p$. In this connection, the following modifications can be made to the algorithm (29). They can be justified *mutatis mutandis* by the same arguments as in the proof of Theorem 1.

In a tracking problem, the variable to be regulated to zero is the tracking error:

$$\varepsilon_y(k) := y(k) - r(k) \tag{9.4-30}$$

Two alternatives are considered for the choice of $s(k)$. In the first,

$$s(k) := \begin{bmatrix} \varepsilon_y(k-n_y) & \cdots & \varepsilon_y(k) & u(k-n_u) & \cdots & u(k-1) \end{bmatrix}' \tag{9.4-31}$$

and, accordingly, MUSMAR acts as a 1-DOF controller. In the second,

$$s(k) := \begin{bmatrix} \left(y_k^{k-n_y}\right)' & \left(u_{k-1}^{k-n_u}\right)' & \left(r_{k+T}^{k-n_r}\right)' \end{bmatrix}' \tag{9.4-32}$$

and, accordingly, MUSMAR acts as a 2-DOF controller. The choice of the controller–regressor $s(k)$ in (32) is justified by resorting to the controller structure in the steady-state LQS servo problem (cf. Theorem 7.5-1 and Proposition 7.5-1).

The 2-DOF version of MUSMAR has typically a better tracking performance than the 1-DOF version, provided that convergence takes place.

Example 9.4-1 [GGMP92] (MUSMAR Autotuning of PID Controllers for a Two–Link Robot)
We consider a double-input double-output plant consisting of the mathematical model of a two-link robot manipulator in a vertical plane. The model is a continuous-time nonlinear dynamic system and refers to the second and third link of the *Unimation PUMA 560* controlled via the two corresponding joint actuators. The control laws considered hereafter pertain to a sampling time of 10^{-2} seconds. Although the plant is nonlinear, we use as controllers two MUSMAR algorithms acting individually on each joint in a fully decentralized fashion. Specifically, for the ith joint, $i = 1, 2$, the corresponding MUSMAR algorithm selects the three-component feedback-gain vector,

$$F_i(t) := \begin{bmatrix} f_{i1}(t) & f_{i2}(t) & f_{i3}(t) \end{bmatrix}' \tag{9.4-33a}$$

in the control law

$$
\begin{aligned}
\delta u_i(t) \quad &:= \quad u_i(t) - u_i(t-1) \\
&= \quad F_i'(t)s_i(t)
\end{aligned}
\tag{9.4-33b}
$$

with pseudo-state

$$
s_i(t) := \left[\ \varepsilon_{y_i}(t) \quad \varepsilon_{y_i}(t-1) \quad \varepsilon_{y_i}(t-2)\ \right]'
\tag{9.4-33c}
$$

and tracking error

$$
\varepsilon_{y_i}(t) := r_i(t) - y_i(t)
\tag{9.4-33d}
$$

where $r_i(t)$ is the reference for the ith joint. $F_i(t)$ is selected so as to minimize, as indicated after (12a), the following performance index with a 10-step prediction horizon:

$$
C_{10}^i(t) \quad = \quad \mathcal{E}\left\{J_{10}^i(t) \mid s_i(t-10)\right\}
\tag{9.4-34a}
$$

$$
J_{10}^i(t) \quad = \quad \frac{1}{10}\sum_{k=t-10}^{t-1}\left[\varepsilon_{yi}^2(k) + 6\cdot 10^{-8}\delta u_i^2(k)\right]
\tag{9.4-34b}
$$

By omitting the argument t in the feedback gain, (33) can also be rewritten as follows:

$$
u_i(t) = \left[K_{P_i} + K_{I_i}T_s\frac{1+d}{2(1-d)} + K_{D_i}\frac{1}{T_s}(1-d)\right]\varepsilon_{yi}(t)
\tag{9.4-35a}
$$

with $T_s = 10^{-2}$ seconds and

$$
2K_{P_i} = f_{i1} - f_{i2} - 3f_{i3} \qquad T_sK_{I_i} = f_{i1} + f_{i2} + f_{i3} \qquad K_{D_i} = T_sf_{i3}
\tag{9.4-35b}
$$

The control law (35a) is a digital version of the classical PID controller obtained by using the Tustin approximation for the integral term and the backward difference for the derivative term [ÅW84]. Thus, MUSMAR in configuration (33) and (34) can be used to adaptively auto-tune the two digital PID controllers of the robot manipulator. To this end, the reference trajectories for the two joints are chosen to be periodic so as to represent repetitive tasks for the robot manipulator. For each joint, a smooth trapezoidal reference trajectory is used (Fig. 1). A payload of 7 kilograms is considered to be picked up at the lower rest position and released at the upper rest position of the terminal link. Figures 2 and 3 show the time evolution of the MUSMAR three feedback components, reexpressed as K_{P_i}, K_{I_i}, and K_{D_i} via (35b), for the two joints over a 200-second run. The feedback gains on which MUSMAR self-tunes are used in a constant-feedback controller, one for each joint. If these two resulting fixed decentralized controllers are used to control the manipulator, we obtain the tracking-error behavior indicated by the solid lines in Figs. 4(a) and 4(b) for the two joints. In the same figures, the dashed lines indicate the tracking-error behavior obtained by two digital decentralized PID controllers whose gains are selected via the classical Ziegler and Nichols trial-and-error tuning method [ÅW84]. Note that the MUSMAR auto-tuned feedback gains yield a definitely better tracking performance. We point out that, because the optimal feedback gains for a restricted-complexity controller are dependent on the selected trajectories, it is usually required to repeat the MUSMAR auto-tuning procedure when the robot task is changed. A final remark concerns the possible use in the two controllers of the common pseudo-state

$$
s(t) = \left[\ s_1'(t) \quad s_2'(t)\ \right]'
$$

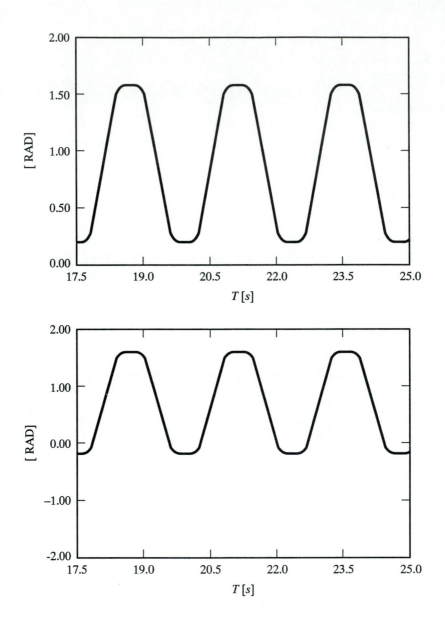

Figure 9.4-1: Reference trajectories for joint 1 (above) and joint 2 (below).

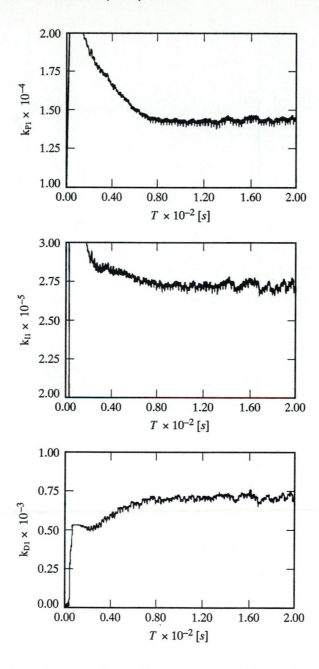

Figure 9.4-2: Time evolution of the three PID feedback-gains K_P, K_I, and K_D adaptively obtained by MUSMAR for the joint 1 of the robot manipulator.

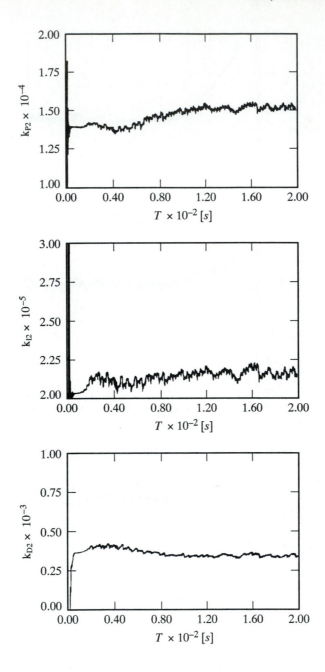

Figure 9.4-3: Time evolution of the three PID feedback-gains K_P, K_I, and K_D adaptively obtained by MUSMAR for the joint 2 of the robot manipulator.

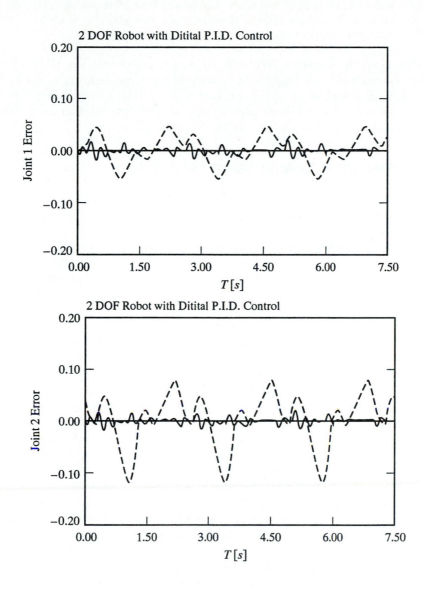

Figure 9.4-4: Time evolution of the tracking errors for the robot manipulator controlled by a digital PID autotuned by MUSMAR (solid lines) or Ziegler and Nichols method (dotted lines): (a) joint 1 error; (b) joint 2 error.

where $s_1(t)$ and $s_2(t)$ are defined as before. In this way, the controller on each joint has some information on the current state of the other link. This generally leads to a further improvement of tracking performance.

Main Points of the Section. Whereas in the previous section, the MUSMAR algorithm was obtained as an implicit TCI-based adaptive regulator using the knowledge of the CARMA plant structure (degrees of the CARMA polynomials), in this section, it is shown that the same algorithm is a candidate for adaptively tuning reduced-complexity control laws for highly uncertain plants.

9.5   MUSMAR LOCAL-CONVERGENCE PROPERTIES

For the implicit adaptive predictive control algorithms introduced in the previous two sections, convergence analysis turns out to be a difficult task. One of the reasons is that the parameters the controller attempts to identify do not depend solely on the plant but also, and in a complicated way, on the past feedback and feedforward gains. In particular, the estimated parameters in the MUSMAR algorithm are in a sense more directly related to the controller than to the plant, particularly if the former is of reduced complexity relatively to the latter.

Under these circumstances, implicit adaptive predictive control algorithms do not appear amenable to global convergence analysis. Further, we may be legitimately concerned about their possible lack of global convergence properties for two main reasons. First, their underlying control laws need not be stabilizing unless some provisions are taken. For example, in Chapter 5 some guidelines were given on the prediction-horizon length in order to make a TCI-based controller stabilizing. Second, the on-line control synthesis is based on parameters that, in turn, depend on the past controller gains. Hence, if at a given time, the latter makes the closed-loop system unstable and regressors of very high magnitude are experienced, the identifier gains can quickly go to zero and the subsequent estimates corresponding to a nonstabilizing set of controller gains will stay constant. It cannot be ruled out that such estimates produce a nonstabilizing new set of controller gains, so saturations may occur such as to prevent subsequent use of the adaptive algorithm. This consideration leads us to conjecture that the implicit predictive controllers of the last two sections, based on the use of standard RLS identifiers, could be improved by equipping their identifiers with constant-trace and data-normalization features. Such a conjecture is in fact confirmed by experimental evidence.

Though global convergence is a very difficult task, local convergence analysis of implicit adaptive predictive controllers, for example, MUSMAR, can be carried out via stochastic averaging methods, such as the ordinary differential equation approach. Such an analysis is important in that it can reveal the possible convergence points even when plant neglected dynamics are present. For example, the derivation of the last section makes MUSMAR a candidate for a reduced-complexity adaptive controller. Hereafter, by using the ODE method, we shall uncover how MUSMAR can behave in the presence of neglected dynamics.

9.5.1   Stochastic Averaging: The ODE Method

The updating formula of many recursive algorithms typically has the form:

$$\theta(t) = \theta(t-1) + \gamma(t)\mathcal{R}^{-1}(t)\varphi(t-1)\varepsilon(t) \qquad (9.5\text{-}1a)$$

$$\mathcal{R}(t) = \mathcal{R}(t-1) + \gamma(t)\left[\varphi(t-1)\varphi'(t-1) - \mathcal{R}(t-1)\right] \qquad (9.5\text{-}1b)$$

where $\theta(t)$ denotes the estimate at time t, $\{\gamma(t)\}$ a scalar-valued gain sequence, $\varphi(t-1)$ a regression vector, and $\varepsilon(t)$ a prediction error. For example, the RLS algorithm (6.3-13) can be rewritten as in (1) if we set

$$\mathcal{R}(t) := \frac{P^{-1}(t)}{t+1}, \qquad \mathcal{R}(0) := P^{-1}(0) \qquad (9.5\text{-}2a)$$

and

$$\gamma(t) := \frac{1}{t+1} \qquad (9.5\text{-}2b)$$

The SG algorithm (8.7-24) can be written as

$$\theta(t) = \theta(t-1) + a\gamma(t)\varrho^{-1}(t)\varphi(t-1)\varepsilon(t) \qquad (9.5\text{-}3a)$$

$$\varrho(t) = \varrho(t-1) + \gamma(t)\left[\|\varphi(t-1)\|^2 - \varrho(t-1)\right] \qquad (9.5\text{-}3b)$$

with $a > 0$,

$$\varrho(t) := \frac{q(t)}{t+1}, \qquad \varrho(0) := q(0) \qquad (9.5\text{-}3c)$$

and $\gamma(t)$ given again as in (2b).

In both cases, $\varepsilon(t)$ and $\varphi(t-1)$ depend in the usual way on the I/O variables of a CARMA plant:

$$A(d)y(t) = B(d)u(t) + C(d)e(t) \qquad (9.5\text{-}4)$$

with e a stationary zero-mean sequence of independent random variables such that all moments exist.

A rigorous proof of the ODE method for analyzing the stochastic recursive algorithm is given in [Lju77b], and it turns out to be a quite formidable problem. We shall give a heuristic derivation of the method together with statements, without proof, of the main results of interest to us.

From (1a), write for $k \in \mathbb{Z}_+$

$$\theta(t+k) = \theta(t) + \sum_{i=t+1}^{t+k} \gamma(i)\mathcal{R}^{-1}(i)\varphi(i-1)\varepsilon(i)$$

or

$$\frac{\theta(t+k) - \theta(t)}{k} = \frac{1}{k}\sum_{i=t+1}^{t+k} \gamma(i)\mathcal{R}^{-1}(i)\varphi(i-1)\varepsilon(i)$$

$$\cong \mathcal{R}^{-1}(t)\gamma(t)\frac{1}{k}\sum_{i=t+1}^{t+k} \varphi(i-1)\varepsilon(i)$$

$$\cong \mathcal{R}^{-1}(t)\gamma(t)\mathcal{E}\left\{\varphi(t-1,\theta)\varepsilon(t,\theta)\right\}_{\theta=\theta(t)}$$

In the last equation, we have used the following approximations. First, assuming t large enough, we see from (1b) and (2b) that $\mathcal{R}(i)$ and $\gamma(i)$ for $t + 1 \leq i \leq t + k$ are slowly time-varying and hence can be well approximated by $\mathcal{R}(t)$ and $\gamma(t)$, respectively. Second, the regressor $\varphi(i - 1)$ and the prediction error $\varepsilon(i)$ depend on the previous estimates $\theta(j), j = i - 1, i - 2, \cdots$, via an underlying control law that has not yet been made explicit. Moreover, by (1a), $\theta(j)$ is slowly time-varying. Hence,

$$
\begin{aligned}
\frac{1}{k} \sum_{i=t+1}^{t+k} \varphi(i-1)\varepsilon(i) &\cong \frac{1}{k} \sum_{i=t+1}^{t+k} \varphi\left(i-1, \theta(t)\right) \varepsilon(i, \theta(t)) \\
&\cong \mathcal{E}\left\{\varphi(t-1, \theta)\varepsilon(t, \theta)\right\}_{\theta=\theta(t)} \\
&=: f(\theta(t)) \qquad\qquad (9.5\text{-}5a)
\end{aligned}
$$

where with the second approximation we have replaced the time average over the interval $[t+1, t+k]$ with the indicated expectation. This is a plausible approximation because, by the presence of the innovations process e in (3), both $\varphi(i - 1, \theta(t))$ and $\varepsilon(i, \theta(t))$ are processes exhibiting fast time variations in their sample paths. The expectation $\mathcal{E}\{\varphi(t - 1, \theta)\varepsilon(t, \theta)\}$ has to be taken w.r.t. the probability density function induced on u and y by e, assuming that the system is in the stochastic steady state corresponding to the constant estimate θ.

From the previous discussion, we see that the asymptotic behavior of the stochastic recursive algorithm can be described in terms of the system of ODEs:

$$
\frac{d\theta(t)}{dt} = \gamma(t)\mathcal{R}^{-1}(t)f(\theta(t)) \qquad\qquad (9.5\text{-}5b)
$$

$$
\frac{d\mathcal{R}(t)}{dt} = \gamma(t)\left[\mathcal{G}(\theta(t)) - \mathcal{R}(t)\right] \qquad\qquad (9.5\text{-}5c)
$$

where the latter equation is obtained similarly to the first with

$$
\mathcal{G}(\theta(t)) := \mathcal{E}\left\{\varphi(t-1, \theta)\varphi'(t-1, \theta)\right\}_{\theta=\theta(t)} \qquad\qquad (9.5\text{-}5d)
$$

where the expectation $\mathcal{E}$ has to be carried out as indicated before. It is now convenient to further simplify the foregoing ODEs by operating the following time-scale change: Substitute t with a new independent variable $\tau = \tau(t)$ such that

$$
\frac{d\tau}{dt} = \gamma(t) \simeq \frac{1}{t} \qquad \text{or} \qquad \tau = \ln t \qquad\qquad (9.5\text{-}6a)
$$

and then (5) become

$$
\frac{d\theta(\tau)}{d\tau} = \mathcal{R}^{-1}(\tau)f(\theta(\tau)) \qquad\qquad (9.5\text{-}6b)
$$

$$
\frac{d\mathcal{R}(\tau)}{d\tau} = -\mathcal{R}(\tau) + \mathcal{G}(\theta(\tau)) \qquad\qquad (9.5\text{-}6c)
$$

These are called the ODEs associated to the stochastic recursive algorithm (1).

These considerations make plausible Result 1, which follows. For the sake of simplicity, the validity conditions stated there are not the most general, but specifically tailored for our needs.

Result 9.5-1.   *Consider the stochastic recursive system (1) and (4), to-gether with a linear regulation law*

$$u(t) = F'(\theta(t))s(t) + \eta(t) \tag{9.5-7}$$

where $s(t)$ is a vector with components from $\varphi(t)$, and $\eta(t)$ has the same interpretation as in (4-18). Further, assume the following:

- $\theta := \begin{bmatrix} \Theta_1' & \cdots & \Theta_T' & M_1' & \cdots & M_T' \end{bmatrix}'$ *and $F(\theta)$ are as in (3-17) with $\rho > 0$ so that $F(\theta)$ is a bounded rational function of the components of θ*

- $\theta(t)$ *belongs to $\mathcal{D}_s$ for infinitely many t, a.s., $\mathcal{D}_s$ being a compact subset of $\mathbb{R}^{n_\theta}$ in which every vector defines an asymptotically stable closed-loop system (4) and (7)*

- $\|\varphi(t-1)\|$ *is bounded for infinitely many t, a.s.*

Then:

(i) *The trajectories of the ODEs (6) within $\mathcal{D}_s$ are the asymptotic paths of the estimates generated by the stochastic system (1) and (4)*

(ii) *The only possible convergence points of the stochastic recursive algorithm (1) are the locally stable equilibrium points of the associated ODEs (6). Specifically, if, with nonzero probability,*

$$\lim_{t\to\infty} \theta(t) = \theta_* \qquad and \qquad \lim_{t\to\infty} \mathcal{R}(t) = \mathcal{R}_* > 0$$

(with $\theta_* \in \mathcal{D}_s$, necessarily) then $(\theta_*, \mathcal{R}_*)$ is a locally stable equilibrium point of the associated ODEs (6), viz.,

$$f(\theta_*) = O_{n_\theta} \qquad and \qquad \mathcal{G}(\theta_*) = \mathcal{R}_* \tag{9.5-8}$$

and the matrix

$$\mathcal{R}_*^{-1} \frac{\partial f(\theta)}{\partial \theta}\bigg|_{\theta=\theta_*} \tag{9.5-9}$$

has all its eigenvalues in the closed left half-plane.

Remark 9.5-1.

- Note that the time-scale transformation $t \mapsto \tau = \ln t$ in (6a) yields a "time compression" in that events at large values of t take place earlier in τ. This is an important advantage of investigating convergence of (1) through simulation of the associated ODE rather than running the recursive algorithm (1) itself.

- From the validity conditions of Result 1, in particular the role played there by $\mathcal{D}_s$, we see that stability in the ODE method must be assumed from the outset and never comes as a consequence of ODE analysis. This clearly limits the importance of the ODE method in adaptive control. Nonetheless, the method is widely applicable to determine necessary conditions for algorithm convergence to desirable points.

In the next example, we apply the ODE method to analyzing the implicit RLS+MV ST regulator introduced in Sec. 8.7. We commented there that global convergence analysis of such an adaptive regulator is a difficult task and were only able to establish global convergence of the SG+MV regulator. In the example that follows it is shown that, though global convergence cannot be addressed, ODE analysis is by all means valuable in that it pinpoints some positive real properties on the $C(d)$ polynomial that must be satisfied so as to possibly achieve convergence.

Example 9.5-1 [Lju77a]   (Implicit RLS+MV ST Regulation)

Consider the plant (8.7-1) with I/O delay equal to 1, along with (8.7-6) to (8.7-9) corresponding to MV regulation. In particular, hereafter we assume that b_1 is a priori known and hence is not estimated. The implicit model used for RLS estimation is then

$$y(t) = b_1 u(t-1) + \varphi'(t-1)\theta_{\mathrm{MV}} + e(t) \tag{9.5-10a}$$

$$\varphi(t-1) := \left[\ \left(y_{t-\hat{n}}^{t-1} \right)' \quad \left(u_{t-\hat{n}}^{t-2} \right)' \ \right]' \tag{9.5-10b}$$

$$\theta_{\mathrm{MV}} := \left[\ c_1 - a_1 \quad \cdots \quad c_{\hat{n}} - a_{\hat{n}} \quad b_2 \quad \cdots \quad b_{\hat{n}} \ \right]' \in \mathbb{R}^{n_\theta} \tag{9.5-10c}$$

with $n_\theta = 2\hat{n} - 1$. Consequently, the RLS algorithm is given by (1) and (2) with

$$\varepsilon(t) = y(t) - b_1 u(t-1) - \varphi'(t-1)\theta(t-1)$$

Further, the plant input is

$$u(t) = -\frac{\theta'(t)}{b_1}\varphi(t) \tag{9.5-11}$$

Hence, by (11),

$$
\begin{aligned}
\varepsilon(t,\theta) &= y(t,\theta) \\
&= b_1 u(t-1,\theta) + \theta^{o'}\varphi(t-1,\theta) + C(d)e(t) \\
&= (\theta^o - \theta)'\,\varphi(t-1,\theta) + C(d)e(t) \\
&= (\theta^o - \theta_{\mathrm{MV}})'\,\varphi(t-1,\theta) + (\theta_{\mathrm{MV}} - \theta)'\,\varphi(t-1,\theta) + C(d)e(t) \quad (9.5\text{-}12a)
\end{aligned}
$$

where

$$\theta^o := \left[\ -a_1 \quad \cdots \quad -a_{\hat{n}} \quad b_2 \quad \cdots \quad b_{\hat{n}} \ \right]' \tag{9.5-12b}$$

Observe that

$$\theta^o - \theta_{\mathrm{MV}} = -\begin{bmatrix} c_1 & \cdots & c_{\hat{n}} & O_{1\times(\hat{n}-1)} \end{bmatrix}'$$

$$\begin{aligned}(\theta^o - \theta_{\mathrm{MV}})'\,\varphi(t-1,\theta) &= -c_1 y(t-1,\theta) - \cdots - c_{\hat{n}} y(t-\hat{n},\theta) \\ &= [1 - C(d)]y(t,\theta) \\ &= [1 - C(d)]\varepsilon(t,\theta)\end{aligned}$$

Therefore, using this in (12a),

$$\varepsilon(t,\theta) = [1 - C(d)]\varepsilon(t,\theta) + (\theta_{\mathrm{MV}} - \theta)'\,\varphi(t-1,\theta) + C(d)e(t)$$

or

$$C(d)\varepsilon(t,\theta) = (\theta_{\mathrm{MV}} - \theta)'\,\varphi(t-1,\theta) + C(d)e(t)$$

Hence,

$$\varepsilon(t,\theta) = (\theta_{\mathrm{MV}} - \theta)'\,\varphi_c(t-1,\theta) + e(t)$$

where $\varphi_c(t-1,\theta)$ is the filtered regressor:

$$\varphi_c(t-1,\theta) := H(d)\varphi(t-1,\theta) \tag{9.5-13a}$$

$$H(d) := \frac{1}{C(d)} \tag{9.5-13b}$$

Thus, (5a) becomes

$$\begin{aligned}f(\theta) &= \mathcal{E}\left\{\varphi(t-1,\theta)\varepsilon(t,\theta)\right\} \\ &= \mathcal{E}\left\{\varphi(t-1,\theta)\varphi_c'(t-1,\theta)\right\}(\theta_{\mathrm{MV}} - \theta) + \mathcal{E}\left\{\varphi(t-1,\theta)e(t)\right\} \\ &= \mathcal{E}\left\{\varphi(t-1,\theta)\varphi_c'(t-1,\theta)\right\}(\theta_{\mathrm{MV}} - \theta) \end{aligned} \tag{9.5-13c}$$

from which the associated ODE follows:

$$\begin{aligned}\dot{\theta}(\tau) &= -\mathcal{R}^{-1}(\tau)M(\theta(\tau))\left[\theta(\tau) - \theta_{\mathrm{MV}}\right] \tag{9.5-14a} \\ \dot{\mathcal{R}}(\tau) &= -\mathcal{R}(\tau) + \mathcal{G}(\theta(\tau)) \tag{9.5-14b}\end{aligned}$$

where the dot denotes the derivative,

$$M(\theta) := \mathcal{E}\left\{\varphi(t-1,\theta)\varphi_c'(t-1,\theta)\right\} \tag{9.5-14c}$$

and $\mathcal{G}(\theta)$ as in (5d).

The equilibrium points $\theta_*, \mathcal{R}_*$, of (14) are the solutions of the following algebraic equations:

$$M(\theta_*)\,(\theta_* - \theta_{\mathrm{MV}}) = O_{n_\theta} \tag{9.5-15a}$$

$$\mathcal{R}_* = \mathcal{G}(\theta_*) > 0 \tag{9.5-15b}$$

A comment on (15b) is in order. If $\hat{n}$ is larger than the minimum plant order and hence $\varphi(t-1,\theta)$ need not be a full-rank process, some measure must be taken so as to make $\mathcal{G}(\theta(t))$ positive definite, for example, a small positive definite matrix can be added within the brackets in (1b).

To proceed, we have to avail of the following lemma.

Lemma 9.5-1. *Let $H(d)$ be SPR (cf. (6.4-34)). Then, for $\ell \in \mathbb{R}^{n_\theta}$,*

$$M(\theta)\ell = O_{n_\theta} \qquad \Longrightarrow \qquad \varphi'(t-1,\theta)\ell = 0$$

Proof: $M(\theta)\ell = O_{n_\theta}$ implies $0 = \ell' M(\theta)\ell = \mathcal{E}\{z(t-1,\theta)z_c(t-1,\theta)\}$, where

$$z(t-1,\theta) := \ell' \varphi(t-1,\theta) \qquad \text{and} \qquad z_c(t-1,\theta) := H(d)z(t-1,\theta)$$

Thus, if $\Psi_z(d)$ denotes the spectral density function of $z(t-1,\theta)$, from Problem 7.3-5, it follows that

$$
\begin{aligned}
0 = \ell' M(\theta)\ell &= \langle H(d)\Psi_z(d) \rangle = \frac{1}{2}\left[\langle H(d)\Psi_z(d) + H^*(d)\Psi_z^*(d)\rangle\right] \\
&= \frac{1}{2}\langle (H(d) + H^*(d))\,\Psi_z(d)\rangle \qquad [(7.3\text{-}24b)] \\
&= \frac{1}{2\pi}\int_{-\pi}^{\pi} \mathrm{Re}\left[H\left(e^{j\omega}\right)\right]\Psi_z\left(e^{j\omega}\right)d\omega \qquad [(3.1\text{-}43)]
\end{aligned}
$$

Because $\Psi_z\left(e^{j\omega}\right) = \Psi_z\left(e^{-j\omega}\right) \geq 0$, $H(d)$ SPR implies that $\Psi_z\left(e^{j\omega}\right) \equiv 0$.

Then, by Lemma 1, assuming $1/C(d)$ SPR or, equivalently, $C(d)$ SPR, (15a) implies that $\varphi'(t-1,\theta_*)\,(\theta_* - \theta_{\mathrm{MV}}) = 0$. Hence, from (12a),

$$
\begin{aligned}
y(t,\theta_*) &= \varphi'(t-1,\theta_*)(\theta^o - \theta_{\mathrm{MV}}) + C(d)e(t) \\
&= [1 - C(d)]\,y(t,\theta_*) + C(d)e(t)
\end{aligned}
$$

that is,

$$C(d)y(t,\theta_*) = C(d)e(t) \qquad \text{or} \qquad y(t,\theta_*) = e(t) \qquad (9.5\text{-}16)$$

We can conclude that

$$
\left.\begin{array}{c}
\mathrm{Re}\left[C\left(e^{j\omega}\right)\right] > 0 \\
\forall \omega \in [-\pi,\pi)
\end{array}\right\}
\Longrightarrow
\left\{\begin{array}{l}
\text{all the equilibrium points of the ODE} \\
\text{associated to the implicit RLS+MV} \\
\text{regulator yield the MV regulation law}
\end{array}\right. \qquad (9.5\text{-}17)
$$

This means that the implicit RLS+MV ST regulator is weakly self-optimizing w.r.t. the MV criterion. Assume now that $\hat{n}$ is not only an upper bound of the minimum plant order, but that it equals the latter, viz.,

$$\hat{n} = \max\{n_a, n_b, n_c\} \qquad (9.5\text{-}18)$$

Then, under (18), (16) implies $\theta_* = \theta_{\mathrm{MV}}$. We next study local stability of θ_{MV} under the assumption (18). We have from (13c)

$$f(\theta) = -M(\theta)\,(\theta - \theta_{\mathrm{MV}})$$

and

$$
\begin{aligned}
\mathcal{R}_{\mathrm{MV}}^{-1}\frac{\partial f(\theta)}{\partial \theta}\bigg|_{\theta=\theta_{\mathrm{MV}}} &= -\mathcal{R}_{\mathrm{MV}}^{-1}M\left(\theta_{\mathrm{MV}}\right) \\
&= -\mathcal{G}^{-1}\left(\theta_{\mathrm{MV}}\right)M\left(\theta_{\mathrm{MV}}\right) \qquad (9.5\text{-}19)
\end{aligned}
$$

According to Result 1, in order to show that θ_{MV} is a possible convergence point, we have to prove that all the eigenvalues of (19) are in the closed left half-plane. To this end, we shall use the following lemma.

Lemma 9.5-2.   *Let S and M be two real-valued square matrices such that*

$$S = S' > 0 \qquad and \qquad M + M' \geq 0$$

Then, $-SM$ is a stability matrix.

Proof: Consider the linear differential equation $\dot{x}(t) = -SMx(t)$. For $P = S^{-1}$, we have

$$P(-SM) + (-SM)'P + (M + M') = 0$$

Hence, $P = P' > 0$, so it follows that $(-SM, H)$, $H'H = M + M' \geq 0$, is an observable pair and that $-SM$ is a stability matrix. (cf. Problem 2.4-3.)

We can now prove that in the implicit RLS+MV ST regulator w.s. self-tuning occurs.

Proposition 9.5-1.   *Consider the implicit RLS+MV ST regulator applied to the given CARMA plant. Let $\hat{n}$ be set as in (18). Then, provided that $C(d)$ is SPR, the only possible convergence point is θ_{MV}.*

Proof: By the proof of Lemma 1, we have for $H(d) = 1/C(d)$

$$\ell'(M + M')\ell = \frac{1}{\pi} \int_{-\pi}^{\pi} \mathrm{Re}\left[H\left(e^{j\omega}\right)\right] \Psi_z \left(e^{j\omega}\right) d\omega \geq 0$$

whenever $C(d)$ is SPR. Then, applying Lemma 2, we conclude that (19) is a stability matrix.

Via the ODE method, it can be also shown [Lju77a] that, with the choice (18) and under the validity conditions of Result 1, the previous implicit RLS+MV does indeed converge to θ_{MV} provided that the same SPR condition as in Theorem 6.4-36 holds:

$$\frac{1}{C(d)} - \frac{1}{2} \text{ is SPR} \tag{9.5-20}$$

Note that (20) is a stronger condition than $C(d)$ being SPR.

In [Lju77a], it is also shown via the ODE method that the variant of the adaptive regulator discussed before with RLS substituted by the SG identifier (3), with $a = 1$, under the general validity conditions of Result 1 does indeed converge to the MV control law provided that $C(d)$ is SPR. Note that this conclusion agrees with Theorem 8.7-1, where global convergence of the SG+MV ST regulator was established via the stochastic Lyapunov function method. It has to be pointed out that the condition $\theta_{\mathrm{MV}} \epsilon \mathcal{D}_s$ implies that the plant must be minimum-phase.

Problem 9.5-1 (ODEs for SG-Based Algorithms)   Following a heuristic derivation similar to the one used for RLS-based algorithms, show that the ODEs associated with SG-based algorithms are given by

$$\frac{d\theta(\tau)}{d\tau} = \frac{a}{r(\tau)} f(\theta(\tau))$$

$$\frac{dr(\tau)}{d\tau} = -r(\tau) + g(\theta(\tau))$$

with

$$g(\theta) = \mathcal{E}\left\{\|\varphi(t - 1, \theta)\|^2\right\}$$

and $f(\theta)$ again as in (5a).

Problem 9.5-2 (ODE Analysis of the SG+MV Regulator) Use the ODE associated to the SG-based algorithms in Problem 1 and conditions similar to (8) and (9) in order to find the possible converging points of the SG+MV ST regulator (8.7-24) and (8.7-25).

9.5.2 MUSMAR ODE Analysis

We now consider a SISO CARMA plant:

$$A(d)y(t) = B(d)u(t - \ell) + C(d)e(t) \tag{9.5-21}$$

with all the properties as in (2-1). In (21), $\ell + 1$ is the I/O delay, and e is a stationary zero-mean sequence of independent random variables with variance σ_e^2 and such that all moments exist. Moreover, let

$$n = \max\{\partial A(d), \partial B(d) + \ell, \partial C(d)\}$$

Associated with (21), a quadratic cost function is considered:

$$\mathcal{C}_T := \frac{1}{T}\mathcal{E}\{L_T\} \tag{9.5-22a}$$

$$L_T := \frac{1}{2}\sum_{i=1}^{T}\left[y^2(t + i + \ell) + \rho u^2(t + i - 1)\right] \tag{9.5-22b}$$

where $\rho \geq 0$. We recall that in previous sections the MUSMAR algorithm has been introduced in order to adaptively select a feedback-gain vector F such that in stochastic steady state,

$$u(t) = F's(t) + \eta(t) \tag{9.5-23}$$

minimizes in a receding-horizon sense the cost (22) for the unknown plant. In (23), η is a stationary zero-mean white dither sequence of independent random variables with variance σ_η^2 independent of e, and $s(t)$ denotes the pseudo-state made up by past I/O and, possibly, dither samples. In order to deal with a causal control strategy, the pseudo-state $s(t)$ is made up by any finite subset of the available data:

$$I^t = \{y^t, u^{t-1}, \eta^{t-1}\}$$

Problem 3, which follows, shows that the control law of the form

$$u(t) = \bar{u}(t) + \eta(t) \tag{9.5-24a}$$

minimizing in stochastic steady state cost $\mathcal{C}_\infty$, with $\bar{u}(t) \in \sigma\{I^t\}$, is given by

$$R_0(d)u(t) = -S_0(d)y(t) + C(d)\eta(t) \tag{9.5-24b}$$

Problem 9.5-3 Show that the control law of the form (24a), minimizing in stochastic steady state cost $\mathcal{C}_\infty$, is given by (24b). (*Hint:* Consider any causal regulation law (cf. (7.3-72))

$$R_0(d)u(t) = -S_0(d)y(t) + \eta(t) + v(t)$$

where $R_0(d)$ and $S_0(d)$ are the LQS regulation polynomials:

$$A(d)R_0(d) + d^\ell B(d)S_0(d) = C(d)E(d)$$

$$E^*(d)E(d) = \rho A^*(d)A(d) + B^*(d)B(d)$$

and v any wide-sense stationary process such that $v(t) \in \sigma\left\{I^t\right\}$. Following the same lines used after (7.3-72), show that under the previous control law

$$\mathcal{C}_\infty = \mathcal{C}_{\mathrm{LQS}} + \mathcal{E}\left\{\left[\frac{1}{C(d)}\left[v(t) + \eta(t)\right]\right]^2\right\}$$

where $\mathcal{C}_{\mathrm{LQS}}$ is the LQS cost for $\eta(t) = v(t) \equiv 0$. Take into account the constraint $v(t) \in \sigma\left\{I^t\right\}$ to show that $\mathcal{C}_\infty$ is minimized for

$$v(t) = -C(d)\mathcal{E}\left\{\frac{\eta(t)}{C(d)} \,\middle|\, I^t\right\}$$

Hence, conclude that $v(t) = [C(d) - 1]\eta(t)$.)

In the absence of dither, that is, $\sigma_\eta^2 = 0$, (24b) coincides with the optimal regulation law for the LQS regulation problem under consideration. Equation (24b) can be written in the form (23) with $f \in \mathbb{R}^{2n+m}$ and

$$s(t) := \left[\ (y_t^{t-n+1})' \quad (u_{t-1}^{t-m})' \quad (\eta_{t-1}^{t-n})'\ \right]' \qquad (9.5\text{-}25a)$$

with (cf. Problem 7.3-16)

$$m = \begin{cases} n-1, & \rho = 0 \\ n, & \rho > 0 \end{cases} \qquad (9.5\text{-}25b)$$

This result shows that in order to make the MUSMAR algorithm compatible with the steady-state LQS regulator, past dither samples should be included in the regressor. In practice, an unknown input dither component is unavoidably present due to the finite word length of the digital processor implementing the algorithm. Consequently, in order to deal with this more realistic situation, we assume hereafter that the pseudo-state vector only includes past I/O samples and no dither, viz.,

$$s(t) := \left[\ (y_t^{t-\hat{n}+1})' \quad (u_{t-1}^{t-\hat{m}})'\ \right]' \qquad (9.5\text{-}26a)$$

where $\hat{n}$ and $\hat{m}$ are two nonnegative integers. In case $\hat{n}$ is the presumed order of the plant, $\hat{m}$ can be chosen according to the rule

$$\hat{m} = \begin{cases} \hat{n}-1, & \rho = 0 \\ \hat{n}, & \rho > 0 \end{cases} \qquad (9.5\text{-}26b)$$

This situation will be referred to as the *unknown dither case*. Problem 4, which follows, shows that when the pseudo-state is given by (26a), with $\hat{n} \geq n$ and $\hat{m}$ as in (26b), the control law (23), minimizing in stochastic steady state cost C_∞ for the CARMA plant (21), is given by

$$R_0(d)u(t) = -S_0(d)y(t) + \eta(t) - C(d)\mathcal{E}\left\{\left.\frac{\eta(t)}{C(d)}\ \right|\ s(t)\right\} \tag{9.5-27}$$

Because, for $\sigma_\eta^2/\sigma_e^2 \ll 1$,

$$\mathcal{E}\left\{\left.\frac{\eta(t)}{C(d)}\ \right|\ s(t)\right\} \propto \frac{\sigma_\eta^2}{\sigma_e^2}$$

(27) tends to the steady-state LQS optimal regulation law as $\sigma_\eta^2/\sigma_e^2 \to 0$.

Problem 9.5-4   Show that when the pseudo-state is given by (26a), with $\hat{n} \geq n$ and $\hat{m}$ as in (26b), the control law (23), minimizing in stochastic steady state cost C_∞ for the CARMA plant (21), is given by (27). (*Hint*:   Proceed as indicated in the hint of Problem 3 by replacing I^t with $s(t)$.)

Problem 9.5-5   Prove that in the unknown dither case, for any $\hat{n}$ and $\hat{m}$, and under a stabilizing regulation law, the pseudo-state covariance matrix $\Psi_s := \mathcal{E}\left\{s(t)s'(t)\right\}$ is always positive definite. (*Hint*:   Construct a proof by contradiction.)

   Remark 9.5-2.   As Problem 5 shows, in the unknown dither case, for any $\hat{n}$ and under a stabilizing regulation law, the pseudo-state covariance matrix $\Psi_s := \mathcal{E}\{s(t)s'(t)\}$ is always strictly positive definite. This property makes it convenient for the ODE analysis to assume the presence of a nonzero dither in (23). However, in practice, the dither need not be used provided that the identifiers are equipped with a covariance resetting logic fix or variants thereof.

   MUSMAR is based on the set of $2T$ prediction models (3-11), one for each output and input variable included in the cost function (22), viz.,

$$\left.\begin{array}{rcl} y(t+i+\hat{\ell}) & = & \theta_i u(t) + \Gamma_i' s(t) + \epsilon_i(t+i) \\ u(t+i-1) & = & \mu_i u(t) + \Lambda_i' s(t) + \nu_i(t+i-1) \end{array}\right\} \tag{9.5-28}$$

where $\hat{\ell} \leq \ell$. Notice that all the models in (28) share the same regressor:

$$\varphi(t) := \left[\begin{array}{cc} u(t) & s'(t) \end{array}\right]'$$

In the following, we set for simplicity $\hat{\ell} = 0$. The parameters of the prediction models in (28) are separately estimated via standard RLS identifiers, as shown in Fig. 3-4. The estimate updating equations, therefore, are given by (3-16), for $i = 1, \cdots, T$:

$$\left[\begin{array}{c} \theta_i(t) \\ \Gamma_i(t) \end{array}\right] = \left[\begin{array}{c} \theta_i(t-1) \\ \Gamma_i(t-1) \end{array}\right] + K(t-T) \tag{9.5-29a}$$

$$\times \left[y(t-T+i) - \theta_i(t-1)u(t-T) - \Gamma_i'(t-1)s(t-T)\right]$$

and, for $i = 2, \cdots, T$:

$$\begin{bmatrix} \mu_i(t) \\ \Lambda_i(t) \end{bmatrix} = \begin{bmatrix} \mu_i(t-1) \\ \Lambda_i(t-1) \end{bmatrix} + K(t-T) \tag{9.5-29b}$$

$$\times \left[u(t-T+i-1) - \mu_i(t-1)u(t-T) - \Lambda_i'(t-1)s(t-T) \right]$$

In (29), $K(t-T) = P(t-T+1)\varphi(t-T)$ denotes the updating gain. Obviously, $\mu_1(t) \equiv 1$ and $\Lambda_1(t) \equiv 0$. Finally, the control signal is chosen, at each sampling instant t, according to (cf. (3-17))

$$u(t) = F'(t)s(t) + \eta(t) \tag{9.5-30}$$

$$F(t) := -\Xi^{-1}(t) \sum_{i=1}^{T} [\theta_i(t)\Gamma_i(t) + \rho\mu_i(t)\Lambda_i(t)] \tag{9.5-31}$$

$$\Xi(t) := \sum_{i=1}^{T} [\theta_i^2(t) + \rho\mu_i^2(t)] \tag{9.5-32}$$

Conditions on the form of feedback F in (23), under which the regulated CARMA plant (21) admits in stochastic steady state the multiple-prediction model (28), have been given in Theorem 3-1. Hereafter, an ODE local convergence analysis of the MUSMAR algorithm is carried out. No assumption is made on the pseudo-state complexity $\hat{n}$ or the true I/O delay. The main interest is in the possible convergence points of MUSMAR feedback-gain vector F.

Because, if F converges to a stabilizing control law, Ψ_s converges to a strictly positive definite bounded matrix, Theorem 6.4-2 also shows that parameter estimates $\theta(t) := \{\theta_i(t), \Gamma_i(t), \mu_i(t), \Lambda_i(t)\}_{i=1}^{T}$ converge. Thus, the only possible convergence feedback gains are the ones provided by (31) and (32) for a given parameter estimate $\theta(t)$ at convergence.

The multipredictor coefficients of (28) are estimated via standard RLS algorithms, so the asymptotic average evolution of their estimates is described by the following set of ODEs (cf. (6)):

$$\dot{\Theta}_i(\tau) = \begin{bmatrix} \dot{\theta}_i(\tau) \\ \dot{\Gamma}_i(\tau) \end{bmatrix} = \mathcal{R}^{-1}(\tau)f_{\Theta_i}(\tau) \tag{9.5-33a}$$

$$\dot{M}_i(\tau) = \begin{bmatrix} \dot{\mu}_i(\tau) \\ \dot{\Lambda}_i(\tau) \end{bmatrix} = \mathcal{R}^{-1}(\tau)f_{M_i}(\tau) \tag{9.5-33b}$$

$$\dot{\mathcal{R}}(\tau) = -\mathcal{R}(\tau) + \mathcal{R}_\varphi(\tau) \tag{9.5-33c}$$

where

$$f_{\Theta_i}(\tau) := \mathcal{E}\left\{\varphi(t)\left[y(t+i)\right.\right. \tag{9.5-34a}$$

$$\left.\left. - [\theta_i(\tau)F(\tau) + \Gamma_i(\tau)]' s(t) - \theta_i(\tau)\eta(t)\right] \mid F(\tau)\right\}$$

$$f_{M_i}(\tau) \quad := \quad \mathcal{E}\left\{\varphi(t)\Big[u(t+i-1)\right. \tag{9.5-34b}$$

$$\left. - \left[\mu_i(\tau)F(\tau) + \Lambda_i(\tau)\right]' s(t) - \mu_i(\tau)\eta(t)\Big] \mid F(\tau)\right\}$$

$$\mathcal{R}_\varphi(\tau) \quad := \quad \mathcal{E}\left\{\varphi(t)\varphi'(t) \mid F(\tau)\right\} \tag{9.5-34c}$$

$$= \quad \begin{bmatrix} F'(\tau)\mathcal{R}_s(\tau)F(\tau) + \sigma_\eta^2 & F'(\tau)\mathcal{R}_s(\tau) \\ \mathcal{R}_s(\tau)F(\tau) & \mathcal{R}_s(\tau) \end{bmatrix}$$

$$\mathcal{R}_s(\tau) \quad := \quad \mathcal{E}\left\{s(t)s'(t) \mid F(\tau)\right\} \tag{9.5-34d}$$

and $F(\tau)$ is as in (31), with t replaced by τ. In (34), $\mathcal{E}\{\cdot \mid F(\tau)\}$ denotes the expectation w.r.t. the probability density function induced on u and y by e and η, assuming that the system is in the stochastic steady state corresponding to the constant control law

$$u(t) = F'(\tau)s(t) + \eta(t) \tag{9.5-35}$$

It is now convenient to derive the ODE for $F(\tau)$ from (33) to (35).

Lemma 9.5-3.   *Consider ODEs (33) to (35) associated with the MUSMAR algorithm (29) to (32). Then the ODE associated with the MUSMAR feedback-gain vector can be written as follows:*

$$\dot{F}(\tau) = -\Xi^{-1}(\tau)\mathcal{R}_s^{-1}(\tau)p(\tau) + o(\|\tilde{F}(\tau)\|) \tag{9.5-36}$$

where $\tilde{F}(\tau) := F(\tau) - F^*$, F^* denotes any equilibrium point of (36), $o(\|x\|)$ is such that $\lim_{x \to 0}[o(\|x\|)/\|x\|] = 0$,

$$p(\tau) := \sigma_\eta^{-2} \sum_{i=1}^{T} [\mathcal{R}_{y\eta}(i;\tau)\mathcal{R}_{ys}(i;\tau) + \rho\mathcal{R}_{us}(i-1;\tau)\mathcal{R}_{us}(i-1;\tau)] \tag{9.5-37}$$

and

$$\mathcal{R}_{y\eta}(i;\tau) := \mathcal{E}\left\{y(t+i)\eta(t) \mid F(\tau)\right\} \tag{9.5-38a}$$

$$\mathcal{R}_{ys}(i;\tau) := \mathcal{E}\left\{y(t+i)s(t) \mid F(\tau)\right\} \tag{9.5-38b}$$

with similar definitions for $\mathcal{R}_{u\eta}(i-1;\tau)$ and $\mathcal{R}_{us}(i-1;\tau)$.

Proof: Premultiplying both sides of (33a) by $\mathcal{R}(\tau) = \mathcal{R}_\varphi(\tau) - \dot{\mathcal{R}}(\tau)$ and using (35), we have (τ is omitted)

$$\left(\mathcal{R}_\varphi - \dot{\mathcal{R}}\right) \begin{bmatrix} \dot{\theta}_i \\ \dot{\Gamma}_i \end{bmatrix}$$

$$= \quad \mathcal{E}\left\{\begin{bmatrix} F's(t) + \eta(t) \\ s(t) \end{bmatrix} \left[y(t+i) - [\theta_i F + \Gamma_i]' s(t) - \theta_i \eta(t)\right] \mid F(\tau)\right\}$$

$$= \quad \begin{bmatrix} F'\mathcal{R}_{ys}(i) - F'\mathcal{R}_s[\theta_i F + \Gamma_i] + \mathcal{R}_{y\eta}(i) - \sigma_\eta^2 \theta_i \\ \mathcal{R}_{ys}(i) - \mathcal{R}_s[\theta_i F + \Gamma_i] \end{bmatrix} \tag{9.5-39}$$

where $\mathcal{R}_{ys}$ and $\mathcal{R}_{y\eta}$ are defined in (38). Recalling (34c), we also have

$$\mathcal{R}_\varphi \begin{bmatrix} \dot{\theta}_i \\ \dot{\Gamma}_i \end{bmatrix} = \begin{bmatrix} \left[F'\mathcal{R}_s F + \sigma_\eta^2 \right] \dot{\theta}_i + F'\mathcal{R}_s \dot{\Gamma}_i \\ \mathcal{R}_s \left[F\dot{\theta}_i + \dot{\Gamma}_i \right] \end{bmatrix} \tag{9.5-40}$$

If we define

$$\begin{bmatrix} K_i & G_i' \end{bmatrix}' := \dot{\mathcal{R}} \begin{bmatrix} \dot{\theta}_i & \dot{\Gamma}_i' \end{bmatrix}' \tag{9.5-41}$$

and

$$g_i := \mathcal{R}_{ys}(i) - \mathcal{R}_s \left[\theta_i F + \Gamma_i \right] \tag{9.5-42}$$

and taking into account (40), (39) can be rewritten as follows:

$$F'\mathcal{R}_s \left[F\dot{\theta}_i + \dot{\Gamma}_i \right] + \sigma_\eta^2 \dot{\theta}_i = F'g_i + \mathcal{R}_{y\eta}(i) - \sigma_\eta^2 \theta_i + K_i$$

$$\mathcal{R}_s \left[F\dot{\theta}_i + \dot{\Gamma}_i \right] = g_i + G_i$$

Substituting the latter equation into the former, we get

$$F' \left[g_i + G_i \right] + \sigma_\eta^2 \dot{\theta}_i = F'g_i + \mathcal{R}_{y\eta}(i) - \sigma_\eta^2 \theta_i + K_i$$

Thus, (39) can be rewritten as

$$\dot{\theta}_i = -\theta_i + \sigma_\eta^{-2} \mathcal{R}_{y\eta}(i) + \sigma_\eta^{-2} \left[K_i - F'G_i \right] \tag{9.5-43a}$$

$$\mathcal{R}_s \left[F\dot{\theta}_i + \dot{\Gamma}_i \right] = g_i + G_i \tag{9.5-43b}$$

In a similar way, (33b) can be rewritten as

$$\dot{\mu}_i = \mu_i + \sigma_\eta^{-2} \mathcal{R}_{u\eta}(i-1) + \sigma_\eta^{-2} \left[V_i - F'H_i \right] \tag{9.5-44a}$$

$$\mathcal{R}_s \left[F\dot{\mu}_i + \dot{\Lambda}_i \right] = h_i + H_i \tag{9.5-44b}$$

where

$$\begin{bmatrix} V_i & H_i' \end{bmatrix}' := \dot{\mathcal{R}} \begin{bmatrix} \dot{\mu}_i & \dot{\Lambda}_i' \end{bmatrix}' \tag{9.5-45}$$

$$h_i := \mathcal{R}_{us}(i-1) - \mathcal{R}_s \left[\mu_i F + \Lambda_i \right] \tag{9.5-46}$$

By taking into account (31), the corresponding ODE for $F(\tau)$ is obtained:

$$\begin{aligned} \dot{F}(\tau) &= -\frac{1}{\Xi(\tau)} \sum_{i=1}^{T} \Big\{ \theta_i \left[\dot{\theta}_i F + \dot{\Gamma}_i \right] \\ &\qquad + \rho\mu_i \left[\dot{\mu}_i F + \dot{\Lambda}_i \right] + \dot{\theta}_i \left[\Gamma_i + \theta_i F \right] + \rho\dot{\mu}_i \left[\varphi_i + \mu_i F \right] \Big\} \\ &= -\frac{1}{\Xi(\tau)} \left[\mathcal{R}_s^{-1}(\tau) p(\tau) - r(\tau) \right] \end{aligned} \tag{9.5-47}$$

where the last equality follows from (43) and (44) if $p(\tau)$ is as in (37) and

$$\begin{aligned} r(\tau) &:= \sum_{i=1}^{T} \Big\{ \sigma_\eta^{-2} \left[K_i(\tau) - F'(\tau)G_i(\tau) \right] \mathcal{R}_s^{-1}(\tau) \left[\mathcal{R}_{ys}(i;\tau) + G_i(\tau) \right] \\ &\qquad + \rho\sigma_\eta^{-2} \left[V_i(\tau) - F'(\tau)H_i(\tau) \right] \mathcal{R}_s^{-1}(\tau) \left[\mathcal{R}_{us}(i-1;\tau) + H_i(\tau) \right] \\ &\qquad + \sigma_\eta^{-2} \mathcal{R}_s^{-1} \left[\mathcal{R}_{y\eta}(i;\tau)G_i(\tau) + \rho\mathcal{R}_{u\eta}(i-1;\tau)H_i(\tau) \right] \\ &\qquad - \dot{\theta}_i(\tau) \left[\theta_i(\tau)F(\tau) + \dot{\Gamma}_i(\tau) \right] - \rho\dot{\mu}_i(\tau) \left[\mu_i(\tau)F(\tau) + \dot{\Lambda}_i(\tau) \right] \Big\} \end{aligned}$$

Now it is to be noticed that if $\Gamma^* = \{\theta_i^*, \Gamma_i^*, \mu_i^*, \Lambda_i^*\}_{i=1}^T$ denotes any equilibrium point of (33) and, according to (31), F^* is the corresponding feedback-gain vector and $\tilde{F}(\tau) :=$ $F(\tau) - F^*$, then

$$K_i(\tau) - F'(\tau)G_i(\tau) = \sigma_\eta^2 o(\|\tilde{F}(\tau)\|)$$

$$V_i(\tau) - F'(\tau)H_i(\tau) = \sigma_\eta^2 o(\|\tilde{F}(\tau)\|)$$

$$G_i(\tau) = o(\|\tilde{F}(\tau)\|)$$

$$H_i(\tau) = o(\|\tilde{F}(\tau)\|)$$

$$\dot{\theta}_i \left[\dot{\theta}_i F + \Gamma_i\right] = o(\|\tilde{F}(\tau)\|)$$

$$\dot{\mu}_i \left[\dot{\mu}_i F + \Lambda_i\right] = o(\|\tilde{F}(\tau)\|)$$

Consequently,

$$r(\tau) = o(\|\tilde{F}(\tau)\|) \tag{9.5-48}$$

In order to give a convenient interpretation to (37), the following lemma is introduced.

Lemma 9.5-4.   *Let $\chi(d; \tau) = \chi(d; F(\tau))$ be the closed-loop d-characteristic polynomial corresponding to $F(\tau)$. Then*

$$\sigma_\eta^{-2} \mathcal{R}_{y\eta}(i; \tau) = \left[\frac{d^\ell B(d)}{\chi(d; \tau)}\right]_i \tag{9.5-49a}$$

$$\sigma_\eta^{-2} \mathcal{R}_{u\eta}(i; \tau) = \left[\frac{A(d)}{\chi(d; \tau)}\right]_i \tag{9.5-49b}$$

where $[H(d)]_i$ denotes the ith sample of the impulse response associated with transfer function $H(d)$.

Proof: In a closed loop, (35) can be written as

$$R(d; \tau)u(t) = -S(d; \tau)y(t) + \eta(t)$$

Consequently, if

$$\chi(d; \tau) = A(d)R(d; \tau) + d^\ell B(d)S(d; \tau)$$

we find

$$y(t) = \frac{d^\ell B(d)}{\chi(d; \tau)}\eta(t) + \frac{C(d)R(d; \tau)}{\chi(d; \tau)}e(t)$$

$$u(t) = \frac{A(d)}{\chi(d; \tau)}\eta(t) - \frac{C(d)S(d; \tau)}{\chi(d; \tau)}e(t)$$

Because η and e are uncorrelated, (49) easily follow.

According to (49), (37) can be rewritten as follows:

$$
\begin{aligned}
p(\tau) &= \sum_{i=1}^{T} \mathcal{E} \left\{ \left[\frac{d^{\ell} B(d)}{\chi(d;\tau)} \right]_i y(t+i)s(t) \right. \\
&\quad \left. + \rho \left[\frac{A(d)}{\chi(d;\tau)} \right]_{i-1} u(t+i-1)s(t) \,\Big|\, F(\tau) \right\} \\
&= \mathcal{E} \left\{ y(t) \left[\left[\frac{d^{\ell} B(d)}{\chi(d;\tau)} \right]_{|T} s(t) \right] \right. \\
&\quad \left. + \rho u(t) \left[\left[\frac{A(d)}{\chi(d;\tau)} \right]_{|T-1} s(t) \right] \,\Big|\, F(\tau) \right\}
\end{aligned}
\tag{9.5-50}
$$

where $H(d)_{|T}$ denotes the truncation to the Tth power of the power series expansion in d of transfer function $H(d)$, viz.,

$$
H(d)_{|T} = \sum_{i=0}^{T} h_i d^i \quad \text{if} \quad H(d) = \sum_{i=0}^{\infty} h_i d^i
$$

It will now be shown that (50) can be interpreted as the gradient of cost (22) in a receding-horizon sense. In order to see this, let us introduce the following *receding-horizon variant* of the cost (22):

$$
\mathcal{C}_T(F,l) := T^{-1} \mathcal{E} \left\{ L_T(F,l) \right\}
\tag{9.5-51a}
$$

$$
L_T(F,l) := \frac{1}{2} \sum_{i=1}^{T} \left[y^2(t+i) + \rho u^2(t+i-1) \,\Big|\,
\right.
$$

$$
\left. u(k \neq t) = F's(k); u(t) = l's(t) \right]
\tag{9.5-51b}
$$

The idea is to evaluate the cost assuming that all inputs, except the first included in the cost, are given by a constant stabilizing feedback F for all times and since the remote past. Then denoting by $\nabla_l \mathcal{C}_T(F,l)$ the value taken on at $l = F$ by the gradient of $\mathcal{C}_T(F,l)$ w.r.t. l,

$$
\nabla_l \mathcal{C}_T(F,l) := \left. \frac{\partial \mathcal{C}_T(F,l)}{\partial l} \right|_{l=F}
\tag{9.5-52}
$$

we get the following.

Lemma 9.5-5. *Let $F(\tau)$ be a stabilizing feedback for the plant (1), and $p(\tau)$ as in (50). Then*

$$
p(\tau) = T \nabla_l \mathcal{C}_T(F,l)
\tag{9.5-53}
$$

Proof: Let

$$u(t) = l's(t) = F's(t) + (l' - F')s(t)$$

Thus, for all k,

$$u(k) = F's(k) + v(t)\delta_{t,k}$$

or

$$R(d; F)u(k) = -S(d; F)y(k) + v(t)\delta_{t,k}$$

where $v(t) := (l - F)'s(t)$. Consequently, if

$$\chi(d; F) = A(d)R(d; F) + d^\ell B(d)S(d; F)$$

we find

$$y(k) = \frac{d^\ell B(d)}{\chi(d; F)}v(t)\delta_{t,k} + \frac{C(d)R(d; F)}{\chi(d; F)}e(k)$$

$$u(k) = \frac{A(d)}{\chi(d; F)}v(t)\delta_{t,k} - \frac{C(d)S(d; F)}{\chi(d; F)}e(k)$$

Thus,

$$\left. \frac{\partial \mathcal{C}_T(F, l)}{\partial l} \right|_{l=F}$$

$$= \frac{1}{T}\sum_{i=1}^{T}\mathcal{E}\left\{ \left[y(t+i)\frac{\partial y(t+i)}{\partial v(t)}s(t) + \rho u(t+i-1)\frac{\partial u(t+i-1)}{\partial v(t)}s(t) \right]_{l=F} \right\}$$

Because

$$\frac{\partial y(t+i)}{\partial v(t)} = \left[\frac{d^\ell B(d)}{\chi(d; F)} \right]_i$$

$$\frac{\partial u(t+i)}{\partial v(t)} = \left[\frac{A(d)}{\chi(d; F)} \right]_i$$

taking into account (50), (53) follows.

Taking into account (36) together with Lemma 5, we get the following result.

Proposition 9.5-2.   *The ODE associated with the MUSMAR feedback-gain vector is given by*

$$\dot{F}(\tau) = -\left[\Xi(\tau)\right]^{-1}\mathcal{R}_s^{-1}(\tau)T\nabla_l\mathcal{C}_T(F(\tau), l) + o(\|\tilde{F}(\tau)\|) \tag{9.5-54}$$

Remark 9.5-3.   Because $\mathcal{R}_s(\tau) > 0$ and $\Xi(\tau) > 0$ for $\rho > 0$, (54) for $\rho > 0$ implies that the equilibrium points $\bar{F}$ of the MUSMAR algorithm are the extrema of the cost $\mathcal{C}_T$ in a receding-horizon sense, viz., $\mathcal{C}_T\left(\bar{F}, u(t) = \bar{F}'s(t)\right)$ is an extremum of $\mathcal{C}_T\left(\bar{F}, u(t) = l'\bar{s}(t)\right)$, where $\bar{s}(t)$ denotes the pseudo-state at time t corresponding in stochastic steady state to feedback $\bar{F}$. Such a conclusion holds true irrespective of the plant I/O delay $\ell + 1$ and the regulator complexity $\hat{n}$.

In order to establish which equilibrium points are possible convergence points of the MUSMAR algorithm, let us consider the cost in stochastic steady state:

$$\mathcal{C}(F) = \frac{1}{2}\mathcal{E}\left\{y^2(t) + \rho u^2(t) \mid u(k) = F's(k)\right\} \tag{9.5-55}$$

as a function of the stabilizing constant feedback F. As indicated in Problem 6,

$$\nabla_F \mathcal{C}(F) \; = \; \frac{\partial \mathcal{C}(F)}{\partial F} \tag{9.5-56}$$

$$= \; \mathcal{E}\left\{y(t)\left[\frac{d^\ell B(d)}{\chi(d;\tau)}s(t)\right] + \rho u(t)\left[\frac{A(d)}{\chi(d;\tau)}s(t)\right]\right\}$$

Problem 9.5-6   Verify that (56) gives the gradient of the stochastic steady-state cost (55) with respect to a stabilizing constant feedback F.

Thus, comparing (50) with (56) and taking into account the dependence of $y(t)$, $u(t)$, and $s(t)$ on e, (50) is seen to be a good approximation to (56) whenever

$$\left|\lambda\left[\chi(d,F)\right]\right|^{2(T-\ell)} \gg 1 \tag{9.5-57}$$

where $\lambda[\chi]$ denotes any root of χ. Therefore, in a neighborhood of any equilibrium point satisfying (57), the ODE (54) can be approximated by

$$\dot{F}(\tau) = -\left[\Xi(\tau)\right]^{-1}\mathcal{R}_s^{-1}(\tau)T\nabla_F\mathcal{C}(F(\tau)) + o(\|\tilde{F}(\tau)\|) \tag{9.5-58}$$

The foregoing results are summarized in the following theorem.

Theorem 9.5-1.   *Consider the MUSMAR algorithm for any I/O transport delay ℓ, an arbitrary strictly Hurwitz $C(d)$ innovations polynomial, and any pseudo-state complexity $\hat{n}$. Then*

(i) *For any $T > \ell$, MUSMAR equilibrium points are the extrema F^* of the receding-horizon variant (51) of the quadratic cost.*

(ii) *Amongst the equilibria F^* giving rise to a closed-loop system with well-damped modes relative to the regulation horizon T such that (50) can be replaced by (56), the only possible MUSMAR converging points for any $\rho > 0$ approach the local minima $(\nabla_F^2\mathcal{C}(F^*) > 0)$ or ridge points $(\nabla_F^2\mathcal{C}(F^*) \geq 0)$ of the cost (55).*

Proof: Part (i) is proved in Remark 3. Part (ii) is proved by Result 1 according to which the only possible convergence points of a recursive stochastic algorithm are the locally stable equilibrium points of the associated ODE. Because in (58), for $\rho > 0$, $\Xi(\tau) > 0$, and $\mathcal{R}_s(\tau) > 0$, the conclusion follows.

Remark 9.5-4. The relevance of Theorem 1 is twofold. First, because no assumption was made on the I/O delay or the pseudo-state complexity $\hat{n}$, it turns out that, if T is large enough, the only possible convergence points of MUSMAR tightly approach the local minima of the criterion even in the presence of unmodeled plant dynamics. Moreover, this holds true irrespective of any positive-real condition ([MZ84] and [MZ87]), although MUSMAR is based on RLS (cf. Proposition 1).

It is difficult to characterize the closed-loop behavior of the plant for $\hat{n} < n$. Conversely, if the feedback control law has enough parameters, then, according to [Tru85], the minima of $\mathcal{C}(F)$ are related to steady-state LQS regulation. More precisely:

Result 9.5-2. *If, in addition to the assumptions in (ii) of Theorem 1, $\hat{n} \geq n$ and the dither intensity is negligible w.r.t. that of the innovation process, $\mathcal{C}(F)$ has a unique minimum coinciding with the steady-state LQS feedback.*

Therefore, from Theorem 1 and Result 2, it follows that, if T is large enough in the sense of (57), $\sigma_\eta^2 \ll \sigma_e^2$, and $\hat{n} \geq n$, MUSMAR has a unique possible convergence point that tightly approximates the steady-state LQS feedback.

9.5.3   Simulation Results

The results of the preceding ODE analysis are important in that they show that if the algorithm converges, then under general assumptions, it converges to desirable points. Thus, the analysis allows us to disprove the existence of possible undesirable convergence points. However, ODE analysis leaves unanswered fundamental queries on the algorithm. Among them, it is of paramount interest to establish if the algorithm converges under realistic conditions. Because, in this respect, any further analysis appears to be prevented, we are forced to resort to simulation experiments.

In all the examples, the estimates are obtained by a factorized U-D version of RLS with no forgetting (cf. Sec. 6.3); the innovations and dither variance are respectively 1 and 10^{-4}; and simulation runs involve 3000 steps.

Example 9.5-2

Consider the plant

$$y(t+1) + 0.9y(t) + \varepsilon y(t-1) = u(t) + e(t+1) - 0.7e(t)$$

where $\varepsilon = -0.5$, and $\rho = 1$. This is a second-order plant. However, it is regulated under the assumption that it is of first-order, the term in ε being considered as a perturbation. Hence, the controller has the structure

$$u(t) = \begin{bmatrix} f_1 & f_2 \end{bmatrix} \begin{bmatrix} y(t) \\ u(t-1) \end{bmatrix}$$

Figure 1 shows the evolution, in feedback parameter space, of the feedback vector $F(t)$ for $T = 1, 2$, and 3, superimposed to the level curves of the unconditional quadratic cost,

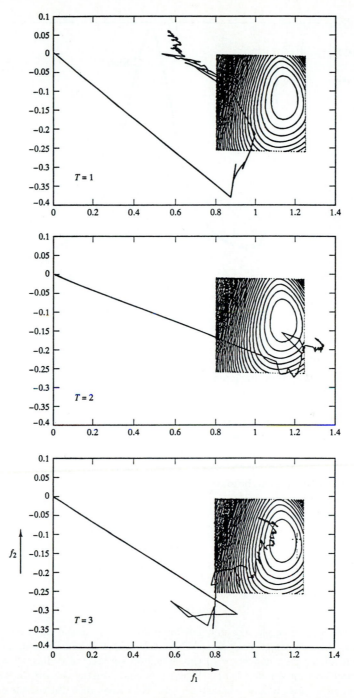

Figure 9.5-1: Time behaviour of the MUSMAR feedback parameters in Example 1 for $T = 1, 2, 3$, respectively.

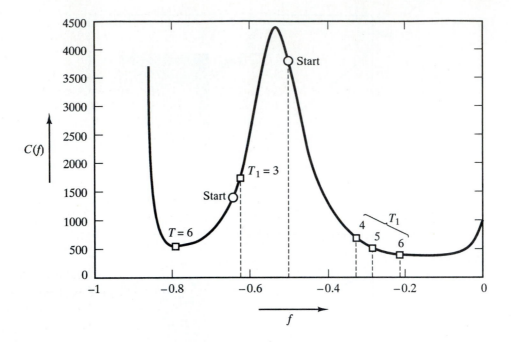

Figure 9.5-2: The unconditional cost $C(F)$ in Example 3 and feedback convergence points for various control horizons T.

$\mathcal{E}\{y^2(t) + u^2(t)\}$, constrained to the chosen regulator structure. For $T = 1$, convergence occurs to a point far from the optimum (where the cost is twice the minimum). For $T = 2$, MUSMAR is already quite close to the optimum and, for $T = 3$, the result is even better.

Example 9.5-3   Consider the sixth-order plant:

$$y(t + 6) - 3.102y(t + 5) + 4.049y(t + 4) - 2.974y(t + 3)$$
$$+ 1.356y(t + 2) - 0.37y(t + 1) + 0.0461y(t)$$
$$= 0.01u(t + 5) + 0.983u(t + 4) - 1.646u(t + 3)$$
$$+ 1.1788u(t + 2) - 0.3343u(t + 1) + 0.0353u(t) + e(t + 6)$$

with the proportional regulator

$$u(t) = fy(t)$$

and $\rho = 0$. In this example, the unconditional cost $C(f)$, as shown in Fig. 2, exhibits a finite maximum between two minima. When f is held constant at -0.5 for the first 100 steps, the feedback gain converges to the indicated squares according to various choices of the control horizon denoted by T_1. When f is held constant at -0.65, it converges for $T = 6$ to a value close to the other minimum. No convergence to the local maximum is observed, even when the initial feedback is close to it.

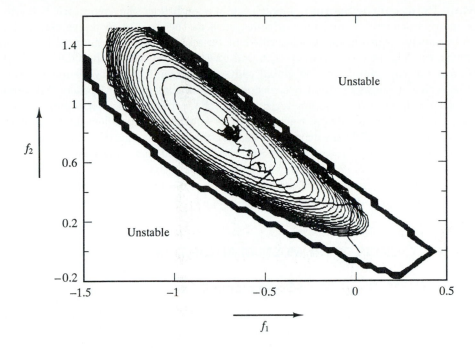

Figure 9.5-3: Time behaviour of MUSMAR feedback parameters of Example 4 for $T = 5$.

Example 9.5-4
Consider the plant

$$y(t+4) - 0.167y(t+3) - 0.74y(t+2) - 0.132y(t+1) + 0.87y(t)$$
$$= 0.132u(t+3) + 0.545u(t+2) + 1.117u(t+1) + 0.262u(t) + e(t+4)$$

This model corresponds to the fexible robot arm described in [Lan85]. It is a nonminimum-phase plant with a high-frequency resonance peak. With a reduced complexity two-term regulator

$$u(t) = \begin{bmatrix} f_1 & f_2 \end{bmatrix} \begin{bmatrix} y(t) \\ y(t-1) \end{bmatrix}$$

and $\rho = 10^{-4}$, the unconditional performance index exhibits a narrow "valley" from $\begin{bmatrix} 0 & 0 \end{bmatrix}$ to the minimum at $\begin{bmatrix} -0.787 & 0.86 \end{bmatrix}$. Figure 3 shows that MUSMAR, for $T = 5$, converges slowly but steadily to the point $\begin{bmatrix} -0.677 & 0.753 \end{bmatrix}$, close to the minimum (a loss of 1.28, against 1.26 at the optimum).

For both plants of Examples 3 and 4, the use of $T - \ell = 1$ yields unstable closed-loop systems.

Main Points of the Section. Under stability conditions, the asymptotic behavior of many stochastic recursive algorithms, such as the ones of recursive estimation and adaptive control, can be described in terms of a set of ordinary differential equations. The method of analysis, based on this result, called the ODE method, though not capable of yielding global convergence results, it is by all means valuable in that it allows us to uncover necessary conditions for convergence. Although the feedback-dependent parameterization of the implicit closed-loop plant model makes MUSMAR global convergence analysis a formidable problem, ODE analysis enables us to establish local convergence properties of the algorithm. These results reveal that, even in the presence of any structural mismatching condition, MUS-MAR equilibrium points coincide with the extrema of the cost in a receding-horizon sense. Further, as the length of the prediction horizon increases, MUSMAR possible convergence points approach the minima of the adopted quadratic criterion.

9.6   EXTENSIONS OF THE MUSMAR ALGORITHM

9.6.1   MUSMAR with Mean-Square Input Constraint

In all control applications, actuator power is limited. It is therefore important to explicitly take into account such a restriction in the controller design specifications. This can be done by adopting either a hard-limit input constraint or a mean-square (MS) input costraint approach. These are two possible alternatives; the most convenient to use depends on the application at hand. A hard-limit input constraint leads to a difficult nonlinear optimization problem. In this connection, approximate solutions are proposed in [TC88], [Toi83a], and [Böh85]. In [TC88], an adaptive GPC yielding an approximate solution to a quadratic programming problem is considered. In [Toi83a], an approximation to the probability density function of the plant input is used. In [Böh85], spread-in-time Riccati iterations are considered. For the alternative approach, [Toi83b] considered an input MS constraint. Specifically, an algorithm was proposed by combining the GMV self-tuning regulator with a stochastic approximation scheme. Though appealing for its simplicity, this algorithm has drawbacks (nonminimum-phase, unstable plants, and/or time-varying I/O delay) inherited from the one-step-ahead cost.

In this section, we study an MS input constrained adaptive control algorithm whose underlying control law is capable of overcoming the foregoing drawbacks. The algorithm is obtained by combining conveniently the MUSMAR algorithm discussed in the previous sections with the stochastic approximation scheme of [Toi83b]. Hereafter, this algorithm is referred to as CMUSMAR (*constrained MUS-MAR*). The main interest is in its convergence properties. Local convergence results are obtained. The strongest of them asserts that the isolated *constrained* minima of the underlying steady-state quadratic cost are possible convergence points of CMUSMAR. This conclusion also holds true in the presence of plant unmodeled dynamics and unknown I/O transport delay. The study is carried out by using the ODE convergence analysis of Sec. 5 and singular perturbation theory of ODEs ([KKO86] and [Was65]). The actual convergence of CMUSMAR to the possible equilibrium gains predicted by the theory is explored by means of simulation examples.

Formulation of the problem.   Consider the CARMA plant

$$A(d)y(t) = B(d)u(t) + C(d)e(t) \qquad (9.6\text{-}1)$$

with all the properties in (2-1), and

$$n = \max\{\partial A(d), \partial B(d), \partial C(d)\}$$

Also let e be a sequence of zero-mean independent identically distributed random variables such that all moments exist. A linear control regulation law

$$R(d)u(t) = -S(d)y(t) \qquad (9.6\text{-}2)$$

is considered for the plant (1). In (2), $R(d)$ and $S(d)$ are polynomials, with $R(d)$ monic. Equation (2) can be equivalently rewritten as

$$u(t) = F's(t) \qquad (9.6\text{-}3)$$

where F is a vector whose entries are the coefficients of $R(d)$ and $S(d)$, and $s(t)$ the pseudo-state (cf. Remark 2-1 and 5-2), given by

$$s(t) = \left[\ \left(y_t^{t-n+1}\right)'\quad \left(u_{t-1}^{t-m}\right)'\ \right]' \qquad (9.6\text{-}4)$$

The following problem is considered.

Problem 9.6-1   Given $c^2 > 0$, find in (3) an F solving

$$\min_{F}\ \lim_{t\to\infty}\ \mathcal{E}\left\{y^2(t)\right\} \qquad (9.6\text{-}5)$$

subject to the constraint

$$\lim_{t\to\infty}\ \mathcal{E}\left\{u^2(t)\right\} \leq c^2 \qquad (9.6\text{-}6)$$

According to the Kuhn–Tucker theorem [Lue69], Problem 1 is converted to the following *unconstrained* minimization problem.

Problem 9.6-2   Given $c^2 > 0$, find in (3) an F solving

$$\min_{F} \mathcal{L}(F, \rho) \qquad (9.6\text{-}7)$$

where $\mathcal{L}$ is the Lagrangian function given by the unconditional cost

$$\mathcal{L}(F,\rho) := \lim_{t\to\infty}\ \mathcal{E}\left\{y^2(t) + \rho u^2(t)\right\} \qquad (9.6\text{-}8)$$

and the Lagrange multiplier ρ satisfies the Kuhn–Tucker complementary condition

$$\rho\left(\lim_{t\to\infty}\ \mathcal{E}\left\{u^2(t)\right\} - c^2\right) = 0 \qquad (9.6\text{-}9)$$

For an unknown plant, Problem 1, or, equivalently, Problem 2, is to be solved by an adaptive control algorithm capable of selecting ρ and approximating an F that minimizes (8) under the constraint (9).

Remark 9.6-1. Let r be the output set point and $\tilde{y}(t) := y(t) - r$ the tracking error whose MS value $\mathcal{E}\{\tilde{y}^2(t)\}$ has to be minimized in stochastic steady state under the constraint (6). This problem can be cast into the previous formulation by changing $y(t)$ into $\tilde{y}(t)$. In case the circumstances are such that $\mathcal{E}\{\delta u^2(t)\} \le c^2$, $\delta u(t) := u(t) - u(t-1)$, is more suitable than (6), one can use the pseudo-state

$$s_\delta(t) := \begin{bmatrix} \tilde{y}(t) & \cdots & \tilde{y}(t - \hat{n}) & \delta u(t-1) & \cdots & \delta u(t - \hat{m}) \end{bmatrix}'$$

and the control variable $\delta u(t) = F' s_\delta(t)$ at the input of a CARMA plant:

$$A(d)\Delta(d)\tilde{y}(t) = B(d)\delta u(t) + C(d)e(t) \tag{9.6-10}$$

$\Delta(d) := 1 - d$. This is an integral action variant of (1) to (6) with $y(t)$ changed into $\tilde{y}(t)$ and $s(t)$ into $s_\delta(t)$, capable of ensuring in stochastic steady-state rejection of constant disturbances.

MS Input Constrained MUSMAR. As a candidate algorithm for solving the problem stated before, the stochastic approximation approach proposed in [Toi83b], combined with the MUSMAR algorithm, is considered.

At each sampling time t, the MUSMAR algorithm selects, via the enforced certainty equivalence procedure, $u(t)$ so as to minimize in stochastic steady state and in a receding-horizon sense the multistep quadratic cost (5-51). Next, the Lagrange multiplier $\rho = \rho(t)$ is updated via the following recurrent scheme:

$$\rho(t) = \rho(t-1) + \varepsilon\gamma(t)\rho(t-1)\left[u^2(t-1) - c^2\right] \tag{9.6-11}$$

where ε is a positive real, and $\{\gamma(t)\}$ a sequence of real numbers, whose selection will be made precise in (15). CMUSMAR is, then, obtained as detailed in what follows.

Algorithm 9.6-1 (CMUSMAR Algorithm) At each step t, recursively execute the following steps:

(i) Update RLS estimates of the closed-loop system parameters θ_i, μ_i, Γ_i, and Λ_i, $i = 1, \cdots, T$

$$\begin{bmatrix} \theta_i(t) \\ \Gamma_i(t) \end{bmatrix} = \begin{bmatrix} \theta_i(t-1) \\ \Gamma_i(t-1) \end{bmatrix} + K(t-T)\Big[y(t - T + i)$$
$$- \theta_i(t-1)u(t-T) - \Gamma_i'(t-1)s(t-T)\Big] \tag{9.6-12}$$

and, for $i = 2, \cdots, T$,

$$\begin{bmatrix} \mu_i(t) \\ \Lambda_i(t) \end{bmatrix} = \begin{bmatrix} \mu_i(t-1) \\ \Lambda_i(t-1) \end{bmatrix} + K(t-T)\Big[u(t-T+i-1)$$

$$- \mu_i(t-1)u(t-T) - \Lambda_i'(t-1)s(t-T) \Big] \qquad (9.6\text{-}13)$$

In (12) and (13), $K(t-T) = P(t-T+1)\varphi(t-T)$ denotes the RLS updating gain associated with the regressors

$$\varphi(j) := [\ u(j) \quad s'(j)\ ]' \qquad (9.6\text{-}14)$$

and $\mu_1(t) \equiv 1$ and $\Lambda_0(t) \equiv 0$.

(ii) Update the control cost weight $\rho(t)$ by using (11) with

$$\gamma(t) = [K'(t-T)K(t-T)]^{1/2} \qquad (9.6\text{-}15)$$

(iii) Update the vector of feedback gains F by

$$\Xi(t) = \sum_{i=1}^{T} \theta_i^2(t) + \rho(t)\left[1 + \sum_{i=2}^{T} \mu_i^2(t) \right] \qquad (9.6\text{-}16)$$

$$F(t) = -\frac{1}{\Xi(t)}\left[\sum_{i=1}^{T} \theta_i(t)\Gamma_i(t) + \rho(t)\left(\sum_{i=2}^{T} \mu_i(t)\Lambda_i(t) \right) \right] \qquad (9.6\text{-}17)$$

(iv) Apply to the plant an input given by

$$u(t) = F'(t)s(t) + \eta(t) \qquad (9.6\text{-}18)$$

where η is a zero-mean independent identically distributed dither noise independent of e and such that all moments exist.

The dither presence is introduced so as to guarantee persistency of exitation (cf. Problem 5-5). For $T = 1$, the Algorithm 1 reduces to the constrained MV self-tuner given in [Toi83b].

ODE Convergence Analysis. The previous algorithm is now analyzed using the ODE method. We can associate to CMUSMAR the following set of ODEs as in (5-33) ($i = 1, \cdots, T$; and $j = 2, \cdots, T$):

$$\begin{bmatrix} \dot{\theta}_i(\tau) \\ \dot{\Gamma}_i(\tau) \end{bmatrix} = \mathcal{R}^{-1}(\tau) \qquad (9.6\text{-}19)$$

$$\times \mathcal{E}\left\{ \varphi(t)\left[y(t+i) - \theta_i(\tau)u(t) - \Gamma_i'(\tau)s(t) \right] \Big| F(\tau) \right\}$$

$$\begin{bmatrix} \dot{\mu}_j(\tau) \\ \dot{\Lambda}_j(\tau) \end{bmatrix} = \mathcal{R}^{-1}(\tau) \qquad (9.6\text{-}20)$$

$$\times \mathcal{E}\left\{ \varphi(t)\left[u(t+j-1) - \mu_j(\tau)u(t) + \Lambda_j'(\tau)s(t) \right] \Big| F(\tau) \right\}$$

$$\dot{\mathcal{R}}(\tau) = -\mathcal{R}(\tau) + \mathcal{R}_\varphi(\tau) \tag{9.6-21a}$$
$$\mathcal{R}_\varphi(\tau) = \mathcal{E}\left\{\varphi(t)\varphi'(t) \mid F(\tau)\right\} \tag{9.6-21b}$$

$$\dot{\rho}(\tau) = \varepsilon\rho(\tau)\left[\mathcal{E}\left\{u^2(t)\right\} - c^2\right] \tag{9.6-22}$$

where $\varphi(t)$ is as in (14). In (19) and (20), a dot denotes a derivative with respect to τ, and $\mathcal{E}\{\cdot\}$ the expectation w.r.t. the probability density function induced on $u(t)$ and $y(t)$ by e and η, assuming that the system is in stochastic steady state corresponding to the constant control law

$$u(t) = F'(\tau)s(t) + \eta(t) \tag{9.6-23}$$

and a constant $\rho(\tau)$. Hereafter, in order to simplify the notation, the variable τ and the conditioning upon $F(\tau)$ will be omitted wherever no ambiguity can arise. In order to obtain a differential equation for F, differentiate (17) with respect to τ:

$$\dot{F} = \dot{F}_0 - \frac{1}{\Xi}\dot{\rho}\left[F\sum_i \mu_i^2 + \sum_i \mu_i\Lambda_i\right] \tag{9.6-24}$$

where $\dot{F}_0$ denotes the derivative of F assuming ρ constant. In Proposition 5-2, it has been shown that the following ODE holds:

$$\dot{F}_0 = -\frac{1}{\Xi}\mathcal{R}_s^{-1}T\nabla_T\mathcal{L} + o(\|\tilde{F}\|) \tag{9.6-25}$$

where $\tilde{F} := F_0 - F^*$, F^* denotes any equilibrium point, $o(\|x\|)$ is such that $\lim_{x\to 0}[o(\|x\|)/\|x\|] = 0$, and $\mathcal{R}_s := \mathcal{E}\left\{s(t)s'(t) \mid F_0(\tau)\right\}$. Finally, $\nabla_T\mathcal{L}$ is an approximation to the gradient of $\mathcal{L}$ w.r.t. F_0, which becomes increasingly tighter as $T \to \infty$. Thus, the ODE for F associated with CMUSMAR is

$$\dot{F} = -\frac{1}{\Xi}\mathcal{R}_s^{-1}T\nabla_T\mathcal{L} - \frac{1}{\Xi}\dot{\rho}\left[F\sum_i \mu_i^2 + \sum_i \mu_i\Lambda_i\right] + o(\|\tilde{F}\|) \tag{9.6-26}$$

and

$$\dot{\rho} = \varepsilon\rho\left[\mathcal{E}\left\{u^2(t)\right\} - c^2\right] \tag{9.6-27}$$

If F converges to a stabilizing control law, $\mathcal{R}_s$ converges to a strictly positive definite bounded matrix and ρ converges to a positive number; as pointed out in the previous section also parameter estimates $\theta(t) := \{\theta_i(t), \Gamma_i(t), \mu_i(t), \Lambda_i(t)\}$ converge. Therefore, the only possible convergence points of CMUSMAR are given by the stable equilibrium points of (26) and (27). These are given by

$$\nabla_T\mathcal{L} = 0, \quad \rho = 0 \tag{A}$$
$$\nabla_T\mathcal{L} = 0, \quad \mathcal{E}\{u^2(t)\} - c^2 = 0 \tag{B}$$

(A) equilibria correspond to the extrema of the receding-horizon variant of the MS output cost for which the corresponding MS input is less than c^2. (B) equilibria correspond to the extrema of the receding-horizon variant of the MS output cost on the boundary of the feasibility region $\mathcal{E}\{u^2(t)\} \le c^2$.

Stable (A) Equilibria. We have the following result:

Proposition 9.6-1. *Consider CMUSMAR (Algorithm 1) with any controller complexity and any plant I/O delay smaller than T. Then if T is large enough in the sense of* (ii) *of Theorem 5-1, among the (A) equilibria, the only possible convergence points of CMUSMAR are the minima or ridge points of the MS output value in the feasibility region $\mathcal{E}\{u^2(t)\} \le c^2$.*

Proof: Equations (26) and (27) are of the form

$$\dot{F} = G(F, \rho) \tag{9.6-28}$$

$$\dot{\rho} = \varepsilon H(F, \rho) \tag{9.6-29}$$

In order to find the possibly locally stable equilibria of (28) and (29), the following Jacobian matrix is considered

$$
J = \begin{bmatrix} \dfrac{\partial G}{\partial F} & \dfrac{\partial G}{\partial \rho} \\[2ex] \varepsilon \dfrac{\partial H}{\partial F} & \varepsilon \dfrac{\partial H}{\partial \rho} \end{bmatrix} \tag{9.6-30}
$$

The entries of the Jacobian matrix at the (A) equilibria are given by

$$
J = \begin{bmatrix} -\dfrac{1}{\Xi} \mathcal{R}_s^{-1} T \nabla_T^2 \mathcal{L} & \dfrac{\partial G}{\partial \rho} \\[2ex] 0 & \varepsilon \left[\mathcal{E}\{u^2(t)\} - c^2 \right] \end{bmatrix}
$$

This, being upper-block triangular with $\Xi > 0$, $\varepsilon > 0$, and $\mathcal{R}_s > 0$, corresponds to possibly locally stable equilibria, whenever $\nabla_T^2 \mathcal{L} \ge 0$ and $\mathcal{E}\{u^2(t)\} \le c^2$.

Stable (B) Equilibria. Stability analysis of (B) equilibria appears to be a difficult task because, in this case, the Jacobian matrix (30) need not be block diagonal. In such a case, we consider (28) and (29) for small positive reals ε. Then (28) and (29) can be regarded as a singularly perturbed differential system ([KKO86] and [Was65]) of which (28) and (29) describe the "fast" and the "slow" states, respectively.

Hereafter, the interest is directed to the (B) equilibria at which $\nabla^2 \mathcal{L} > 0$, viz., (B) equilibria that are isolated minima of the cost (8) for a fixed ρ_0. Any such a (B) equilibrium point will be denoted by $\beta = \begin{bmatrix} F_0' & \rho_0 \end{bmatrix}'$.

Because the plant to be regulated is linear and time-invariant, and, at every β point, the closed-loop system is asymptotically stable, the next property holds.

Property 9.6-1 Functions G and H in (28) and (29), respectively, are continuously differentiable in a neighborhood of β.

Because, for every β,

$$\frac{\partial G}{\partial F}\bigg|_{\beta} \propto -\nabla_T^2 \mathcal{L}\big|_{\beta} + O(\varepsilon)$$

where $\lim_{\varepsilon \to 0} O(\varepsilon) = 0$, for ε small enough,

$$\frac{\partial G}{\partial F}\bigg|_{\beta} < 0 \qquad\qquad (9.6\text{-}31)$$

Then the implicit function theorem [Zei85] assures that the following property is satisfied:

Property 9.6-2   In a neighborhood of ρ_0, the equation $G(f, \rho) = 0$ has an isolated solution $F = F(\rho)$, with $F(\cdot)$ continuously differentiable.

Setting $t := \varepsilon \tau$, (28) and (29) become

$$\varepsilon \frac{dF}{dt} = G(F, \rho) \qquad \text{and} \qquad \frac{d\rho}{dt} = H(F, \rho) \qquad (9.6\text{-}32)$$

Property 9.6-3   Consider the "reduced system"

$$\frac{d\rho}{dt} = H(F(\rho), \rho) = \rho \left[\bar{u}^2(\rho) - c^2 \right] \qquad (9.6\text{-}33)$$

Then ρ_0, such that $\bar{u}^2(\rho_0) =: \mathcal{E}\{u^2(t)\} = c^2$, is an isolated equilibrium point at which (33) is exponentially stable.

In order to prove Property 3, it will be shown by next Lemma 1 and Property 1 that the following input *MS monotonicity property* holds:

$$\frac{\partial H}{\partial \rho}\bigg|_{\rho_0} = \frac{\partial \bar{u}^2(\rho)}{\partial \rho}\bigg|_{\rho_0} < 0$$

Lemma 9.6-1.   *Let $\bar{u}^2(\rho)$ be the input MS value $\bar{u}^2(\rho) := \mathcal{E}\{u^2(\rho)\}$ corresponding to an isolated minimum of the stochastic steady-state quadratic cost $\mathcal{L}(F, \rho)$ for a given ρ. Then, $\bar{u}^2(\rho)$ is a strictly decreasing function of ρ in a neighborhood of ρ_0, ρ_0 being specified as in Property 2.*

Proof: Let ρ_1 and ρ_2, $\rho_2 > \rho_1$, be in a suitably small neighborhood of ρ_0. Let, according to Property 2, $F_i = F(\rho_i) = \arg\min_F \mathcal{L}(F, \rho_i)$, $i = 1, 2$. Further let $\bar{u}_i^2$, $\bar{y}_i^2$ denote the corresponding stochastic steady-state MS values of the input and the output, respectively. Then, proceeding as in the proof of Theorem 7.4-1, we get that $\rho_2 > \rho_1$ implies $\bar{u}_2^2 - \bar{u}_1^2 < 0$.

Property 9.6-4   Consider, for fixed ρ, the "boundary layer system":

$$\frac{dF}{d\tau} = G(F, \rho) \qquad\qquad (9.6\text{-}34)$$

Then (34) is exponentially stable at $F = F(\rho)$ uniformly in ρ in a suitable neighborhood of ρ_0.

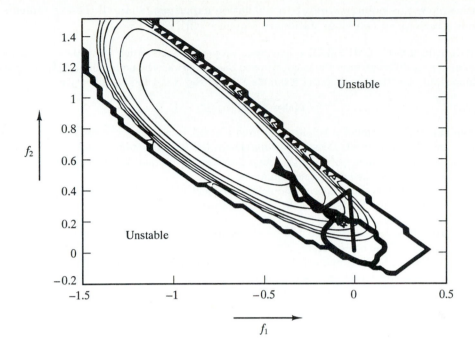

Figure 9.6-1: Superposition of the feedback time-evolution over the constant level curves of $\mathcal{E}\{y^2(t)\}$ and the set boundary of the allowed $\mathcal{E}\{u^2(t)\} \leq 0.1$ for CMUSMAR with $T = 5$ and the plant in Example 1.

Property 4 is fulfilled by virtue of (31). In fact, (31) implies, by Theorem 9.3 of [BN66], exponential stability at $F_0 = F(\rho)$. Next, (31), together with Property 2, implies that there exists a suitably small neighborhood of ρ_0 where the exponential stability referred to before is uniform in ρ.

Taking into account Properties 1 to 4, stability theory of singularly perturbed ODEs (cf. Corollary 7.2.3 of [KKO86]) yields the following conclusion.

Theorem 9.6-1. *Let the control horizon T of CMUSMAR be large enough w.r.t. the time constants of the closed-loop system, and $s(t)$ chosen so as to yield isolated minima of the cost. Then there exists an $\bar{\varepsilon} > 0$ such that, for every positive $\varepsilon < \bar{\varepsilon}$, any feedback gain solving Problem 1 (viz., minimizing the MS output value inside or along the boundary of the feasibility region $\mathcal{E}\{u^2(t)\} \leq c^2$) is a possible convergence point of CMUSMAR.*

Simulation Results. Proposition 1 and Theorem 1 suggest that CMUSMAR *may* possess nice convergence properties. However, there is no guarantee that CMUSMAR will actually be capable of converging to the desired points. In order to explore this point, we resort to simulation experiments. In all the experiments, e is a zero-mean Gaussian stationary sequence with $\mathcal{E}\{e^2\} = 1$.

Example 9.6-1   CMUSMAR convergence properties are studied when the constrained minimum is different from the unconstrained one. Consider the nonminimum-phase open-loop stable fourth-order plant of Example 5-4 and the restricted complexity controller

$$u(t) = f_1 y(t) + f_2 y(t-1)$$

The Lagrange multiplier ρ is initialized from a small value, viz., $\rho = 10^{-4}$, and $T = 5$ is the control horizon used. Because ρ grows slowly, the feedback gain initially approach the unconstrained minimum of $\mathcal{E}\{y^2(t)\}$. As ρ converges to its final value, the gains converge to a point close to the constrained minimum.

Figure 1 shows the superposition of the feedback gains with the constant-level curves of $\mathcal{E}\{y^2(t)\}$ and the boundary of the region defined by the restriction (6) with $c^2 = 0.1$.

Example 9.6-2   As referred before, to ensure that CMUSMAR has the constrained local minima of the steady-state LQS regulation cost as possible convergence points, the horizon T must be large enough. In this example, the plant of Example 7.4-1 is used, for which the control based on a single-step cost function ($T = 1$) greatly differs from the steady-state LQS regulation. Consider the plant:

$$y(t+3) - 2.75y(t+2) + 2.61y(t+1) + 0.855y(t)$$
$$= u(t+2) - 0.5u(t+1) + e(t+3) - 0.2e(t+2) + 0.5e(t+1) - 0.1e(t)$$

and the full complexity controller defined by

$$s(t) = \begin{bmatrix} y(t) & y(t-1) & y(t-2) & u(t-1) & u(t-2) & u(t-3) \end{bmatrix}'$$

As shown in Example 7.4-1, using $T = 1$ and taking ρ as a parameter, this plant gives rise to a relationship between $\mathcal{E}\{u^2(t)\}$, and $\mathcal{E}\{y^2(t)\}$, which is not monotone (Fig. 7.4-1). Instead, for steady-state LQS regulation, the relationship is monotone as guaranteed by Theorem 7.4-1. It can be seen from Fig. 7.4-1 that, in this example, the single-step-ahead constrained self-tuner has two possible equilibrium points denoted by B and D in Fig. 7.4-1. Both of them correspond to the same value of the input variance but to quite different values of the output variance. These equilibria can be attained depending on the initial conditions. This unpleasant phenomenon is eliminated by increasing the value of T in CMUSMAR (Fig. 2). In fact, the dashed line in Fig. 7.4-1, which exhibits a monotonic behavior and corresponds to steady-state LQS regulation, is already tightly approached for $T = 2$.

9.6.2   Implicit Adaptive MKI: MUSMAR-∞

So far no extension of the celebrated weak self-tuning property of the RLS+MV adaptive regulator [ÅW73] was shown to exactly hold for steady-state LQS regulation. In this connection, however, the MUSMAR algorithm represents almost an exception, because it exhibits approximately the weak self-tuning property, the approximation becoming sharper as $T \rightarrow \infty$. Hereafter, we pose the following question:

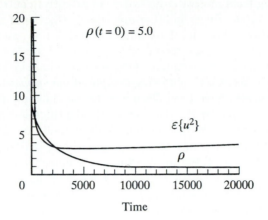

Figure 9.6-2: Time evolution of ρ and $\mathcal{E}\{u^2(t)\}$ for CMUSMAR with $T = 2$ and the plant of Example 2.

Is it possible to adaptively get the steady-state LQS regulation for any CARMA plant by using a finite number of predictors whose parameters are estimated by standard RLS?

An adaptive regulation algorithm solving this problem is considered. It embodies a standard RLS separate identification of the parameters of $T \geq n + 1$ predictors of the I/O joint process, n being the order of the CARMA plant, together with an appropriate control synthesis rule. The proposed algorithm turns out to be a modified version of MUSMAR performing spread-in-time MKI (cf. Sec. 5.7).

In Sec. 5 it was shown that MUSMAR possible convergence points are close approximations to the minima of the adopted unconditional quadratic cost, even under mismatching conditions, provided that the prediction horizon T is chosen large enough. More precisely, T should be chosen such that

$$\left| \lambda_M^{2(T-\ell)} \right| \ll 1$$

where $\ell + 1$ is the plant I/O delay, and λ_M the eigenvalue with maximum modulus of the closed-loop system (cf. (5-57)). This implies that when $|\lambda_M|$ is only slightly less than 1, T must be very large so as to let all the transients decay within the prediction horizon. When $|\lambda_M|$ is a priori unknown, there is no definite criterion for a suitable a priori choice of T. In practice, T is chosen as a compromise between the algorithm computational complexity, which increases with T, and the stabilization requirement for generic unknown plants. In fact, the latter would impose, in principle, $T = \infty$, and hence an unacceptable computational load, as well as an irrealizable implementation.

The foregoing facts motivate the search for adaptive control algorithms based on a finite number of identifiers and that may yield a tight approximation to steady-state LQS regulation. For the deterministic case, [SF81] and [OK87] proposed schemes in which a state-space model of the plant is built upon estimates of the one-step-ahead predictor. An estimate of the state provided by an adaptive observer is then fed back, the feedback gain being computed via spread-in-time Riccati iterations. Similar schemes have been developed for stochastic plants [Pet86]. When CARMA plants are considered, RELS or RML identification algorithms must be used. This has the drawback that the inherent simplicity of standard RLS is lost. Here, "simplicity" refers not only to the computational burden, but mainly to the fact that both RELS and RML involve highly nonlinear operations in that their regressor depends not only on the current experimental data, but also on previous estimates. Along this line, it is interesting to establish whether the tuning properties of the classical RLS+MV self-tuning regulator can be extended to RLS+LQS adaptive regulation.

Given the previous motivations, the problem that we shall consider is the following:

> *Is it possible to suitably modify the MUSMAR algorithm so as to adaptively get steady-state LQS regulation for any CARMA plant by using a small number of predictors whose parameters are separately estimated via standard RLS estimators?*

Problem Formulation.   The SISO CARMA plant (1) is considered. As usual, the order of the plant is denoted by n:

$$n = \max\{\partial A(d), \partial B(d), \partial C(d)\}$$

Associated with the plant, we consider a quadratic cost defined over an NT-step horizon:

$$\mathcal{E}\left\{L_N(t)\right\} = \frac{1}{NT}\mathcal{E}\left\{\sum_{k=t}^{t+NT-1}\left[y^2(k+1) + \rho u^2(k)\right]\right\} \qquad (9.6\text{-}35)$$

with $\rho > 0$, to be minimized in a receding-horizon sense. In the sequel, it will become clear why in (35) the prediction horizon is denoted by NT. In fact, it will be convenient to increase the regulation horizon by holding T fixed and letting N become larger.

Our goal is to find a convenient procedure by which to adaptively select the input $u(t)$ to the plant (1) minimizing the cost (35) as $N \to \infty$, subject to the following requirements:

- The feedback is updated on the grounds of the predictor parameters estimated by standard RLS algorithms.

- The number of operations per single cycle does not grow with N.

The following receding-horizon scheme, for any $T \geq n+1$, is considered to possibly achieve the stated goals.

Algorithm 9.6-2 (MUSMAR-∞ Algorithm)

(i) Given all I/O data up to time t, compute RLS estimates of the parameter matrices Ψ and Θ in the following set of prediction models:

$$z(t) = \Psi z(t-T) + \Theta u(t-T) + \zeta(t) \tag{9.6-36}$$

where $\zeta(t)$ denotes a residual vector uncorrelated with both $z(t-T)$ and $u(t-T)$,

$$z(t) \quad := \quad \left[\begin{array}{cc} s'(t) & \gamma'(t) \end{array} \right]'$$

$$s(t) \quad := \quad \left[\begin{array}{cc} (y_{t-n+1}^t)' & (u_{t-n}^{t-1})' \end{array} \right]'$$

$$\gamma(t) \quad := \quad \left[\begin{array}{cc} (y_{t-T+1}^{t-n})' & (u_{t-T}^{t-n-1})' \end{array} \right]'$$

Ψ is a $2T \times 2T$ matrix, and Θ is a $2T \times 1$ vector such that the bottom row of Ψ is zero, the bottom element of Θ is 1, and the last $2(T-n)$ columns of Ψ are zero.

(ii) Update the matrix of weights P by the difference pseudo-Riccati equation (cf. (5.7-9)):

$$P = \Psi_F' \bar{P} \Psi_F + \Omega \tag{9.6-37}$$

where

$$\Omega := \operatorname{diag} \{ \underbrace{1 \cdots 1}_{n}, \ \underbrace{\rho \cdots \rho}_{n}, \ \underbrace{1 \cdots 1}_{T-n}, \ \underbrace{\rho \cdots \rho}_{T-n} \}$$

$$\Psi_F := \Psi + \Theta \bar{F}' \tag{9.6-38}$$

and $\bar{P}$ and $\bar{F}$ are respectively the matrix of weights and the feedback vector used at time $t-1$.

(iii) Update the augmented vector of feedback gains by

$$F = -\left(\Theta' P \Theta\right)^{-1} \Psi' P \Theta \tag{9.6-39}$$

with Θ and Ψ replaced by their current estimates, and then apply the control at time t given by

$$u(t) = F_s' s(t) \tag{9.6-40}$$

where F_s is made up by the first $2n$ components of F.

(iv) Set $\bar{P} = P$, $\bar{F} = F$, sample the output of the plant, and go to step (i) with t replaced by $t+1$.

Remark 9.6-2. The estimation of the parameters in (36) is performed by first updating RLS estimates of the parameters θ_i, Γ_i, $i = 1, \cdots, T$; and μ_i, Λ_i, $i = 2, \cdots, T$, in the following set of prediction models (cf. (3-11)):

$$
\begin{aligned}
y(t - T + i) &= \theta_i u(t - T) + \Gamma_i' s(t - T) + \epsilon_i(t - T + i) & \text{(9.6-41a)} \\
u(t - T + i - 1) &= \mu_i u(t - T) + \Lambda_i' s(t - T) + \nu_i(t - T + i - 1) & \text{(9.6-41b)}
\end{aligned}
$$

This is accomplished with the formulas

$$
\begin{bmatrix} \theta_i(t) \\ \Gamma_i(t) \end{bmatrix} = \begin{bmatrix} \theta_i(t-1) \\ \Gamma_i(t-1) \end{bmatrix} + K(t - T) \qquad\qquad \text{(9.6-42a)}
$$
$$
\times \; [y(t - T + i) - \theta_i(t - 1)u(t - T) - \Gamma_i'(t - 1)s(t - T)]
$$

$$
\begin{bmatrix} \mu_i(t) \\ \Lambda_i(t) \end{bmatrix} = \begin{bmatrix} \mu_i(t-1) \\ \Lambda_i(t-1) \end{bmatrix} + K(t - T) \qquad\qquad \text{(9.6-42b)}
$$
$$
\times \; [u(t - T + i - 1) - \mu_i(t - 1)u(t - T) - \Lambda_i(t - 1)s(t - T)]
$$

$$
\varphi(t - T) = \begin{bmatrix} u(t - T) & s'(t - T) \end{bmatrix}'
$$
$$
K(t - T) = P(t - T + 1)\varphi(t - T) \qquad\qquad\qquad \text{(9.6-42c)}
$$
$$
P^{-1}(t - T + 1) = P^{-1}(t - T) + \varphi(t - T)\varphi'(t - T) \qquad \text{(9.6-42d)}
$$

In the previous equations, θ_i and μ_i are scalars, Γ_i and Λ_i are column vectors of dimension $2n$, and $\epsilon_i(t + i)$ and $\nu_i(t + i)$ are uncorrelated with both $u(t)$ and $s(t)$. Note that because the regressor $\varphi(t - T)$ is the same for all the models in the RLS algorithms, there is only the need to update *one* covariance matrix of dimension $2n + 1$. As pointed out for MUSMAR, this considerably reduces the numerical complexity of the algorithm. The estimates of matrices Ψ and Θ are of the form

$$
\Psi' = \qquad\qquad\qquad\qquad\qquad\qquad\qquad\qquad\qquad\qquad\qquad \text{(9.6-43a)}
$$

$$
\overbrace{\begin{bmatrix} \Gamma_T \cdots \Gamma_{T-n+1} & \Lambda_T \cdots \Lambda_{T-n+1} & \Gamma_{T-n} \cdots \Gamma_1 & \Lambda_{T-n} \cdots \Lambda_2 & 0 \\ \hline & & 0 & & \end{bmatrix}}^{2T} \begin{array}{l} \left.\right\} 2n \\[1.5em] \left.\right\} 2(T-n) \end{array}
$$

$$
\Theta = \begin{bmatrix} \theta_T \cdots \theta_{T-n+1} & \mu_T \cdots \mu_{T-n+1} & \theta_{T-n} \cdots \theta_1 & \mu_{T-n} \cdots \mu_2 & 1 \end{bmatrix}' \quad \text{(9.6-43b)}
$$

Remark 9.6-3. Vector F has dimension $2T$. Given the structure of Ψ, with zeros on the last $2(T - n)$ columns, the last $2(T - n)$ entries of F are also zero.

Justification of MUSMAR-∞. We show that, under suitable assumptions, the steady-state LQS regulation feedback is an equilibrium point of MUSMAR-∞. Some required results drawn from Theorem 3-1 are summed up in the following lemma.

Lemma 9.6-2. *Let the inputs to the CARMA plant (1) be given by*

$$u(k) = F's(k) \tag{9.6-44}$$

or, equivalently, for suitable polynomials $R(d)$ and $S(d)$ by

$$R(d)u(k) = -S(d)y(k) \tag{9.6-45}$$

Let R and S be coprime and such that the closed-loop d-characteristic polynomial $\chi(d) = A(d)R(d) + B(d)S(d)$ be strictly Hurwitz and divided by $C(d)$:

$$C(d) \mid \chi(d) \tag{9.6-46}$$

Then if the foregoing conditions are fulfilled for

$$k = t - n, \cdots, t - 1, t + 1, \cdots, t + T - 1 \tag{9.6-47}$$

$z(t + T)$ has the following representation:

$$z(t + T) = \Psi z(t) + \Theta u(t) + \zeta(t + T) \tag{9.6-48}$$

where

$$\zeta(t + T) \in \text{Span} \left\{ e_{t+T}^{t+1} \right\} \tag{9.6-49}$$

Remark 9.6-4. The parameters in (48) characterize the dynamics of the closed-loop system. Therefore, they depend on the feedback-gain polynomials $R(d)$ and $S(d)$, as well as on the plant and disturbance dynamics, defined by polynomials $A(d)$, $B(d)$, and $C(d)$.

The interest in (48) is that (35), which may be written as

$$\mathcal{E}\{J_N(t)\} = \frac{1}{NT}\mathcal{E}\left\{ \sum_{i=1}^{N} \|z(t + iT)\|_{\Omega}^2 \right\} \tag{9.6-50}$$

can be easily minimized w.r.t. $u(t)$ provided that Ψ and Θ are known and suitable assumptions, to be discussed next, are made on the magnitude of T and past and future inputs. In fact, if (44) to (47) are assumed, (48) expresses $z(t+T)$ in terms of $u(t)$. Next,

$$z(t + iT) = \Psi z(t + (i-1)T) + \Theta u(t + (i-1)T) + \zeta(t + iT) \tag{9.6-51}$$

also for all $i \geq 2$ if inputs $u(k)$ are given by the previous feedback law for

$$k = t - n + (i-1)T, \cdots, t + iT - 1 \tag{9.6-52}$$

Because (50) has to be minimized w.r.t. $u(t)$, $u(t)$ must be left unconstrained. By taking $i = 2$, it is seen that all the inputs between $t + (-n + T)$ and $t + 2T - 1$

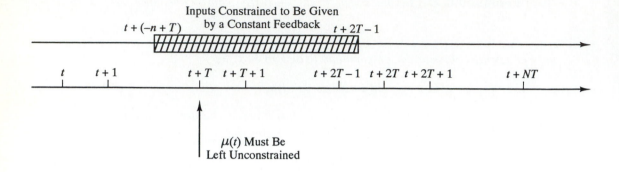

Figure 9.6-3: Illustration of the minorant imposed on T.

(the shaded interval in Fig. 3) must be given by a constant feedback. Because $u(t)$ must be left unconstrained, this implies (cf. Fig. 3)

$$T \geq n+1 \tag{9.6-53}$$

Clearly, according to the definition of n, (53) already comprises the plant I/O delay. The previous considerations are summed up in the following lemma.

     Lemma 9.6-3.   *Let assumptions (44) to (46) be fulfilled for*

$$k = t - n, \cdots, t - 1, t + 1, \cdots, t + NT - 1$$

Then, if T satisfies (53), $z(t+iT)$, $i = 1, 2, \cdots, N$, has the state-space representation (51) irrespective of the plant $C(d)$ innovations polynomial and the value taken on by $u(t)$.

     Remark 9.6-5.   Inequality (53) specifies in terms of the plant order n, the minimal dimension of the state z required to carry out in a correct way the minimization procedure under consideration.

     Thus, assuming (53), (51) can be used for all $i \geq 1$. For $i \geq 2$, using (44) in (51), one has

$$
\begin{aligned}
z(t + iT) &= \Psi_F z(t + (i - 1)T) + \zeta(t + iT) \\
&= \bar{z}(t + iT) + \tilde{z}(t + iT)
\end{aligned}
\tag{9.6-54}
$$

where Ψ_F is as in (38), $\bar{z}(t + iT)$ is the zero-input response from the initial state $z(t+2T)$, and $\tilde{z}(t+iT)$ is the response due to $\zeta(t+iT)$ from the zero state at time $t + 2T$. Thus, taking into account that

$$\mathcal{E}\left\{\bar{z}(t + iT)\tilde{z}'(t + iT)\right\} = 0 \tag{9.6-55}$$

and denoting (50) by $\mathcal{C}_N(t, F)$, so as to point out that past and future inputs are given by a constant feedback, one has

$$\mathcal{C}_N(t, F) = \frac{1}{NT}\mathcal{E}\left\{\|z(t+T)\|_\Omega^2 + \|z(t+2T)\|_{P(N)}^2 + V_N(t, F)\right\} \qquad (9.6\text{-}56)$$

where

$$V_N(t, F) = \sum_{i=3}^{N} \|\tilde{z}(t+iT)\|_\Omega^2$$

is not affected by $u(t)$, and

$$P(N) = \Omega + \sum_{i=1}^{N-2} \left(\Psi_F'\right)^i \Omega \Psi_F^i$$

satisfies the following Lyapunov-type equation:

$$P(N) = \Psi_F' P(N)\Psi_F + \Omega - \Delta(N) \qquad (9.6\text{-}57)$$

with $\Delta(N) := \left(\Psi_F^{N-1}\right)' \Omega \Psi_F^{N-1}$. Thus, the first two additive terms in (56) equal

$$\mathcal{E}\left\{\|z(t+T)\|_\Omega^2 + \|\Psi_F z(t+T)\|_{P(N)}^2\right\} = \mathcal{E}\left\{\|z(t+T)\|_{P(N)+\Delta(N)}^2\right\}$$

Consequently, $\arg\min_{u(t)} \mathcal{C}_N(t, F) = \hat{F}'(N)z(t)$ with

$$\hat{F}'(N) = -\left\{\Theta'\left[P(N) + \Delta(N)\right]\Theta\right\}^{-1} \Theta'\left[P(N) + \Delta(N)\right]\Psi \qquad (9.6\text{-}58)$$

We now consider the minimization of $\mathcal{C}_N(t, F)$ w.r.t. $u(t)$ as $N \to \infty$. With Ψ_F a stability matrix, we can define

$$\mathcal{C}_\infty(t, F) := \lim_{N \to \infty} \mathcal{C}_N(t, F) \qquad (9.6\text{-}59)$$

Because

$$P(N) + \Delta(N) > P(N) > \Omega \geq \epsilon I$$

with $0 < \epsilon \leq \min(1, \rho)$, $\hat{F}(N)$ is a continuous function of $P(N) + \Delta(N)$. Consequently, because $P(N) + \Delta(N) \to P$ as $N \to \infty$, one has

$$\hat{F}' := \lim_{N \to \infty} \hat{F}'(N) = -\left[\Theta' P\Theta\right]^{-1} \Theta' P\Psi \qquad (9.6\text{-}60)$$

with P the solution of the following Lyapunov equation:

$$P = \Psi_F' P\Psi_F + \Omega \qquad (9.6\text{-}61)$$

Theorem 9.6-2.   *Under the same assumptions as in Lemma 3, the input at time t minimizing $\mathcal{C}_\infty(t, F)$ in (59) is given by $u(t) = \hat{F}'z(t)$, with $\hat{F}$ specified by (60) and (61). Further, if the procedure used to generate $\hat{F}$ from F is iterated, the steady-state LQS regulation feedback is an equilibrium point for the resulting iterations.*

Proof: The validity of the last assertion follows from the properties of Kleinman iterations (cf. Sec. 2.5).

By the structure of matrix Ψ, the last $T-n$ elements of $\hat{F}$ are zero, and thus (40) holds. Further, in order to circumvent difficulties associated with possible feedback vectors temporarily making the closed-loop system unstable, and hence the Lyapunov equation (61) meaningless, in MUSMAR-∞, P is updated via the pseudo-Riccati difference equation (37). This change makes MUSMAR-∞ underlying control law a stochastic variant of spread-in-time MKI, as discussed in Sec. 5.7.

Algorithmic Considerations.   Matrix Ψ and vector Θ can be partitioned in the following blocks:

$$\Psi = \begin{bmatrix} \Psi_s & 0 \\ \Psi_\gamma & 0 \end{bmatrix} \begin{matrix} \scriptstyle 2n \\ \scriptstyle 2(T-n) \end{matrix}$$

$$\Theta = \begin{bmatrix} \Theta_s \\ \Theta_\gamma \end{bmatrix} \begin{matrix} \scriptstyle 2n \\ \scriptstyle 2(T-n) \end{matrix}$$

with the bottom elements of Ψ_γ and Θ_γ 0 and 1, respectively, that is, the prediction model (36) does not impose any constraint on $u(t)$.

Let P be partitioned accordingly:

$$P = \begin{bmatrix} P_s & \tilde{P} \\ \tilde{P}' & P_\gamma \end{bmatrix} \begin{matrix} \scriptstyle 2n \\ \scriptstyle 2(T-n) \end{matrix} \qquad (9.6\text{-}62)$$

and the same for $\bar{P}$. Then a simple calculation shows that (37) and (39) can be simplified as follows:

$$P_s = \left(\Psi_s + \Theta_s \bar{F}_s\right)' \bar{P}_s \left(\Psi_s + \Theta_s \bar{F}_s\right) + \Omega_s \qquad (9.6\text{-}63)$$

and

$$F_s' = -\frac{\Theta_s' P_s \Psi_s + \Theta_\gamma' \Omega_\gamma \Psi_\gamma}{\Theta_s' P_s \Theta_s + \Theta_\gamma' \Omega_\gamma \Theta_\gamma} \qquad (9.6\text{-}64)$$

with

$$\Omega_s := \operatorname{diag}\{\underbrace{1\cdots 1}_{n}, \underbrace{\rho\cdots\rho}_{n}\}$$

$$\Omega_\gamma := \operatorname{diag}\{\underbrace{1\cdots 1}_{T-n}, \underbrace{\rho\cdots\rho}_{T-n}\}$$

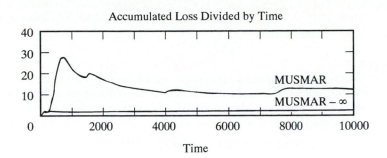

Accumulated Loss Divided by Time

Figure 9.6-4: The accumulated loss divided by time when the plant of Example 3 is regulated by MUSMAR $(T = 3)$ and MUSMAR-∞ $(T = 3)$.

and $\bar{P}_s$ initialized by $\bar{P}_s = \Omega_s$.

The algorithm assumes $\rho > 0$. This in practice constitutes no restriction because ρ may be made as small as needed.

Simulation Results. Some examples are considered in order to show the features of the MUSMAR-∞ algorithm. Comparisons are made with an indirect steady-state LQS adaptive controller (ILQS) based on the same underlying control problem as MUSMAR-∞, the difference is that the former identifies the usual one-step-ahead prediction model and computes the steady-state LQS regulation law indirectly. Both full and reduced complexity controllers are considered, the aim being to show that MUSMAR-∞ is capable of stabilizing plants requiring long prediction horizons, a feature due to its underlying regulation law, and still capable of retaining good performance robustness thanks to its parallel identification scheme.

Example 9.6-3

A second-order nonminimum-phase plant is adaptively regulated by MUSMAR and MUSMAR-∞. The plant to be regulated is

$$y(t + 1) - 1.5y(t) + 0.7y(t - 1) = u(t) - 1.01u(t - 1) + e(t + 1)$$

with input weight $\rho = 0.1$ in the performance index (35). Here, and in the following examples, e is a sequence of independent Gaussian random variables with zero mean and unit variance.

Figure 4 compares the accumulated loss divided by time, viz.,

$$\frac{1}{t} \sum_{i=1}^{t} \left[y^2(i) + \rho u^2(i - 1) \right]$$

for MUSMAR $(T = 3)$ and MUSMAR-∞ $(T = 3)$. In both cases, a full complexity controller is used. Because this plant has a nonminimum-phase zero very close to the stability boundary, the prediction horizon T must be very large in order for MUSMAR to behave close to the optimal performance. For $T = 3$ (a value chosen according to the rule $T = n + 1$), MUSMAR-∞ yields a much smaller cost than MUSMAR, exhibiting a loss very close to the optimal one.

Example 9.6-4
Because MUSMAR-∞ is based on a state-space model built upon separately estimated predictors, it turns out, according to Sec. 3, not to be affected by a $C(d)$ polynomial different from 1 in the CARMA plant representation. In this example, the following plant with a highly correlated equation error is considered:

$$y(t + 1) = u(t) + e(t + 1) - 0.99e(t) \qquad (9.6\text{-}65)$$

For $\rho = 1$, MUSMAR-∞ converges to $F_s = [\ 0.492 \quad 0.494\ ]'$, with the optimal feedback $F_s^* = [\ 0.495 \quad 0.495\ ]$.

Example 9.6-5
MUSMAR was shown to be robustly self-optimizing in the sense that, if T is large enough, and MUSMAR converges, it converges to the minima of the cost constrained to the chosen regulator regressor. MUSMAR-∞ is expected to inherit this robustness property because it is based on a set of implicit prediction models whose parameters are separately esti-mated. Consider the open-loop unstable plant:

$$y(t + 1) + 0.9y(t) - 0.5y(t - 1) = u(t) + e(t + 1) - 0.7e(t) \qquad (9.6\text{-}66)$$

Although the plant is of second-order, and hence its pseudo-state is

$$\begin{bmatrix} y(t) & y(t - 1) & u(t - 1) & u(t - 2) \end{bmatrix}$$

$s(t)$ is instead chosen to be

$$s(t) = \begin{bmatrix} y(t) \\ u(t - 1) \end{bmatrix} \qquad (9.6\text{-}67)$$

The optimal feedback constrained to this choice of $s(t)$ is, for $\rho = 1$,

$$F_s^* = [\ 1.147 \quad -0.109\ ]$$

Figure 5 shows the time evolution of the feedback when this plant is regulated by MUSMAR-∞ on the space $\begin{bmatrix} f_1 & f_2 \end{bmatrix}$, superimposed on the level curves of the quadratic cost, con-strained to the chosen regulator regressor. As is apparent, MUSMAR-∞ is able to tune closely on the minimum of the underlying cost, despite the presence of unmodeled plant dynamics.

Example 9.6-6
This example shows the importance of the separate estimation of the predictor parameters in MUSMAR-∞. A comparison is made with ILQS.
    Consider the fourth-order nonminimum-phase open-loop stable plant:

$$y(t + 4) - 0.167y(t + 3) - 0.74y(t + 2) - 0.132y(t + 1) + 0.87y(t)$$
$$= 0.132u(t + 3) + 0.545u(t + 2) + 1.117u(t + 1) + 0.262u(t) + e(t + 4)$$

Figures 6 and 7 show the results obtained when a reduced complexity regulator is used for this plant:
$$u(t) = f_1 y(t) + f_2 y(t - 1) + f_3 u(t - 1) + f_4 u(t - 2)$$

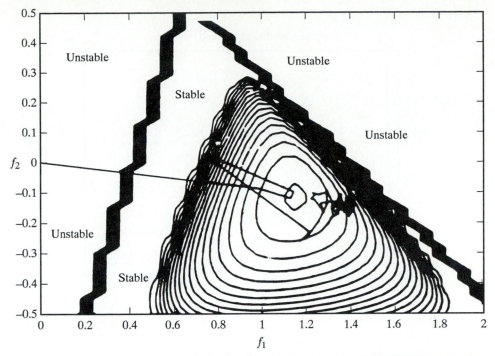

Figure 9.6-5: The evolution of the feedback calculated by MUSMAR-∞ in Example 5, superimposed to the level curves of the underlying quadratic cost.

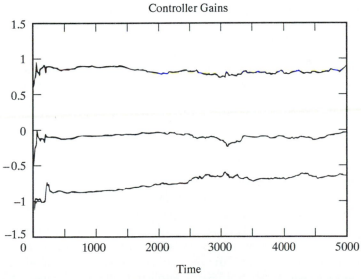

Figure 9.6-6: Convergence of the feedback when the plant of Example 6 is controlled by MUSMAR-∞.

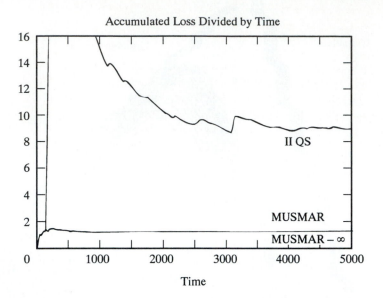

Figure 9.6-7: The accumulated loss divided by time when the plant of Example 6 is controlled by ILQS, standard MUSMAR ($T = 3$) and MUSMAR-∞ ($T = 3$).

and $\rho = 10^{-4}$. Figure 6 shows the time evolution of the first three components of the feedback when MUSMAR-∞ is used. Figure 7 shows the accumulated loss divided by time when ILQS, MUSMAR ($T = 3$), and MUSMAR-∞ ($T = 3$) are used. Although both MUSMAR-∞ and ILQS yield the steady-state LQS feedback under model-matching conditions, due to the presence of unmodeled dynamics, ILQS presents a large detuning. MUSMAR-∞, instead, being based on a multipredictor model, tends to be insensitive to plant unmodeled dynamics.

Main Points of the Section. Implicit modeling theory can be further exploited so as to construct extensions of the MUSMAR algorithm. The first extension, CMUSMAR, embodies a mean-square input value constraint. ODE analysis and singular perturbation methods show that when suitable provisions are taken, the local *constrained* minima of the underlying quadratic performance index are possible convergence points of CMUSMAR, also in the presence of unmodeled dynamics. In the second extension, MUSMAR-∞, implicit modeling theory is blended with spread-in-time MKI so as to realize an implicit adaptive regulator for which the steady-state LQS regulation feedback is an equilibrium point.

NOTES AND REFERENCES

Adaptive LQ controllers have been considered in [Gri84], [OK87], [Pet84], [Sam82], and [SF81]. The basic pole assignment approach to self-tuning has been discussed in [PW81], [WEPZ79], and [WPZ79]. In contrast with CC and MV control, both LQ and pole assignment control require the fulfillment of a stabilizability condition ([LG85] and [DL84]) for the identified model. This can be ensured by using a projection facility to constrain the estimated parameters in a convex set containing the unknown parameters and such that every element of the set satisfies the stabilizability condition required for computing the control law. Although the existence of such a set can be postulated for theoretical developments, in practice, the definition of such convex sets for higher-order plants is complicated and sometimes unfeasible. An alternative approach to deal with the stabilizability condition is to suitably enforce persistency of excitation as in [Cri87], [ECD85], and [Kre89]. We mainly borrow similar ideas for constructing, by using a self-excitation mechanism, the globally convergent adaptive SIORHC for the ideal case described in Sec. 1 ([MZ93] and [MZB93]). Along similar lines, we construct the robust adaptive SIORHC for the bounded disturbance case by using a constant trace normalized RLS with a dead zone and a suitable self-excitation mechanism.

In the neglected dynamics case, combinations of data normalization, relative dead zones, and projection of estimates onto convex sets have been proposed by many authors. For example, see [CMS91] and [MGHM88]. In [WH92], it is shown that in the presence of neglected dynamics, projection of the estimates suffices to get global boundedness for adaptive pole assignment. An attempt to use the persistency of excitation approach in the neglected dynamics case is described in [GMDD91]. The use of both low-pass prefiltering of I/O data for identification and high-pass dynamic weights for control synthesis is in practice of vital importance in the presence of neglected dynamics. For a description of the first of these concepts expressly tailored for GPC, see [RC89], [SMS91], and [SMS92].

The presentation in Sec. 2 of implicit multistep prediction models of linear-regression type is a simplified version of [MZ85] and [MZ89c], where the main results on this topics were first presented. See also [CDMZ87] and [CMP91]. For the difficulties with the self-tuning approach for general cost criteria, see also [LKS85]. Though we use the notion of implicit linear-regression prediction models so as to provide a motivated derivation of MUSMAR, the introduction of the latter ([GMMZ84], [MM80], [Mos83], and [MZ83]) preceded the discovery of the preceding implicit modeling property. Ever since its introduction, a great deal of simulative and application experience in case studies [GIM+90] has revealed MUSMAR self-optimizaing properties as a reduced-complexity adaptive controller. The local analysis of MUSMAR self-optimizing properties in Sec. 5 appeared in [MZL89]. See also [Mos83], [MZ84], and [MZ87]. This study is based on the ODE method for analyzing stochastic recursive algorithms ([Lju77a] and [Lju77b]). See also [ABJ+86] for nonstochastic averaging methods.

The MUSMAR algorithm with mean-square input constraint was reported in [MLMN92] and its analysis is based on some results of singular perturbation theory of ODEs ([KKO86] and [Was65]). MUSMAR-∞, the implicit adaptive LQG algorithm based on spread-in-time modified Kleinman iterations was introduced in [ML89].

SOME RESULTS FROM LINEAR SYSTEMS THEORY

In this appendix, we briefly review some results from linear systems theory used in this book. For more extensive treatments, standard textbooks, for example, [Bro70], [CD91], [Che70], [Des70a], [Kai80], [PA74], and [ZD63], should be consulted.

A.1 STATE-SPACE REPRESENTATIONS

A discrete-time dynamic linear system is represented in state-space form as follows:

$$
\left.
\begin{aligned}
x(k+1) &= \Phi(k)x(k) + G(k)u(k) \\
y(k) &= H(k)x(k) + J(k)u(k)
\end{aligned}
\right\}
\tag{A.1-1}
$$

where $k \in \mathbb{Z}$; $x(k) \in \mathbb{R}^n$ is the system *state* at time k; $u(k) \in \mathbb{R}^m$ and $y(k) \in \mathbb{R}^p$ are the system *input* and *output*, respectively, at time k; n is the system *dimension*; and $\Phi(k)$, $G(k)$, $H(k)$, and $J(k)$ are matrices with real entries of compatible dimensions.

The basic idea of state is that it contains all the information on the past history of the system relevant to describe its future behavior in terms of the present and future inputs only. In fact from (1), we can compute the state $x(j)$ given the state $x(k)$, $k \leq j$, in terms of inputs $u_{[k,j)}$ only:

$$
\begin{aligned}
x(j) &= \varphi\left(j, k, x(k), u_{[k,j)}\right) \\
&:= \Phi(j,k)x(k) + \sum_{i=k}^{j-1} \Phi(j,i+1)u(i)
\end{aligned}
\tag{A.1-2}
$$

where

$$
\Phi(j,k) :=
\begin{cases}
I_n, & j = k \\
\Phi(j-1)\cdots\Phi(k), & j > k
\end{cases}
\tag{A.1-3}
$$

is the system *state-transition matrix*, and $\varphi\left(j, k, x(k), u_{[k,j)}\right)$ is the *global state-transition function*. Note that by linearity of the system, the latter is the superposition of the *zero-input response* $\varphi\left(j, k, x(k), O_{\Omega}\right)$ with the *zero-state response* $\varphi\left(j, k, O_X, u_{[k,j)}\right)$

$$\varphi\left(j, k, x(k), u_{[k,j)}\right) \;\; = \;\; \varphi\left(j, k, x(k), O_{\Omega})\right) + \varphi\left(j, k, O_X, u_{[k,j)}\right) \tag{A.1-4}$$

$$\varphi\left(j, k, x(k), O_{\Omega}\right) \;\; := \;\; \Phi(j,k)x(k) \tag{A.1-5}$$

$$\varphi\left(j, k, O_X, u_{[k,j)}\right) \;\; := \;\; \sum_{i=k}^{j-1} \Phi(j, 1+1)u(i) \tag{A.1-6}$$

In the previous equations, O_{Ω} denotes the zero input sequence.

Similar superposition properties hold for the system output response. In particular, if the system is time-invariant, that is,

$$\Phi(k) \equiv \Phi \qquad G(k) \equiv G \qquad H(k) \equiv H \qquad J(k) \equiv J \quad , \quad \forall k \in \mathbb{Z} \tag{A.1-7}$$

we have, for $i \in \mathbb{Z}_+$:

$$y(k+i) = S_i x(k) + \sum_{j=0}^{i} w(j)u(k+i-j) \tag{A.1-8}$$

where

$$S_i := H\Phi^i \tag{A.1-9}$$

and

$$w(j) := \begin{cases} J, & j = 0 \\ H\Phi^{j-1}G, & j \ge 1 \end{cases} \tag{A.1-10}$$

is the jth sample of the system *impulse response* $W := \{w(j)\}_{j=1}^{\infty}$.

For the time-invariant system, $\Sigma = (\Phi, G, H, J)$, the following definitions and results apply.

- A state $\hat{x}$ is *reachable* from a state $\tilde{x}$ if there exists an input sequence $u_{[0,N)}$ of finite length N that can drive the system from the initial state $\tilde{x}$ to the final state $\hat{x}$:

$$\hat{x} = \varphi\left(N, 0, \tilde{x}, u_{[0,N)}\right)$$

- Σ is said to be *completely reachable*, or (Φ, G) a *reachable pair*, if every state is reachable from any other state. This happens to be true if and only if every state is reachable from state vector O_X.

Theorem A.1-1. $\Sigma = (\Phi, G, H, J)$ *is completely reachable if and only if*

$$\text{rank}\, R = n = \dim \Sigma \tag{A.1-11}$$

where

$$R := \begin{bmatrix} G & \Phi G & \cdots & \Phi^{n-1}G \end{bmatrix} \tag{A.1-12}$$

is the reachability matrix *of* Σ.

Theorem A.1-2. (GK Canonical Reachability Decomposition)
Consider system Σ with reachability matrix R such that

$$\operatorname{rank} R = n_r < n = \dim \Sigma$$

Then there exist nonsingular matrices M that transform Σ into $\hat{\Sigma}$ of the form

$$Mx(t) =: \hat{x}(t) = [\ x_r'(t)\quad x_{\bar{r}}'(t)\ ]', \qquad \dim x_r(t) = n_r$$

$$\begin{bmatrix} x_r(t+1) \\ x_{\bar{r}}(t+1) \end{bmatrix} = \begin{bmatrix} \Phi_r & \Phi_{r\bar{r}} \\ O & \Phi_{\bar{r}} \end{bmatrix} \begin{bmatrix} x_r(t) \\ x_{\bar{r}}(t) \end{bmatrix} + \begin{bmatrix} G_r \\ O \end{bmatrix} u(t) \qquad \text{(A.1-13a)}$$

$$y(t) = [\ H_r \quad H_{\bar{r}}\ ] \begin{bmatrix} x_r(t) \\ x_{\bar{r}}(t) \end{bmatrix} + Ju(t) \qquad \text{(A.1-13b)}$$

with $\Sigma_r = (\Phi_r, G_r, H_r, J)$ completely reachable. $\hat{\Sigma}$ is said to be obtained from Σ via a Gilbert–Kalman (GK) canonical reachability decomposition.

- A state x is said to be *controllable* if there exists an input sequence $u_{[0,N)}$ of finite length that drives the system state to O_X:

$$\varphi(N, 0, x, u_{[0,N)}) = O_X$$

- Σ is said to be *completely controllable*, or (Φ, G) a *controllable pair*, if every state is controllable.

Theorem A.1-3. *The system Σ is completely controllable if and only if either Σ is completely reachable or the matrix $\Phi_{\bar{r}}$ resulting from a GK canonical reachability decomposition of Σ is nilpotent.*

- A state x is said to be *unobservable* if the system output response, from the state x and for the zero input, is zero at all times:

$$y(k) = H\Phi^k x = O_Y\ , \qquad \forall k \in \mathbb{Z}_+ \qquad \text{(A.1-14)}$$

- Σ is said to be *completely observable*, or (Φ, H) an *observable pair*, if the only unobservable state is the zero state O_X.

Theorem A.1-4. $\Sigma = (\Phi, G, H, J)$ *is completely observable if and only if*

$$\operatorname{rank} \Theta = n = \dim \Sigma \qquad \text{(A.1-15)}$$

where

$$\Theta := \begin{bmatrix} H \\ H\Phi \\ \vdots \\ H\Phi^{n-1} \end{bmatrix} \qquad \text{(A.1-16)}$$

is the observability matrix *of Σ.*

Theorem A.1-5.   **(GK Canonical Observability Decomposition)**
Consider the system Σ with observability matrix Θ such that

$$\text{rank}\,\Theta = n_o < n = \dim \Sigma$$

Then there exist nonsingular matrices M that transform Σ into $\hat{\Sigma}$ of the form

$$Mx(t) =: \hat{x}(t) = \begin{bmatrix} x_o'(t) & x_{\bar{o}}'(t) \end{bmatrix}', \qquad \dim x_o(t) = n_o$$

$$\begin{bmatrix} x_o(t+1) \\ x_{\bar{o}}(t+1) \end{bmatrix} = \begin{bmatrix} \Phi_o & O \\ \Phi_{\bar{o}o} & \Phi_{\bar{o}} \end{bmatrix} \begin{bmatrix} x_o(t) \\ x_{\bar{o}}(t) \end{bmatrix} + \begin{bmatrix} G_o \\ G_{\bar{o}} \end{bmatrix} u(t) \qquad (\text{A.1-17a})$$

$$y(t) = \begin{bmatrix} H_o & O \end{bmatrix} \begin{bmatrix} x_o(t) \\ x_{\bar{o}}(t) \end{bmatrix} + Ju(t) \qquad (\text{A.1-17b})$$

with $\Sigma_o = (\Phi_o, G_o, H_o, J)$ completely observable. $\hat{\Sigma}$ is said to be obtained from Σ via a GK canonical observability decomposition.

- Σ is said to be *completely reconstructible*, or (Φ, H) a *reconstructible pair*, if every final state of Σ can be uniquely determined by knowing the final output and the previous input and output sequences over intervals of finite but arbitrary length.

Theorem A.1-6.   *A system Σ is completely reconstructible if and only if either Σ is completely observable or the matrix $\Phi_{\bar{o}}$ resulting from a GK canonical observability decomposition of Σ is nilpotent.*

A.2   STABILITY

- The dynamic linear system (1-1) is said to be *exponentially stable* if there exist two positive reals α, $\lambda < 1$, such that

$$\|\varphi(t, t_0, x(t_0), O_\Omega)\| \leq \alpha \lambda^{(t-t_0)} \|x(t_0)\| \qquad (\text{A.2-1})$$

for all $x(t_0) \in \mathbb{R}^n$ and $t \geq t_0$. The system is *asymptotically stable* if

$$\lim_{t \to \infty} \Phi(t, t_0, x(t_0), O_\Omega) = O_X$$

for every $x(t_0) \in \mathbb{R}^n$. If the system is time-invariant, asymptotic stability is equivalent to exponential stability. A square matrix Φ is said to be a *stability matrix* if all its eigenvalues have modulus less than 1:

$$\text{sp}(\Phi) \subset \mathbb{C}_s$$

where $\text{sp}(\Phi)$, the *spectrum* of Φ, is the set of the eigenvalues of Φ, and $\mathbb{C}_s$ is the unit open disc in the complex plane.

Theorem A.2-1. *The time-invariant dynamic linear system (1-1), (1-7) is asymptotically stable if and only if its state-transition matrix Φ is a stability matrix or the d-characteristic polynomial of Φ, $\chi_\Phi(d) := \det(I - d\Phi)$, is strictly Hurwitz.*

- Σ is said to be *stabilizable*, or (Φ, G) a *stabilizable pair*, if there exist matrices $F \in \mathbb{R}^{m \times n}$ such that $\Phi + GF$ is a stability matrix:

$$\mathrm{sp}\,(\Phi + GF) \subset \mathbb{C}_s$$

- Σ is said to be *detectable*, or (Φ, H) a *detectable pair*, if there exists matrices $K \in \mathbb{R}^{n \times p}$ such that $\Phi - KH$ is a stability matrix:

$$\mathrm{sp}\,(\Phi - KH) \subset \mathbb{C}_s$$

Theorem A.2-2. *Σ is stabilizable (detectable) if and only if either Σ is completely reachable (observable) or the matrix $\Phi_{\bar{r}}$ ($\Phi_{\bar{o}}$) resulting from a GK canonical reachability (observability) decomposition of Σ is a stability matrix.*

The following stability property of slowly time-varying systems is frequently used in the global convergence analysis of adaptive systems.

Theorem A.2-3. [Des70b] *Consider the linear dynamic system*

$$x(k + 1) = \Phi(k)x(k) \quad , \qquad k \in \mathbb{Z}_+ \qquad \text{(A.2-2)}$$

where the $\Phi(k)$ are bounded stability matrices, viz.,

$$\|\Phi(k)\| < M < \infty \qquad and \qquad |\lambda\,[\Phi(k)]| < 1 - \epsilon \quad , \qquad \forall k \in \mathbb{Z}_+$$

with $\epsilon > 0$ and $\lambda\,[\Phi(k)]$ any eigenvalue of $\Phi(k)$. Then, provided that

$$\lim_{k \to \infty} \|\Phi(k) - \Phi(k-1)\| = 0 \qquad \text{(A.2-3)}$$

the system (2) is exponentially stable.

Note that (3) does not imply convergence of $\Phi(k)$.

A.3   STATE-SPACE REALIZATIONS

Given the transfer function

$$H(z) = \frac{b_1 z^{n-1} + \cdots + b_n}{z^n + a_1 z^{n-1} + \cdots + a_n} \qquad \text{(A.3-1)}$$

we can find a state-space representation $\Sigma = (\Phi, G, H)$, such that $H(z) = H(zI_n - \Phi)^{-1} G$, in the following straightforward way:

$$
\Phi = \left[\begin{array}{c|c}
0 & I_{n-1} \\
\hline
-a_n & -a_{n-1} \cdots - a_1
\end{array} \right]
\qquad
G = \left[\begin{array}{c}
0 \\
\hline
1
\end{array} \right]
\qquad
H = \left[\begin{array}{ccc} b_n \cdots b_1 \end{array} \right]
\qquad \text{(A.3-2)}
$$

Σ is said to be a *realization* of $H(z)$. In particular, (A.3-2) is the so-called *canonical reachable realization* of (1).

It is more difficult to find a realization Σ of the impulse-response sequence $W = \{w_k\}_{k=1}^{\infty}$, viz., a triplet (Φ, G, H) such that

$$
w_k = H\Phi^{k-1}G \quad , \qquad \forall k \in \mathbb{Z}_1 \qquad \text{(A.3-3)}
$$

In this connection, a key point consists of considering the following *Hankel matrices*:

$$
H_N = \left[\begin{array}{cccc}
w_1 & w_2 & \cdots & w_N \\
w_2 & w_3 & \cdots & w_{N+1} \\
\vdots & \vdots & \ddots & \vdots \\
w_N & w_{N+1} & \cdots & w_{2N-1}
\end{array} \right] \quad , \qquad N \in \mathbb{Z}_1 \qquad \text{(A.3-4)}
$$

Then the *minimal dimension* of the realizations of W equals the integer N, for which $\det H_N \neq 0$ and $\det H_{N+i} = 0$, $\forall i \in \mathbb{Z}_1$. Realizations of minimal dimension are called *minimal*. A realization Σ is minimal if and only if Σ is completely reachable and completely observable.

SOME RESULTS OF POLYNOMIAL MATRIX THEORY

This appendix provides a quick review of those results of polynomial matrix theory used in this book. For more extensive treatments, standard textbooks, for example, [Bar83], [BY83], [Kai80], [Ros70], and [Wol74], should be consulted.

B.1 MATRIX-FRACTION DESCRIPTIONS

Polynomial matrices arise naturally in linear system theory. Consider the $p \times m$ transfer matrix

$$H(z) := H(zI_n - \Phi)^{-1}G \qquad \text{(B.1-1)}$$

associated with the finite-dimensional linear time-invariant dynamical system (Φ, G, H). $H(z)$ is a *rational* matrix in the indeterminate z, viz., a matrix whose elements are rational functions of z. Let $\ell(z)$ be the least-common multiple of the denominator polynomials of the $H(z)$ entries. Then we can write

$$H(z) = \frac{N(z)}{\ell(z)} \qquad \text{(B.1-2)}$$

where $N(z)$ is a $p \times m$ *polynomial* matrix.

Equation (2) can be also rewritten as a *right matrix fraction*:

$$\left. \begin{array}{rcl} H(z) & = & N(z)M^{-1}(z) \\ M(z) & := & \ell(z)I_m \end{array} \right\} \qquad \text{(B.1-3)}$$

or a *left* matrix fraction:

$$\left. \begin{array}{rcl} H(z) & = & \tilde{M}^{-1}(z)N(z) \\ \tilde{M}(z) & := & \ell(z)I_p \end{array} \right\} \qquad \text{(B.1-4)}$$

Equations (3) and (4) are two examples of right and left *matrix-fraction descriptions (MFDs)* of $H(z)$. There are then many MFDs of a given transfer matrix. We are interested in finding MFDs that are irreducible in a well-defined sense. In order to introduce the concept, we define the *degree of an MFD* $N(z)M^{-1}(z)$ as the degree of the determinantal polynomial of its *denominator* matrix:

$$\text{the degree of the MFD} := \partial[\det M(z)] \qquad (\text{B.1-5})$$

Referring to (3), we find $\partial[\det M(z)] = \partial[\ell^m(z)] = m\partial\ell(z)$. Likewise, for the degree of the left MFD (4), we find $\partial[\det \tilde{M}(z)] = \partial[\ell^p(z)] = p\partial[\ell(z)]$. Given an MFD, we shall see how to obtain MFDs of minimal degree. One reason for this interest is that MFDs of minimal degree are intimately related to minimal state-space representations.

B.1.1   Divisors and Irreducible MFDs

From now on, we shall only consider right MFDs. All the material can be easily duplicated to cover, *mutatis mutandis*, the case of left MFDs.

Given a pair $(M(z), N(z))$ of polynomial matrices with equal number of columns and $M(z)$ square and nonsingular, viz., $\det M(z) \not\equiv 0$, we call $\Delta(z)$, $\dim \Delta(z) = \dim M(z)$, a *common right divisor (crd)* of $M(z)$ and $N(z)$ if there exist polynomial matrices $\hat{M}(z)$ and $\hat{N}(z)$ such that

$$M(z) = \hat{M}(z)\Delta(z) \quad \text{and} \quad N(z) = \hat{N}(z)\Delta(z) \qquad (\text{B.1-6})$$

Because

$$\partial[\det M(z)] = \partial[\det \hat{M}(z)] + \partial[\det \Delta(z)] \qquad (\text{B.1-7})$$

it follows that

$$\partial[\det M(z)] \geq \partial[\det \hat{M}(z)] \qquad (\text{B.1-8})$$

Further, $N(z)M^{-1}(z) = \hat{N}(z)\hat{M}^{-1}(z)$. Therefore, the degree of a MFD can be reduced by removing right divisors of the numerator and denominator matrices.

A square polynomial matrix $\Delta(z)$ is called *unimodular* if its determinant is a nonzero constant, independent of z. For instance,

$$\Delta(z) = \begin{bmatrix} z+1 & z \\ z & z-1 \end{bmatrix}$$

is unimodular because $\det \Delta(z) = -1$. A polynomial matrix $\Delta(z)$ is unimodular if and only if its inverse $\Delta^{-1}(z)$ is polynomial.

We see from (7) that equality holds in (8) if and only if $\Delta(z)$ is unimodular.

We say that $\Delta(z)$ is *a greatest common right divisor (gcrd)* of $M(d)$ and $N(d)$ if for any crd, $\hat{\Delta}(z)$ of $M(z)$ and $N(z)$, there exists a polynomial matrix $X(z)$ such that

$$\Delta(z) = X(z)\hat{\Delta}(z)$$

Let $\Delta(z)$ and $\tilde{\Delta}(z)$ be two gcrd's of $M(z)$ and $N(z)$. Then, for some polynomial matrices $X(z)$ and $Y(z)$,

$$\left.\begin{array}{rcl} \Delta(z) & = & X(z)\tilde{\Delta}(z) \\ \tilde{\Delta}(z) & = & Y(z)\Delta(z) \end{array}\right\} \qquad \Rightarrow \qquad \Delta(z) = X(z)Y(z)\Delta(z)$$

Hence, $X(z) = Y^{-1}(z)$. It follows that

1. All gcrd's can only differ by a unimodular (left) factor.

2. If a gcrd is unimodular, then all gcrd's must be unimodular.

3. All gcrd's have determinant polynomials of the same degree.

$M(z)$ and $N(z)$ as in (6) are *relatively right prime* or *right coprime* if their gcrd's are unimodular. In such a case, the right MFD $N(z)M^{-1}(z)$ is said to be *irreducible* because it has the minimum possible degree.

B.1.2 Elementary Row (Column) Operations for Polynomial Matrices

1. Interchange of any two rows (columns).

2. Addition to any row (column) of a polynomial multiple of any other row (column).

3. Scaling any row (column) by any nonzero real number.

These operations can be represented by *elementary matrices*, in that we can carry out elementary row (column) operations on a polynomial matrix by premultiplying (postmultiplying) the latter by an elementary matrix. Some examples are

$$\begin{bmatrix} 0 & 1 & 0 \\ 1 & 0 & 0 \\ 0 & 0 & 1 \end{bmatrix} A = \begin{bmatrix} A_{2.} \\ A_{1.} \\ A_{3.} \end{bmatrix}$$

$$\begin{bmatrix} 1 & \alpha(z) & 0 \\ 0 & 1 & 0 \\ 0 & 0 & 1 \end{bmatrix} A = \begin{bmatrix} A_{1.} + \alpha(z)A_{2.} \\ A_{2.} \\ A_{3.} \end{bmatrix}$$

where $A_{i.}$ denotes the ith row of A, and $\alpha(z)$ is a polynomial.

All these elementary matrices are unimodular. Conversely, premultiplication (postmultiplication) by a unimodular matrix corresponds to the actions of a sequence of elementary row (column) operations.

B.1.3   A Construction for a gcrd

Given $m \times m$ and $p \times m$ polynomial matrices $M(z)$ and $N(z)$, form the matrix $[M'(z)N'(z)]'$. Next, find elementary row operations (or, equivalently, a unimodular matrix $U(z)$) such that p of the bottom rows of the matrix on the RHS of the following equation are identically zero:

$$
U(z) \underbrace{\left[\begin{array}{c} M(z) \\ \hline N(z) \end{array} \right] \begin{array}{c} \}m \\ \}p \end{array}}_{m} = \underbrace{\left[\begin{array}{c} \Delta(z) \\ \hline 0 \end{array} \right] \begin{array}{c} \}m \\ \}p \end{array}}_{m} \tag{B.1-9}
$$

Then the square matrix denoted $\Delta(z)$ in (9) is a gcrd of $M(z)$ and $N(z)$. In particular, to this end, we can use the procedure to construct the Hermite form [Kai80].

B.1.4   Bezout Identity

$M(z)$ and $N(z)$ are right coprime if and only if there exists polynomial matrices $X(z)$ and $Y(z)$ such that

$$
X(z)M(z) + Y(z)N(z) = I_n \tag{B.1-10}
$$

B.2   COLUMN-REDUCED AND ROW-REDUCED MATRICES

A rational transfer matrix $H(z)$ is said to be *proper* if $\lim_{z\to\infty} H(z)$ is finite, and *strictly proper* if $\lim_{z\to\infty} H(z) = 0$. Properness or strict properness of a transfer matrix can be verified by inspection. If we refer to a MFD $N(z)M^{-1}(z)$, the situation is not so simple. We need the following definition:

$$
\left. \begin{array}{c} \text{the degree of a} \\ \text{polynomial vector} \end{array} \right\} = \left\{ \begin{array}{c} \text{the highest degree of} \\ \text{all the entries of the vector} \end{array} \right.
$$

Let

$$
k_j := \partial M_j(z): \text{ the degree of the } j\text{th column of } M(z) \tag{B.2-1}
$$

Then, clearly,

$$
\partial[\det M(z)] \le \sum_{j=1}^{m} k_j \tag{B.2-2}
$$

If equality holds in (2), we shall say that $M(z)$ is *column-reduced*. In general, we can always write

$$
M(z) = M_{hc}S(z) + L(z) \tag{B.2-3}
$$

where

$$
\begin{aligned}
S(z) &:= \text{diag } \{z^{k_j}, = 1, 2, \cdots, m\} \\
M_{hc} &:= \text{highest-column-degree coefficient matrix of } M(z)
\end{aligned}
$$

with M_{hc} the matrix whose jth column comprises the coefficient of k_j in the jth column of $M(z)$. Finally, $L(z)$ denotes the *remaining* terms and is a polynomial matrix with column degrees strictly less than those of $M(z)$. For example,

$$M(z) = \begin{bmatrix} z^3 + 1 & z^2 + 2 \\ z^2 + z & 1 \end{bmatrix} = \underbrace{\begin{bmatrix} 1 & 1 \\ 0 & 0 \end{bmatrix}}_{M_{hc}} \underbrace{\begin{bmatrix} z^3 & 0 \\ 0 & z^2 \end{bmatrix}}_{S(z)} + \underbrace{\begin{bmatrix} 1 & 2 \\ z^2 + z & 1 \end{bmatrix}}_{L(z)}$$

Then

$$\det M(z) = (\det M_{hc}) z^{\sum_j k_j} + \text{ terms of lower degree in } z \qquad \text{(B.2-4)}$$

Therefore, it follows:

A nonsingular polynomial matrix is column-reduced if and only if its highest-column-degree coefficient matrix is nonsingular.

Properness of $N(z)M^{-1}(z)$ can be established provided that $M(z)$ is column (row)-reduced.

If $M(z)$ is column-reduced, then $N(z)M^{-1}(z)$ is strictly proper (proper) if and only if each column of $N(z)$ has degree less than (less than or equal to) the degree of the corresponding column of $M(z)$.

Any nonsingular polynomial matrix can be made column (row)-reduced by using elementary column (row) operations to successively reduce the individual column (row) degrees until column (row) reduction is achieved. For example, taking $M(z)$ as in the equation preceding (4),

$$M(z) \begin{bmatrix} 1 & 0 \\ -z & 1 \end{bmatrix} = \underbrace{\begin{bmatrix} -2z + 1 & z^2 + 2 \\ -z^3 + z^2 - 1 & 1 \end{bmatrix}}_{\tilde{M}(z)}$$

$$= \underbrace{\begin{bmatrix} 0 & 1 \\ -1 & 0 \end{bmatrix}}_{M_{hc}} \underbrace{\begin{bmatrix} z^3 & 0 \\ 0 & z^2 \end{bmatrix}}_{S(z)} + \underbrace{\begin{bmatrix} -2z + 1 & 2 \\ z^2 - z & 1 \end{bmatrix}}_{L(z)}$$

and we see that $\tilde{M}(z)$ is column-reduced. Then, given a nonsingular $M(z)$, there exist unimodular matrices $W(z)$ such that $\tilde{M}(z) = M(z)W(z)$ is column-reduced. Therefore, any right MFD $N(z)M^{-1}(z)$ of $H(z)$ can be transformed into a right MFD with a column-reduced denominator matrix. In fact, $H(z) = N(z)W(z)[M(z)W(z)]^{-1} = \tilde{N}(z)\tilde{M}^{-1}(z)$ with $\tilde{N}(z) := N(z)W(z)$ and $\tilde{M}(z) := M(z)W(z)$.

B.3   REACHABLE REALIZATIONS FROM RIGHT MFDS

W.l.o.g., we assume that the right MFD $H(z) = N(z)M^{-1}(z)$ has a column-reduced denominator matrix. It is also assumed that $H(z)$ is strictly proper. We note that a system having transfer matrix $H(z)$ can be represented in terms of a system of difference equations as follows:

$$y(t) = N(z)M^{-1}(z)u(t) \quad \sim \quad \begin{cases} M(z)\xi(t) &= u(t) \\ y(t) &= N(z)\xi(t) \end{cases} \tag{B.3-1}$$

where z is now to be interpreted as the forward-shift operator $zy(t) := y(t+1)$; and $\xi(t) \in \mathbb{R}^m$ is called the *partial state*. Let $\xi_i(t)$, $i = 1, \cdots, m$, be the ith component of $\xi(t)$. Define

$$x(t) := [\xi_1(t) \cdots \xi_1(t + k_1) \cdots \xi_m(t) \cdots \xi_m(t + k_m)]' \in \mathbb{R}^{\sum_i k_i} \tag{B.3-2}$$

Then a state-space realization (Φ, G, H) with state $x(t)$ of dimension (cf. (2-4))

$$\dim \Phi = \sum_i k_i = \partial \det M(z) \tag{B.3-3}$$

can be constructed (cf. [Kai80]) with the following properties:

1. (Φ, G) is completely reachable.

2. (Φ, H) is completely observable if and only if $N(z)$ and $M(z)$ are right co-prime.

3. $\bar{\chi}_\Phi(z) := \det(zI - \Phi) = (\det M_{hc})^{-1} \det M(z).$ $\tag{B.3-4}$

B.4   RELATIONSHIP BETWEEN z AND d MFDS

Let

$$\begin{aligned} \bar{H}(z) &= \bar{N}(z)\bar{M}^{-1}(z) \\ &= H(zI_n - \Phi)^{-1}G \end{aligned}$$

with $\bar{M}(z) = \bar{M}_{hc}\,\bar{S}(z) + \bar{L}(z)$ column-reduced and such that

$$n := \dim \Phi = \partial \det \bar{M}(z)$$

and

$$\bar{\chi}_\Phi(z) := \det(zI - \Phi) = (\det \bar{M}_{hc})^{-1} \det \bar{M}(z)$$

We have

$$\bar{M}(z)\bar{S}(z^{-1})\bar{M}_{hc}^{-1} = I_m + \bar{L}(z)\bar{S}(z^{-1})\bar{M}_{hc}^{-1} =: M(d)|_{d=z^{-1}} \tag{B.4-1}$$

Similarly,

$$\bar{N}(z)\bar{S}(z^{-1})\bar{M}_{hc}^{-1} =: N(d)|_{d=z^{-1}} \tag{B.4-2}$$

In the previous equations, we have used the fact that $\bar{S}(z^{-1}) = \bar{S}^{-1}(z)$. We can see that, with $\bar{M}(z)$ column-reduced, $M(d)$ and $N(d)$ are polynomial matrices in the indeterminate d. Further,

$$
\begin{aligned}
N(d)M^{-1}(d) \quad = \quad \bar{H}(d^{-1}) \quad &= \quad H(d^{-1}I_n - \Phi)^{-1}G \\
&= \quad H(I_n - d\Phi)^{-1}dG
\end{aligned}
\tag{B.4-3}
$$

Then $N(d)M^{-1}(d)$ is a right MFD of the d-transfer matrix $H(I_n - d\Phi)^{-1}dG$ associated with (Φ, G, H) (cf. (3.1-28)). Further, we find

$$
\begin{aligned}
\det M(d) \quad &= \quad (\det \bar{M}_{hc})^{-1} \det \bar{M}(d^{-1}) \det \bar{S}(d) && [(1)] \\
&= \quad d^{\sum_j k_i} \det(d^{-1}I_n - \Phi) && [(3\text{-}4)] \quad (\text{B.4-4}) \\
&= \quad \det(I_n - d\Phi) =: \chi_\Phi(d)
\end{aligned}
$$

Finally, if $\bar{M}(z)$ and $\bar{N}(z)$ are right coprime, $M(d)$ and $N(d)$ are such. Then, if $H(d) = H(I_n - d\Phi)^{-1}dG$ has an irreducible right MFD $N(d)M^{-1}(d)$, Fact 3.1-1 follows.

B.5 DIVISORS AND SYSTEM-THEORETIC PROPERTIES
PBH RANK TESTS [Kai80, p. 136.]

1. A pair (Φ, G) is reachable if and only if

$$\text{rank}\begin{bmatrix} zI_n - \Phi & G \end{bmatrix} = n \quad \text{for all } z \in \mathbb{C} \tag{B.5-1}$$

2. A pair (H, Φ) is observable if and only if

$$\text{rank}\begin{bmatrix} H \\ zI_n - \Phi \end{bmatrix} = n \quad \text{for all } z \in \mathbb{C} \tag{B.5-2}$$

By setting $d := z^{-1}$,

$$\begin{bmatrix} zI_n - \Phi & G \end{bmatrix} = z\begin{bmatrix} I_n - d\Phi & dG \end{bmatrix}$$

Then we have the following results:

3. $\text{rank}\begin{bmatrix} I_n - d\Phi & dG \end{bmatrix} = n \quad \text{for all} \quad d \in \mathbb{C} \tag{B.5-3}$
   if and only if the pair (Φ, G) is controllable, that is, it has no nonzero unreachable eigenvalue.

It is to be pointed out that (3) is equivalent to right coprimeness of polynomial matrices $A(d) := I_n - d\Phi$ and $B(d) := dG$.

4. rank $\begin{bmatrix} H \\ I_n - d\Phi \end{bmatrix} = n \quad$ for all $\quad d \in \mathbb{C}$ $\hspace{2cm}$ (B.5-4)

   if and only if the pair (H, Φ) is reconstructible, that is, it has nonzero unobservable eigenvalue.

It is to be pointed out that (4) is equivalent to left coprimeness of the polynomial matrices $I_n - d\Phi$ and H.

The following properties, which can be easily verified via GK canonical decompositions, relate reducible MFDs to system-theoretic attributes of the triplet (H, Φ, G).

5. The gcld's of $I_n - d\Phi$ and dG are strictly Hurwitz if and only if the pair (Φ, G) is stabilizable.

6. The gcrd's of $I_n - d\Phi$ and H are strictly Hurwitz if and only if the pair (H, Φ) is detectable.

SOME RESULTS ON LINEAR DIOPHANTINE EQUATIONS

This appendix provides a quick review of those results on linear Diophantine equations used in this book. For a more extensive treatment, the monograph [Kuč79] should be consulted. Diophantus of Alexandria studied in the third century A.D. the problem, isomorphic to the next equation (1-1), of finding integers (x, y) solving the equation $ax + by = c$ with a, b, and c as given integers.

C.1 UNILATERAL POLYNOMIAL MATRIX EQUATIONS

Let $R_{pm}[d]$ denote the set of $p \times m$ matrices whose entries are polynomials in the indeterminate d. We consider either the equation

$$A(d)X(d) + B(d)Y(d) = C(d) \tag{C.1-1}$$

or the equation

$$X(d)A(d) + Y(d)B(d) = C(d) \tag{C.1-2}$$

In (1), $A(d) \in R_{pp}[d]$ and nonsingular, and $C(d) \in R_{pp}[d]$. In (2), $A(d) \in R_{mm}[d]$ and nonsingular, and $C(d) \in R_{mm}[d]$. In both (1) and (2), $B(d) \in R_{pm}[d]$, and $X(d)$ and $Y(d)$ are polynomial matrices of compatible dimensions. By a solution, we mean any pair of polynomial matrices $X(d)$ and $Y(d)$ satisfying either (1) or (2).

Result C.1-1. *Equation (1) is solvable if and only if the gcld's of $A(d)$ and $B(d)$ are left divisors of $C(d)$. Provided that (1) is solvable, the general solution of (1) is given by*

$$X(d) = X_0(d) - B_2(d)P(d) \tag{C.1-3}$$

$$Y(d) = Y_0(d) + A_2(d)P(d) \tag{C.1-4}$$

where $(X_0(d), Y_0(d))$ is a particular solution of (1); $A_2(d)$ and $B_2(d)$ are right coprime and such that

$$A^{-1}(d)B(d) = B_2(d)A_2^{-1}(d) \tag{C.1-5}$$

and $P(d) \in R_{mp}[d]$ is an arbitrary polynomial matrix.

Result C.1-2. *Equation (2) is solvable if and only if the gcrd's of $A(d)$ and $B(d)$ are right divisors of $C(d)$. Provided that (2) is solvable, the general solution of (2) is given by*

$$X(d) = X_0(d) - P(d)B_1(d) \tag{C.1-6}$$

$$Y(d) = Y_0(d) + P(d)A_1(d) \tag{C.1-7}$$

where $(X_0(d), Y_0(d))$ is a particular solution of (2); $A_1(d)$ and $B_1(d)$ are left coprime and such that

$$B(d)A^{-1}(d) = A_1^{-1}(d)B_1(d) \tag{C.1-8}$$

and $P(d) \in R_{mp}[d]$ is an arbitrary polynomial matrix.

In applications, we are usually interested in solutions of either (1) or (2) with some specific properties. In particular, we consider the *minimum-degree solution* of (1) w.r.t. $Y(d)$. By this, we mean, whenever it exists unique, the pair $(X(d), Y(d))$ solving (1) with minimum $\partial Y(d)$. Here $\partial Y(d)$ denotes the *degree* of $Y(d)$:

$$Y(d) = Y_0 + Y_1 d + \cdots + Y_{\partial Y} d^{\partial Y}, \quad Y_{\partial Y} \neq 0$$

with $Y_0, Y_1, \cdots, Y_{\partial Y}$ constant matrices. We say that a square polynomial matrix

$$Q(d) = Q_0 + Q_1 d + \cdots + Q_{\partial Q} d^{\partial Q}$$

is *regular* if its highest degree matrix coefficient $Q_{\partial Q}$ is nonsingular.

Result C.1-3. *Let (1) be solvable and $A_2(d)$ regular. Then, the minimum-degree solution of (1) w.r.t. $Y(d)$ exists unique and can be found as follows. Use the left division algorithm to divide $A_2(d)$ into $Y_0(d)$:*

$$Y_0(d) = A_2(d)Q(d) + \Gamma(d), \quad \partial \Gamma(d) < \partial A_2(d) \tag{C.1-9}$$

Then (4) becomes

$$Y(d) = A_2(d)[Q(d) + P(d)] + \Gamma(d)$$

Hence, the required minimum-degree solution is obtained by setting

$$P(d) = -Q(d) \tag{C.1-10}$$

to get

$$\begin{aligned} X(d) &= X_0(d) + B_2(d)Q(d) & \text{(C.1-11)} \\ Y(d) &= \Gamma(d) & \text{(C.1-12)} \end{aligned}$$

Result C.1-4. *Let (2) be solvable and $A_1(d)$ regular. Then, the minimum-degree solution of (2) w.r.t. $Y(d)$ exists unique and can be found as follows. Use the right division algorithm to divide $A_1(d)$ into $Y_0(d)$:*

$$Y_0(d) = Q(d)A_1(d) + \Gamma(d) \; , \;\; \partial\Gamma(d) < \partial A_1(d) \qquad (\text{C.1-13})$$

Then (7) becomes

$$Y(d) = [Q(d) + P(d)]A_1(d) + \Gamma(d)$$

Hence, the minimum-degree solution of (2) w.r.t. $Y(d)$ is given by

$$X(d) \;\;=\;\; X_0(d) + Q(d)B_1(d) \qquad (\text{C.1-14})$$
$$Y(d) \;\;=\;\; \Gamma(d) \qquad (\text{C.1-15})$$

C.2   BILATERAL POLYNOMIAL MATRIX EQUATIONS

In this book, we shall find various bilateral polynomial matrix equations of the form

$$\bar{E}(d)X(d) + Z(d)G(d) = C(d) \qquad (\text{C.2-1})$$

where $\bar{E}(d) \in R_{mm}[d]$, $G(d) \in R_{nn}[d]$, $C(d) \in R_{mn}[d]$, and $X(d)$ and $Z(d)$ are unknown polynomial matrices of compatible dimensions. Solvability conditions for (1) are more complicated than the ones for (1-1) and (1-2). However, we shall only encounter (1) in the special case where $G(d)$ is strictly Hurwitz and $\bar{E}(d)$ anti-Hurwitz. This implies that

$$\det \bar{E}(d) \quad \text{and} \quad \det G(d) \quad \text{are coprime polynomials} \qquad (\text{C.2-2})$$

Result C.2-1. *Provided that (2) is fulfilled, (1) is solvable. Further, the general solution of (1) is given by*

$$X(d) \;\;=\;\; X_0(d) + L(d)G(d) \qquad (\text{C.2-3})$$
$$Z(d) \;\;=\;\; Z_0(d) - E(d)L(d) \qquad (\text{C.2-4})$$

where $(X_0(d), \; Z_0(d))$ is a particular solution of (1) and $L(d) \in R_{mn}[d]$ is an arbitrary polynomial matrix.

Result C.2-2. *Let (1) be solvable and $\bar{E}(d)$ regular. Then, the minimum-degree solution of (1) w.r.t. $Z(d)$ exists unique and can be found as follows. Use the left division algorithm to divide $\bar{E}(d)$ into $Z_0(d)$:*

$$Z_0(d) = \bar{E}(d)Q(d) + \Gamma(d) \quad , \quad \partial\Gamma(d) < \partial\bar{E}(d) \qquad (\text{C.2-5})$$

Then (4) becomes

$$Z(d) = \bar{E}(d)[Q(d) - L(d)] + \Gamma(d)$$

Hence, the desired minimum-degree solution is obtained by setting

$$L(d) = Q(d) \tag{C.2-6}$$

to get

$$X(d) = X_0(d) + Q(d)G(d) \tag{C.2-7}$$
$$Z(d) = \Gamma(d) \tag{C.2-8}$$

PROBABILITY THEORY AND STOCHASTIC PROCESSES

This appendix provides a quick review of those concepts and results from probability and the theory of stochastic processes used in this book. For more extensive treatments standard textbooks, for example, [Chu68], [Cra46], [Doo53], [Loè63], and [Nev75], should be consulted.

D.1 PROBABILITY SPACE

A *probability space* is a triple $(\Omega, \mathcal{F}, \mathbb{P})$, where Ω, the *sample space*, is a nonempty set of elements ω; $\mathcal{F}$ is a *σ-field* (or a *σ-algebra*) of subsets of Ω, viz., a collection of subsets containing the empty set $\emptyset$ and closed under complements and countable unions; $\mathbb{P}$ is a *probability measure*, viz., a function $\mathbb{P} : \mathcal{F} \to \mathbb{R}$ satisfying the following axioms:

$$
\begin{aligned}
\mathbb{P}(A) &\geq 0, \quad \forall A \in \mathcal{F} \\
\mathbb{P}\left(\bigcup_{i=1}^{\infty} A_i\right) &= \sum_{i=1}^{\infty} \mathbb{P}(A_i) \quad \text{if } A_i \in \mathcal{F} \text{ and } A_i \cap A_j = \emptyset, i \neq j \\
\mathbb{P}(\Omega) &= 1
\end{aligned}
$$

Any element of $\mathcal{F}$ is called an *event*; in particular, Ω and $\emptyset$ are sometimes referred to as the *sure* and the *impossible* event, respectively.

Given a family $\mathcal{S}$ of subsets of Ω, there is a uniquely determined σ-field, denoted $\sigma(\mathcal{S})$, on Ω which is the smallest σ-field containing $\mathcal{S}$. $\sigma(\mathcal{S})$ is called the *σ-field generated by $\mathcal{S}$*.

D.2   RANDOM VARIABLES

Let $(\Omega, \mathcal{F}, \mathbb{P})$ be a probability space. A real-valued function $v(\omega)$ on Ω, $v : \Omega \to \mathbb{R}$ is called a *random variable* if it is *measurable* w.r.t. $\mathcal{F}$, viz., the set $\{\omega \mid v(\omega) \in \mathcal{R}\}$ belongs to $\mathcal{F}$ for every open set $\mathcal{R} \in \mathbb{R}$.

Let v be a random variable such that $\int_{\Omega} |v(\omega)| dP(\omega) < \infty$. Then its *expected value*, or *mean*, is defined as

$$\mathcal{E}\{v\} := \int_{\Omega} v(\omega) d\,\mathbb{P}(\omega)$$

The mean of $v^k(\omega)$ is called the kth *moment* of $v(\omega)$. From the Cauchy–Schwarz inequality [Lue69], it follows that the existence of the second moment $\mathcal{E}\{v^2\}$ of $v(\omega)$ implies that its mean $\mathcal{E}\{v\}$ does exist. The quantity

$$\mathrm{Var}(v) := \mathcal{E}\{v^2\} - [\mathcal{E}\{v\}]^2$$

is called the *variance* of v. Whenever this, or, equivalently, $\mathcal{E}\{v^2\}$, exists, $v(\omega)$ is said to be *square-integrable* or to have *finite variance*.

Consider the set of all real-valued square-integrable random variables on $(\Omega, \mathcal{F}, \mathbb{P})$. This set can be made a vector space over the real field under the usual operation of pointwise sum of functions and multiplication of functions by real numbers. Given two square-integrable random variables u and v, set $\langle u, v \rangle := \mathcal{E}\{uv\}$ and $\|u\| := +\sqrt{\langle u, u \rangle}$. Now let u denote the equivalence class of random variables on $(\Omega, \mathcal{F}, \mathbb{P})$, where v is equivalent to u if $\|u - v\|^2 = \mathcal{E}\{(u - v)^2\} = 0$, that is, u denotes the collection of random variables that are identical to v except on a set of zero probability measure. With such a stipulation, $\langle \cdot, \cdot \rangle$ satisfies all the axioms of an inner product. The foregoing vector space of (equivalence classes of) square-integrable random variables equipped with the inner product $\langle \cdot, \cdot \rangle$ is denoted by $L_2(\Omega, \mathcal{F}, \mathbb{P})$. It is an important result in analysis [Roy68] that $L_2(\Omega, \mathcal{F}, \mathbb{P})$ is a Hilbert space.

Let $v : \Omega \to \mathbb{R}^n$ be a random vector with n finite-variance components. Then v is called a finite-variance random vector, and its mean $\bar{v} := \mathcal{E}\{v\}$ and *covariance* matrix

$$\mathrm{Cov}(v) := \mathcal{E}\left\{ (v - \bar{v}) (v - \bar{v})' \right\}$$

are well-defined. Further, if $V = \mathrm{Cov}(v)$, we have $V = V' \geq 0$. By setting $v = \bar{v} + \tilde{v}$, with $\bar{v} := \mathcal{E}\{v\}$ and $\tilde{v} := v - \bar{v}$, and using $\mathcal{E}\{\tilde{v}\} = O_n$, the following lemma is easily proved.

Lemma D.2-1.   *Let $v : \Omega \to \mathbb{R}^n$ be a finite variance random vector. Let G be an $n \times n$ matrix. Then*

$$\mathcal{E}\{v'Gv\} = \bar{v}'G\bar{v} + \mathrm{Tr}\left[G\,\mathrm{Cov}(v) \right] \qquad \text{(D.2-1)}$$

The *probability distribution (function)* of v is a function $P_v : \mathbb{R} \to [0,1]$ defined as follows:

$$P_v(\alpha) := \mathbb{P}\left(\{\omega \mid v(\omega) \leq \alpha\}\right), \quad \alpha \in \mathbb{R}$$

P_v is clearly nondecreasing, continuous from the right, and

$$\lim_{\alpha \to -\infty} P_v(\alpha) = 0, \qquad \lim_{\alpha \to \infty} P_v(\alpha) = 1$$

If $P_v(\alpha)$ is absolutely continuous w.r.t. the Lebesgue measure [Roy68], then there exists a function $p_v : \mathbb{R} \to \mathbb{R}_+$ called the *probability density (function)* of v such that

$$P_v(\alpha) = \int_{-\infty}^{\alpha} p_v(\beta) d\beta$$

$v(\omega) := \begin{bmatrix} v_1(\omega) & \cdots & v_n(\omega) \end{bmatrix}'$ is an n-dimensional *random vector* if its n components $v_i(\omega)$, $i \in \underline{n}$, are random variables. In such a case, the probability distribution is a function $P_v : \mathbb{R}^n \to [0,1]$:

$$P_v(\alpha) := \mathbb{P}\left(\{\omega \mid v_i(\omega) \leq \alpha_i,\ i \in \underline{n}\}\right) \qquad \alpha = \begin{bmatrix} \alpha_1 & \cdots & \alpha_n \end{bmatrix}' \in \mathbb{R}^n$$

As for a single random variable, if P_v is absolutely continuous, we have

$$P_v(\alpha) = \int_{-\infty}^{\alpha_1} \cdots \int_{-\infty}^{\alpha_n} p_v(\beta) d\beta \qquad \beta = \begin{bmatrix} \beta_1 & \cdots & \beta_n \end{bmatrix}' \in \mathbb{R}^n$$

with p_v the probability density of random vector v.

D.3   CONDITIONAL PROBABILITIES

The events $A_1, \cdots, A_n \in \mathcal{F}$ are *independent* if

$$\mathbb{P}\left(A_1 \cap A_2 \cdots \cap A_n\right) = \mathbb{P}\left(A_1\right) \times \mathbb{P}\left(A_2\right) \times \cdots \times \mathbb{P}\left(A_n\right)$$

The *conditional probability* of A given B, $A, B \in \mathcal{F}$, is defined as

$$\mathbb{P}(A \mid B) := \frac{\mathbb{P}(A \cap B)}{\mathbb{P}(B)} \qquad \text{provided that } \mathbb{P}(B) > 0$$

Note that $\mathbb{P}(A \mid B) = \mathbb{P}(A)$ if and only if A and B are independent. Further, $\mathbb{P}(\cdot \mid B)$ is itself a probability measure. Thus, if v is a random variable defined on $(\Omega, \mathcal{F}, \mathbb{P})$, we can define the *conditional mean* of v given B as

$$\mathcal{E}\{v \mid B\} = \int_{\Omega} v(\omega) d\,\mathbb{P}(\omega \mid B)$$

More generally, let $\mathcal{A}$, $\mathcal{A} \subset \mathcal{F}$, denote a sub-$\sigma$-field of $\mathcal{F}$, viz., a family of elements of $\mathcal{F}$ that also forms a σ-field. The conditional expectation of v w.r.t. $\mathcal{A}$, or given $\mathcal{A}$, denoted $\mathcal{E}\{v \mid \mathcal{A}\}$, is a random variable such that

1. $\mathcal{E}\{v \mid \mathcal{A}\}$ is $\mathcal{A}$-measurable

2. $\int_A \mathcal{E}\{v \mid \mathcal{A}\} d\,\mathbb{P}(\omega) = \int_A v(\omega) d\,\mathbb{P}(\omega)$ for all $A \in \mathcal{A}$

If $\mathcal{A}$ is the σ-field generated by the set of random variables $\{v_1, \cdots, v_n\}$, $\mathcal{A} = \sigma\{v_1, \cdots, v_n\}$, we write

$$\mathcal{E}\{v \mid \mathcal{A}\} = \mathcal{E}\{v \mid v_1, \cdots, v_n\}$$

Properties of conditional expectation are as follows:

1. If $u = \mathcal{E}\{v \mid \mathcal{A}\}$ and $w = \mathcal{E}\{v \mid \mathcal{A}\}$, then $u = w$ a.s. (where "a.s." means *almost surely*, that is, except on a set having probability measure zero)

2. If u is $\mathcal{A}$-measurable, then

$$\mathcal{E}\{uv \mid \mathcal{A}\} = u\mathcal{E}\{v \mid \mathcal{A}\} \quad \text{a.s.}$$

3. $\mathcal{E}\{uv \mid \mathcal{A}\} = \mathcal{E}\{u\}\mathcal{E}\{v \mid \mathcal{A}\}$ if u is independent of every set in $\mathcal{A}$.

4. (*Smoothing properties* of conditional expectations). If $\mathcal{F}_{t-1}$ and $\mathcal{F}_t$ are two sub-σ-fields of $\mathcal{F}$ such that $\mathcal{F}_{t-1} \subset \mathcal{F}_t$, then

$$\mathcal{E}\{\mathcal{E}\{v \mid \mathcal{F}_{t-1}\} \mid \mathcal{F}_t\} = \mathcal{E}\{v \mid \mathcal{F}_{t-1}\} \quad \text{a.s.}$$
$$\mathcal{E}\{\mathcal{E}\{v \mid \mathcal{F}_t\} \mid \mathcal{F}_{t-1}\} = \mathcal{E}\{v \mid \mathcal{F}_{t-1}\} \quad \text{a.s.}$$

D.4   GAUSSIAN RANDOM VECTORS

A random vector v with n components is said to be *Gaussian* if its probability density function $p_v(\alpha)$, $\alpha \in \mathbb{R}^n$, equals

$$p_v(\alpha) = [(2\pi)^n \det V]^{-1/2} \exp\left(-\frac{1}{2}\|\alpha - \bar{v}\|^2_{V^{-1}}\right) \qquad \text{(D.4-1)}$$
$$=: n(\bar{v}, V)$$

Function $n(\bar{v}, V)$ is the Gaussian or normal probability density of v with mean $\bar{v}$ and covariance matrix V, the latter assumed here to be positive definite.

Result D.4-1. *Let v be a Gaussian random vector with probability density $n(\bar{v}, V)$. Then $u(\omega) = Lv(\omega) + \ell$, with L a matrix and ℓ a vector both of compatible dimension, is a Gaussian random vector with probability density $n(\bar{u}, U)$, where*

$$\bar{u} = L\bar{v} + \ell \qquad and \qquad U = LVL'$$

Result D.4-2. *Let v be a Gaussian random vector with probability density $n(\bar{v}, V)$. Let v, $\bar{v}$ and V be partitioned conformably as follows:*

$$v(\omega) = \begin{bmatrix} v_1(\omega) \\ v_2(\omega) \end{bmatrix} \qquad \bar{v} = \begin{bmatrix} \bar{v}_1 \\ \bar{v}_2 \end{bmatrix} \qquad V = \begin{bmatrix} V_{11} & V_{12} \\ V_{12} & V_{22} \end{bmatrix}$$

Then v_i, $i = 1, 2, \cdots$, has (marginal) probability density $n(\bar{v}_i, V_{ii})$. Further, the conditional probability density of v_1 given v_2 is Gaussian and given by

$$p_{v_1 | v_2} = n \left(\bar{v}_1 + V_{12} V_{22}^{-1} (v_2 - \bar{v}_2), V_{11} - V_{12} V_{22}^{-1} V_{21} \right)$$

D.5   STOCHASTIC PROCESSES

A discrete-time *stochastic process* $v = \{v(t, \omega), t \in T\}$, $T \subset \mathbb{Z}$, is an integer-indexed family of random vectors defined on a common underlying probability space $(\Omega, \mathcal{F}, \mathbb{P})$. To indicate a stochastic process, we interchangeably use the notations $\{v(t, \omega)\}$, $\{v(t)\}$, or simply v. For fixed t, $v(t, \cdot)$ is a random variable. For fixed ω, $v(\cdot, \omega)$ is called a *realization* or a *sample path* of the process.

The *mean*, $\bar{v}(t)$, and the *covariance*, $K_v(t, \tau)$, of the process are defined as follows:

$$\begin{aligned} \bar{v}(t) &:= \mathcal{E}\{v(t, \omega)\} \\ K_v(t, \tau) &:= \mathcal{E}\left\{[v(t, \omega) - \bar{v}(t)][v(\tau, \omega) - \bar{v}(\tau)]'\right\} \end{aligned}$$

If $\bar{v}(t) \equiv \bar{v}$ and $K_v(t, \tau) = K_v(t + k, \tau + k)$, $\forall k$, $t + k$, $\tau + k \in T$, that is, the mean and covariance are invariant w.r.t. time shifts, we say that process v is *wide-sense stationary*. In such a case, abusing the notations, we write $K_v(\tau)$ in place of $K_v(t_1, t_2)$, where $\tau := t_1 - t_2$. If $K_v(\tau) = \Psi_v \delta_{\tau, 0}$ we say that the process is *white*.

The sequence of random vectors and σ-fields $\{v(t), \mathcal{F}_t\}$, $t \in \mathbb{Z}_+$, with $\mathcal{F}_t \subset \mathcal{F}$, is called a *martingale* if the following holds:

1. $\mathcal{F}_t \subset \mathcal{F}_{t+1}$ and $v(t)$ is $\mathcal{F}_t$-measurable (the latter condition is referred to by saying that $\{v(t)\}$ is $\{\mathcal{F}_t\}$-*adapted*)

2. $\mathcal{E}\{\|v(t)\|\} < \infty$

3. $\mathcal{E}\{v(t+1) \mid \mathcal{F}_t\} = v(t)$ a.s.

If, instead of the equality in 3., we have

$$\mathcal{E}\{v(t+1) \mid \mathcal{F}_t\} \geq v(t) \quad \text{a.s.}$$

$\{v(t), \mathcal{F}_t\}$ is said to be a *submartingale*. It is called a *supermartingale* if

$$\mathcal{E}\{v(t+1) \mid \mathcal{F}_t\} \leq v(t) \quad \text{a.s.}$$

If 3. is replaced by

$$\mathcal{E}\{v(t+1) \mid \mathcal{F}_t\} = 0 \quad \text{a.s.}$$

$\{v(t), \mathcal{F}_t\}$ is called a *martingale difference*.

D.6 CONVERGENCE

1. A sequence of random vectors $\{v(t, \omega), t \in \mathbb{Z}_+\}$ is said to *converge almost surely* (a.s.), or *with probability one*, to $v(\omega)$ if

$$\mathbb{P}\left\{\omega \mid \lim_{t \to \infty} v(t, \omega) = v(\omega)\right\} = 1$$

2. $\{v(t, \omega), t \in \mathbb{Z}_+\}$ *converges in probability* to $v(\omega)$ if, for all $\varepsilon > 0$, we have

$$\lim_{t \to \infty} \mathbb{P}\left\{\omega \mid \|v(t, \omega) - v(\omega)\| > \varepsilon\right\} = 0$$

3. $\{v(t, \omega), t \in \mathbb{Z}_+\}$ *converges in νth mean* $(\nu > 0)$ to $v(\omega)$ if

$$\mathcal{E}\left\{\|v(t, \omega) - v(\omega)\|^\nu\right\} = 0$$

If $\nu = 2$, we say that convergence is in *quadratic mean*, or *mean-square*.

4. $\{v(t, \omega), t \in \mathbb{Z}_+\}$ *converges in distribution* to $v(\omega)$ if

$$\lim_{t \to \infty} P_{v_t}(\alpha) = P_v(\alpha)$$

at all the continuity points of $P_v(\cdot)$.

We point out that the well-known *Markov inequality*

$$\mathbb{P}\left(\omega \mid \|v(\omega)\| \geq \varepsilon\right) \leq \frac{\mathcal{E}\left\{\|v\|^\nu\right\}}{\varepsilon^\nu}$$

for any ε, $\nu > 0$, shows that the convergence in νth mean implies convergence in probability. The following connections exist between the foregoing types of convergence:

$$\begin{array}{ccc}
\begin{array}{c} v(t) \to v \\ \text{a.s.} \end{array} & \Longrightarrow & \\
& & \begin{array}{c} v(t) \to v \\ \text{in probability} \end{array} \Longrightarrow \begin{array}{c} v(t) \to v \\ \text{in distribution} \end{array} \\
\begin{array}{c} v(t) \to v \\ \text{in } \nu\text{th mean} \end{array} & \Longrightarrow &
\end{array}$$

Some convergence results follow.

Lemma D.6-1. (Martingale Stability) *Let $\{v(t), \mathcal{F}_t\}$ be a martingale difference sequence. Then*

$$\sum_{t=1}^{\infty} \frac{1}{t^p} \mathcal{E}\left\{|v(t)|^p \mid \mathcal{F}_{t-1}\right\} < \infty \qquad \text{a.s.}$$

for some $0 < p \leq 2$, implies that

$$\lim_{N \to \infty} \frac{1}{N} \sum_{t=1}^{N} v(t) = 0 \qquad \text{a.s.}$$

Proposition D.6-1.   *Let $\{v(t), \mathcal{F}_t\}$ be a martingale difference sequence. Then*

$$\mathcal{E}\left\{v^2(t) \mid \mathcal{F}_{t-1}\right\} = \sigma^2 \qquad \text{a.s.}$$

and

$$\mathcal{E}\left\{v^4(t) \mid \mathcal{F}_{t-1}\right\} \leq M < \infty \qquad \text{a.s.}$$

imply that

$$\lim_{N \to \infty} \frac{1}{N} \sum_{t=1}^{N} v^2(t) = \sigma^2 \qquad \text{a.s.}$$

Proof: The result follows from Lemma 1. Set $u(t) := v^2(t) - \mathcal{E}\{v^2(t) \mid \mathcal{F}_{t-1}\} = v^2(t) - \sigma^2$. Then $\mathcal{E}\{u(t) \mid \mathcal{F}_{t-1}\} = 0$. Hence, $\{u(t), \mathcal{F}_t\}$ is a martingale difference sequence. Also

$$\sum_{t=1}^{\infty} \frac{1}{t^2} \mathcal{E}\left\{u^2(t) \mid \mathcal{F}_{t-1}\right\} \quad = \quad \sum_{t=1}^{\infty} \frac{1}{t^2} \left[\mathcal{E}\left\{v^4(t) \mid \mathcal{F}_{t-1}\right\} - 2\sigma^4 + \sigma^4\right]$$

$$\leq \quad \sum_{t=1}^{\infty} \frac{1}{t^2}\left(M - \sigma^4\right) < \infty$$

The following *positive supermartingale convergence* result is important for convergence analysis of stochastic recursive algorithms (cf. Theorem 6.4-3).

Theorem D.6-1.   **(The Martingale Convergence Theorem)** *Let $\{v(t), \alpha(t), \beta(t), t \in \mathbb{Z}_+\}$ be three sequences of positive random variables adapted to an increasing sequence of σ-fields $\{\mathcal{F}_t, t \in \mathbb{Z}_+\}$ and such that*

$$\mathcal{E}\left\{v(t) \mid \mathcal{F}_{t-1}\right\} \leq v(t-1) - \alpha(t-1) + \beta(t-1) \qquad \text{a.s.}$$

with

$$\sum_{t=0}^{\infty} \beta(t) \leq \infty \qquad \text{a.s.}$$

Then $\{v(t), t \in \mathbb{Z}_+\}$ converges a.s. to a finite random variable

$$\lim_{t \to \infty} v(t) = v < \infty \qquad \text{a.s.}$$

and

$$\sum_{t=0}^{\infty} \alpha(t) < \infty \qquad \text{a.s.}$$

The following property is used in Chapter 6 to establish convergence results.

Result D.6-1.   **(Kronecker Lemma)** *Let $\{a(t)\}$ and $\{b(t)\}$ be two real-valued sequences such that*

$$\lim_{k \to \infty} \sum_{t=1}^{k} a(t) < \infty$$

$\{b(t)\}$ *is nondecreasing and* $\lim_{t \to \infty} b(t) = \infty$ *Then*

$$\lim_{k \to \infty} \frac{1}{b(k)} \sum_{t=1}^{k} b(t) a(t) = 0$$

D.7 MINIMUM MEAN-SQUARE-ERROR ESTIMATORS

Consider the square-integrable random variables w and $\{y_i\}_{i=1}^n$ on a common probability space $(\Omega, \mathcal{F}, \mathbb{P})$. Let $\mathcal{A} = \sigma(y)$ be the sub-σ-field of $\mathcal{F}$ generated by the components of $y := [\ y_1 \ \cdots \ y_n\ ]'$. Then $L_2(\mathcal{A}) := L_2(\Omega, \mathcal{A}, \mathbb{P})$ is a closed subspace of $L_2(\mathcal{F}) = L_2(\Omega, \mathcal{F}, \mathbb{P})$. Its elements can be conceived as all square-integrable random variables given by any nonlinear transformation of y. We show that the conditional mean $\mathcal{E}\{w \mid y\} = \mathcal{E}\{w \mid \mathcal{A}\}$ enjoys the following property:

$$\mathcal{E}\{w \mid y\} = \arg \min_{v \in L_2(\mathcal{A})} \mathcal{E}\left\{(w - v)^2\right\} \tag{D.7-1}$$

In fact, setting $\hat{w} = \mathcal{E}\{w \mid y\}$,

$$
\begin{aligned}
\mathcal{E}\left\{(w - v)^2\right\} &= \mathcal{E}\left\{[(w - \hat{w}) + (\hat{w} - v)]^2\right\} \\
&= \mathcal{E}\left\{(w - \hat{w})^2 + (\hat{w} - v)^2\right\} + 2\mathcal{E}\left\{(w - \hat{w})(\hat{w} - v)\right\} \\
&= \mathcal{E}\left\{(w - \hat{w})^2\right\} + \mathcal{E}\left\{(\hat{w} - v)^2\right\} \\
&\geq \mathcal{E}\left\{(w - \hat{w})^2\right\}
\end{aligned}
$$

where the third equality follows because by the smoothing properties of conditional expectations,

$$
\begin{aligned}
\mathcal{E}\left\{(\hat{w} - v)(w - \hat{w})\right\} &= \mathcal{E}\left\{\mathcal{E}\left\{(\hat{w} - v)(w - \hat{w}) \mid y\right\}\right\} \\
&= \mathcal{E}\left\{(\hat{w} - v)\mathcal{E}\left\{w - \hat{w} \mid y\right\}\right\} = 0
\end{aligned}
$$

The RHS of (1) is called the *minimum mean–square error (MMSE) or minimum variance estimator* of w based on y. Hence, (1) shows that the MMSE estimator of w given y is given by the conditional mean $\mathcal{E}\{w \mid y\}$. The latter can be interpreted as the orthogonal projection of $w \in L_2(\mathcal{F})$ onto $L_2(\sigma(y))$.

REFERENCES

[ABJ+86] B. D. O. Anderson, R. Bitmead, C. R. Johnson, P. V. Kokotovic, R. L. Kosut, I. Mareels, L. Praly, and B. Riedle. *Stability of Adaptive Systems Passivity and Averaging Analysis*. The MIT Press, Cambridge, Mass., 1986.

[ÅBLW77] K. J. Åstrom, V. Borisson, L. Ljung, and B. Wittenmark. Theory and application of self-tuning regulators. *Automatica*, 13:457–476, 1977.

[ÅBW65] K. J. Åström, T. Bohlin, and S. Wensmark. Automatic construction of linear stochastic dynamic models for stationary industrial processes with random disturbances using operating records. TP 18.150, IBM Nordic Laboratory, Stockholm, Sweden, June 1965.

[AF66] M. Athans and P. L. Falb. *Optimal Control: An Introduction to the Theory and Its Applications*. McGraw-Hill, New York, 1966.

[AJ83] B. D. O. Anderson and R. M. Johnstone. Adaptive systems and time varying plants. *Int. J. Control*, 37:367–377, 1983.

[AL84] W. F. Arnold and A. J. Laub. Generalized eigenproblem algorithms and software for algebraic Riccati equations. *Proc. IEEE*, 72:1746–1754, 1984.

[AM71] B. D. O. Anderson and J. B. Moore. *Linear Optimal Control*. Prentice Hall, Englewood Cliffs, N.J., 1971.

[AM79] B. D. O. Anderson and J. B. Moore. *Optimal Filtering*. Prentice Hall, Englewood Cliffs, N.J., 1979.

[AM90] B. D. O. Anderson and J. B. Moore. *Optimal Control: Linear Quadratic Methods*. Prentice Hall, Englewood Cliffs, N.J., 1990.

[And67] B. D. O. Anderson. An algebraic solution to the spectral factorization problem. *IEEE Trans. Automat. Control*, 12:410–414, 1967.

[Åst70] K. J. Åström. *Introduction to Stochastic Control Theory*. Academic Press, New York, 1970.

[Åst87]   K. J. Åström. Adaptive feedback control. *Proc. IEEE*, 75:185–217, 1987.

[Ath71]   M. Athans (ed.). Special issue on the linear–quadratic–Gaussian problem. *IEEE Trans. Automat. Control*, 16:527–869, 1971.

[ÅW73]   K. J. Åström and B. Wittenmark. On self–tuning regulators. *Automatica*, 9:185–189, 1973.

[ÅW84]   K. J. Åström and B. Wittenmark. *Computer Controlled Systems: Theory and Design*. Prentice Hall, Englewood Cliffs, N.J., 1984.

[ÅW89]   K. J. Åström and B. Wittenmark. *Adaptive Control*. Addison-Wesley, Reading, Mass., 1989.

[Bar83]   S. Barnett. *Polynomial and Linear Control Systems*. Marcel Dekker, New York, 1983.

[BB91]   S. P. Boyd and C. H. Barrat. *Linear Controller Design: Limits of Performance*. Prentice Hall, Englewood Cliffs, N.J., 1991.

[BBC90a]   S. Bittanti, P. Bolzern, and M. Campi. Convergence and exponential convergence of identification algorithms with directional forgetting factor. *Automatica*, 26:929–932, 1990.

[BBC90b]   S. Bittanti, P. Bolzern, and M. Campi. Recursive least–squares identification algorithms with incomplete excitation: Convergence analysis and application to adaptive control. *IEEE Trans. Automat. Control*, 35:1371–1373, 1990.

[Bel57]   R. Bellman. *Dynamic Programming*. Princeton University Press, Princeton, 1957.

[Ber76]   D. P. Bertsekas. *Dynamic Programming and Stochastic Control*. Academic Press, New York, 1976.

[BGW90]   R. R. Bitmead, M. Gevers, and V. Wertz. *Adaptive Optimal Control: The Thinking Man's GPC*. Prentice Hall, Englewood Cliffs, N.J., 1990.

[BH75]   A. E. Bryson and Y. C. Ho. *Applied Optimal Control*. Hemisphere, New York, 1975.

[Bie77]   G. J. Bierman. *Factorization Methods for Discrete Sequential Estimation*. Academic Press, New York, 1977.

[Bit89]   S. Bittanti (ed.). *Preprints Workshop on the Riccati Equation in Control, Systems, and Signals*. Pitagora, Bologna, Italy, 1989.

[BJ76]   G. E. P. Box and G. Jenkins. *Time Series Analysis, Forecasting and Control*, 2d ed. Holden-Day, San Francisco, California, 1976.

[BN66]   F. Brauer and J. Nohel. *Ordinary Differential Equations*. W. A. Benjamin, New York and Amsterdam, 1966.

[Böh85]   J. Böhm. LQ self-tuners with signal level constraints. In *Preprints of the 7th IFAC Symp. Ident. Syst. Param. Est.*, pp. 131–135. Pergamon, Oxford, UK, 1985.

[Bro70]   R. W. Brockett. *Finite Dimensional Linear Systems*. Wiley, New York, 1970.

[BS78]   D. Bertsekas and S. E. Shreve. *Stochastic Optimal Control: The Discrete Time Case*. Academic Press, New York, 1978.

[BST74]   Y. Bar-Shalom and E. Tse. Dual effect, certainty equivalence and separation in stochastic control. *IEEE Trans. Automat. Control*, 19:494–500, 1974.

[But92]   H. Butler. *Model Reference Adaptive Control: Bridging the Gap between Theory and Practice*. Prentice Hall, Englewood Cliffs, N.J., 1992.

[BY83]   H. Blomberg and R. Ylinen. *Algebraic Theory for Multivariable Linear Systems*. Academic Press, New York, 1983.

[Cai76]   P. E. Caines. Prediction error identification methods for stationary stochastic processes. *IEEE Trans. Automat. Control*, 21:500–506, 1976.

[Cai88]   P. E. Caines. *Linear Stochastic Systems*. Wiley, New York, 1988.

[CD89]   B. S. Chen and T. Y. Dong. LQG optimal control system design under plant perturbation and noise uncertainty: A state-space approach. *Automatica*, 25:431–436, 1989.

[CD91]   F. M. Callier and C. A. Desoer. *Linear System Theory*. Springer-Verlag, New York, 1991.

[CDMZ87]   G. Casalino, F. Davoli, R. Minciardi, and G. Zappa. On implicit modelling theory: Basic concepts and application to adaptive control. *Automatica*, 23:189–201, 1987.

[CG75]   D. W. Clarke and P. J. Gawthrop. Self-tuning controller. *Proc. IEE*, 122, D:929–934, 1975.

[CG79]   D. W. Clarke and P. J. Gawthrop. Self-tuning control. *Proc. IEE*, 126, D:633–640, 1979.

[CG88]   H. F. Chen and L. Guo. A robust adaptive controller. *IEEE Trans. Automat. Control*, 33:1035–1043, 1988.

[CG91]   H. F. Chen and L. Guo. *Identification and Stochastic Adaptive Control*. Birkhäuser, Boston, 1991.

[CGMN91]   A. Casavola, M. Grimble, E. Mosca, and P. Nistri. Continuous-time LQ regulator design by polynomial equations. *Automatica*, 25:555–558, 1991.

[CGS84]   S. W. Chan, G. C. Goodwin, and K. S. Sin. Convergence properties of the Riccati difference equation in optimal filtering of nonstabilizable systems. *IEEE Trans. Automat. Control*, 29:110–118, 1984.

[Cha87]   V. V. Chalam. *Adaptive Control Systems*. Marcel Dekker, New York, 1987.

[Che70]   C. T. Chen. *Introduction to Linear System Theory*. Holt, Rinehart & Winston, New York, 1970.

[Che81]   H. F. Chen. Quasi-least-squares identification and its strong consistency. *Int. J. Control*, 34:921–936, 1981.

[Che85]   H. F. Chen. *Recursive Estimation and Control for Stochastic Systems*. Wiley, New York, 1985.

[Chu68]   K. L. Chung. *A Course in Probability Theory*. Hartcourt, Brace, & World, New York, 1968.

[CM89]   D. W. Clarke and C. Mohtadi. Properties of generalized predictive control. *Automatica*, 25:859–875, 1989.

[CM91]   A. Casavola and E. Mosca. Polynomial LQG regulator design for general systems configurations. In *Proceedings of the 30th IEEE Conference on Decision and Control*, pp. 2307–2312. Brighton, UK, 1991.

[CM92a]   L. Chisci and E. Mosca. Adaptive predictive control of ARMAX plants with unknown deadtime. In Preprints *IFAC Symposium on Adaptive Systems in Control and Signal Processing*, pp. 199–204. ENSIEG, Grenoble, France, 1992.

[CM92b]   L. Chisci and E. Mosca. Polynomial equations for the linear MMSE state estimation. *IEEE Trans. Automat. Control*, 37:623–626, 1992.

[CM93]   L. Chisci and E. Mosca. Stabilizing predictive control: The singular transition matrix case. In *Proceedings of the Conference on Advances in Model-Based Predictive Control*, Vol. 1:130–138, Dept. of Engineering Science, Oxford University, Oxford, UK, 1993.

[CMP91]   G. Casalino, R. Minciardi, and T. Parisini. Development of a new self-tuning control algorithm for finite and infinite horizon quadratic adaptive optimization. *Int. J. Adapt. Control and Signal Processing*, 5:505–525, 1991.

[CMS91]   D. W. Clarke, E. Mosca, and R. Scattolini. Robustness of an adaptive predictive controller. In *Proceedings of the 30th IEEE Conference on Decision and Control*, pp. 1788–1789. Brighton, UK, 1991.

[CMT85]   D. W. Clarke, P. S. Tuffs, and C. Mohtadi. Self-tuning control of a difficult process. In *Preprints of the 7th IFAC Symp. on Identification and System Parameter Estimation*, pp. 1009–1014. York, UK, 1985.

[CMT87a] D. W. Clarke, C. Mohtadi, and P. S. Tuffs. Generalized predictive control–Part I: The basic algorithm. *Automatica*, 23:137–148, 1987.

[CMT87b] D. W. Clarke, C. Mohtadi, and P. S. Tuffs. Generalized predictive control–Part II: Extensions and interpretations. *Automatica*, 23:149–160, 1987.

[CR80]   C. R. Cutler and B. L. Ramaker. Dynamic matrix control: a computer control algorithm. In *Proceedings of the Joint American Control Conference*. San Francisco, 1980.

[Cra46]   H. Cramér. *Mathematical Methods of Statistics*. Princeton University Press, Princeton, 1946.

[Cri87]   R. Cristi. Internal persistency of excitation in indirect adaptive control. *IEEE Trans. Automat. Control*, 32:1101–1103, 1987.

[CS82]   C. C. Chen and L. Shaw. On receding horizon control. *Automatica*, 18:349–352, 1982.

[CS91]   D. W. Clarke and R. Scattolini. Constrained receding horizon predictive control. *Proc. IEE*, 138, D:347–354, 1991.

[DC75]   C. A. Desoer and M. C. Chan. The feedback interconnection of lumped linear time-invariant systems. *J. Franklin Inst.*, 300:335–351, 1975.

[Des70a] C. A. Desoer. *Notes for a Second Course on Linear Systems*. Van Nostrand Reinhold, New York, 1970.

[Des70b] C. A. Desoer. Slowly varying discrete system $x_{i+1} = A_i x_i$. *Electron. Lett.*, 7:339–340, 1970.

[DG91]   H. Demircioğlu and P. J. Gawthrop. Continuous-time generalized predictive control (CGPC). *Automatica*, 27:55–74, 1991.

[DG92]   H. Demircioğlu and P. J. Gawthrop. Multivariable continuous-time generalized predictive control (MCGPC). *Automatica*, 28:697–713, 1992.

[dKvC85]   R. M. C. de Keyser and A. R. van Cauvenberghe. Extended prediction self-adaptive control. In *Preprints of the 7th IFAC Symposium on Identification and System Parameter Estimation*, pp. 1255–1260. York, UK, 1985.

[DL84]   P. De Larminat. On the stabilizability condition in indirect adaptive control. *Automatica*, 20:793–795, 1984.

[DLMS80]   C. A. Desoer, R. W. Liu, J. Murray, and R. Saeks. Feedback system design: The fractional representation approach to analysis and synthesis. *IEEE Trans. Automat. Control*, 25:399–412, 1980.

[Doo53]   J. L. Doob. *Stochastic Processes*. Wiley, New York, 1953.

[DS79]   J. C. Doyle and G. Stein. Robustness with observers. *IEEE Trans. Automat. Control*, 24:607–611, 1979.

[DS81]   J. C. Doyle and G. Stein. Multivariable feedback design: Concepts for a classical/modern synthesis. *IEEE Trans. Automat. Control*, 26:4–61, 1981.

[dS89]   C. E. de Souza. Monotonicity and stabilizability results for the solution of the Riccati difference equations. In *Proceedings of the Workshop on the Riccati Equation in Control, Systems and Signals*, S. Bittanti (ed.), pp. 38–41. Pitagora, Bologna, Italy, 1989.

[dSGG86]   C. E. de Souza, M. Gevers, and G. C. Goodwin. Riccati equations in optimal filtering of nonstabilizable systems having singular state transition matrices. *IEEE Trans. Automat. Control*, 31:831–839, 1986.

[DUL73]   J. D. Deshpande, T. N. Upadhyay, and P. G. Lainiotis. Adaptive control of linear stochastic systems. *Automatica*, 9:107–115, 1973.

[DV75]   C. A. Desoer and M. Vidyasagar. *Feedback Systems: I/O Properties*. Academic Press, New York, 1975.

[DV85]   M. H. A. Davis and R. B. Vinter. *Stochastic Modelling and Control*. Chapman and Hall, London, 1985.

[ECD85]   H. Elliot, R. Cristi, and M. Das. Global stability of adaptive pole placement algorithms. *IEEE Trans. Automat. Control*, 30:348–356, 1985.

[Ega79]   B. Egardt. *Stability of Adaptive Controllers*. Springer-Verlag, 1979.

[Eme67]   S. V. Emelyanov. *Variable Structure Control Systems*. Oldenburger Verlag, Munich, F.G.R., 1967.

[Eyk74] P. Eykhoff. *System Identification: Parameter and State Estimation.* Wiley, New York, 1974.

[Fel65] A. A. Feldbaum. *Optimal Control Systems.* Academic Press, New York, 1965.

[FFH+91] C. Foias, B. Francis, J. W. Helton, H. Kwakernaak, and J. B. Pearson. *$\mathcal{H}_\infty$-Control Theory.* E. Mosca and L. Pandolfi (eds.). Springer-Verlag, Berlin, 1991.

[FKY81] T. R. Fortescue, L. S. Kershenbaum, and B. E. Ydstie. Implementation of self-tuning regulators with variable forgetting factors. *Automatica*, 17:831–835, 1981.

[FM78] A. Feuer and S. Morse. Adaptive control for single-input single-output linear systems. *IEEE Trans. Automat. Control*, 23:557–570, 1978.

[FN74] Y. Funahashi and K. Nakamura. Parameter estimation of discrete-time systems using short periodic pseudo-random sequences. *Int. J. Control*, 19:1101–1110, 1974.

[FR75] W. H. Fleming and R. W. Rishel. *Deterministic and Stochastic Optimal Control.* Springer-Verlag, New York, 1975.

[Fra64] J. S. Frame. Matrix functions and applications—Part IV. *Spectrum*, 1:123–131, 1964.

[Fra87] B. A. Francis. *A Course in $\mathcal{H}_\infty$ Control Theory.* Springer-Verlag, New York, 1987.

[Fra91] B. A. Francis. Lectures on $\mathcal{H}_\infty$ control and sampled-data systems. In *$\mathcal{H}_\infty$-Control Theory.* E. Mosca and L. Pandolfi (eds.), pp. 37–105. Springer-Verlag, New York, 1991.

[FW76] B. A. Francis and W. M. Wonham. The internal model principle of control theory. *Automatica*, 12:457–465, 1976.

[Gau63] K. F. Gauss. *Teoria Motus Corporum Cœlestium in Sectionibus Conicus Solem Ambientium,* 1809. Reprinted as *Theory of the Motion of the Heavenly Bodies Moving about the Sun in Conic Sections.* Dover, New York, 1963.

[Gaw87] P. J. Gawthrop. *Continuous-Time Self-Tuning Control.* Wiley, New York, 1987.

[GC91] L. Guo and H. F. Chen. The Åstrom-Wittenmark self-tuning regulator revisited and ELS-based adaptive trackers. *IEEE Trans. Automat. Control*, 36:802–812, 1991.

[GdC87]   J. C. Geromel and J. J. da Cruz. On the robustness of optimal regulators for nonlinear discrete-time systems. *IEEE Trans. Automat. Control*, 32, 1987.

[Gel74]   A. Gelb (ed.). *Applied Optimal Estimation*. The MIT Press, Cambridge, Mass.,1974.

[GGMP92]   L. Giarré, R. Giusti, E. Mosca, and M. Pacini. Adaptive digital PID autotuning for robotic applications. In *Proceedings of the 36th ANIPLA Annual Conference*, vol. III, pp. 1–11. Pirella, Genoa, Italy, 1992.

[GIM⁺90]   M. Galanti, F. Innocenti, S. Magrini, E. Mosca, and V. Spicci. An innovative adaptive control approach to complex high performance servos: An Officine Galileo case study. In *Modelling the Innovation: Communications, Automation and Information Systems,* M. Carnevale, M. Lucertini, and S. Nicosia (eds.), pp. 499–506. Elsevier-North Holland, Amsterdam, Holland, 1990.

[GJ88]   M. J. Grimble and M. A. Johnson. *Optimal Control and Stochastic Estimation*, vols. 1 and 2. Wiley, Chichester, UK, 1988.

[GMDD91]   F. Giri, M. M'Saad, J. M. Dion, and L. Dugard. On the robustness of discrete-time indirect adaptive (linear) controllers. *Automatica*, 27:153–159, 1991.

[GMMZ84]   C. Greco, G. Menga, E. Mosca, and G. Zappa. Performance improvements of self-tuning controllers by multistep horizons: the MUSMAR approach. *Automatica*, 20:681–699, 1984.

[GN85]   P. J. Gawthrop and M. T. Nihtila. Identification of time-delays using a polynomial identification method. *Syst. Control Lett.*, 5:267–271, 1985.

[GP77]   G. C. Goodwin and R. L. Payne. *Dynamic System Identification: Experiment Design and Data Analysis*. Academic Press, New York, 1977.

[GPM89]   C. E. García, D. M. Prett, and M. Morari. Model predictive control: Theory and practice—A survey. *Automatica*, 25:335–348, 1989.

[GRC80]   G. C. Goodwin, P. J. Ramadge, and P. Caines. Discrete-time multivariable control. *IEEE Trans. Automat. Control*, 25:449–456, 1980.

[GRC81]   G. C. Goodwin, P. J. Ramadge, and P. E. Caines. Discrete time stochastic adaptive control. *SIAM J. Control Optim.*, 19:829–853, 1981.

[Gri84]   M. J. Grimble. Implicit and explicit LQG self-tuning controllers. *Automatica*, 20:661–669, 1984.

[Gri85]   M. J. Grimble. Polynomial system approach to optimal linear filtering and prediction. *Int. J. Control*, 41:1545–1564, 1985.

[Gri87]   M. J. Grimble. Relationship between polynomial and state-space solutions of the optimal regulator problem. *Syst. Control Lett.*, 8:411–416, 1987.

[Gri90]   M. J. Grimble. LQG predictive optimal control for adaptive applications. *Automatica*, 26:949–961, 1990.

[GS84]   G. C. Goodwin and K. S. Sin. *Adaptive Filtering Prediction and Control*. Prentice Hall, Englewood Cliffs, N.J., 1984.

[GVL83]   G. H. Golub and C. F. Van Loan. *Matrix Computations*. The Johns Hopkins University Press, Baltimore, 1983.

[Hag83]   T. Hagglund. The problem of forgetting old data in recursive estimation. In *Preprints of the 1st IFAC Workshop on Adaptive Systems in Control and Signal Processing*. San Francisco, 1983.

[Han76]   E. J. Hannan. The convergence of some time-series recursions. *Ann. Stat.*, 4:1258–1270, 1976.

[HD88]   E. J. Hannan and M. Deistler. *The Statistical Theory of Linear Systems*. Wiley, New York, 1988.

[Hew71]   G. A. Hewer. An iterative technique for the computation of the steady state gains for the discrete time optimal regulator. *IEEE Trans. Automat. Control*, 16:382–384, 1971.

[Hij86]   O. Hijab. *The Stabilization of Control Systems*. Springer-Verlag, New York, 1986.

[HKŠ92]   J. Hunt, V. Kučera, and M. Šebek. Optimal regulation using measurement feedback: A polynomial approach. *IEEE Trans. Automat. Control*, 37:682–685, 1992.

[HŠG87]   K. J. Hunt, M. Šebek, and M. J. Grimble. Optimal multivariable LQG control using a single Diophantine equation. *Int. J. Control*, 46:1445–1453, 1987.

[HŠK91]   K. J. Hunt, M. Šebek, and V. Kučera. Polynomial approach to $\mathcal{H}_2$-optimal control: The multivariable standard problem. In *Proceedings of the 30th IEEE Conference on Decision and Control*, pp. 1261–1266. Brighton, UK, 1991.

[ID91]   P. Ioannou and A. Datta. Robust adaptive control: A unified approach. *Proc. IEEE*, 79:1736–1767, 1991.

[IK83]   P. Ioannou and P. V. Kokotovic. *Adaptive Systems with Reduced Models*. Springer-Verlag, New York, 1983.

[IK84]   P. Ioannou and P.V. Kokotovic. Instability analysis and improvement of robustness of adaptive control. *Automatica*, 20:583–594, 1984.

[ILM92]   R. Isermann, K.-H. Lachmann, and D. Matko. *Adaptive Control Systems*. Prentice Hall, New York, 1992.

[IS88]   P. Ioannou and J. Sun. Theory and design of robust direct and indirect adaptive control schemes. *Int. J. Control*, 47:775–813, 1988.

[IT86a]   P. Ioannou and K. Tsakalis. A robust direct adaptive controller. *IEEE Trans. Automat. Control*, 31:1033–1043, 1986.

[IT86b]   T. Ishihara and H. Takeda. Loop transfer recovery techniques for discrete-time optimal regulators using prediction estimators. *IEEE Trans. Automat. Control*, 31:1149–1151, 1986.

[Itk76]   U. Itkis. *Control Systems of Variable Structure*. Wiley, New York, 1976.

[Jaz70]   A. H. Jazwinski. *Stochastic Processes and Filtering Theory*. Academic Press, New York, 1970.

[JJBA82] R. M. Johnstone, C. R. Johnson, R. R. Bitmead, and B. D. O. Anderson. Exponential convergence of recursive least squares with exponential forgetting factor. *Syst. Control Lett.*, 2:77–82, 1982.

[JK85]   J. Ježek and V. Kučera. Efficient algorithm for matrix spectral factorization. *Automatica*, 21:663–669, 1985.

[Joh88]   C. R. Johnson Jr. *Lectures on Adaptive Parameter Estimation*. Prentice Hall, Englewood Cliffs, N.J., 1988.

[Joh92]   R. Johansson. Supermartingale analysis of minimum variance adaptive control. In *Preprints of the 4th IFAC International Symposium on Adaptive Systems in Control and Signal Processing*, pp. 521–526, ENSIEG, Grenoble, France, 1992.

[KA86]   G. Kreisselmeir and B. D. O. Anderson. Robust model reference adaptive control. *IEEE Trans. Automat. Control*, 31:127–134, 1986.

[Kai68]   T. Kailath. An innovations approach to least-squares estimation. Part 1: Linear filtering in additive white noise. *IEEE Trans. Automat. Control*, 13:646–654, 1968.

[Kai74]   T. Kailath. A view of three decades of linear filtering theory. *IEEE Trans. Inf. Theory*, 20:145–181, 1974.

[Kai76]   T. Kailath. *Lectures on Linear Least Squares Estimation*. Springer-Verlag, New York, 1976.

[Kai80]   T. Kailath. *Linear Systems*. Prentice Hall, Englewood Cliffs, N.J., 1980.

[Kal58]   R. E. Kalman. Design of a self-optimizing control system. *Trans. ASME*, 80:468–478, 1958.

[Kal60a]   R. E. Kalman. Contributions to the theory of optimal control. *Boletin de la Societad Matematica Mexicana*, 5:102–119, 1960.

[Kal60b]   R. E. Kalman. A new approach to linear filtering and prediction problems. *ASME Trans., Series D, J. Basic Eng.*, 82:35–45, 1960.

[KB61]   R. E. Kalman and R. S. Bucy. New results in linear filtering and prediction theory. *ASME Trans., Series D, J. Basic Eng.*, 83:95–108, 1961.

[KBK83]   W. H. Kwon, A. N. Bruckstein, and T. Kailath. Stabilizing state–feedback design via the moving horizon method. *Int. J. Control*, 37, 1983.

[KG85]   S. S. Keerthi and E. G. Gilbert. An existence theorem for discrete-time infinite-horizon optimal control problems. *IEEE Trans. Automat. Control*, 30:907–909, 1985.

[KG86]   S. S. Keerthi and E. G. Gilbert. Optimal infinite-horizon control and the stabilization of linear discrete-time systems: state-control constraints and nonquadratic cost functions. *IEEE Trans. Automat. Control*, 31:264–266, 1986.

[KG88]   S. S. Keerthi and E. G. Gilbert. Optimal infinite-horizon feedback laws for a general class of constrained discrete-time systems: stability and moving-horizon approximations. *J. Optimization Theory and Applications*, 57:265–293, 1988.

[KHB+85]   M. Kárný, A. Halousková, J. Böhm, R. Kulavý and P. Nedoma. Design of linear quadratic adaptive control: Theory and algorithms for practice. *Kybernetika*, 21:Supplement, 1985.

[KK84]   R. Kulhavý and M. Kárný. Tracking of slowly varying parameters by directional forgetting. In *Preprints of the 9th IFAC World Congress*, vol. X, pp. 178–183. Budapest, 1984.

[KKO86]   P. V. Kokotović, H. K. Khalil, and J. O'Reilly. *Singular Perturbation Methods in Control: Analysis and Design*. Academic Press, New York, 1986.

[Kle68]   D. L. Kleinman. On an iterative tecnique for Riccati equation computation. *IEEE Trans. Automat. Control*, 13:114–115, 1968.

[Kle70]   D. L. Kleinman. An easy way to stabilize a linear constant system. *IEEE Trans. Automat. Control*, 15:692, 1970.

[Kle74]   D. L. Kleinman. Stabilizing a discrete constant linear system with application to iterative methods for solving the Riccati equation. *IEEE Trans. Automat. Control*, 19:252–254, 1974.

[Kol41]   A. N. Kolmogorov. Stationary sequences in Hilbert space (in Russian). *Bull. Math. Univ. Moscow*, 2(6), 1941. English translation in *Linear Least Squares*, T. Kailath (ed.). Dowden, Hutchinson & Ross, New York, 1977.

[KP75]   W. H. Kwon and A. E. Pearson. On the stabilization of a discrete constant linear system. *IEEE Trans. Automat. Control*, 20:800–801, 1975.

[KP78]   W. H. Kwon and A. E. Pearson. On feedback stabilization of time-varying discrete linear systems. *IEEE Trans. Automat. Control*, 23:479–481, 1978.

[KP87]   P. R. Kumar and L. Praly. Self-tuning trackers. *SIAM J. Control and Optim.*, 25:1053–1071, 1987.

[KR76]   R. L. Kashyap and A. R. Rao. *Dynamic Stochastic Models from Empirical Data*. Academic Press, New York, 1976.

[Kre89]   G. Kreisselmeir. An indirect adaptive controller with a self-excitation capability. *IEEE Trans. Automat. Control*, 34:524–528, 1989.

[KS72]   H. Kwakernaak and R. Sivan. *Linear Optimal Control Systems*. Wiley, New York, 1972.

[KS79]   M. G. Kendall and A. Stuart. *The Advanced Theory of Statistics*, vol. 2, 4th ed. Griffin, London, 1979.

[Kuč75]   V. Kučera. Algebraic approach to discrete linear control. *IEEE Trans. Automat. Control*, 20:116–120, 1975.

[Kuč79]   V. Kučera. *Discrete Linear Control: The Polynomial Equation Approach*. Wiley, Chichester, 1979.

[Kuč81]   V. Kučera. New results in state estimation and regulation. *Automatica*, 17:745–748, 1981.

[Kuč83]   V. Kučera. Linear quadratic control, state space vs. polynomial equations. *Kybernetica*, 19:185–195, 1983.

[Kuč91]   V. Kučera. *Analysis and Design of Discrete Linear Control Systems*. Prentice Hall, Englewood Cliffs, N.J., 1991.

[Kul87]   R. Kulhavý. Restricted exponential forgetting in real-time identification. *Automatica*, 23:589–600, 1987.

[KV86]   P. R. Kumar and P. Varaiya. *Stochastic Systems*. Prentice Hall, Englewood Cliffs, N.J., 1986.

[Kwa69]   H. Kwakernaak. Optimal low sensitivity linear feedback systems. *Automatica*, 5:279–286, 1969.

[Kwa85]   H. Kwakernaak. Minimax frequency domain performance and robustness optimization of linear feedback systems. *IEEE Trans. Automat. Control*, 30:994–1004, 1985.

[Lan79]   Y. D. Landau. *Adaptive Control—The Model Reference Approach*. Marcel Dekker, New York, 1979.

[Lan85]   Y. D. Landau. Adaptive control techniques for robotic manipulators: the status of the art. In *Preprints of the IFAC Syroco—85*. Barcelona, 1985.

[Lan90]   I. D. Landau. *System Identification and Control Design*. Prentice Hall, Englewood Cliffs, N.J., 1990.

[LDU72]   P. G. Lainiotis, J. D. Deshpande, and T. N. Upadhhay. Optimal adaptive control: A nonlinear separation theorem. *Int. J. Control*, 15:877–888, 1972.

[Lew86]   F. L. Lewis. *Optimal Control*. Wiley, New York, 1986.

[LG85]   R. Lozano–Leal and G. C. Goodwin. A globally convergent adaptive pole placement algorithm without a persistency of excitation requirement. *IEEE Trans. Automat. Control*, 30:795–798, 1985.

[LH74]   C. L. Lawson and R. J. Hanson. *Solving Least Squares Problems*. Prentice Hall, Englewood Cliffs, N.J., 1974.

[Lju77a]   L. Ljung. On positive real functions and the convergence of some recursive schemes. *IEEE Trans. Automat. Control*, 22:539–551, 1977.

[Lju77b]   L. Ljung. Analysis of recursive stochastic algorithms. *IEEE Trans. Automat. Control*, 22:551–575, 1977.

[Lju78]   L. Ljung. Convergence analysis of parametric identification methods. *IEEE Trans. Automat. Control*, 23:770–783, 1978.

[Lju87]   L. Ljung. *System Identification: Theory for the User*. Prentice Hall, Englewood Cliffs, N.J., 1987.

[LKS85]   W. Lin, P. R. Kumar, and T. I. Seidman. Will the self-tuning approach work for general cost criteria? *Syst. Control Lett.*, 6:77–85, 1985.

[LM76]   A. Luvison and E. Mosca. Development of recursive deconvolution algorithms via innovations analysis with applications to identification by PRBS's. In *Preprints of the 4th IFAC Symp. Identification and System Parameter Estimation*, vol. 3, pp. 604–614, Institute of Control Sciences, Tbilisi, Moscow, 1976.

[LM85]   J. M. Lemos and E. Mosca. A multipredictor-based LQ self-tuning controller. In *Preprints of the 7th IFAC Symposium on Identification and System Parameter Estimation*, pp. 137–142. York, UK, 1985.

[LN88]   T. H. Lee and K. S. Narendra. Robust adaptive control of discrete-time systems using persistent excitation. *Automatica*, 24:781–788, 1988.

[Loè63]   M. Loève. *Probability Theory*, 3d ed. Van Nostrand, New York, 1963.

[Loz89]   R. Lozano-Leal. Robust adaptive regulation without persistent excitation. *IEEE Trans. Automat. Control*, 34:1260–1267, 1989.

[LPVD83] D. P. Looze, H. V. Poor, K. S. Vastola, and J. C. Darragh. Minimax control of linear stochastic systems with noise uncertainty. *IEEE Trans. Automat. Control*, 28:882–896, 1983.

[LS83]   L. Ljung and T. Söderström. *Theory and Practice of Recursive Identification*. The MIT Press, Cambridge, Mass., 1983.

[LSA81]   N. A. Lehtomaki, N. R. Sandell Jr., and M. Athans. Robustness results in linear quadratic Gaussian based multivariable control design. *IEEE Trans. Automat. Control*, 26:75–92, 1981.

[Lue69]   D. G. Luenberger. *Optimization by Vector Space Methods*. Wiley, New York, 1969.

[LW82]   T. L. Lai and C. Z. Wei. Least squares estimates in stochastic regression models with application to identification and control of dynamic systems. *Ann. Statist.*, 10:154–166, 1982.

[LW86]   T. L. Lai and C. Z. Wei. Extended least squares and their application to adaptive control and prediction in linear systems. *IEEE Trans. Automat. Control*, 31:898–906, 1986.

[Mac85]   J. M. Maciejowski. Asymptotic recovery for discrete-time systems. *IEEE Trans. Automat. Control*, 30:602–605, 1985.

[Mar76a] J. M. Martín Sánchez. *Adaptive Predictive Control System*. U.S. Patent No. 4, 196, 576. Priority date August 4, 1976.

[Mar76b] J. M. Martín Sánchez. A new solution to adaptive control. *Proc. IEEE*, 64:1209–1218, 1976.

[Mår85] B. Mårtensson. The order of any stabilizing regulator is sufficient information for adaptive stabilization. *Syst. Control Lett.*, 6:87–91, 1985.

[May65] D. Q. Mayne. Parameter estimation. *Automatica*, 3:245–255, 1965.

[May79] P. S. Maybeck. *Stochastic Models Estimation and Control*, vol. 1. Academic Press, New York, 1979.

[May82a] P. S. Maybeck. *Stochastic Models Estimation and Control*, vol. 2. Academic Press, New York, 1982.

[May82b] P. S. Maybeck. *Stochastic Models Estimation and Control*, vol. 3. Academic Press, New York, 1982.

[MC85] S. P. Meyn and P. E. Caines. The zero divisor problem in multivariable stochastic adaptive control. *Syst. Control Lett.*, 6:235–238, 1985.

[McC69] N. H. McClamroch. Duality and bounds for the matrix Riccati equation. *J. Math. Anal. Appl.*, 25:622–627, 1969.

[MCG90] E. Mosca, A. Casavola, and L. Giarrè. Minimax LQ stochastic tracking and servo problems. *IEEE Trans. Automat. Control*, 35:95–97, 1990.

[Med69] J. S. Meditch. *Stochastic Optimal Linear Estimation and Control.* McGraw-Hill, New York, 1969.

[Men73] J. M. Mendel. *Discrete Techniques for Parameter Estimation: Equation Error Formulation.* Marcel Dekker, New York, 1973.

[MG90] R. H. Middleton and G. C. Goodwin. *Digital Control and Estimation: A Unified Approach.* Prentice Hall, Englewood Cliffs, N.J., 1990.

[MG92] E. Mosca and L. Giarré. A polynomial approach to the MIMO LQ servo and disturbance rejection problems. *Automatica*, 28:209–213, 1992.

[MGHM88] R. H. Middleton, G. C. Goodwin, D. J. Hill, and D. Q. Mayne. Design issues in adaptive control. *IEEE Trans. Automat. Control*, 33:50–58, 1988.

[ML76] R. K. Mehra and D. G. Lainiotis (eds.). *System Identification—Advances and Case Studies.* Academic Press, New York, 1976.

[ML89] E. Mosca and J. M. Lemos. A semi–infinite horizon LQ self-tuning regulator for ARMAX plants based on RLS. In *Preprints of the 3rd IFAC Symposium on Adaptive Systems in Control and Signal Processing*, pp. 347–352, Glasgow, UK, 1989. Also in *Automatica*, 28:401–406, 1992.

[MLMN92] E. Mosca, J. M. Lemos, T. Mendonça, and P. Nistri. Adaptive predictive control with mean-square input constraint. *Automatica*, 28:593–597, 1992.

[MLZ90]   E. Mosca, J. M. Lemos, and J. Zhang. Stabilizing I/O receding-horizon control. In *Proceedings of the 29th IEEE Conference on Decision and Control*, pp. 2518–2523. Honolulu, 1990.

[MM80]   G. Menga and E. Mosca. MUSMAR: Multivariable adaptive regulators based on multistep cost functionals. In *Advances in Control*, D. G. Lainiotis and N. S. Tzannes (eds.), pp. 334–341. Reidel, Dordrecht, Holland, 1980.

[MM85]   D. R. Mudgett and A. S. Morse. Adaptive stabilization of linear systems with unknown high frequency gains. *IEEE Trans. Automat. Control*, 30:549–554, 1985.

[MM90a]   D. Q. Mayne and H. Michalska. An implementable receding horizon controller for stabilization of nonlinear systems. In *Proceedings of the 29th IEEE Conference on Decision and Control*, pp. 3396–3397. Honolulu, 1990.

[MM90b]   D. Q. Mayne and H. Michalska. Receding horizon control of nonlinear systems. *IEEE Trans. Automat. Control*, 35:814–824, 1990.

[MM91a]   D. Q. Mayne and H. Michalska. Receding horizon control of constrained nonlinear systems. In *Proceedings of the 1st European Control Conference*, pp. 2037–2042. Hermés, Paris, 1991.

[MM91b]   D. Q. Mayne and H. Michalska. Robust receding horizon control. In *Proceedings of the 30th IEEE Conference on Decision and Control*, pp. 64–69, Brighton, 1991.

[MN89]   E. Mosca and P. Nistri. A direct polynomial approach to LQ regulation. In *Proceedings of the Workshop on the Riccati Equation in Control, Systems and Signals,* S. Bittanti (ed.), pp. 8–9. Pitagora, Bologna, Italy, 1989.

[Mon74]   R. V. Monopoli. Model reference adaptive control with an augmented error signal. *IEEE Trans. Automat. Control*, 19:474–484, 1974.

[Mor80]   A. S. Morse. Global stability of parameter-adaptive control. *IEEE Trans. Automat. Control*, 25:433–439, 1980.

[Mos75]   E. Mosca. An innovations approach to indirect sensing measurement problems. *Ricerche di Automatica*, 6:1–24, 1975.

[Mos83]   E. Mosca. Multivariable adaptive regulators based on multistep cost functionals. In *Nonlinear Stochastic Problems,* R. S. Bucy and J. M. F. Moura (eds.), pp. 187–204. Reidel, Dordrecht, Holland, 1983.

[MS82]   P. E. Möden and T. Söderström. Stationary performance of linear stochastic systems under single step optimal control. *IEEE Trans. Automat. Control*, 27:214–216, 1982.

[MZ83]   E. Mosca and G. Zappa. A MV adaptive controller for plants with time-varying I/O transport delay. In *Proceedings of the 1st IFAC Workshop on Adaptive Systems*, pp. 207–211. Pergamon, New York, 1983.

[MZ84]   E. Mosca and G. Zappa. Removal of a positive realness condition in minimum variance adaptive regulators by multistep horizons. *IEEE Trans. Automat. Control*, 29:844–846, 1984.

[MZ85]   E. Mosca and G. Zappa. ARX modeling of controlled ARMAX plants and its application to robust multipredictor adaptive control. In *Proceedings of the 24th IEEE Conference on Decision and Control*, pp. 856–861. Fort Lauderdale, 1985.

[MZ87]   E. Mosca and G. Zappa. On the absence of positive realness conditions in self-tuning regulators based on explicit criterion minimization. *Automatica*, 23:259–260, 1987.

[MZ88]   M. E. Magama and S. H. Zak. Robust state feedback stabilization of discrete-time uncertain dynamical systems. *IEEE Trans. Automat. Control*, 33, 1988.

[MZ89a]   M. Morari and E. Zafiriou. *Robust Process Control*. Prentice Hall, Englewood Cliffs, N.J., 1989.

[MZ89b]   E. Mosca and G. Zappa. Matrix fraction solution to the discrete-time LQ stochastic tracking and servo problems. *IEEE Trans. Automat. Control*, 34:240–242, 1989.

[MZ89c]   E. Mosca and G. Zappa. ARX modeling of controlled ARMAX plants and LQ adaptive controllers. *IEEE Trans. Automat. Control*, 34:371–375, 1989.

[MZ91]   E. Mosca and J. Zhang. Globally convergent predictive adaptive control. In *Proceedings of the 1st European Control Conference*, pp. 2169–2179. Hermès, Paris, 1991.

[MZ92]   E. Mosca and J. Zhang. Stable redesign of predictive control. *Automatica*, 28:1229–1233, 1992.

[MZ93]   E. Mosca and J. Zhang. Adaptive 2-DOF tracking with reference-dependent self-excitation. In *Proceedings of the 12th IFAC World Congress*, pp. 369–372, The Institution of Engineers, Sidney, 1993.

[MZB93]   E. Mosca, J. Zhang, and D. Borrelli. Adaptive self-excited predictive tracking based on a constant trace normalized RLS. In *2nd European Control Conference*, pp. 2207–2215, ECCA, Groningen, Holland, 1993.

[MZL89]   E. Mosca, G. Zappa, and J. M. Lemos. Robustness of multipredictor adaptive regulators: MUSMAR. *Automatica*, 25:521–529, 1989.

[NA89]   K. S. Narendra and A. M. Annaswamy. *Stable Adaptive Systems*. Prentice Hall, Englewood Cliffs, N.J., 1989.

[NDD92]   A. T. Neto, J. M. Dion, and L. Dugard. On the robustness of LQ regulators for discrete-time systems. *IEEE Trans. Automat. Control*, 37:1564–1568, 1992.

[Nev75]   J. Neveu. *Discrete Parameter Martingales*. North Holland, Amsterdam, 1975.

[Nor87]   J. P. Norton. *System Identification*. Academic Press, New York, 1987.

[Nus83]   R. D. Nussbaum. Some remarks on a conjecture in parameter adaptive control. *Syst. Control Lett.*, 3:243–246, 1983.

[NV78]   K. S. Narendra and L. Valavani. Stable adaptive controllers. Direct control. *IEEE Trans. Automat. Control*, 23:570–583, 1978.

[OK87]   K. A. Ossman and E. W. Kamen. Adaptive regulation of MIMO linear discrete-time systems without requiring a persistent excitation. *IEEE Trans. Automat. Control*, 32:397–404, 1987.

[OY87]   R. Ortega and T. Yu. Theoretical results on robustness of direct adaptive controllers: A survey. In *Preprints of the 10th IFAC World Congress*, vol. 10, pp. 1–15, IFAC, Munich, 1987.

[PA74]   L. Padulo and M. A. Arbib. *System Theory*. W. B. Saunders, San Francisco, 1974.

[Pan68]   V. Panuska. A stochastic approximation method for identification of linear systems using adaptive filtering. In *Proceedings of the Joint Automatic Control Conference*, pp. 1014–1021. Ann Arbor, 1968.

[Par66]   P. C. Parks. Lyapunov redesign of model reference adaptive control systems. *IEEE Trans. Automat. Control*, 11:362–365, 1966.

[Pet70]   V. Peterka. Adaptive digital regulation of noisy systems. In *Preprints of the 2nd IFAC Symposium on Identification and System Parameter Estimation*, paper 6.2. Academia, Prague, 1970.

[Pet72]   V. Peterka. On steady state minimum variance control strategy. *Kybernetika*, 8:219–232, 1972.

[Pet81]   V. Peterka. Bayesian approach to system identification. In *Trends and Progress in System Identification,* P. Eykhoff (ed.). Pergamon, New York, 1981.

[Pet84]   V. Peterka. Predictor-based self-tuning control. *Automatica,* 20:39–50, 1984.

[Pet86]   V. Peterka. Control of uncertain processes: Applied theory and algorithms. *Kybernetica,* 22:Supplement, 1986.

[Pet89]   I. R. Petersen. The matrix Riccati equation in state feedback $\mathcal{H}_\infty$ control and in the stabilization of uncertain systems with norm bounded uncertainties. In *Proceedings of the Workshop on the Riccati Equation in Control, Systems and Signals,* S. Bittanti (ed.), pp. 51–56. Pitagora, Bologna, Italy, 1989.

[PK92]   Y. Peng and M. Kinnaert. Explicit solution to the singular LQ regulation problem. *IEEE Trans. Automat. Control,* 37:633–636, 1992.

[PLK89]   L. Praly, S. F. Lin, and P. R. Kumar. A robust adaptive minimum variance controller. *SIAM J. Control Optim.,* 27:235–266, 1989.

[POF89]   M. P. Polis, A. W. Olbrot, and M. Fu. An overview of recent results on the parametric approach for robust stability. In *Proceedings of the 28th IEEE Conference on Decision and Control,* pp. 23–29. Tampa, 1989.

[Pra83]   L. Praly. Robustness of indirect adaptive control based on pole-placement design. In *Preprints of the IFAC Workshop on Adaptive Systems.* San Francisco, 1983.

[Pra84]   L. Praly. Robust model reference adaptive controllers—Part I: Stability analysis. In *Proceedings of the 23rd IEEE Conference on Decision and Control,* pp. 1009–1014. Las Vegas, 1984.

[PW71]   W. W. Peterson and E. J. Weldon. *Error-Correcting Codes.* The MIT Press, Cambridge, Mass., 1971.

[PW81]   D. L. Prager and P. E. Wellstead. Multivariable pole assignment regulator. *Proc. IEE,* 128, D:9–18, 1981.

[Rao73]   C. R. Rao. *Linear Stochastic Inference and Its Applications.* Wiley, New York, 1973.

[RC89]   B. D. Robinson and D. W. Clarke. Robustness effects of a prefilter in generalized predictive control. *Proc. IEE,* 138, D:2–8, 1989.

[RM92]   M. S. Radenkovic and A. N. Michel. Robust adaptive systems and self-stabilization. *IEEE Trans. Automat. Control,* 37:1355–1369, 1992.

[Ros70]   H. H. Rosenbrock. *Multivariable and State-Space Theory.* Wiley, New York, 1970.

[Roy68]   H. L. Royden. *Real Analysis.* Macmillan, New York, 1968.

[RPG92]   D. E. Rivera, J. F. Pollard, and C. E. Garcia. Control-relevant pre-filtering: A systematic design approach and case study. *IEEE Trans. Automat. Control,* 37:964–974, 1992.

[RRTP78] J. Richalet, A. Rault, J. L. Testud, and J. Papon. Model predictive heuristic control: Application to industrial processes. *Automatica,* 14:413–428, 1978.

[RT90]   J. Richalet and S. Tzafestas (eds.). *Proceedings of the CIM—Europe Workshop on Computer Integrated Design of Controlled Industrial Systems.* ESPRIT CIM, Paris, 1990.

[RVAS81] C. E. Rohrs, L. Valavani, M. Athans, and G. Stein. Analytical verification of undesirable properties of direct model reference adaptive control algorithm. In *Proceedings of the 20th IEEE Conference on Decision and Control,* pp. 1272–1284. San Diego, 1981.

[RVAS82] C. E. Rohrs, L. Valavani, M. Athans, and G. Stein. Robustness of adaptive control algorithms in the presence of unmodelled dynamics. In *Proceedings of the 21st IEEE Conference on Decision and Control,* pp. 3–11. Orlando, 1982.

[RVAS85] C. E. Rohrs, L. Valavani, M. Athans, and G. Stein. Robustness of continuous-time adaptive control algorithms in the presence of unmodelled dynamics. *IEEE Trans. Automat. Control,* 30:881–889, 1985.

[Sam82]   C. Samson. An adaptive LQ controller for non-minimum-phase systems. *Int. J. Control,* 35:1–28, 1982.

[SB89]   S. Sastry and M. Bodson. *Adaptive Control: Stability, Convergence and Robustness.* Prentice Hall, Englewood Cliffs, N.J., 1989.

[SB90]   R. Scattolini and S. Bittanti. On the choice of the horizon in long-range predictive control—Some simple criteria. *Automatica,* 26:915–917, 1990.

[SE88]   H. Selbuz and V. Elden. Kleinman's controller: a further stabilizing property. *Int. J. Control,* 48:2297–2301, 1988.

[SF81]   C. Samson and J. J. Fuchs. Discrete adaptive regulation of not-necessarily minimum-phase systems. *Proc. IEE,* 128, D:102–108, 1981.

[Sha79]   L. Shaw. Nonlinear control of linear multivariable systems via state-dependent feedback gains. *IEEE Trans. Automat. Control*, 24:108–112, 1979.

[Sha86]   U. Shaked. Guaranteed stability margins for the discrete-time linear quadratic optimal regulator. *IEEE Trans. Automat. Control*, 31:162–165, 1986.

[Sim56]   H. A. Simon. Dynamic programming under uncertainty with a quadratic criterion function. *Econometrica*, 24:74–81, 1956.

[SK86]   V. Shaked and P. R. Kumar. Minimum variance control of multivariable ARMAX plants. *SIAM J. Control Optim.*, 24:396–411, 1986.

[SL91]   J. J. Slotine and W. Li. *Applied Nonlinear Control*. Prentice Hall, Englewood Cliffs, N.J., 1991.

[SMS91]   D. S. Shook, C. Mohtadi, and S. L. Shah. Identification for long range predictive control. *Proc. IEE*, 140, D:75–84, 1991.

[SMS92]   D. S. Shook, C. Mohtadi, and S. L. Shah. A control-relevant identification strategy for GPC. *IEEE Trans. Automat. Control*, 37:975–980, 1992.

[Söd73]   T. Söderström. An on-line algorithm for approximate maximum likelihood identification of linear dynamic systems. Report 7308, Dept. of Automatic Control, Lund Institute of Technology, Lund, Sweden, 1973.

[Soe92]   R. Soeterboek. *Predictive Control. A Unified Approach*. Prentice Hall, Englewood Cliffs, N.J., 1992.

[Sol79]   V. Solo. On the convergence of AML. *IEEE Trans. Automat. Control*, 24:958–962, 1979.

[Sol88]   V. Solo. *Time Series Analysis*. Springer–Verlag, New York, 1988.

[SS89]   T. Söderström and P. Stoica. *System Identification*. Prentice Hall, Englewood Cliffs, N.J., 1989.

[TAG81]   Y. Z. Tsypkin, E. D. Avedyan, and O. V. Galinskij. On the convergence of recursive identification algorithms. *IEEE Trans. Automat. Control*, 26:1009–1017, 1981.

[TC88]   T. T. C. Tsang and D. W. Clarke. Generalized predictive control with input constraints. *Proc. IEE*, 135, D:451–460, 1988.

[Tho75]   Y. A. Thomas. Linear quadratic optimal estimation and control with receding horizon. *Electron. Lett.*, 11:19–21, 1975.

[Toi83a]   H. T. Toivonen. Suboptimal control of discrete stochastic amplitude constrained systems. *Int. J. Control*, 37:493–502, 1983.

[Toi83b]   H. T. Toivonen. Variance constrained self-tuning control. *Automatica*, 19:415–418, 1983.

[Tru85]   E. Trulsson. Uniqueness of local minima for linear quadratic control design. *Syst. Control Lett.*, 5:295–302, 1985.

[TSS77]   Y. A. Thomas, D. Sarlat, and L. Shaw. A receding horizon approach to the synthesis of nonlinear multivariable regulators. *Electron. Lett.*, 13-11:329, 1977.

[UR87]   H. Unbehauen and G. P. Rao. *Identification of Continuous Systems*. North-Holland, Amsterdam, 1987.

[Utk77]   V. I. Utkin. Variable structure systems with sliding modes. *IEEE Trans. Automat. Control*, 22:212–222, 1977.

[Utk87]   V. I. Utkin. Discontinuous control systems: State of the art in theory and applications. In *Preprints of the 10th IFAC World Congress*, pp. 75–94. IFAC, Munich, 1987.

[Utk92]   V. I. Utkin. *Sliding Modes in Control Optimization*. Springer–Verlag, Berlin, 1992.

[Vid85]   M. Vidyasagar. *Control System Synthesis: A Factorization Approach*. The MIT Press, Cambridge, Mass., 1985.

[VJ84]   M. Verma and E. Jonckheere. L_∞–compensation with mixed sensitivity as a broadband matching problem. *Syst. Control Lett.*, 4:125–130, 1984.

[Was65]   W. Wasow. *Asymptotic Expansions for Ordinary Differential Equations*. Wiley, New York, 1965.

[WB84]   J. C. Willems and C. I. Byrnes. Global adaptive stabilization in the absence of information on the sign of the high frequency gain. In *Proceedings of the INRIA Conference on Analysis and Optimization of Systems*, pp. 49–57, Springer, Berlin, 1984.

[WEPZ79]   P. E. Wellstead, J. M. Edmunds, D. L. Prager, and P. H. Zanker. Pole zero assignment self-tuning regulator. *Int. J. Control*, 30:11–26, 1979.

[WH92]   C. Wen and D. J. Hill. Global boundedness of discrete-time adaptive control just using estimator projection. *Automatica*, 28:1143–1157, 1992.

[Whi81]   P. Whittle. *Optimization Over Time: Dynamic Programming and Stochastic Control*, vols. 1 and 2. Wiley, New York, 1981.

[Wie49]   N. Wiener. *Extrapolation, Interpolation and Smoothing of Stationary Time Series, with Engineering Applications.* Technology Press and Wiley, New York, 1949. Originally issued in February 1942 as a classified National Defense Research Concil Report. Now available as *Time Series*, the MIT Press, Cambridge, Mass., 1977.

[Wil71]   J. C. Willems. Least squares stationary optimal control and the algebraic Riccati equation. *IEEE Trans. Automat. Control*, 16:621–634, 1971.

[Wit71]   H. S. Witsenhausen. Separation of estimation and control for discrete-time systems. *Proc. IEEE*, 59:1557–1566, 1971.

[WM57]   N. Wiener and P. Masani. The prediction theory of multivariate stochastic processes—Part 1: The regularity condition. *Acta Math.*, 98:111–150, 1957.

[WM58]   N. Wiener and P. Masani. The prediction theory of multivariate stochastic processes—Part 2: The linear predictor. *Acta Math.*, 99:93–137, 1958.

[Wol38]   H. Wold. *Study in the Analysis of Stationary Time Series.* Almquist and Wicksell, Uppsala, Sweden, 1938.

[Wol74]   W. A. Wolovich. *Linear Multivariable Systems.* Springer-Verlag, New York, 1974.

[Won68]   W. H. Wonham. On the separation theorem of stochastic control. *SIAM J. Control*, 6:312–326, 1968.

[Won70]   E. Wong. *Stochastic Processes in Information and Dynamical Systems.* McGraw-Hill, New York, 1970.

[WPZ79]   P. E. Wellstead, D. L. Prager, and P. H. Zanker. A pole assignment self-tuning regulator. *Proc. IEE*, 126, D:781–787, 1979.

[WR79]   B. Wittenmark and P. K. Rao. Comments on single step versus multi-step performance criteria for steady-state SISO systems. *IEEE Trans. Automat. Control*, 24:140–141, 1979.

[WS85]   B. Widrow and S.D. Stearns. *Adaptive Signal Processing.* Prentice Hall, Englewood Cliffs, N.J., 1985.

[WW71]   J. Wieslander and B. Wittenmark. An approach to adaptive control using real time identification. *Automatica*, 7:211–217, 1971.

[WZ91]   P. E. Wellstead and M. B. Zarrop. *Self-Tuning Systems.* Wiley, New York, 1991.

[Yaz89]   E. Yaz. Equivalence of two stochastic stabilizability conditions and its implications. In *Int. J. System Science*, 20:1745–1751, 1989.

[Yds84]   B. E. Ydstie. Extended horizon adaptive control. In *Preprints of the 9th IFAC World Congress*, vol. VII, pp. 133–137. IFAC, Budapest, 1984.

[Yds91]   B. E. Ydstie. Stability of the direct self-tuning regulator. In *Foundations of Adaptive Control*, P. Kokotovic (ed.). Springer-Verlag, New York, 1991.

[YJB76]   D. C. Youla, H. A. Jabar, and J. J. Bongiorno. Modern Wiener–Hopf design of optimal controllers—Part II: The multivariable case. *IEEE Trans. Automat. Control*, 21:319–338, 1976.

[You61]   D. C. Youla. On the factorization of rational matrices. *IRE Trans. Inform. Theory*, 7:172–189, 1961.

[You68]   P. C. Young. The use of linear regression and related procedures for the identifications of dynamic processes. In *Proceedings of the 7th IEEE Symposium on Adaptive Processes*, pp. 501–505. San Antonio, Texas, 1968.

[You74]   P. C. Young. Recursive approaches to time series analysis. *Bull. Inst. Math. Appl.*, 10:209–224, 1974.

[You84]   P. C. Young. *Recursive Estimation and Time–Series Analysis*. Springer–Verlag, New York, 1984.

[Zam81]   G. Zames. Feedback and optimal sensitivity: Model reference transformations, multiplicative seminorms and approximate inverses. *IEEE Trans. Automat. Control*, 26:301–320, 1981.

[ZD63]   L. A. Zadeh and C. A. Desoer. *Linear System Theory*. McGraw-Hill, New York, 1963.

[Zei85]   E. Zeidler. *Nonlinear Functional Analysis and its Applications*, vol. I. Springer-Verlag, New York, 1985.

INDEX